Mesoscale Eddies in the South China Sea

Advanced Series on Physical Oceanography

Series Editors: Ruixin Huang (*Woods Hole Oceanographic Institution, USA*)
Quanan Zheng (*University of Maryland, USA*)

Published:

Vol. 1 *Mesoscale Eddies in the South China Sea*
by Quanan Zheng, Lingling Xie, Chung-Ru Ho and Jianyu Hu

Advanced Series on Physical Oceanography – Volume 1

Mesoscale Eddies in the South China Sea

Quanan Zheng
University of Maryland, USA

Lingling Xie
Guangdong Ocean University, China

Chung-Ru Ho
National Taiwan Ocean University, Taiwan

Jianyu Hu
Xiamen University, China

World Scientific

NEW JERSEY • LONDON • SINGAPORE • BEIJING • SHANGHAI • HONG KONG • TAIPEI • CHENNAI • TOKYO

Published by

World Scientific Publishing Co. Pte. Ltd.
5 Toh Tuck Link, Singapore 596224
USA office: 27 Warren Street, Suite 401-402, Hackensack, NJ 07601
UK office: 57 Shelton Street, Covent Garden, London WC2H 9HE

Library of Congress Cataloging-in-Publication Data
Names: Zheng, Quanan, 1943– author | Xie, Lingling author | Ho, Chung-Ru author | Hu, Jianyu author
Title: Mesoscale eddies in the South China Sea / authors, Quanan Zheng, University of Maryland, USA; Lingling Xie, Guangdong Ocean University, China; Chung-Ru Ho, National Taiwan Ocean University, Taiwan; and Jianyu Hu, Xiamen University, China.
Description: Hackensack, NJ : World Scientific Publishing Co. Pte. Ltd., [2026] | Series: Advanced Series on Physical Oceanography ; volume 1 | Includes bibliographical references and index.
Identifiers: LCCN 2025049808 | ISBN 9789819827329 hardcover | ISBN 9789819827336 ebook for institutions | ISBN 9789819827343 ebook for individuals
Subjects: LCSH: Eddies--South China Sea | Ocean currents--South China Sea
Classification: LCC GC289.J3 Z44 2026 | DDC 551.461/472--dc23/eng/20260529
LC record available at https://lccn.loc.gov/2025049808

British Library Cataloguing-in-Publication Data
A catalogue record for this book is available from the British Library.

For any available supplementary material, please visit
https://www.worldscientific.com/worldscibooks/10.1142/14704#t=suppl

Desk Editors: Soundararajan Raghuraman/Amanda Yun

Typeset by Stallion Press
Email: enquiries@stallionpress.com

Preface

The discovery of mesoscale eddies by a research team led by oceanographer C. Iselin from the Woods Hole Oceanographic Institution, USA in 1930s marked a great breakthrough of physical oceanography. Referring to calculations by Richman and Hogan, the eddy kinetic energy on scales from 50 to 500 km accounts for 76% of total mean surface kinetic energy in the global ocean. Mesoscale eddies are a key link of the energy cascade chain from the large-scale ocean circulation to sub-mesoscale and small-scale processes. Meanwhile, mesoscale eddies behave like heart pulses of the ocean, which link closely to the variability of ocean circulation and the global climate change. Thus, ocean eddy science has become a central theme of modern oceanography during the past few decades.

The South Chia Sea (SCS) is a semi enclosed marginal sea of the western North Pacific. Its circulation system is characterized by energetic mesoscale eddies occurring all year round and comprises an ideal test site for mesoscale eddy research. Since the beginning of the 21st century, a great number of research results have been published. This book aims to integrating the most important discoveries and research results of mesoscale eddies in the SCS and delivering a novel tool book containing systematically updated research results to the oceanography community.

This monograph consists of five parts with 18 chapters. Part 1, Background of Ocean Eddy Science, provides a macroscopic review for progresses in the world-wide research of ocean mesoscale eddies. It consists of four chapters. Chapter 1, An Overview, Chapter 2,

Roles of Mesoscale Eddies in Ocean Circulation, Climate and Ecology,Chapter 3, Generation, Development and Dissipation of Mesoscale Eddies and Chapter 4, Detection and Observation Methods for Mesoscale Eddies. Part 2, Generation Mechanisms of Eddies in the South China Sea, integrates theoretical and observational results of geneses and dissipation of mesoscale eddies. Part 2 contains Chapter 5, Local Basin Processes for Eddy Genesis, Chapter 6, Disturbance Processes from the Pacific for Eddy Genesis in the SCS, Chapter 7, Intermediary role of the Kuroshio Current, Chapter 8, Typhoon Forced Ocean Mesoscale Eddies. Part 3, Cruise Observations of Eddies in the South China Sea includes three chapters. Chapter 9, Cruise Observed Mesoscale Eddies in the South China Sea, Chapter 10, Three-dimensional Structures and Evolution of Mesoscale Eddies and Chapter 11, Eddy-induced Transports in the South China Sea. Part 4, Statistical Features of Eddies in the South China Sea, undertakes tasks to collect the statistical results mainly using satellite data products, numerical model simulation data, and satellite tracked Argo float data. Part 4 includes two chapters. Chapter 12, Statistics of Mesoscale Eddies in the SCS and Chapter 13, Statistics of Eddies on the Northern SCS Continental Shelf. Part 5, Inter-scale Interactions, summarizes the research results of inter-scale and interdisciplinary interactions of eddies with co-existing dynamic processes. Chapter 14, Mesoscale Eddy Modulation to Sea Surface Waves, Chapter 15, Mesoscale Eddy Modulation to Near Inertial Waves, Chapter 16, Mesoscale Eddy Modulation to Internal Solitary Waves, Chapter 17, Atmospheric Responses to Ocean Mesoscale Eddies and Chapter 18, Mesoscale Eddies in Nearby Marginal Seas, including the Andaman Sea of the Indian Ocean, the Sulu Sea, the Celebes Sea and the Banda Sea in the tropic West Pacific.

Major materials used for preparing the book contents are taken from published works of authors' research teams and others, some unpublished new analysis results are also included. Individual contributions of the authors are as follows: Quanan Zheng: Chapters 1, 3, 4, 7, 8, 10, 11, 12, 15 and 16; Lingling Xie: Chapters 2, 5, 14 and 18; Chung-Ru Ho: Chapters 6 and 17; and Jianyu Hu: Chapters 9 and 11. Zheng integrates individual contributions into an entire book.

My special thanks to co-authors, Professor Xie, Professor Ho and Professor Hu, for their long-term sincere collaborations that make the book a great success, and to Guangdong Ocean University for providing SOE, technical assistances and travel supports during the period from book preparation to finalization.

Quanan Zheng, Ph.D,
Research Scientist Emeritus,
University of Maryland College Park,
The United States.
May 2026

About the Authors

Quanan Zheng received his Ph.D. degree in physical oceanography from the Institute of Oceanology, Chinese Academy of Sciences, China in 1987. He is a Research Scientist Emeritus at the Department of Atmospheric and Oceanic Science, University of Maryland, College Park, MD, USA. His research interests include physical oceanography, satellite oceanography, and nonlinear wave dynamics. He has published 226 journal papers and 5 books. Dr. Zheng is an Associate Editor-in-Chief of the journals, *Acta Oceanological Sinica* and *Ocean-Land-Atmosphere Research,* and an editor for Volume 8 of *Comprehensive Remote Sensing* published by Elsevier. Dr. Zheng was named a 2024 Top Scholar by ScholarGPS (top 0.5% of all scholars worldwide). Dr. Zheng is a member of the American Geophysical Union.

Lingling Xie earned her Ph.D. degree in physical oceanography from the Ocean University of China in 2009. She currently serves as a Professor and Dean in the College of Oceanography and Meteorology at Guangdong Ocean University, China. Her research focuses on multi-scale ocean dynamic processes and ocean responses to typhoons, with an emphasis on ocean eddies in the western Pacific marginal seas. She has authored more than 100 peer-reviewed papers and two books, and received awards from the Chinese Society of Oceanography and local governments. She also serves as an editor for *Acta Oceanological Sinica*, *Ocean-Land-Atmosphere Research*, *Marine Science*, and *Journal of Guangdong Ocean University*.

Chung-Ru Ho earned his Ph.D. in Applied Ocean Science from the University of Delaware, USA, in 1994. He is a Distinguished Professor in the Department of Marine Environmental Informatics at National Taiwan Ocean University. Additionally, he has held positions as Deputy Director of the National Museum of Marine Science and Technology in Taiwan, as well as Associate Vice President for Research and Development, and Department Chair at National Taiwan Ocean University. His research interests focus on utilizing satellite remote sensing data with numerical modeling techniques to enhance our understanding of ocean processes. Specifically, he studies the upper ocean's response to typhoons and the complex interactions between the Kuroshio Current and ocean eddies. Throughout his career, Professor Ho has presented at numerous international conferences, authored over 100 peer-reviewed papers, and received multiple awards for his contributions to the field. Looking ahead, his research will concentrate on addressing the challenges posed by global warming. He aims to apply his expertise to develop practical solutions in coastal management, fisheries, and related areas, particularly in light of the increasing intensity and frequency of unusual natural phenomena. Professor Ho currently represents Taiwan in the International Association for the Physical

Sciences of the Oceans (IAPSO) and serves on the editorial boards of several international journals, including the *Journal of Marine Science and Technology*, *Remote Sensing*, and *Contemporary Science, Engineering, and Technology.*

Jianyu Hu received his Ph.D. in Physical Oceanography from Tohoku University, Japan, in 2001 and a Ph.D. in Engineering in Environmental Science from Xiamen University, China, in 2002. He is currently a Professor at State Key Laboratory of Marine Environmental Science, Center for Marine Meteorology and Climate Change, College of Ocean and Earth Sciences, Xiamen University, China. His main research interest is in regional environmental oceanography, especially in the South China Sea and the Taiwan Strait. He, together with Drs. Chung-Ru Ho, Lingling Xie and Quanan Zheng, have published one book, *Regional Oceanography of the South China Sea* (World Scientific).

Acknowledgements

Special thanks to Dr. Changming Dong for sharing their review paper published in *Ocean-Land-Atmosphere Research* that was cited as the major contents of Chapters 2–4. We appreciate helps of Drs. Yu-Hao Tseng, Ching-Yuan Lu, Po-Chun Hsu, Yu-Hsin Cheng, Jia-Yi Lin and Zhe-Wen Zheng to Chapters 5–8 and 17; Zhiyuan Hu, Zhenyu Sun, Longqi Yang, Jia Zhu, Zhaozhang Chen, Zhida Huang, Xiaoqing Chu, Feng Nan, Xia Wang, Xiaolin Bai, Hang Zhao, Chun Wu and Chunhua Qiu to Chapters 9 and 11. Thanks to Ying Xiong for assistance in editing and formatting of entire manuscript.

Quanan Zheng

Contents

Part 1

Background of Ocean Eddy Science

Chapter 1

An Overview

1.1 Brief History of Discovery and Expenditure of Mesoscale Eddies in Global Ocean

Owing to intrinsic fluid properties of sea water, vortices of various horizontal length scales as a sort of flow patterns are broadly distributed in the global ocean. Among them, vortices with horizontal diameters ranging from 100 to 500 km and lifetime scales ranging from days to months are generally defined as the mesoscale eddy (Zheng *et al.*, 2017). Referring to calculations by Richman and Hogan (2011), the mean eddy kinetic energy (EKE) on scales from 50 km to 500 km in the global ocean (70°S–70°N) accounts for 76% (68%) of total mean surface (abyssal ocean) kinetic energy. Mesoscale eddy is a key link of the energy cascade chain from the large-scale ocean circulation to sub-mesoscale and small-scale processes (Zhang *et al.*, 2014, 2016; Chen and Han, 2019). Mesoscale eddies behave like heart pulses of the ocean, understanding of their behavior would reveal the variability of ocean circulation, which is closely linked with the global climate change. Thus, eddy science has become a central theme of modern oceanography during the past few decades (Fu *et al.*, 2010; Huang, 2013; see details in Chapter 2).

The South China Sea (SCS) is a semi enclosed marginal sea of the western North Pacific (WNP). Its dynamic environments provide favorable conditions for genesis, growth, propagation, and dissipation of mesoscale eddies. In fact, observations and investigations have shown that the SCS circulation system is indeed characterized by energetic mesoscale eddies occurring all year round (Lin and Sun, 2020). Thus, the SCS comprises an ideal test site for mesoscale eddy research, from which a great number of research results have been generated and published particularly since the beginning of the 21st century. This book is designed with the aim of integrating the most important discoveries and research results of mesoscale eddies in the SCS (SCSMEs) and delivering a novel tool book containing systematically new knowledge to the oceanography community. This chapter serves as an overview of contents in the book, including brief history of discovery and expenditure of mesoscale eddies in the global ocean and the SCS, mesoscale dynamic environments in the SCS, and brief introduction to five parts of the book.

1.1.1 *Initial discovery era (1930s–1960s)*

The earliest scientific record on discovery of mesoscale eddies in the global ocean might be traced back to 1930s. At that time a research team led by oceanographer C. Iselin from the Woods Hole Oceanographic Institution, USA, observed a large vortex in the Gulf Stream region of the Northwest Atlantic, and referred to it as the internal solitary wave (Greene, 2006). Later, similar phenomena were observed in the ocean off Southern California (McEwen, 1948) and in the south of Kuroshio (Masuzawa, 1957). In 1950s–1960s, Russian (former Soviet Union) oceanographers made great affords to exploration of ocean mesoscale eddies supported by the Soviet POLYGON Experiment started from 1935 (Brekhovskikh *et al.*, 1971; Koshlyakov and Monin, 1978). Yasui (1961) detected eddy signals from long-term mooring observed data in the northern Philippine Sea (24°N 135°E) from 1950 to 1953. Koshlyakov *et al.* (1970) observed cyclonic eddies with the horizontal length scale of $O(200)$ km in the Arabian Sea (10°–15°N 64°–66°E) from hydrographic survey data taken in 1967 and defined them as "synoptic eddies" (Koshlyakov and Monin, 1978). Meanwhile, more cyclonic cold vortices concentrated in the Gulf Stream region were observed by US oceanographers using hydrographic survey. The vortices were defined as mesoscale rings (in 1970s) or mesoscale eddies (after 1980s) (Iselin and Fuglister, 1948; Greenspan, 1968; Fuglister, 1972). The initial discovery era achieved the confirmation of mesoscale eddies as a new oceanographic phenomenon and understandings of their basic physical features.

1.1.2 *Prosperous era (1970s–1980s)*

1970s and 1980s were a prosperous era for the research of ocean mesoscale eddies, which comprised a hotspot of worldwide oceanographic research. Particularly, the early 1970s were a great discovery period of mesoscale eddies. From 1974 to 1976, the US oceanographers recorded 93 cases of mesoscale eddies in the Gulf Stream region (Lai and Richardson, 1977). The great discovery relied on the advances in satellite remote sensing technologies. For example, since NOAA 2 launched in 1973, the very high resolution radiometer (VHRR) onboard US National Oceanic and Atmospheric Administration (NOAA) meteorological satellite series has provided visible and infrared (IR) band sea surface images with a coverage or swatch as wide as 2500 km, a spatial resolution as high as 1 km and a repeatable observation time interval as frequent as 1 d (OSCAR, 2016). On meteorological satellite cloud free IR images, all patterns of mesoscale eddies may show up, so that the important data and information on eddies may be extracted, such as sea surface temperature (SST) distribution, occurrence time, geographic location, 2-D horizontal geometry and fine structures (Rao *et al.*, 1971; DeRycke and Rao, 1973; Strong and DeRycke, 1973; Gotthardt and Potocsky, 1974; Worthington, 1977; McDowell and Rossby, 1978; Olson and Spence, 1978; Lee *et al.*, 1981). Meanwhile, time series satellite images were used to observe propagation and evolution processes of mesoscale eddies (Mied and Lindemann, 1979; Richardson, 1980). Zheng *et al.* (1984) developed the early manual feature tracking methods, which used the movable position information of cold-water wedges surrounding Gulf Stream warm eddies on time series satellite images to calculate the eddy rotating velocities as shown in Figure 1.1. Later the methods and the measured data were used for the dynamic analysis of viscous decay mechanisms of eddies (Zheng *et al.*, 1989).

During the two decades, intensive cruise survey and direct measurement programs were carried out with the aim of understanding 3-D structure and dynamic features of mesoscale eddies. Cheney and Richardson (1976) investigated the decay process of a Gulf Stream cold eddy using cruise measurements of the conductivity-temperature-depth profilers (CTDs) and the expendable bathythermographs (XBTs). McCartney *et al.* (1978) observed large cold eddies originating from the northeastern Sargasso Sea using CTDs

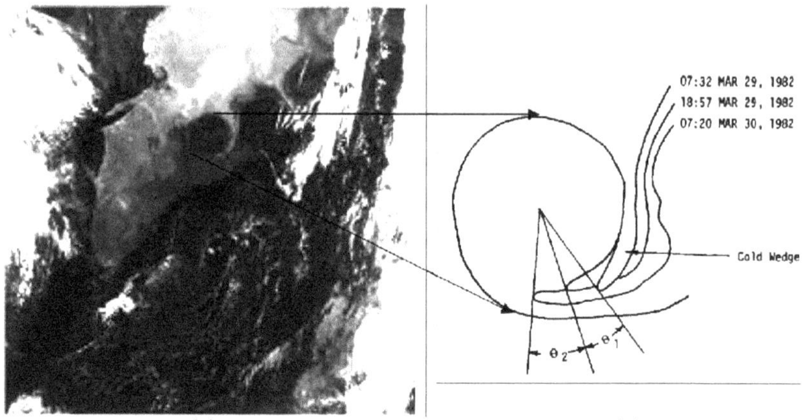

(a) (b)

Fig. 1.1. (a) IR image over the Gulf Stream and the northwestern continental shelf water of the Atlantic taken by NOAA-5 VHRR on 16 June 1978 at 13:53 UTC, Orbit 8502. Dark area on left is the US east coast region. Bright-dark gray levels of the ocean are dependent on cold and warm SST. Three dark near-circular patterns in the upper middle are imagery of warm mesoscale eddies with a mean radius of about 100 km. Cotton-like bright imagery shows the clouds. (b) Interpretation map of (a) showing early pattern recognition methods to measure eddy rotating velocity [$\omega = d\theta/dt$] using moving positions of cold-water wedges of warm eddies.

Source: Taken and redrawn from Zheng *et al.* (1984).

and the aircraft expendable bathythermographs (AXBTs). Richardson *et al.* (1979) explored the physical structure and life history of Gulf Stream cold eddy "Allen" north of Bermuda to track its continuous movement over its life from September 1976 to April 1977 using CTDs, XBTs, satellite tracking drifters and velocity profile measurements. From December 1976 to November 1977, The Ring Group, a group of oceanographers from US multi-institutions, organized the largest scale expenditure for Gulf Stream cold eddies. Four multidisciplinary cruises were conducted by the state-of-the art technologies, including CTDs, XBTs and satellite tracking drifters. During the expenditure period, six cold eddies were tracked and profiled. The data analysis results revealed physics, chemistry and biology of Gulf Stream cold eddies (The Ring Group, 1981). The prosperous era underwent the great discovery period and intensive cruise surveys and achieved understandings of basic laws of eddy movement and evolution, as well as their roles in the ocean circulation.

1.1.3 *Globalized era (1990s–present)*

Since 1990s, the research of ocean mesoscale eddies has entered a globalized era with supports of new generation satellite data with global ocean coverage, such as geostationary meteorological satellites and radar altimeter satellites, as well as great advances in satellite tracking drifter, high resolution ocean modeling and advanced data processing technologies. During the globalized era, the expenditure ranges have spread to the global ocean (Chelton *et al.*, 2007). New hotspot regions have been formed, including the Atlantic deep basin (Grundlingh, 1995; Duncombe Rae *et al.*, 1996; Aguedjou *et al.*, 2019), the Pacific (Flament *et al.*, 1996; Qiu, 1999; Willett *et al.*, 2006; Yan *et al.*, 2019), the Indian Ocean (Ogata and Masumoto, 2010; Jia *et al.*, 2011; Yang *et al.*, 2019; Azis Ismail *et al.*, 2021), the Southern Ocean (Frenger *et al.*, 2013; Lamont *et al.*, 2020), boundary current regions, such as the Kuroshio Current (Nakata, 1992; Kimura *et al.*, 1997; Tsai *et al.*, 2015; Jan *et al.*, 2017, Yan *et al.*, 2019; Mensah *et al.*, 2020), the Brazil Current (Johns *et al.*, 1990; Richardson *et al.*, 1994; Goni and Johns, 2001), the Australian Current (Nilsson and Cresswell, 1980; Ribbe *et al.*, 2018; Oke *et al.*, 2019) and the Agulhas current (Lutjeharms *et al.*, 1989; Biastoch and Krauss, 1999; Penven *et al.*, 2001). As one of the Pacific marginal seas, the SCS is characterized by energetic mesoscale eddies, which are the theme of this book.

Meanwhile, the research directions have shifted more toward global distribution features and the long-term variability of mesoscale eddies (Chelton *et al.*, 2007), their roles in ocean circulation and ocean kinetic energy cascade, global climate change and ecologic effects (Zhang *et al.*, 2014; see details in Chapter 2). Theories and model simulations of dynamics of mesoscale eddies have obtained great progresses (Fu *et al.*, 2010; Qiu and Chen, 2010; Shi *et al.*, 2021; see details in Chapter 3).

In recent years, cut-edge data analysis technologies, such as Big Data and Artificial intelligence (AI) technologies, have been applied to global mesoscale eddy research and obtained important breakthroughs on the understanding of global ocean circulation variation trends. Figure 1.2 shows an example of 2-D distributions of global ocean abnormal mesoscale eddies derived from the fusion of satellite sea surface height (SSH) data and satellite tracking Argo's drifter

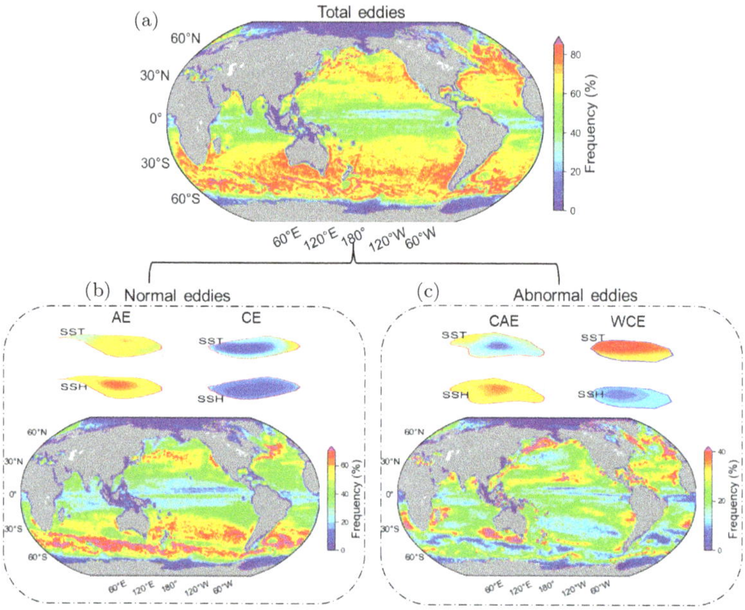

Fig. 1.2. Spatial distributions of (a) total, (b) normal and (c) abnormal mesoscale eddy occurrence frequency in the global ocean from 1993 to 2015 by AI technology. Image resolution is 0.25° by 0.25°.
Source: Taken from Liu *et al.* (2021).

temperature data by deep learning technology (Liu *et al.*, 2021). Abnormal mesoscale eddies refer to warm cyclonic eddies (WCEs) and cold anticyclonic eddies (CAEs). From Figure 1.2, one can see distributions of abnormal eddies in the global ocean derived from 23 years of SSH and SST data from 1993–2015. Further analysis reveals a surprising trend that the annual mean velocity of ocean currents in the global ocean has been getting slowly with the yearly increase rate of $-4.0 \times 10^{-3}\,\text{a}^{-1}$ from 1993 to 2015, which induces the decrease of abnormal eddy numbers in the global ocean.

1.2 Favorable Dynamic Environments in the South China Sea

The SCS is the largest marginal sea of the WNP. It is a near-enclosed ocean basin, which is surrounded by the lands on north and west sides, separated from the Pacific by a generally south-north running island chain consisting of the Taiwan Island and the Philippines Archipelago on east side and from the Indian Ocean by the Indonesian Archipelago on the south side. Since an SCSME was firstly observed in 1956, a great quantity of observations and investigations have shown that the SCS is the typical marginal sea with energetic mesoscale eddies. The reason is because there are favorable dynamic environments for genesis and evolution of mesoscale eddies in the SCS.

1.2.1 *Intrinsic conditions*

The SCS occupies the south-north range from 2°30′N to 25°25′N latitude and the west-east range from 99°10′E to 121°50′E longitude within tropic and subtropic regions (International Hydrographic Organization, 1986), where the atmospheric and oceanic dynamic processes are active and powerful due to the heat influx from solar radiation. Its total surface area reaches 3500×10^3 km^2 with the average depth of 1212 m and the maximum depth of 5377 m. The central deep basin with the depth deeper than 1000 m has the total area of 1000×10^3 km^2 with the length and the width greater than 1000 km. Thus, the huge water body of the SCS can provide enough space for genesis and evolution of mesoscale eddies with horizontal length scale of $O(200)$ km as shown in Figure 1.3.

The SCS bottom topography is featured by the central deep basin surrounded by broad continental shelves. Forced by the topography and the geostrophic effect, the ocean circulation in the deep basin behaves more like the boundary currents. Furthermore, the current velocity shear strengthened by the viscous dissipation of side boundaries and the current vorticity enhanced by flow direction deflected by the topography comprise favorable dynamic conditions for genesis and growth of mesoscale eddies (Qiu and Chen, 2010; see details in Chapter 3). On the other hand, the continental shelves and westward shoaling of deep basin bottom provide broad spaces for eddy kinetic energy (EKE) dissipation of mesoscale eddies.

As a near enclosed water body, the SCS has the intrinsic property for the genesis of Rossby normal modes, which are 2-D waves (Pedlosky, 1987). On the SST and SSH fields, the wave modes appear as coherent grouping mesoscale eddies in the form of check-board-like

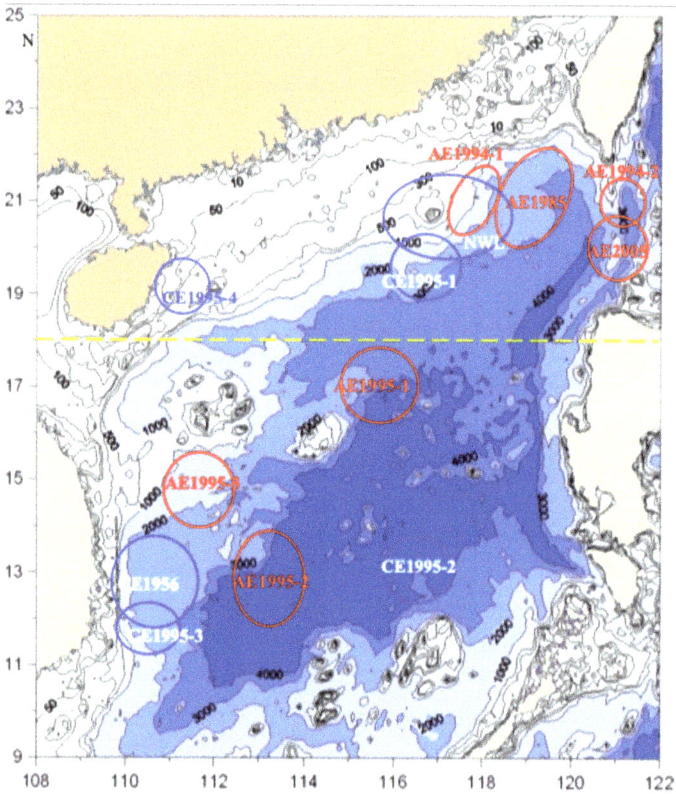

Fig. 1.3. Bottom topographic map of the SCS deep basin and mesoscale eddies discovered at early stage. Red (blue) circles represent warm (cold) eddies with positions and ranges. CE1956 was discovered by Dale (1956) NML by Nitani (1970) AE1985 by Wang and Chen (1987a, 1987b) AE1994-1 and -2 by Li *et al.* (1995) AE1995-1, -2, -3 and CE1995-1, -2, -3, and -4 by Chu *et al.* (1998a, 1998b) and AE2005 by Xie *et al.* (2011). Yellow dash line is 18°N latitude.

Source: Taken and redrawn from Zheng *et al.* (2017).

patterns distributed over the entire basin (Zheng *et al.*, 2014; Xie *et al.*, 2018; see details in Chapter 5).

1.2.2 *External conditions*

The SCS deep basin connects the WNP through the Luzon Strait with a south-north width of about 400 km, which sits between the Taiwan Island and the Luzon Island as shown in Figure 1.3. The Luzon Strait serves as a main channel for water exchange between the SCS and the Pacific with the mass transport being estimated from 2 to 11 Sv (Chen *et al.*, 2011a). On the Pacific side, there is a latitudinal belt between 20°N and 25°N, in which mesoscale wave-like eddy trains propagate westward all year round (Cheng *et al.*, 2017). The Luzon Strait and the east coast of the Taiwan Island are just located at the target section of westward propagating eddies. Thus, it is inevitable that powerful disturbances from the Pacific propagate through the Luzon Strait and enter the SCS in the forms of waves or eddies or jet currents, which is the strong dynamic conditions for the genesis and growth of mesoscale eddies in west of the Luzon Strait (Hu *et al.*, 2012; Xie *et al.*, 2016).

The Kuroshio current, a western boundary current of the North Pacific, flows northward, passing by east of the Luzon Strait from south to north. It behaves most like a dynamic buffer sandwiched by the SCS and the Pacific. Forced by westward propagating eddy train from the Pacific, the Kuroshio intrudes into the SCS in the forms of branching and meandering or loop current. The Kuroshio intrusions are enhanced by its flow path blocked by cold eddies, which have two origins: one is from the Pacific, and another is from the SCS (Zheng *et al.*, 2011). Thus, the Kuroshio can be treated as a contributor and a modulator to the SCS mesoscale eddy fields. See details in Chapter 7.

The SCS sits at the hallway of East Asia monsoon. Driven by seasonally reversal monsoon winds, the circulation in the SCS deep basin is also seasonally reversed: anticyclonic in summer and cyclonic in winter. The instability of ocean circulation is a favorable condition for the genesis and growth of mesoscale eddies. The SCS is the birthplace for the SCS typhoons and sits at the pathway of the Pacific typhoons. Powerful disturbances from typhoons comprise an important dynamic condition for the genesis and evolution of mesoscale eddies (Ho *et al.*, 2009; Zheng *et al.*, 2015; Lu *et al.*, 2020).

1.3 History of Research of Mesoscale Eddies in the South China Sea

1.3.1 *Discovery stage (1956–1990s)*

Dale (1956) discovered a cool pool in the western SCS off the middle coast of Vietnam in summer (CE1956 in Figure 1.3). It was the earliest report on SCSME, about 20 years later than discovery of the Gulf Stream eddies. Later, the discovery was confirmed by Uda and Nakao (1972). In 1970, Nitani (1970) discovered a cyclonic cold eddy in northwest of the Luzon Island (NWL in Figure 1.3). Xu *et al.* (1982) speculated multi-vortex structures of the SCS circulation and their seasonal variability by the objective analysis of 50 years of historical hydrographic data from 1921 to 1970. It was a concept innovation of the understanding of the SCS circulation under those historical conditions.

In 1985, the South China Sea Institute of Oceanography, Chinese Academy of Sciences, reported discovery of a warm eddy in the central SCS, which existed in winter and summer (SCSIO, 1985). Later, Wang and Chen (1987a, 1987b) discovered an anticyclonic mesoscale eddy in the northern SCS based on cruise hydrographic survey data in 1985. The eddy centered at 119°E, 21°N had a horizontal diameter of about 200 km (AE1985). Li *et al.* (1995) confirmed existence of multiple eddies in the SCS by discoveries of two anticyclonic eddies with horizontal scales of 100–150 km derived from cruise hydrographic survey data in the northern SCS (AE1994-1 and -2). Soong *et al.* (1995) was a pioneer using satellite altimeter data to study SCSME. They detected a cyclonic cold mesoscale eddy in the northern SCS from TOPEX/POSEIDON satellite along-track SSH data from 1993 to 1994. Chu *et al.* (1998a, 1998b) distinguished 7 SCSMEs from AXBT data and historical salinity data in 1995 and further confirmed multi-eddy structure of the SCS (AE1995-1, -2, -3, CE1995-1, -2, -3, and -4).

1.3.2 *Globalized stage (2000–present)*

The SCSME discovery attracted attentions from international and local oceanographic communities and became a hotspot of the SCS circulation research, particularly after entering the 21st century, driven by requirements of resources developments and strategic interests for the innovative knowledge of the SCS circulation (Song *et al.*, 2020), the SCSME research is benefited by the requirements, so that all kinds of the state-of-the art ocean survey technologies rushed into SCSME research fields at almost the same time, such as CTDs (Hu *et al.*, 2011, 2012; Xie *et al.*, 2011), mooring acoustic doppler current profiler (ADCP) (Chu and Fan, 2001; Wang and Gan, 2014; Lin *et al.*, 2016), Lagrangian drifter (Zheng *et al.*, 2008;

Li *et al.*, 2011; Zu *et al.*, 2013), glider (Shu *et al.*, 2016; Li *et al.*, 2019), satellite radar altimeter (Hwang and Chen, 2000; Wang *et al.*, 2000, 2003; Yuan *et al.*, 2007; Nan *et al.*, 2011), satellite synthetic aperture radar (SAR) (Zheng *et al.*, 2008) and various numerical models (Yang *et al.*, 2000; Xiu *et al.*, 2010; Wang and Gan, 2014; He *et al.*, 2015; Lin *et al.*, 2015; Xia *et al.*, 2016; Zhao *et al.*, 2016). Following the successful experiences obtained by the research of global ocean mesoscale eddies, the great technologic advances played a key role in promoting quick progresses in the SCSME research.

1.3.2.1 *Targeted individual eddy*

This sub-stage is featured by that investigations aim at individual eddy. The investigation areas were mostly concentrated in the northern SCS, particularly the water area west of the Luzon Strait. Metzger and Hurlburt (2001) found an anticyclonic eddy in the Luzon Strait west of Babuyan Islands. Wang *et al.* (2003) analyzed the earliest satellite altimeter SSH data from 1993 to 2000 and found that the northern SCS, particularly the west of the Luzon Strait, was an eddy-concentrated region. Wang *et al.* (2008) investigated propagation of two anticyclonic warm eddies in the northern SCS in 2003–2004 winter using satellite SSH data, satellite-tracking drifter trajectories and cruise hydrographic survey data. They found that the two warm eddies propagated southwestward along the northern SCS continental slope with an average speed of $10\,\mathrm{cm\,s^{-1}}$, which was close to the phase speed of the Rossby waves. Xie *et al.* (2011) examined 3-D structure of an anticyclonic warm eddy with a diameter of about 200 km and depths from 500 to 900 m sitting just at the Luzon Strait center using cruise hydrographic survey data in 2005 (AE2005 in Figure 1.3). They found that the eddy played a role of local water blender, i.e., it brought the SCS middle layer water out off the northern Luzon Strait, carried the water southward and feedbacked the water to the SCS through the southern Luzon Strait with a volume transport as great as 1.9 Sv. Nan *et al.* (2011) analyzed 17 years of satellite SSH data from 1992 to 2009 and found that cyclonic and anticyclonic eddies often co-existed but their behavior was different from each other. Zu *et al.* (2013) used satellite SSH data and re-analysis products to examine 3-D features and evolution of an anticyclonic eddy occurred in the area southwest of the Taiwan Island in fall 2006 and found that the eddy was trapped at birthplace for longer than a month and finally dissipated locally. Zhang *et al.* (2013) used long-term mooring observations, satellite SSH data and re-analysis products to investigate a cyclonic-anticyclonic eddy pair in the area southwest of the Taiwan Island and their influence on the deep-layer circulation. They found that cyclonic (anticyclonic) eddy strengthened the upper layer current velocity to 50 (100) cm s^{-1} and the temperature anomaly to $-3.0°C$ ($7.5°C$). The eddy pair penetrated whole water body to the deep layer and had great influence on the local deep circulation. Meanwhile, their energetic analysis revealed that both barotropic and baroclinic instabilities play important roles in eddy formation and growth, while contribution of the wind stress was minor one. Cheng *et al.* (2015) analyzed 2.5 years of continuous current velocity data measured by mooring ADCP deployed in the northeastern SCS and found that the interseason variability of 30–90 d of ocean current was closely related with mesoscale eddy propagation. Zhang *et al.* (2016) used mooring array ADCP data observed during the South China Sea Mesoscale Eddy Experiment (S-MEE) from 2013 to 2014 to construct 3-D horizontal current fields for one cyclonic and two anticyclonic eddies in the northeastern SCS west of the Luzon Strait as shown in Figure 1.4.

Figure 1.4 reveals important dynamic features of eddies as follows. (1) Horizontally, eddies of two polarities rotate toward opposite directions just following their definitions. The maximum amplitude of current vectors reaches $40\,cm\,s^{-1}$ at the depth of 40 m. (2) Vertically, the amplitudes of current vectors decrease with the depth and still reach $3.5 \sim 5.0\,cm\,s^{-1}$ at the depth of about 2000 m near the sea floor, implying that eddies are at the growing stage characterized by high intensity and integrality. (3) The detailed structures of eddy current fields show asymmetricity of horizontal currents with respect to the eddy centers and tilting central axes with respect to the vertical line to the sea bottom.

The western boundary of the SCS deep basin is another eddy-frequently-occurring region as shown in Figure 1.3. Hu *et al.* (2011) used cruise CTD sectional observation to investigate a cyclonic eddy in the southwestern SCS during local late summer and found that the genesis of the eddy was closely related with the coastal baroclinic jet off the middle Vietnam. Wang *et al.* (2015) observed movement features of mesoscale eddies near the Xisha Islands and found that eddies propagated northward in summer but reversed in winter. Chu

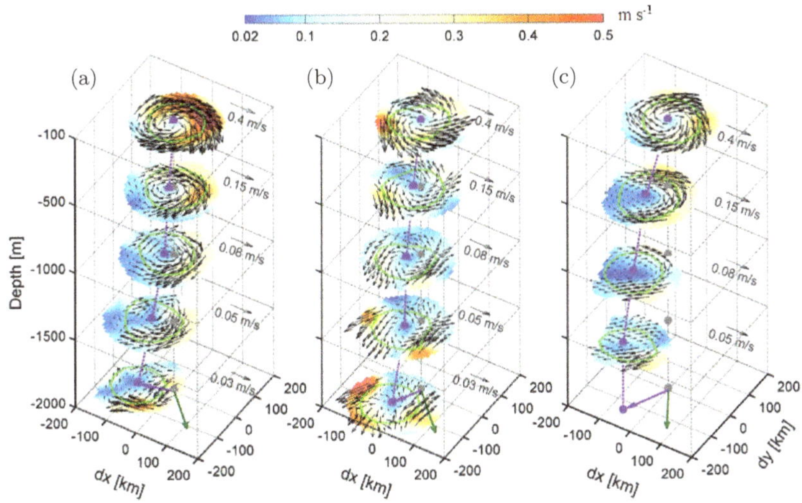

Fig. 1.4. (a) and (c) 3-D structures of two anticyclonic eddies and (b) one cyclonic eddy in the northeastern SCS west of the Luzon Strait. Depths of current field sections are 100, 500, 1000, 1500 and 2000 m. Color codes of current velocity are in $\mathrm{m\,s^{-1}}$.

Source: Taken from Zhang *et al.* (2016).

et al. (2014) found a strong anticyclonic eddy in the western SCS in summer 2010. Cruise hydrographic survey data revealed that the eddy was characterized by 20 cm of the sea level anomaly (SLA) and 7.7°C of the temperature at the eddy center higher than that of surrounding water. Its horizontal length scale is as large as 400 km, and the lifetime is as lengthy as 8 months. They suggested that the genesis of this strong eddy may be attributed to 2009–2010 El Niño event. During the event, enhanced southerly winds and northward currents along the SCS western boundary provided extra energy to strengthen the eddy.

1.3.2.2 *Targeted coherent eddies*

Coherence of eddies is a prominent feature of the SCS because of its closeness. As above mentioned, in as early as 1998, Chu *et al.* (1998a, 1998b) showed co-existence of six SCSMEs. Zheng *et al.* (2008) found a vortex train consisting of five vortexes with a horizontal scale of about 20 km west of the Luzon Strait on European ENVISAT Advanced Synthetic Aperture Radar (ASAR) image on 19

November 2006. It was the first time to treat SCSMEs as a coherent motion process called sub-mesoscale ocean vortex train. Since then, the coherent property of SCSMEs has been laid on the table. Up to now, four scenarios for coherence of SCSMEs have been proposed, i.e., long-lived eddy row, standing wave modes, Rossby normal modes and periodic mesoscale eddies.

Long-lived eddy row Nan *et al.* (2011) identified three anticyclonic eddies ACE1—3 from satellite altimeter SLA images of the SCS in summer 2007 as shown in Figure 1.5. Lifetimes of eddies were 147, 168 and 210 d much longer than 61.6 d of the mean of SCSMEs (Chen *et al.*, 2011a). Three eddies with an average horizontal length

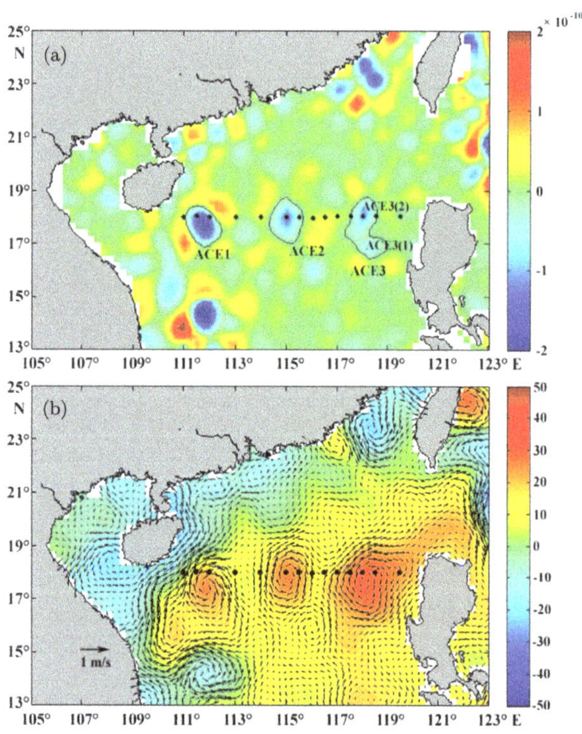

Fig. 1.5. Satellite altimeter SLA images of the northern SCS on 22 August 2007 (a) Okubo-Weiss parameter image (color codes in s^{-2}) shows three anticyclonic eddies ACE1-3. (b) Corresponding SLA image. Color codes and arrows represent SLA in cm and sea surface geostrophic current vectors in $m\,s^{-1}$. Solid dots show positions of cruise hydrographic survey stations along 18°N.

Source: Taken from Nan *et al.* (2011).

scale of $O(200)$ km were arranged along 18°N with the spacing of about 320 km and co-existed for 5 months, implying that three eddies comprised a long-lived eddy row, i.e., a coherent eddy group. Nan *et al.* (2011) speculated that three eddies shared the same mechanism, i.e., the frontal instability of the SCS western boundary current branch was enhanced by southwesterly monsoon winds.

Standing wave modes: Zheng *et al.* (2014) further analyzed cruise hydrographic survey data simultaneously observed with the long-lived eddy row along 18°N as solid dots shown in Figure 1.5. They found that the long-lived eddy row corresponded to strong wave motions in the interior ocean with the maximum amplitude as great as 100 m at the depth of 600 m. The wave disturbance reached the depth of 1000 m as shown in Figure 1.6. Interestingly, the phase speed of the waves was near zero during 5 months of existence. Based

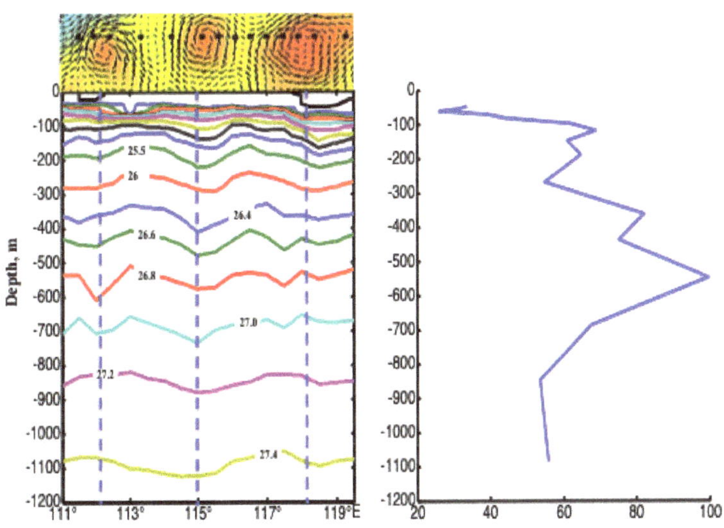

(a) (b)

Fig. 1.6. (a) Corresponding relation of vertical movement of isopycnals $[\sigma_\theta]$ derived from CTD data measured along 18°N as shown in Figure 1.5 from 20 to 24 August 2007 to three anticyclonic eddies on simultaneous satellite altimeter SLA image and (b) vertical distribution of wave motion amplitudes derived from (a). *Source*: Taken and redrawn from Zheng *et al.* (2014).

on these dynamic features, Zheng *et al.* (2014) proposed the scenario that the long-lived eddy row was a component of 2-D horizontal standing wave modes in the closed water body (Lamb, 1945; Raichlen, 1966; Rabinovich, 2009). Standing wave modes are suitable to explain 2-D horizontal distribution patterns of SCSMEs but excluding time-dependent features.

Rossby normal modes: To overcome limitations of standing wave modes, Xie *et al.* (2018) proposed the scenario of Rossby normal modes, which are derived from a linear vorticity equation (Pedlosky, 1987). The solutions consist of two components: westward propagating traveling wave components $\cos\left(\frac{\beta}{2\sigma_{mn}}x + \sigma_{mn}t\right)$ and 2-D standing waves $\sin\left(\frac{m\pi}{L}x\right)\sin\left[\frac{n\pi}{l}y\right]$ in a Cartesian coordinate system, x-axis positive eastward, y-axis positive northward, L and l are horizontal scales of rectangular basin, σ_{mn} is the mode frequency, in which $(m, n) = 1, 2, 3, \ldots$ The Rossby normal modes transform to standing wave modes if and only if $t = 0$. Figure 1.7 shows an example of comparison of Rossby normal modes of the SCS deep basin ζ_{mn} for

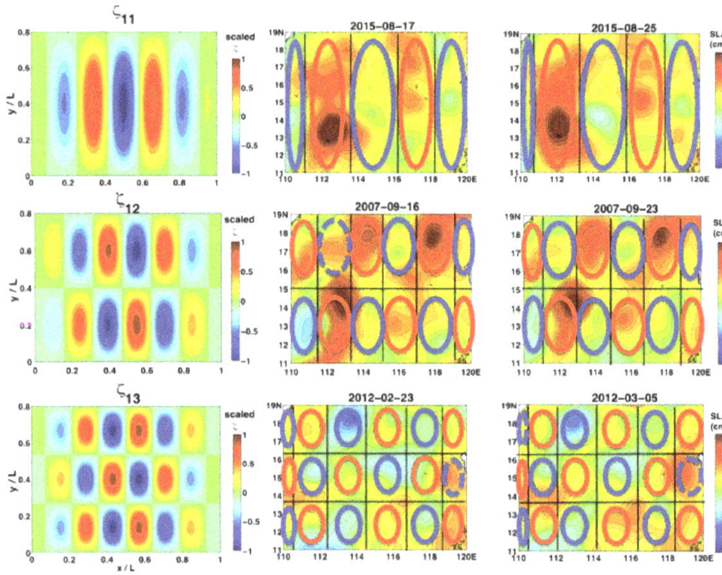

Fig. 1.7. Comparison of the Rossby normal modes ($m = 1$, $n = 1$, 2 and 3, left) to 2-D satellite SLA (in cm) images in the SCS deep basin (right).
Source: Taken from Xie *et al.* (2018).

$m = 1$, $n = 1, 2$ and 3 to 2-D SLA images. Their similarities are 6/6, 6/6, 11/12, 12/12, 17/18 and 16/18, implying a high existence possibility of the Rossby normal modes in the SCS (see Chapter 5 for details).

Periodic mesoscale eddies: Above three scenarios on coherence of SCSMEs were proposed mainly based on their spatial distribution features, i.e., the spatial coherence. On the other hand, Chu *et al.* (2020) proposed a new concept of periodic mesoscale eddies based on long-term cruise hydrographic survey data and satellite altimeter SLA data of the SCS. According to their definition, periodic eddies are a type of eddies that occur nearly annually at fixed time frames with similar horizontal patterns and moving trajectories. Thus, they are temporally coherent eddies. Chu *et al.* (2020) identified 13 periodic eddies and one periodic eddy train of three anticyclonic eddies, which was long-lived eddy row in the SCS as shown in Figure 1.8. One can see that periodic eddies are majorly distributed along the western boundary and around latitudinal section of 17°N, with an anticyclonic "phase" in summer and a cyclonic "phase" in winter. Statistical analysis revealed that the annual repeatability of eddies is directly driven by seasonally reversal monsoon winds and closely associated with large-scale climate variation. Thus, the Niño 3.4 index could be used as a potential predictor for its interannual variation.

1.4 Structures of the Volume

This book is a monograph containing major and updated research results of SCSMEs. Taking on missions, the book is designed to contain 18 chapters distributed in five parts, which are briefly introduced as follows.

Part 1, Background of Ocean Eddy Science, aims to providing a macroscopic review for progresses in the world-wide scientific research of ocean mesoscale eddies. It consists of four chapters.

Chapter 1, An Overview, serves as an introduction of the book, including brief history of discovery and expenditure of mesoscale eddies in the global ocean and the SCS, favorable mesoscale dynamic environments for the eddy genesis and evolution in the SCS.

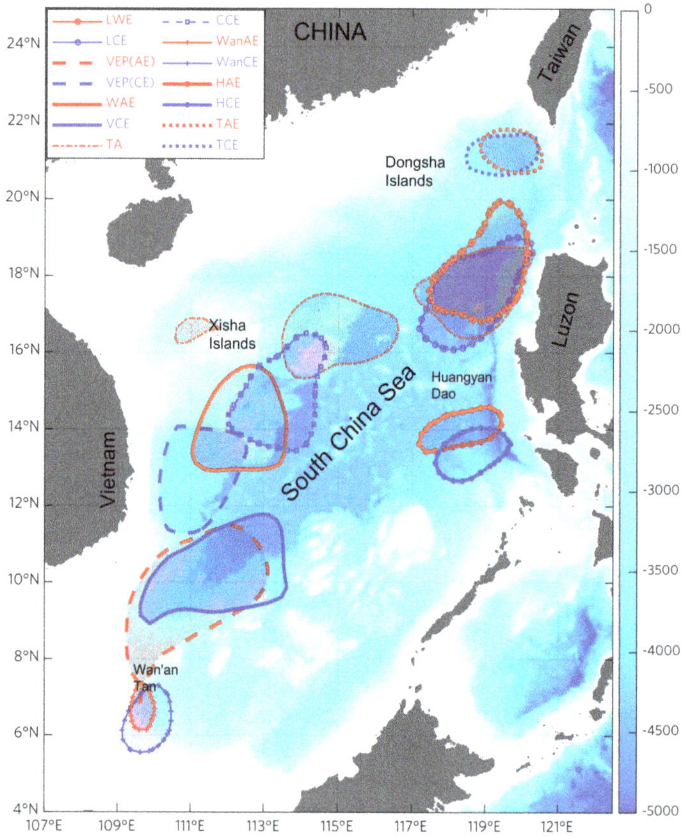

Fig. 1.8. Sketch of 13 periodic mesoscale eddies and Three Anticyclones (TA, i.e., long-lived eddy row) in the SCS. LWE: Luzon Warm Eddy, LCE: Luzon Cold Eddy, VEP (AE): Vietnam Eddy Pair (anticyclonic), VEP (CE): Vietnam Eddy Pair (cyclonic), WAE: Western SCS Anticyclonic Eddy, VCE: Vietnam wintertime Cyclonic Eddy, CCE: Center Cyclonic Eddy, WanAE: Wan'an Tan Anticyclonic Eddy, WanCE: Wan'an Tan Cyclonic Eddy, HAE: Huangyan Dao Anticyclonic Eddy, HCE: Huangyan Dao Cyclonic Eddy, TAE: Taiwan Anticyclonic Eddy, TCE: Taiwan Cyclonic Eddy. Color codes of isobaths are in m.

Source: Taken from Chu *et al.* (2020).

Chapter 2, Roles of Mesoscale Eddies in Ocean Circulation, Climate and Ecology, overviews major achievements and updated progresses in the research of global ocean mesoscale eddies. From the point of view of the global scale, mesoscale eddies occupy a high percentage of the global ocean EKE (Richman and Hogan, 2011), so that it is a critical issue to clarify their roles in affecting the ocean circulation, climate change and ecology.

Chapter 3, Generation, Development and Dissipation of Mesoscale Eddies, gives theoretical research results on generation, growth, and dissipation of mesoscale eddies derived from geophysical fluid dynamics. The solutions have the universal applicability for analyzing dynamics of mesoscale eddies observed anywhere in the global ocean.

Chapter 4, Detection and Observation Methods for Mesoscale Eddies, contains a comprehensive review of recent advances in survey technologies of mesoscale eddies, focusing on automated eddy identification and detection methods mainly from satellite data, as well as cruise observation technologies used by typical observation research programs.

Part 2, Generation Mechanisms of Eddies in the South China Sea, integrates theoretical and observational results of geneses and dissipation of SCSMEs. Distinguished from eddies in the open ocean, generation mechanisms of SCSMEs are complicated, because of superposition of complicated dynamic systems and various side and bottom boundary conditions. Part 2 contains four Chapters 5–8.

Chapter 5, Local Basin Processes for Eddy Genesis, solves theoretical solutions of Rossby normal modes (RNMs) in a rectangular ocean basin with a uniform depth. The complex and variable chessboard-like SLA patterns in the SCS basin, consisting of alternatingly distributed AEs and CEs, are explained well by the RNMs with different spatiotemporal scales.

Chapter 6, Disturbance Processes from the Pacific for Eddy Genesis in the SCS, summarizes the research results of contribution of the Pacific disturbances to the genesis of SCSMEs. In the WNP, wave-like mesoscale eddy trains within the latitudinal belt between 20°N and 25°N propagate westward all year round. The Luzon Strait and the Taiwan Island just sit at their target section. Thus, it is inevitable that the powerful Pacific disturbances propagate through the Luzon Strait and enter the SCS in the forms of waves or eddies or jet currents, which comprise the strong dynamic conditions for the genesis and growth of SCSMEs.

Chapter 7, Intermediary role of the Kuroshio Current, summarizes the research results of the role of the Kuroshio Current that plays in the genesis of SCSMEs. As mentioned in sub-section 1.2.2, the Kuroshio behaves most like a dynamic buffer sandwiched by the SCS and the Pacific. Forced by westward propagating eddy trains from the Pacific, the Kuroshio intrudes into the SCS in the forms of branching and meandering or loop current, which are enhanced by blockage of its flow path by cold eddies originating from the Pacific and the SCS. Thus, the Kuroshio can be treated as a contributor and a modulator for SCSME fields.

Chapter 8, Typhoon Forced Ocean Mesoscale Eddies. The SCS is a hallway for typhoons originating from the NWP and its own deep basin. The ocean inexorably has multiscale responses to episodical injection of a huge amount of momentum and energy from typhoons. Thus, it is reasonable to treat typhoons as a component of localized generation mechanisms of mesoscale eddies. This chapter specifies the role of typhoon played in generation of mesoscale eddies in the SCS.

Part 3, Cruise Observations of Eddies in the South China Sea, is featured by integrating research results mainly derived from cruise hydrographic survey data, of which the amount has dramatically built up, because updated ocean survey technologies have rushed into the SCS since entering the 21st century. The cruise observed data are treated as the ground truth and used to verify theoretical solutions, validate model simulations and calibrate satellite data products. Part 3 includes three Chapters 9–11.

Chapter 9, Cruise Observed Mesoscale Eddies in the South China Sea, presents the observational characteristics of mesoscale eddies in the SCS. Based on extensive in situ data, the spatiotemporal distribution and thermohaline structure of 11 typical mesoscale eddies, eddy pairs, or eddy rows are specified along with their potential dynamic mechanisms.

Chapter 10, Three-dimensional Structures and Evolution of Mesoscale Eddies, focuses on the research results of detailed 3-D structures of SCSMEs and their evolution. High accuracies and high spatial resolutions of cruise hydrographic data assure such research possible.

Chapter 11, Eddy-induced Transports in the South China Sea, reviews advances in understanding eddy-induced transport in the SCS, particularly the role of eddies in heat, salt and volume transports across the continental slope and the Luzon Strait.

Part 4, Statistical Features of Eddies in the South China Sea, undertakes tasks to collect the statistical results of SCSMEs mainly using satellite data products, numerical model simulation data and satellite tracked Argo float data. Part 4 includes two Chapters 12 and 13.

Chapter 12, Statistics of Mesoscale Eddies in the SCS. Since the early statistical result of mesoscale eddies in the SCS was published in 2003, two decades have passed. During this period, technical advances promote continuous generation of refreshed and more reliable statistical results. This chapter aims to summarize the most updated research results and look for the best answers to the questions and uncertainty of eddy statistics such as eddy number per year, ratio of CE number to AE number and eddy responses to climate change.

Chapter 13, Statistics of Eddies on the Northern SCS Continental Shelf, collects recent research results on statistics of dynamic features of mesoscale eddies entering the northern continental shelf of South China Sea (NCSS) derived from Archiving, Validation and Interpretation of Satellite Oceanographic data (AVISO) mesoscale eddy motion trajectory atlas product from 1993 to 2016.

Part 5, Inter-scale Interactions, summarizes the research results of inter-scale and interdisciplinary interactions of SCSMEs with co-existing dynamic processes in Chapters 14–17. Chapter 18 integrates the research results of mesoscale eddies in nearby marginal seas of the SCS used for comparison studies.

Chapter 14, Mesoscale Eddy Modulation to Sea Surface Waves, overviews progress in theoretical research of modulation of mesoscale eddies to the sea surface waves and updated results derived from satellite observations. The baseline includes 2-D horizontal SLA structures of eddies derived from satellite altimeter observations, and the sea surface wave fields from the Surface Wave Investigation and Monitoring (SWIM) wave spectrometer onboard China-France Ocean satellite (CFOSAT).

Chapter 15, Mesoscale Eddy Modulation to Near Inertial Waves, and Chapter 16, Mesoscale Eddy Modulation to Internal Solitary Waves are of physical analogue, i.e., modulation of mesoscale eddy to sub-mesoscale internal waves. The SCS is a favorable test area for those subjects, because of co-existences of strong mesoscale eddies and those sub-mesoscale waves. Here dividing them into two chapters is propitious to specify the characteristics of those modulation processes.

Chapter 17, Atmospheric Responses to Ocean Mesoscale Eddies, belongs to interdisciplinary (atmospheric and oceanic dynamics) and inter-scale (large-scale and mesoscale) subjects. This chapter summarizes the research results of effects of SCSMEs on the overlying atmosphere.

Chapter 18, Mesoscale Eddies in Nearby Marginal Seas, contains the research results of mesoscale eddies in marginal seas nearby the SCS, which are useful for comparison studies. The marginal seas include the Andaman Sea of the Indian Ocean, the Sulu Sea, the Celebes Sea and the Banda Sea in the tropic West Pacific.

1.5 Summaries

Discovery of Gulf Stream mesoscale eddies in 1930s was a great milestone in the world's physical oceanographic research history. Since then, the development history of mesoscale eddy research has undergone three eras: initial discovery era (1930s–1960s), prosperous era (1970s–1980s) and globalized era (1990s–present). The initial discovery era achieved the confirmation of eddies as a new oceanographic phenomenon and the understandings of their basic physical features. The prosperous era underwent the great discovery period and intensive cruise surveys and achieved the understandings of basic laws of eddy movement and evolution, as well as their roles in the ocean circulation. The globalized era expanded the geographic ranges of mesoscale eddy research to the global ocean and achieved the understandings of the universalities of eddy spatiotemporal distributions and their roles in global ocean circulation and global climate change. Thus, up to now, the mesoscale eddy research has gone beyond the significance of eddies' selves and become a unique window, through which one can gain an insight into the variability of global ocean and global climate.

This chapter defines the period from 1956 to 1990s as the discovery stage of SCSMEs, which achieved early discoveries and phenomenon existence confirmations. The major research progresses can be summarized as follows. (1) It was mainly based on cruise hydrographic survey data to discover and confirm the existence of SCSMEs, which occurred all year round. CEs and AEs have generally the same occurrence possibilities. (2) The geographic distribution of SCSMEs almost covered the entire SCS deep basin. Particularly, the northeastern deep basin west of the Luzon Strait and western boundary region of the deep basin are eddy concentrated areas. (3) The 3-D scales of SCSMEs were generally clarified. (4) It was observed that SCSMEs are characterized by multi-eddy-grouping occurrence. These innovative discoveries and understandings contribute a solid basis to further investigations of dynamics of SCSMEs and recognition of the SCS circulation.

Since entering the 21st century, the research of SCSMEs has caught up advances in global ocean mesoscale eddy research and entered the globalized stage. The leap forward mainly relies on applications of the state-of-the art ocean survey technologies. In 2000s, the research mainly targeted physical and dynamic features, evolution processes and fine structures of individual eddy and achieved many important results. Since 2011, coherence of SCSMEs, which seems to be a prominent property for an enclosed basin, has been laid on the table, and four scenarios on occurrence mechanisms have been proposed. The updated research results will be specified in Parts 2–5 of this book.

1.6 Prospects

Since the first mesoscale eddy in the SCS was discovered in 1956, the research of SCSMEs has experienced two development stages: the discovery stage (1956–1990s) and the globalized stage (2000-present). Up to now, the research of SCSMEs has caught up advances in the mesoscale eddy research of the global ocean. The abundant achievements and innovative results presented in this book indicate that the SCS has become a marginal sea with the most active investigations of mesoscale eddies. As mentioned in subsection 1.3.2, requirements of resources developments and strategic interests for knowledges of the SCS circulation comprise the major powers to push the SCS investigations forward. As an important component of the SCS circulation, the SCSME investigation would certainly benefit from the requirements. This macroscopic situation would be sustainable in the expectable future. Meanwhile, updated ocean survey technologies, including hardware and software, would provide powerful tools for 3-D, high-resolution, and high-accuracy ocean data collection, data processing and data analysis. Thus, the SCSME research would keep prosperous in the expectable future.

References

Aguedjou, H. M. A., Dadou, I., Chaigneau, A., Morel, Y. and Alory, G. (2019). Eddies in the Tropical Atlantic Ocean and their seasonal variability, *Geophys. Res. Lett.*, 46(21), pp. 12156–12164, doi:10.1029/2019GL083925.

Azis Ismail, M. F., Ribbe, J., Arifin, T., Taofiqurohman, A. and Anggoro, D. (2021). A census of eddies in the tropical eastern boundary of the Indian Ocean, *J. Geophys. Res. Oceans*, 126, e2021JC017204, doi:10.1029/2021JC017204.

Biastoch, A. and Krauss, W. (1999). The role of mesoscale eddies in the source regions of the Agulhas Current, *J. Phys. Oceanogr.*, 29, pp. 2303–2317, doi:10.1175/1520-0485(1999)029<2303:TROMEI>2.0.CO;2.

Brekhovskikh, L. M., Fedorov, K. N., Fomin, L. M., Koshlyakov, M. N. and Yampolsky, A. D. (1971). Large-scale multibuoy experiment in the Tropical Atlantic, *Deep-Sea Res.*, 18, pp. 1189–1206, doi:10.1016/0011-7471(71)_90026-X.

Chelton, D. B., Schlax, M. G., Samelson, R. M. and de Szoeke, R. A. (2007). Global observations of large oceanic eddies, *Geophys. Res. Lett.*, 34, doi:10.1029/2007gl030812.

Chen, G. and Han, G. (2019). Contrasting short-lived with long-lived mesoscale eddies in the global ocean, *J. Geophys. Res. Oceans*, 124, pp. 3149–3167, doi:10.1029/2019JC014983.

Chen, G., Hou, Y. and Chu, X. (2011a). Water exchange and circulation structure near the Luzon Strait in early summer, *Chin. J. Ocean. Limn.*, 29(2), pp. 470–481, doi:10.1007/s00343-011-0198-0.

Chen, G., Hou, Y. and Chu, X. (2011b). Mesoscale eddies in the South China Sea: Mean properties, spatiotemporal variability, and impact on thermohaline structure, *J. Geophys. Res.*, 116(C6), pp. 102–108, doi:10.1029/2010JC006716.

Cheney, R. E. and Richardson, P. L. (1976). Observed decay of a cyclonic Gulf Stream ring, *Deep-Sea Res.*, 23, pp. 143–155, doi:10.1016/S0011-7471(76)_80023-X.

Cheng, L., Zhang, Z., Zhao, W. and Tian, J. (2015). Temporal variability of the current in the northeastern South China Sea revealed by

2.5-year-long moored observations, *J. Oceanogr.*, 71(4), pp. 361–372, doi:10.1007/s10872-015-0295-3.

Cheng, Y.-H., Ho, C.-R., Zheng, Q., Qiu, B., Hu, J. and Kuo, N.-J. (2017). Statistical features of eddies approaching the Kuroshio east of Taiwan Island and Luzon Island, *J. Oceanogr.*, 73(4), doi:10.1007/s10872-017-0411-7.

Chu, P. C. and Fan, C. (2001). Low salinity, cool-core cyclonic eddy detected northwest of Luzon during the South China Sea Monsoon Experiment (SCSMEX) in July 1998, *J. Oceanogr.*, 57(5), pp. 549–563, doi:10.1023/A:1021251519067.

Chu, P. C., Fan, C., Lozano, C. and Kerling, J. (1998a). An airborne expandable bathythermograph (AXBT) survey of the South China Sea, May 1995, *J. Geophys. Res. Oceans*, 103(C10), pp. 21637–21652, doi:10.1029/98JC02096.

Chu, P. C., Chen, Y. C. and Lu, S. H. (1998b). Wind-driven South China Sea deep basin warm-core/cool-core eddies, *J. Oceanogr.*, 54, pp. 347–360, doi:10.1007/BF02742619.

Chu, X., Chen, G. and Qi, Y. (2020). Periodic mesoscale eddies in the South China Sea, *J. Geophys. Res. Oceans*, 125, e2019JC015139, doi:10.1029/2019JC015139.

Chu, X., Xue, H., Qi, Y., Chen, G., Mao, Q., Wang, D. and Chai, F. (2014). An exceptional anticyclonic eddy in the South China Sea in 2010, *J. Geophys. Res. Oceans*, 119(2), pp. 881–897, doi:10.1002/2013JC009314.

Dale, W. L. (1956). Winds and drift currents in the South China Sea, *Malay Trop. Geogr.*, 8, pp. 1–31.

DeRycke, R. J. and Rao, P. K. (1973). Eddies along a Gulf Stream boundary viewed from a very high resolution radiometer, *J. Phys. Oceanogr.*, 3, pp. 490–492, doi:10.1175/1520-0485(1973)_003<0490:EAAGSB>2.0.CO;2.

Duncombe Rae, C. M., Garzoli, S. L. and Gordon, A. L. (1996). The eddy field of the South-East Atlantic Ocean: A statistical census from the BEST project, *J. Geophys. Res.*, 101, pp. 11949–11964, doi:10.1029/95JC03360.

Flament, P., Kennan, A., Knox, R., Niiler, P. and Bernstein, R. (1996). The three dimensional structure of an upper ocean vortex in the tropical Pacific Ocean, *Nature*, 383, pp. 610–613, doi:10.1038/383610a0.

Frenger, I., Gruber, N., Knutti, R. and Munnich, M. (2013). Imprint of Southern Ocean eddies on winds, clouds and rainfall, *Nat. Geosci.*, 6, pp. 608–612, doi:10.1038/Ngeo1863.

Fu, L.-L., Chelton, D. B., Le Traon, P.-Y. and Morrow, R. (2010). Eddy dynamics from satellite altimetry, *Oceanography*, 23(4), pp. 14–25, doi:10.5670/oceanog.2010.02.

Fuglister, F. C. (1972). Cyclonic rings formed by the Gulf Stream 1965-66, in A. L. Gordon (ed.), *Studies of Physical Oceanography*, Gordon and Breach, New York, pp. 137–168, doi:10.1575/1912/70843.

Goni, G. J. and Johns, W. E. (2001). A census of North Brazil Current rings observed from TOPEX/POSEIDON altimetry: 1992–1998, *Geophys. Res. Lett.*, 28, pp. 1–4, doi:10.1029/2000GL011717.

Gotthardt, G. A. and Potocsky, G. J. (1974). Life cycle of a Gulf Stream anticyclonic eddy observed from several oceanographic platforms, *J. Phys. Oceanogr.*, 4, pp. 131–134, doi:10.1175/1520-0485(1974)_004< 0131:LCOAGS>2.0.CO;2.

Greene, C. H. (2006). Review of down to the sea for science: 75 years of ocean research, education, and exploration at the Woods Hole Oceanographic Institution by Cullen V., *Oceanography*, 19(1), pp. 189–191, doi:10.5670/oceanog.2006.105.

Greenspan, H. P. (1968). *The Theory of Rotating Fluids*, Cambridge University Press, Cambridge, p. 327.

Grundlingh, M. L. (1995). Tracking eddies in the southeast Atlantic and southwest Indian oceans with TOPEX/POSEIDON, *J. Geophys. Res.*, 100, pp. 24977–24986, doi:10.1029/95JC01985.

He, Y., Cai, S., Wang, D. and He, J. (2015). A model study of Luzon cold eddies in the northern South China Sea, *Deep Sea Res. I*, 97, pp. 107–123, doi:10.1016/j.dsr.2014.12.007.

Ho, C.-R., Zheng, Q., Zheng, Z.-W., Kuo, N.-J., Tai, C.-K. and Su, F.-C. (2009). Reply to comment by A. Wada *et al.* on "Importance of pre-existing oceanic conditions to upper ocean response induced by Super Typhoon Hai-Tang", *Geophy. Res. Lett.*, 36(9), L09604, doi:10.1029/2009GL037443.

Hu, J., Gan, J., Sun, Z., Zhu, J. and Dai, M. (2011). Observed three-dimensional structure of a cold eddy in the southwestern South China Sea, *J. Geophys. Res.*, 116(C05016), doi:10.1029/2010JC006810.

Hu, J., Zheng, Q., Sun, Z. and Tai, C.-K. (2012). Penetration of nonlinear Rossby eddies into South China Sea evidenced by cruise data, *J. Geophys. Res.*, 117(C3), doi:10.1029/2011JC007525.

Huang, R. X. (2013). Evolution of oceanic circulation theory: From gyres to eddies, *J. Tropical Oceanogr.*, 32(2), pp. 1–14, doi:10.3969/j.issn.1009-5470.2013.02.001.

Hwang, C. and Chen S. (2000). Circulations and eddies over the South China Sea derived from TOPEX/Poseidon altimetry, *J. Geophys. Res.*, 105(C10), pp. 23943–23965, doi:10.1029/2000JC900092.

International Hydrographic Organization. (1986). *Limits of Oceans and Seas*, 4th edn., IHO Publication S-23, *International Hydrographic Bureau*, pp. 108–109.

Iselin, C. and Fuglister, F. (1948). Some recent developments in the study of the Gulf Stream, *J. Mar. Res.*, 7(3), pp. 317–329.

Jan, S., Mensah, V., Andres, M., Chang, M.-H. and Yang, Y. J. (2017). Eddy-Kuroshio interaction: Local and remote effects, *J. Geophys. Res. Oceans*, 122, pp. 9744–9764, doi:10.1002/2017JC013476.

Jia, F., Wu, L. and Qiu, B. (2011). Seasonal modulation of eddy kinetic energy and its formation mechanism in the southeast Indian Ocean, *J. Phys. Oceanogr.*, 41(4), pp. 657–665, https://doi.org/10.1175/2010JP O4436.1.

Johns, W. E., Lee, T. N., Schott, F. A., Zantopp, R. and Evans, R. H. (1990). The North Brazil Current retroflection: Seasonal structure and eddy variability, *J. Geophys. Res.*, 95, pp. 22103–22120, doi:10.1029/ JC095iC12p22103.

Kimura, S., Kasai, A., Nakata, H., Sugimoto, T., Simpson, J. H. and Cheok, J. V. S. (1997). Biological productivity of meso-scale eddies caused by frontal disturbances in the Kuroshio, *J. Mar. Sci.*, 54, pp. 179–192, doi:10.1006/jmsc.1996.0209.

Koshlyakov, M. N., Galerkin, L. I. and Truong, D. H. (1970). On the mesostructure of the open-ocean geostrophic currents, *Okeanologiya*, 10, pp. 805–814.

Koshlyakov, M. N. and Monin, A. S. (1978). Synoptic eddies in the ocean, *Ann. Rev. Earth Planet. Sci.*, 6, pp. 495–523, doi:10.1146/annurev.ea. 06.050178.002431.

Lai, D. Y. and Richardson, P. L. (1977). Distribution and movement of Gulf Stream rings, *J. Phys. Oceanogr.*, 7(5), pp. 670–683, doi:10.1175/1520-0485(1977)_007<0670:DAMOGS>2.0.CO;2.

Lamb, H. (1945). *Hydrodynamics*, 6th ed. Cambridge University Press, New York, pp. 1–738.

Lamont, T. and van den Berg, M. A. (2020). Mesoscale eddies influencing the sub-Antarctic Prince Edward Islands Archipelago: Origin, pathways, and characteristics, *Cont. Shelf Res.*, 210, pp. 1–13, doi:10.1016/j.csr.2020.104257.

Lee, T., Atkinson, L. and Legeckis, R. (1981). Observations of a Gulf Stream frontal eddy on the Georgia continental shelf, April 1977, *Deep-Sea Res.*, 28, pp. 347–378, doi:10.1016/0198-0149(81)_90004-2.

Li, J. R., Zhang, R. and Jin, B. (2011). Eddy characteristics in the northern South China Sea as inferred from Lagrangian drifter data, *Ocean Sci. Disc.*, 7(5), pp. 661–669, doi:10.5194/os-7-661-2011.

Li, L., Nowlin, W. D. and Su, J. L. (1995). Anticyclonic rings from the Kuroshio in the South China Sea, *Deep Sea Res. Part I*, 45(9), pp. 1469–1482, doi:10.1016/S0967-0637(98)_00026-0.

Li, S., Wang, S., Zhang, F. and Wang, Y. (2019). Constructing the three-dimensional structure of an anticyclonic eddy in the South China Sea using multiple underwater gliders. *J. Atmos. Ocean. Tech.*, 36(12), 2449–2470, doi:10.1175/JTECH-D-19-0006.1.

Lin, H. and Sun, Z. (2020). Mesoscale eddies the South China Sea, in J. Hu *et al.* (eds.), *Regional Oceanography of the South China Sea*, World Scientific, Singapore, pp. 101–131, doi:10.1142/9789811206917_0005.

Lin, X., Dong, C., Chen, D., Liu, Y., Yang, J., Zou, B. and Guan, Y. (2015). Three-dimensional properties of mesoscale eddies in the South China Sea based on eddy-resolving model output, *Deep Sea Res. Part I*, 99, pp. 46–64, doi:10.1016/j.dsr.2015.01.007.

Lin, Y.-C., Oey, L.-Y., Wang, J. and Liu, K.-K. (2016). Rossby waves and eddies observed at a temperature mooring in northern South China Sea, *J. Phys. Oceanogr.*, 46(2), pp. 517–535, doi:10.1175/JPO-D-15-0094.1.

Liu, Y., Zheng, Q. and Li, X. (2021). Characteristics of global ocean abnormal mesoscale eddies derived from the fusion of sea surface height and temperature data by deep learning, *Geophys. Res. Lett.*, 48, e2021GL094772, doi:10.1029/2021GL094772.

Lu, Z., Wang, G. and Shang, X. (2020). Strength and spatial structure of the perturbation induced by a tropical cyclone to the underlying eddies, *J. Geophys. Res. Oceans*, 125(5), e2020JC016097, doi:10.1029/2020JC016097.

Lutjeharms, J. R. E., Catzel, R. and Valentine, H. R. (1989). Eddies and other border phenomena of the Agulhas Current, *Cont. Shelf Res.*, 9, pp. 597–616, doi:10.1016/0278-4343(89)_90032-0.

Masuzawa, J. (1957). An example of cold eddies south of the Kuroshio, *Rec. Oceanogr. Works Jpn.*, 3, pp. 1–7.

McCartney, M. S., Worthington, L. V. and Schmitz, W. J., Jr. (1978). Large cyclonic rings from the northeast Sargasso Sea, *J. Geophys. Res.*, 83, pp. 901–914, doi:10.1029/JC083iC02p00901.

McDowell, S. E. and Rossby, H. T. (1978). Mediterranean water: an intense mesoscale eddy off the Bahamas, *Science*, 202, pp. 1085–1087, doi:10.1126/science.202.4372.1085.

McEwen, G. F. (1948). The dynamics of large horizontal eddies (axes vertical) in the ocean off Southern California, *J. Mar. Res.*, 7, pp. 188–216.

Mensah, V., Jan, S., Andres, M. and Chang, M.-H. (2020). Response of the Kuroshio east of Taiwan to mesoscale eddies and upstream variations, *J. Oceanogr.*, 76, pp. 271–288, doi:10.1007/s10872-020-00544-8.

Metzger, E. J. and Hurlburt, H. E. (2001). The nondeterministic nature of Kuroshio penetration and eddy shedding in the South China Sea, *J. Phys. Oceanogr.*, 31(7), pp. 1712–1732, doi:10.1175/1520-0485 (2001)_031<1712:TNNOKP>2.0.CO;2.
Mied, R. P. and Lindemann, G. J. (1979). The propagation and evolution of cyclonic Gulf Stream rings, *J. Phys. Oceanogr.*, 9, pp. 1183–1206, doi:10.1175/1520-0485(1979)009<1183:TPAEOC>2.0.CO;2.
Nakata, H. (1992). Effect of the Kuroshio frontal eddies on egg and larval transport, *Kaiyo Monthly*, 24, pp. 295–299.
Nan, F., He, Z., Zhou, H. and Wang, D. (2011). Three long-lived anti-cyclonic eddies in the northern South China Sea, *J. Geophys. Res.*, 116(C5), C05002, doi:10.1029/2010JC006790.
Nilsson, C. S. and Cresswell, G. R. (1980). The formation and evolution of East Australian current warm-core eddies, *Progr. Oceanogr.*, 9(3), pp. 133–183, doi:10.1016/0079-6611(80)_90008-7.
Nitani, H. (1970). Oceanographic conditions in the sea east of Philippines and Luzon Strait in summer of 1965 and 1966, *The Kuroshio: A Symposium on the Japan Current*, East West Center Press, Honolulu, pp. 213–232, doi:10.1515/9780824885830-022.
Ogata, T. and Masumoto, Y. (2010). Interactions between mesoscale eddy variability and Indian Ocean dipole events in the Southeastern tropical Indian Ocean—Case studies for 1994 and 1997/1998, *Ocean Dyn.*, 60(3), pp. 717–730, doi:10.1007/s10236-010-0304-4.
Oke, P. R., Roughan, M., Cetina-Heredia, P., Pilo, G. S., Ridgway, K. R., Rykova, T., Archer, M. R., Coleman, R. C., Kerry, C. G., Rocha, C., Schaeffer, A. and Vitarelli, E. (2019). Revisiting the circulation of the East Australian Current: Its path, separation, and eddy field, *Progr. Oceanogr.*, 176, 102139, doi:10.1016/j.pocean.2019.102139.
Olson, D. B. and Spence, T. W. (1978). Asymmetric disturbances in the frontal zone of a Gulf Stream ring, *J. Geophys. Res.*, 83, pp. 4691–4695, doi:10.1029/JC083iC09p04691.
OSCAR (2016). *User Manual for OSCAR/Space and OSCAR/ Requirements*, Doc.ID:04289/2016.ver.1.0.
Pedlosky, J. (1987). *Geophysical Fluid Dynamics*, Springer-Verlag, New York, pp. 374–379, doi:10.1007/978-1-4612-4650-3.
Penven, P., Lutjieharms, J. R. E., Marchesiello, P., Roy, C. and Weeks, S. J. (2001). Generation of cyclonic eddies by the Agulhas current in the lee of the Agulhas Bank, *Geophys. Res. Lett.*, 27, pp. 1055–1058, doi:10.1029/2000GL011760.
Qiu, B. (1999). Seasonal eddy field modulation of the North Pacific Subtropical Countercurrent: TOPEX/Poseidon observations and theory, *J. Phys. Oceanogr.*, 29(10), pp. 2471–2486, doi:10.1175/1520-0485.

Qiu, B. and Chen, S. (2010). Interannual variability of the North Pacific Subtropical Countercurrent and its associated mesoscale eddy field, *J. Phys. Oceanogr.*, 40, pp. 213–225, doi:10.1175/2009JPO4285.1.

Rabinovich, A. B. (2009). Seiches and harbor oscillations, in Y. C. Kim (ed.), *Handbook of Coastal and Ocean Engineering*, World Scientific, Singapore, pp. 193–236, doi:10.1142/9789812819307_0009.

Raichlen, H. (1966). Harbor resonance, in A. T. Ippen (ed.), *Estuary and Coastline Hydrodynamics*, McGraw Hill, New York, pp. 281–340.

Rao, P. K., Strong, A. E. and Koffler, R. (1971). Gulf Stream meanders and eddies as seen in satellite infrared imagery, *J. Phys. Oceanogr.*, 1, pp. 237–239, doi:10.1175/1520-0485(1971)001<0237:GSMAEA>2.0.CO;2.

Ribbe, J., Toaspern, L., Wolff, J.-O. and Ismail, M. F. A. (2018). Frontal eddies along a western boundary current, *Cont. Shelf Res.*, 165, pp. 51–59, doi:10.1016/j.csr.2018.06.007.

Richardson, P. L. (1980). Gulf Stream ring trajectories, *J. Phys. Oceanogr.*, 10, pp. 90–104, doi:10.1175/1520-0485(1980)010<0090:GSRT>2.0.CO;2.

Richardson, P. L., Hufford, G. E., Limeburner, R. and Brown, W. S. (1994). North Brazil Current retroflection eddies, *J. Geophys. Res.*, 99(C3), pp. 5081–5093, doi:10.1029/93JC03486.

Richardson, P. L., Maillard, C. and Stanford, T. B. (1979). The physical structure and life history of cyclonic Gulf Stream Ring Allen, *J. Geophys. Res.*, 84, pp. 7727–7741, doi:10.1029/JC084iC12p07727.

Richman, J. G. and Hogan, P. J. (2011). Energetics of a global ocean circulation model compared to observations, *Geophys. Res. Lett.*, 38, L15607, doi:10.1029/2011GL048347.

SCSIO. (1985). Report of comprehensive investigation in the South China Sea (II), China Science Publishing House, Beijing, pp. 183–231.

Shi, Y., Yang, D. and He, Y. (2021). Numerical study on interaction between eddies and the Kuroshio Current east of Taiwan, China, *J. Ocean. Limn.*, 39(2), pp. 388–402, doi:10.1007/s00343-020-0010-0.

Shu Y., Xiu, P., Xue, H., Yao, J. and Yu, J. (2016). Glider-observed anticyclonic eddy in northern South China Sea, *Aqua. Eco. Health Manage.*, 19(3), pp. 233–241, doi:10.1080/14634988.2016.1208028.

Song, H., Zhang, K. and Sun, S. (2020). An introduction to the South China Sea, in J. Hu *et al.* (eds.), *Regional Oceanography of the South China Sea*, World Scientific, Singapore, pp. 19–42, doi:10.1142/9789811206917_0002.

Soong, Y. S., Hu, J. H., Ho, C.-R. and Niiler, P. P. (1995). Cold-core eddy detected in South China Sea, *EOS Trans. AGU*, 76, pp. 345–347, doi:10.1029/95EO00209.

Strong, A. E. and DeRycke, R. J. (1973). Ocean current monitoring employing a new satellite sensing technique, *Science*, 182, pp. 482–484, doi:10.1126/science.182.4111.482.

The Ring Group (1981). Gulf Stream cold-core rings: Their physics, chemistry and biology, *Science*, 212, pp. 1091–1100, doi:10.1126/science.212.4499.1091.

Tsai, C.-J., Andres, M., Jan, S., Mensah, V., Sanford, T. B., Lien, R.-C. and Lee, C. M. (2015). Eddy-Kuroshio interaction processes revealed by mooring observations off Taiwan and Luzon, *Geophys. Res. Lett.*, 42, pp. 8098–8105, doi:10.1002/2015GL065814.

Uda, M. and Nakao, T. (1972). Water masses and currents in the South China Sea and their seasonal changes, In *The Kuroshio-Proceedings of the 3rd CSK Symposium*, Bangkok, Thailand, pp. 161–188.

Wang, D., Xu, H., Lin, J. and Hu, J. (2008). Anticyclonic eddies in the northeastern South China Sea during winter 2003/2004, *J. Oceanogr.*, 64(6), pp. 925–936, doi:10.1007/s10872-008-0076-3.

Wang, G., Su, J. and Chu, P. C. (2003). Mesoscale eddies in the South China Sea observed with altimeter data, *Geophys. Res. Lett.*, 30(21), 2121, doi:10.1029/2003GL018532.

Wang, J. and Chen, C. S. (1987a). The warm-core eddy in the northern South China Sea. I. Preliminary observations on the warm-core eddy, *Acta Oceanogr. Taiwan.*, 18, pp. 92–103.

Wang, J. and Chen, C. S. (1987b). The warm-core eddy in the northern South China Sea. II. A simple mechanism for the establishment and development of the warm-core eddy, *Acta Oceanogr. Taiwan.*, 18, pp. 104–113.

Wang, L. and Gan, J. (2014). Delving into three-dimensional structure of the West Luzon Eddy in a regional ocean model, *Deep Sea Res. Part I*, 90(8), pp. 48–61, doi:10.1016/j.dsr.2014.04.011.

Wang, L., Koblinsky, C. J. and Howden, S. (2000). Mesoscale variability in the South China Sea from the TOPEX/Poseidon altimetry data, *Deep Sea Res.*, 47(4), pp. 681–708, doi:10.1016/S0967-0637(99)_00068-0.

Wang, Q., Zeng, L., Zhou, W., Xie, Q., Cai, S., Yao, J. and Wang, D. (2015). Mesoscale eddies cases study at Xisha waters in the South China Sea in 2009/2010, *J. Geophys. Res. Oceans*, 120(1), pp. 517–532, doi:10.1002/2014JC009814.

Willett, C., Leben, R. and Lavin, M. (2006). Eddies and tropical instability waves in the eastern tropical Pacific: A review, *Prog. Oceanogr.*, 69, pp. 218–238.

Worthington, L. V. (1977). Intensification of the Gulf Stream after the winter of 1976–77, *Nature*, 270, pp. 415–417, https://doi.org/10.1038/270415a0.

Xia, C., Jung, K. J., Wang, G., Yin, X. and Guo, J. (2016). Case study on the three-dimensional structure of meso-scale eddy in the South China Sea based on a high-resolution model, *Acta Ocean. Sin.*, 35(2), pp. 29–38, doi:10.1007/s13131-016-0805-1.

Xie, L., Tian, J., Zhang, S., Zhang, Y. and Yang, Q. (2011). An anticyclonic eddy in the intermediate layer of the Luzon Strait in autumn 2005, *J. Oceanogr.*, 67(1), pp. 37–46, doi:10.1007/s10872-011-0004-9.

Xie, L., Zheng, Q., Tian, J., Zhang, S., Feng, Y. and Yi, X. (2016). Cruise Observation of Rossby waves with finite wavelengths propagating from the Pacific to the South China Sea, *J. Phys. Oceanogr.*, 46(10), pp. 2897–2913, doi:10.1175/JPO-D-16-0071.1.

Xie, L., Zheng, Q., Zhang, S., Hu, J., Li, M., Li, J. and Xu, Y. (2018). Rossby wave normal modes in the South China Sea deep basin evidenced by satellite altimetry, *Intl. J. Rem. Sens.*, 39(2), pp. 399–417, doi:10.1080/01431161.2017.1384591.

Xiu, P., Chai, F., Shi, L., Xue, H. and Chao, Y. (2010). A census of eddy activities in the South China Sea during 1993–2007, *J. Geophys. Res.*, 115, C03012, doi:10.1029/2009JC005657.

Xu, X., Qiu, Z. and Chen, H. (1982). The general description of the horizontal circulation in the South China Sea, *Proceedings of the Symposium of the Chinese Society of Marine Hydrology and Meteorology, Chinese Society of Oceanology and Limnology*, Science Press, Beijing, pp. 119–127.

Yan, X., Kang, D., Curchitser, E. and Pang, C. (2019). Energetics of eddy-mean flow interactions along the western boundary currents in the North Pacific, *J. Phys. Oceanogr.*, 49, pp. 789–810, doi:10.1175/JPO-D-18-0201.1.

Yang, G., Zheng, Q., Xiong, X., Yuan, Y., Zhuang, Z., Hui, Z., Guo, Y.-L., Yu, L., Sun, J., Ju, X., Ma, D.-J. and Hu, X. (2019). Subsurface cyclonic eddies observed in the southeastern tropical Indian Ocean, *J. Geophys. Res. Oceans*, 124, pp. 7247–7260, doi:10.1029/2019JC015381.

Yang, K., Shi, P., Wang, D., You, X. and Li, R. (2000). Numerical study about the mesoscale multi-eddy system in the northern South China Sea in winter, *Haiyang Xuebao*, 22(1), pp. 27–34.

Yasui, M. (1961). Internal waves in the open ocean (an example of internal waves progressing along the oceanic frontal zone), *Oceanogr. Mag.*, 12, pp. 157.

Yuan, D., Han, W. and Hu, D. (2007). Anti-cyclonic eddies northwest of Luzon in summer-fall observed by satellite altimeters, *Geophys. Res. Lett.*, 34(13), pp. 256–260, doi:10.1029/2007GL029401.

Zhang, Z., Li, Y. and Tian, J. (2013). A modified method to estimate eddy diffusivity in the North Pacific using altimeter eddy statistics, *Chin. J. Ocean. Lim.*, 31(4), pp. 925–933, doi:10.1007/s00343-013-2214-2.

Zhang, Z., Tian, J. Qiu, B., Zhao, W., Chang, P., Wu, D. and Wan, X. (2016). Observed 3D structure, generation, and dissipation of oceanic mesoscale eddies in the South China Sea, *Sci. Rep.*, 6, 24349, doi:10.1038/srep24349.

Zhang, Z., Wang, W. and Qiu, B. (2014). Oceanic mass transport by mesoscale eddies, *Science*, 345(6194), pp. 322–324, doi:10.1126/science.1252418.

Zhao, Y., Liang, X. and Gan, J. (2016). Nonlinear multiscale interactions and internal dynamics underlying a typical eddy-shedding event at Luzon Strait, *J. Geophys. Res.*, 121, pp. 8208–8229, doi:10.1002/2016JC012483.

Zheng, Q., Ho, C.-R., Xie, L. and Li, M. (2019). A case study of a Kuroshio main path cut-off event and impacts on the South China Sea in fall-winter 2013–2014, *Acta Oceanol. Sin.*, 38(4), pp. 12–19, doi:10.1007/s13131-019-1411-9.

Zheng, Q, Hu, J., Zhu, B., Feng, Y., Jo, Y.-H., Sun, Z., Zhu, J., Lin, H., Li, J. and Xu, Y. (2014). Standing wave modes observed in the South China Sea deep basin, *J. Geophys. Res.*, 119(7), pp. 4185–4199, doi:10.1002/2014JC009957.

Zheng, Q., Klemas, V. and Huang, N. E. (1984). Dynamics of the slope water off New England and its influence on the Gulf Stream as inferred from satellite IR data, *Rem. Sens. Environ.*, 15(2), pp. 135–153, doi:10.1016/0034-4257(84)_90042-7.

Zheng, Q., Lin, H., Meng, J., Hu, X., Song, Y. T., Zhang, Y. and Li, C. (2008). Sub-mesoscale ocean vortex trains in the Luzon Strait, *J. Geophys. Res.*, 113(C04032), pp. 977–990, doi:10.1029/2007JC004362.

Zheng, Q., Tai, C.-K., Hu, J., Lin, H., Zhang, R., Su, F.-C. and Yang, X. (2011). Satellite altimeter observations of nonlinear Rossby eddy-Kuroshio interaction at the Luzon Strait, *J. Oceanogr.*, 67, pp. 365–376, doi:10.1007/s10872-011-0035-2.

Zheng, Q., Xie, L., Zheng, Z. and Hu, J. (2017). Progress in research of mesoscale eddies in the South China Sea, *Adv. Mar. Sci.*, 35(2), pp. 131–158, (in Chinese with English abstract).

Zheng, Q. and Yuan, Y. (1989). Study on the analytical model of decay of mesoscale eddy on the continental shelf, *Science In China Series B-Chemistry, Life Sciences & Earth Sciences*, 32(9), pp. 1135–1143.

Zheng, Q., Zhu, B., Li, J., Sun, Z., Xu, Y. and Hu, J. (2015). Growth and dissipation of typhoonforced solitary continental shelf waves in the northern South China Sea, *Clim Dyn.*, 45(3-4), pp. 853–865, doi:10.1007/s00382-014-2318-y.

Zu, T., Wang, D., Yan, C., Beikin, I., Zhuang, W. and Chen, J. (2013). Evolution of an anticyclonic eddy southwest of Taiwan, *Ocean Dyn.*, 63(5), pp. 519–531, doi:10.1007/s10236-013-0612-6.

Chapter 2

Roles of Mesoscale Eddies in Ocean Circulation, Climate and Ecology

2.1 Introduction

Mesoscale eddies are the intermediate process connecting large-scale and small-scale motions, carrying more than 90% of the ocean total kinetic energy, making them the "kinetic energy reservoir" of the global ocean (McWilliams, 2016). Thus, they play a crucial role in constituting an energy cascade chain of the ocean circulation, i.e., energy transition from large-scale to mesoscale (Callies and Ferrari, 2013; Steinberg *et al.*, 2022), from mesoscale to sub-mesoscale such as near-inertial waves (NIWs) (Lelong *et al.*, 2020) and small-scale such as sea surface waves (SSWs) (Romero *et al.*, 2020; Wang *et al.*, 2022; Tan *et al.*, 2023; Zhang and Qiu, 2020). On the other hand, as a huge, movable heat source in the ocean upper layer, mesoscale eddies must have influences on weather systems and climate through direct heat exchange with overlying atmosphere (Businger *et al.*, 1984). Meanwhile, stirring, trapping and Ekman pumping induced by mesoscale eddies directly affect marine ecological environments, including biogeochemistry, horizontal and vertical transports of phytoplankton and chlorophyll growth (Gaube *et al.*, 2014; Siegel *et al.*, 2011). This chapter specifies the advances in above research area in the following sections.

2.2 Eddy Roles in Ocean Circulation

2.2.1 *Energy cascade from large-scale circulation to mesoscale eddies*

Recent observations indicate that 90% of the ocean kinetic energy is stored in mesoscale eddies (Wunsch and Ferrari, 2004). The generation of mesoscale eddies often involves cross-scale energy transfer between large-scale and mesoscale processes. For example, barotropic instability promotes the forward transfer of large-scale ocean kinetic energy to the mesoscale, while baroclinic instability releases large-scale available potential energy and converts it into the kinetic energy of mesoscale eddies. Additionally, mesoscale eddy kinetic energy (EKE) can transfer inversely through nonlinear energy cascades. Understanding the complex energy transfer between mesoscale eddies and large-scale circulation is essential for understanding the ocean energy cycle, though it remains a challenging task.

For the energy cascade problem of mesoscale kinetic energy, although barotropic instability can cause energy to transfer forward from large to mesoscale, ocean eddies primarily originate from baroclinic instability (Gill *et al.*, 1974; Feng *et al.*, 2021, 2022). Theoretical predictions suggest that mesoscale kinetic energy generated by baroclinic instability will ultimately transfer back to large scales: the release of background potential energy is converted into mesoscale kinetic energy, which eventually transfers back to large scales. Various observations (mainly from satellite data) indicate the presence of inverse cascading characteristics in the kinetic energy of mesoscale eddies in the ocean (Scott and Wang, 2005; Callies and Ferrari, 2013; Steinberg *et al.*, 2022). In early studies, the most work was regional, such as Scott and Wang (2005) focusing on the South Pacific, Callies and Ferrari (2013) on the Gulf Stream and North Pacific. Steinberg *et al.* (2022) used a long-track sea surface height data to discuss the global characteristics of inverse energy cascades in mesoscale and revealed their seasonal variations. In addition to direct quantitative analysis of energy cascades, the spatial scale characteristics of ocean eddies also imply the existence of inverse energy cascade to some extent. Oceanographers have already observed that the fastest-growing mode of mesoscale eddies predicted by linear instability is significantly smaller than the observed eddy scale. The inconsistency

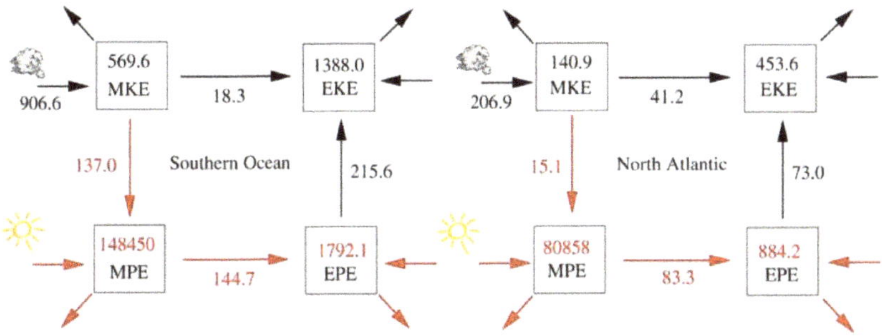

Fig. 2.1. Energy cycle estimated using the Lorenz energy cycle in the North Atlantic and South Atlantic. The units corresponding to energy density and energy exchange terms are 1015 J and 109 W, respectively. The budget range for the North Atlantic is from 10°S to 68°N and from 100°W to 0°W, and for the South Atlantic, the range is from 78°S to 32°S and from 180°W to 180°E.

between the two scales implies that mesoscale flows in the ocean are not simply linear processes, and complex nonlinear processes lead to mutual transformation and transfer of energy, resulting in further growth of eddy scales and indicating the possibility of inverse energy transfer.

However, analyzing inverse energy cascades alone is insufficient to achieve a more quantitative understanding of energy exchange between mesoscale and large scale. The Lorenz Energy Cycle (Figure 2.1) offers a more comprehensive framework for quantifying the transfer of kinetic and potential energy among these scales (Lorenz, 1955). Using this theory, Olbers *et al.* (2012) applied vortex-compatible models to quantitatively estimate the transfer of kinetic and potential energy between large scale and mesoscale in the South and North Atlantic. Their findings suggest tendency for both kinetic and potential energies to flow from large to small scales, indicating that the observed inverse energy cascades at the mesoscale do not cascade back to the Rhines scale as predicted by theory. The Lorenz energy cycle method has been widely used to evaluate energy transfer across different scales (Aiki *et al.*, 2016; Matsuta and Masumoto, 2023; Ssebandeke *et al.*, 2024). However, it is important to note that current estimates are greatly based on model results due to limited observational data, and these results are influenced by the parameterization of sub-grid processes within the models.

2.2.2 *Energy cascade from mesoscale to small scale*

Mesoscale eddies can cascade energy downward from mesoscale to smaller scales such as sub-mesoscale, mainly through two types of processes: sub-mesoscale motions and interactions with topography. Regional observations (Adams *et al.*, 2017; Yang *et al.*, 2017; Zhong *et al.*, 2017) and numerical simulations (Luo *et al.*, 2020) have shown that sub-mesoscale processes at eddy peripheries are important channels for downward energy cascading and dissipation of eddies. Among them, processes such as spiral Rossby waves (Zhang and Qiu, 2020) and instabilities (Luo *et al.*, 2020) play a crucial role in the oceanic energy cascade from mesoscale to small scale. Satellite data show enhancement of sub-mesoscale energy along spiral chlorophyll bands, indicating that mesoscale eddy-trapped Rossby waves can effectively transfer energy to smaller scales in the ocean (Zhang and Qiu, 2020). ADCP velocity observation profiles across anticyclonic eddy (AE) and cyclonic eddy (CE) show that instabilities lead to kinetic energy cascade from mesoscale eddies to internal gravity waves. Model data suggest that enhanced lateral buoyancy gradients and surface buoyancy loss favor triggering of symmetric instability and ageostrophic anticyclonic instability at the edges of mesoscale eddies, which can lead to the consumption of mesoscale energy by sub-mesoscale instability during the sharpening of the eddy periphery fronts (Luo *et al.*, 2020). Both satellite observations and numerical simulations show that under the combined effects of frontogenesis in the West Pacific Ocean and buoyancy loss induced by atmospheric forcing, the enhanced horizontal buoyancy gradient within the eddy region leads to a reduction in potential vorticity. This reduction triggers sub-mesoscale symmetric instability, resulting in strong vertical ageostrophic secondary circulation (Jing *et al.*, 2021) (Figure 2.2). Symmetric instability is the primary mechanism driving the positive cascading of geostrophic energy. Observational profiles based on an AE in the South China Sea (SCS) show that eddy-induced vertical shear can excite symmetric instability, which can extract energy from vertical geostrophic shear and dissipate the kinetic energy of mesoscale eddies to smaller scales. In addition, satellite altimeter and in-situ observation data show that the interaction between mesoscale

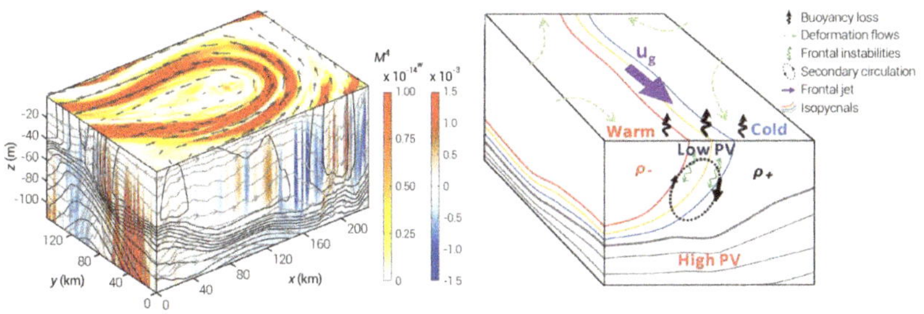

(a) (b)

Fig. 2.2. (a) Typical sub-mesoscale dynamic processes in the mesoscale eddy region. (b) Schematic diagram of a typical sub-mesoscale density front and cross-front ageostrophic secondary circulation that tends to slump isopycnals and re-stratify the surface boundary layer by sub-mesoscale symmetry instability.

eddies and the western boundary topography in the ocean can effectively induce the transfer of mesoscale EKE to smaller scales.

Sub-mesoscale eddies can transfer energy back to mesoscale eddies, enhancing the magnitude of mesoscale eddies and their heat transport (Zhang *et al.*, 2023). A recent observation shows that sub-mesoscale filaments can also transfer energy back to AEs through barotropic instability and gain energy through baroclinic instability.

2.2.3 *Eddy modulation to sea surface waves*

Mesoscale eddies and surface gravity waves are two ubiquitous features of the global ocean, functioning on different temporal and spatial domains, but they are two very important geophysical phenomena. However, their interactions have rarely been investigated. When sea surface wave (SSW) propagation encounters a non-uniform current, several changes occur: the waves become steeper and their energy increases when moving against the current, while they flatten and lose energy when moving with the current. These changes are closely related to the gradient of the current velocity and the angle between the wave propagation direction and the current (Yuan, 1997). Based on the principle of wave action conservation, Wang *et al.*

(2020) derived the wave-current interaction equation that can reflect the spatial variation gradient of ocean current (horizontal shear, convergence and divergence, deformation), and applied it to the study of influence of the Kuroshio on SSWs in the nearby sea area. Their results indicate that gradient of velocity variation is important for effects of ocean currents on SSWs.

The surface current field of mesoscale eddies exhibits a distinct annular structure characterized by a significant velocity gradient and a 360°change in the current direction. When SSWs interact with this current field, an interesting phenomenon occurs: the wave behavior differs on opposite sides of the eddy. The wave height decreases on the side where the wave and current move in the same direction and increases on the side where they move in opposite direction. Romero *et al.* (2020) found that the difference between the significant wave height and the wave steepness under the effect of the eddy is as high as 30%, and the change of the white cap coverage is more than one order of magnitude by the measured data. With the successful launch of the China-France Oceanography SATellite (CFOSAT) in 2018 (Xu *et al.*, 2019; Hauser *et al.*, 2020; Aouf *et al.*, 2021), the along-track resolution of data reaches up to 7 km, providing important data support for the research on eddy modulation effects to SSWs. The data of China-France Oceanography SATellite Surface Wave Investigation and Monitoring (CFOSAT-SWIM) has the advantages of high resolution and wind-wave synchronization. Wang *et al.* (2022) and Tan *et al.* (2023) found that the surface current field of the eddies has a significant modulation effect on the SSWs in the SCS by using the wind wave observation data of CFOSAT-SWIM. The difference in wave height and wave steepness on both sides of the eddy is as high as 20–30%.

At present, the research on the influence of mesoscale eddies on SSWs is mainly based on individual cases. The particularity of the surface current field of mesoscale eddies makes its action process on waves complex, and the energy conversion mechanism between SSWs and eddies has not been clarified. Further study of the mechanism and modulation of mesoscale eddies on SSWs will help to reveal the energy exchange mechanism between ocean phenomena at different scales and to improve the accuracy of wave numerical simulation in the high incidence eddy area. At the meantime, it also provides new ideas for the study of energy exchange in the eddy area.

2.2.4 *Eddy modulation to near-inertial waves*

Near-inertial internal waves (NIW), also known as near-inertial oscillations, are a type of ocean internal wave with frequencies close to the local inertial frequency (Coriolis parameter) and exhibit characteristics like internal tides (Alford *et al.*, 2016). The NIWs contribute significantly to the input of energy into the global ocean, playing a crucial role in driving turbulent mixing. The horizontal spatial scale of near-inertial internal waves is typically on the order of 10–100 km (Fu, 1981). Their group velocities enable strong interactions with mesoscale features in the ocean, such as mesoscale eddies modulating the local effective inertial frequency and influencing the transport of near-inertial energy.

Kunze (1985a, 1985b) and Kunze and Sanford (1984) observed the energy propagation of NIWs along the warm flank of the subtropical front in the North Pacific and derived the dispersion relationship of near-inertial waves within a geostrophic flow. By accounting for the Doppler shift effect, an approximate expression for the effective inertial frequency was obtained,

$$f_{\text{eff}} = [f(f+\zeta)]^{1/2} \approx f_0 + \frac{\zeta}{2}, \tag{2.1}$$

where $\frac{\partial V}{\partial x} - \frac{\partial U}{\partial y}$ is relative vorticity. The vorticity in the background field is proposed to induce a "redshift" or "blueshift" in the near-inertial frequency. Kunze validated this concept through numerical experiments, demonstrating that AEs have the capability to capture NIWs.

Lee and Niiler (1998) investigated the behavior of NIW generation by the surface wind stress on an f-plane using a three-dimensional primitive equation numerical model. They introduced the concept of the "Inertial Chimney" effect through four different eddies, suggesting that AEs can efficiently transfer near-inertial energy from surface layer to deeper layer below thermocline. Zhai *et al.* (2005) utilized the South Pacific Ocean channel model to study the interaction between storm-generated inertial oscillation and mesoscale eddy. Their found that AEs act as conduits leading to the deep sea and significantly enhance the downward propagation of near-inertial energy from the surface layer. Discovered that NIWs forced by typhoons are trapped in regions of negative vorticity and rapidly propagate into the thermocline in those areas. Passage of near-inertial waves through CEs

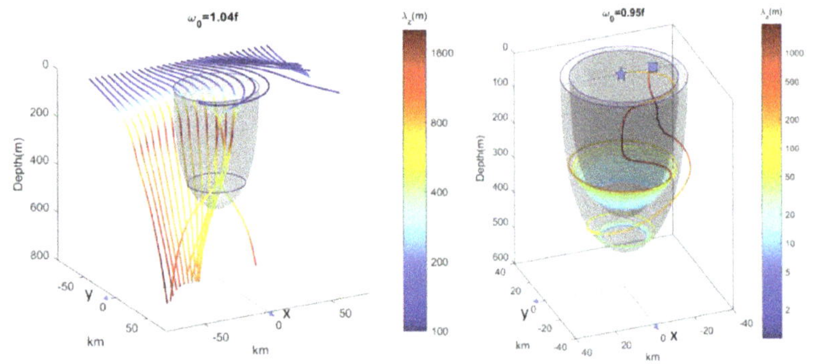

(a) (b)

Fig. 2.3. (a) Ray tracing of super-inertial internal waves ($\omega0 = 1.04f$) emanating from the center of a vortex, with an initial position along $x = -60\,\text{km}$. The color represents the vertical wavelength, and the gray shaded region corresponds to $\zeta = -0.02\,f$, $f_{eff} = 0.98f$. (b) Ray tracing of sub-inertial waves ($\omega0 = 0.95f$) with an initial position along $x = 0\,\text{km}$. The color indicates the vertical wavelength, and the gray shaded regions represent $\zeta = -0.022\,f$, $f_{eff} = 0.978f$ and $\zeta = -0.045\,f$, $f_{eff} = 0.955\,f$. For clarity, only two rays are plotted. The square and star symbols denote the initial positions of rays at $x = -15\,\text{km}$ and $x = 1\,\text{km}$, respectively.

results in their stagnation in the upper ocean, intensifying vertical shear and inducing advection cooling. Through detailed energy budget analysis, Lelong *et al.* (2020) found that AEs play a catalytic role in transferring wind-driven inertial energy into propagating waves (Figure 2.3). In a groundbreaking study, Liu *et al.* (2023) conducted the first global observational estimate of energy transfer between mesoscale eddies and NIWs. They estimated that the energy transfer from mesoscale eddies to NIWs is about 0.025 TW, representing 13% of the global work done by near-inertial winds.

2.3 Ocean Eddy Roles in Mesoscale Air-Sea Interactions

2.3.1 *Atmospheric response to ocean eddies*

Mesoscale eddies are not only an important component of the ocean physical environment but also interact with weather and climate systems. Increasing evidence suggests that the interaction process between ocean mesoscale eddies and the atmosphere is very important, with its influence extending throughout the entire troposphere. Early on, Businger *et al.* (1984), using observational data from the 1978 JASIN (Joint Air-Sea Interaction) experiment, discovered that turbulent heat flux increased above warm eddies. This led to a temperature rise in the atmospheric boundary layer, ultimately resulting in higher wind speed over the warm eddies. With the proliferation of satellite data, scientists have gradually discovered that ocean temperature fluctuations can affect small-scale (tens to hundreds of kilometers) atmospheric circulation. The interactions between the ocean and atmosphere show significant differences at the oceanic mesoscale (tens to hundreds of kilometers) compared to the weather scale (thousands of kilometers) (Chelton *et al.*, 2004, Chelton and Xie, 2010). In contrast to the basin-scale where SST is driven by atmospheric wind fields, showing a significant negative correlation with surface wind speed (Wallace *et al.*, 1990), at smaller ocean scales, SST anomaly is greatly associated with mesoscale processes like ocean fronts, mesoscale eddies, and upwellings, exhibiting a strong positive correlation between SST and surface wind speed. This reflects the strong forcing of the ocean on the atmosphere at meso- and small scales (Small *et al.*, 2003, 2008; Song and Zhou, 2013; Vecchi *et al.*, 2004; White and Annis, 2003; Xie, 2004; Xu *et al.*, 2008).

With the rapid development of observational techniques and technology, a large body of research has found that ocean mesoscale eddies can leave mesoscale imprints on meteorological parameters such as sea surface wind speed, cloud cover and precipitation. Nonaka and Xie (2003) found that the "meandering" and "straight" paths of the Kuroshio Extension stream axis significantly affect the wind speed, cloud cover, and precipitation in that area. The difference between "meandering" and "straight" paths manifests in the appearance of CEs south of Japan, emphasizing the strong forcing effect of larger horizontal oceanic eddies on the atmosphere. Park *et al.* (2006) studied the relationship between sea surface winds and ocean mesoscale eddies in the Gulf Stream area and found that the sea surface winds above AEs increased by 10%, while that above CEs decreased by 15%. Subsequently, used data from the U.S. Navy's Coupled Boundary Layer Air-Sea Transfer experiment to confirm the impact of mesoscale SST changes on the atmospheric boundary layer.

Koseki and Watanabe (2010) conducted a numerical model study on the effects of eddies in the Kuroshio Extension region on the atmospheric boundary layer, with results primarily showing a significant positive correlation between anomalies of sea surface wind speed and mesoscale SST.

Based on satellite observations, Frenger *et al.* (2013) conducted a composite analysis of over 600,000 mesoscale eddies in the Southern Ocean, studying the response anticyclonic (cyclonic) eddies, and anomalies in cloud liquid water content and precipitation rates exhibited maximum values at the eddy centers. However, compared to the Southern Ocean heat flux, the sea-air temperature and humidity difference in the Kuroshio Extension region is significant, releasing a large amount of heat to the atmosphere. The degrees of atmospheric parameter response to ocean mesoscale eddies also vary slightly. Ma *et al.* (2016) examined the seasonal variation in the atmospheric response to mesoscale ocean eddies in the Kuroshio Extension region. They found that the impact of ocean eddies on the atmosphere is strongest in winter and spring, and weaker in summer and autumn, and attributed the differences in atmospheric response primarily to seasonal variations in atmospheric stability. Shan *et al.* (2020) utilized idealized WRF (Weather Research and Forecasting) model experiments to explore the effects of mesoscale ocean eddies on the atmosphere. Their results showed that as background wind speed increased, the latent and sensible heat fluxes from AEs and CEs increased, and the 10 m wind speed, potential for precipitation, and boundary layer height all decreased correspondingly. Using idealized numerical model experiments, examined the effects of mesoscale oceanic eddies on the atmospheric boundary layer and the variations in these effects between coupled and uncoupled models. The results from the experiments indicate that eddies significantly impact various parameters of the atmospheric boundary layer. Cyclonic (anticyclonic) eddies lead to an increase (decrease) in sea-level pressure (SLP). Concurrently, CEs decrease latent and sensible heat fluxes, weaken surface wind speeds, lower the Marine Atmospheric Boundary Layer (MABL) height and diminish atmospheric potential precipitation, while AEs exhibit opposite effects. Moreover, the research found that eddies had a more significant effect on atmospheric parameters in uncoupled experiments. In comparison with coupled experiments, uncoupled experiments demonstrate that eddies could cause a 50% increase in precipitation and a 20–30% increase in wind speed and heat flux. These findings indicate that when considering the coupling effects between the ocean and atmosphere, the impact of eddies on the atmosphere varies.

Ji *et al.* (2020) investigated eddies in four actives but geographically separated regions in the North Pacific, Kuroshio Extension (KE), Subtropical Front (SF), California Current (CC) and Aleutian Islands (AI) that do not influence each other, focusing on the differences in their impact on the atmosphere across regions and Seasons and using observational data and reanalysis products. Using a large sample synthesis method, they assessed the variations in eddy characteristics (EKE and SSTA) and atmospheric stability (potential temperature, TKE flux) and investigated the causes and mechanisms for regional and seasonal differences in eddy impacts on the atmosphere. The results derived from a large-sample synthesis of SSTA, sea surface wind speed, precipitation rate, and heat flux anomalies indicate that AEs in all regions are associated with positive SST anomaly and increase sea surface wind speed, precipitation rate, and heat flux, whereas CEs have opposite effects.

The research results indicate that eddy influence on the atmosphere varies greatly with region and exhibit remarkable seasonal differences. Excluding precipitation rate, eddy effects on SST, sea surface wind speed and heat flux are consistently most pronounced in the KE region, followed by CC and SF and least in the AI region, with a seasonal pattern of being stronger in winter and weaker in summer. Furthermore, the study used linear fitting results of SSTA with wind speed and precipitation rate to confirm that the wind speed anomaly caused by a 1°C of SSTA are strongest in the KE region, followed by the CC and SF regions with the weakest response in the AI region, and distribution of being stronger in winter and weaker in summer. Moreover, the distribution results of precipitation rate anomaly induced by a 1°C of SSTA indicate that the strongest response in winter is still observed in the KE region, followed by AI and CC with the weakest response in the SF region. In contrast, the area with the weakest impact during summer is the CC region, differing from the SF region in winter. The primary reason is that precipitation in the SF region is affected by the monsoon, displaying a seasonal variation opposite to the other three regions, characterized as stronger in summer and weaker in winter. Therefore, the locality and seasonal differences in precipitation across regions are the main reasons for the differences in the distribution of other atmospheric parameters affected by eddies across regions and seasons. Finally, based on the composite results of potential temperature is the smallest in the KE region, followed by AI, CC and SF. In all regions, the vertical change in potential temperature is smaller in winter than that in summer, indicating a weaker stratification during the winter months. Composite results for TKE flux indicate that TKE flux is predominantly positive (negative) over anticyclonic (cyclonic) eddies. When comparing winter and summer, the maximum TKE flux values are observed in the KE region, followed by SF and CC, with the smallest values in AI. The seasonal distribution shows stronger TKE flux in winter and weaker in summer. Thus, the influence of eddies on the atmosphere shows regional variations and seasonal changes, primarily due to differences in the characteristics of eddies and atmospheric stability across regions. Additionally, the composite results for TKE flux and SLP demonstrate that the vertical momentum mixing mechanism is the predominant mechanism, through which eddies influence the atmosphere in these four regions.

The influence of mesoscale ocean eddies on the atmosphere extends beyond local atmospheric transient effects, impacting broader temporal and spatial scales as well. Recent high-resolution satellite observations and atmospheric-oceanic coupled models have revealed that mesoscale ocean eddies have a cumulative effect on air-sea fluxes and significantly affect the climate. Using a high-resolution regional ocean-atmosphere coupled model, Ma *et al.* (2015a, 2015b), indicated that the overall effect of mesoscale ocean eddies in the Kuroshio Extension region can influence North American climate through teleconnection processes. Zhang *et al.* (2020) found that with climate warming, the frequency of strong typhoons increases, and the positive feedback mechanism between typhoons and climate warming can be measured by the energy of ocean eddies induced by typhoons and changes in potential vorticity (PV).

The research identified that the dissipation of numerous eddies, particularly the substantial dissipation of AEs, is directly related to typhoons. Over the past 20 years, the intensity of typhoons in the waters east of Taiwan Island has significantly increased, corresponding to positive anomaly in PV over the ocean. As eddies carrying positive PV anomalies move westward and merge with the Kuroshio, they accelerate the Kuroshio towards the north. This opposes the decreased Kuroshio flow associated with reduced wind stress on a basin scale, potentially exerting a significant warming impact on temperate oceans and climate. Furthermore, it effectively sustains the northward heat transport by the Kuroshio, enhancing the extent of warming in that region and its downstream marine and atmospheric environments. Additionally, the rise in SST is set to significantly increase the frequency of strong typhoons in this maritime zone.

The Southern Ocean connects the global three major oceans, playing a significant role in global ocean circulation, biogeochemical cycles, and climate change. As one of the most powerful oceanic current systems globally, the Antarctic Circumpolar Current-generated mesoscale eddies actively impact air-sea heat exchange and the distribution and storage of ocean heat, influencing extreme ocean temperatures that pose a threat to ecosystem stability. Research indicates that global warming results in more frequent extreme high-temperature events at the sea surface, yet knowledge of subsurface extreme temperature events remains limited. He *et al.* (2023), through analysis of temperature-salinity profiles and satellite remote sensing data in the Southern Ocean, discovered that while AEs are fewer, the frequency of extreme high temperatures within these eddies are elevated. The contribution of eddies to extreme temperatures increases with the depth and the magnitude of temperature anomaly, showing regional differences. The Southern Ocean significantly absorbs heat and carbon dioxide generated from human activities, exhibiting a pronounced rate of warming. The enhanced instability of the Antarctic Circumpolar Current front results in increased EKE, more active eddy movements and intensified ocean stratification, amplifying average and extreme temperature disturbances within eddies.

It indicates that mesoscale eddies significantly contribute to the occurrence and intensification of extreme temperature, potentially enhancing the transfer of near-surface heat to the deep ocean, thereby mitigating the warming of the upper layers.

2.3.2 *Effects of wind input on ocean dynamics via eddies*

The wind is the primary driving force behind internal mixing in the ocean (Munk and Wunsch, 1998; Wunsch and Ferrari, 2004; Huang *et al.*, 2006). It is the wind to generate the wind stress on the sea surface which transfers kinetic energy and momentum into the ocean. Consequently, this process gives rise to various phenomena, including gravity waves, Ekman drift, Langmuir circulation, and turbulent mixing (Wunsch, 1998).

The wind power entering the ocean is calculated by multiplying the wind stress and sea surface velocity (Faller, 1966; Oort *et al.*, 1994; Wunsch, 1998), i.e.,

$$P = \vec{\tau} \cdot \vec{U}, \tag{2.2}$$

where $\vec{\tau}$ is the sea surface wind stress, and $\vec{U}$ the sea surface velocity.

The power exerted by the wind on the ocean is a surface flux from the atmosphere to the sea, serving as a key metric for understanding the interactions and impacts of the ocean on weather and climate systems (Josey *et al.*, 2013). However, measuring surface flux poses challenges, as it is influenced not only by the motion of the sea surface but also by the properties of the MABL layers on both sides of the interface. There are two algorithms commonly used to calculate sea surface wind stress. One is the relative wind stress, which is the wind stress relative to ocean current velocity (Eq. (2.4)). The other is the absolute wind stress, which is calculated using the absolute wind speed without considering ocean current velocity (Eq. (2.2)). Currently, the quadratic calculation of relative wind speed is commonly used (Pacanowski, 1987; Scott and Xu, 2009),

$$\vec{\tau}_{\text{rel}} = \rho_a C_D(\vec{U}_a - \vec{U})|\vec{U}_a - \vec{U}|. \tag{2.3}$$

where ρ_a is the air density above the sea surface (taken as $1.25\,\text{kg}\,\text{m}^{-3}$), $\vec{U}_a$ is the atmospheric wind speed at a height of 10 m above the ocean surface, $\vec{U}$ is the ocean current velocity and C_D is the sea surface drag coefficient, representing the momentum transfer efficiency between the atmosphere and the ocean, and depends on factors such as the wind speed, ocean currents, waves, sea surface roughness and stability (Garratt, 1977; Wu, 1980; Powell *et al.*, 2003; Makin, 2005; Zhao *et al.*, 2015). Another approach is based on absolute wind speed without considering the ocean current velocity. This approach recognizes that ocean velocity is usually less than $1\,\mathrm{m\,s^{-1}}$, which is much smaller than the sea surface wind speed. To simplify the calculation, the surface wind stress exerted by the atmosphere on the ocean can be approximated as a function of atmospheric wind, which is the absolute wind stress (Bunker, 1976; Bye, 1985),

$$\vec{\tau}_{\mathrm{abs}} = \rho_a C_D \vec{U}_a |\vec{U}_a|. \tag{2.4}$$

Recent studies have found that mesoscale eddies play a crucial role in the input of wind energy into the ocean. For instance, Zhao *et al.* (2015) found that relative wind stress can suppress the development of mesoscale eddies, resulting in a reduction of EKE by approximately 20%. This damping effect is primarily caused by the negative wind energy input. Xu *et al.* (2016) estimated the work done by winds on over 1,200,000 mesoscale eddies. They highlighted that winds significantly suppress mesoscale ocean eddies, especially in regions with active eddy motion, such as the western boundary current region and the Southern Ocean. Additionally, Renault *et al.* (2017) conducted a study on the effect of ocean-air coupling on the wind stress based on satellite observations. They found that the coupling coefficient between the sea surface velocity and the surface wind stress varies, representing the transport of eddy energy due to ocean current feedback. This variation generally appears as a linear function of average wind speed. Renault *et al.* (2019), using different wind stresses for numerical simulation, found that using relative wind stress resulted in slightly smaller numbers and intensities of eddies compared to using absolute wind stress. Consequently, they aptly referred to wind stress as the "eddy killer." Moreover, compared the global monthly EKE obtained using absolute and relative wind stress methods in an ocean model. They observed that the EKE obtained using relative wind stress was reduced by 37% compared to using absolute wind stress in the ocean. Based on satellite data, Rai *et al.* (2021) found that wind only inputs kinetic energy into the geostrophic flow at scales larger than 260 km. However, at scales smaller than 260 km, the wind carries energy away from mesoscale eddies in the ocean at an average rate of

50 GW. They concluded that the negative work of relative wind stress on eddies is the primary mechanism for mesoscale energy dissipation, showing significant seasonal variation, with the peak occurring in winter.

Nevertheless, it is crucial to question whether relative wind stress is the sole determinant in eddy suppression. Müller and Frankignoul (1981) found that direct wind stress forcing is one of the main energy sources for ocean mesoscale eddies, particularly in regions far from strong current areas. Although Stammer *et al.* (2001) emphasized that baroclinic instability is the main mechanism for generating mesoscale eddies in the global ocean, strong wind stress vorticity becomes the primary mechanism for eddy generation in areas with low atmospheric forcing. Wang *et al.* (2008) revealed the importance of the northeast monsoon in generating eddies in the eastern SCS during winter, based on a high-resolution reduced gravity model. Jia *et al.* (2011) utilized a numerical model to identify the relationship between intensity variations of eddies and strong winds above the leeward side of Hawaii Island. They found that ocean upwelling and downwelling induced by strong surface wind shear there are the main mechanisms for generating small eddies in the sea area west of Hawaii Islands. Teng *et al.* (2021) designed a simplified idealized model that considers the momentum transfer effect of the wind field on an ideal CE. They calculated the integral of relative wind stress time variation on the work done by the eddy. Their results indicate that wind stress initially does negative work on the eddy.

However, after the eddy adapts to the adjusted wind stress, the structure of the eddy flow field changes, resulting in wind stress doing positive work on the eddy. Moreover, Teng *et al.* (2023), through analyzing remote sensing satellite data, identified significant positive wind power values contributing to eddy generation in the northeastern tropical Pacific Ocean, particularly during the eddy generation stage (Figure 2.4). Low-frequency winds are identified as the primary mechanism for eddy generation in this region, and eddies therein have effectively adapted to wind stress. Consequently, it is evident that the input of wind energy, whether positive or negative, cannot be generalized, as it exhibits spatiotemporal variations, thus necessitating further in-depth research.

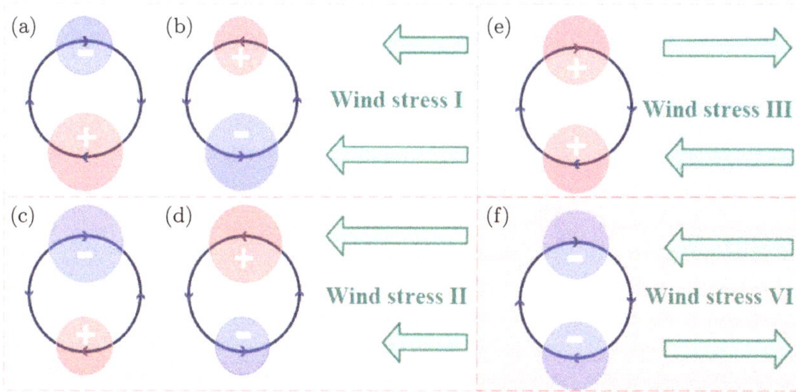

Fig. 2.4. Schematic diagram showing the behavior of eddies under different wind stresses. The wind stresses with negative vorticity impact AE (a) and CE (b), while the wind stresses with positive vorticity act upon AE (c) and CE (d). Notably, AE experiences wind stress from both positive and negative eddies (e–f), displaying a stronger curl compared to the wind stress curl observed in (a–d). The magnitude and direction of wind stress are depicted by the light green arrows, and the eddy boundary is represented by the dark blue circles. The eddy rotation directions are marked by the dark blue arrows. The blue (pink) circles situated within the eddy signify the negative (positive) work done by the wind stress on the eddy, where the circle sizes represent the relative magnitude of positive and negative work performed. The background color corresponds to the magnitude of the wind stress curl, with light red (blue) denoting positive (negative) wind stress curl. Moreover, the relative tone of the background colors in (a–d) and (e–f) represent the relative magnitude of wind stress curl.

2.4 Typhoon-Eddy Interaction

Typhoons play a vital role in air-sea interaction (Sun *et al.*, 2014; Song *et al.*, 2018). A thorough understanding of the interaction among eddies and typhoons can help improving the accuracy of typhoon forecasts.

Pre-existing mesoscale eddies can influence water temperature changes during typhoon passages (Ho *et al.*, 2009). Combining observational data with air-sea coupled models, studies have found that CEs can enhance the sea surface cooling induced by typhoons, while AEs inhibit this cooling effect (Ma *et al.*, 2017). Using a dataset of typhoon tracks and eddy trajectories in the SCS over 15 years (2003–2017), Yu *et al.* (2021) comprehensively analyzed the impact of existing eddies on the sea surface cooling induced by typhoons. Their results show that CEs on the right of the typhoon track lead to more significant cooling, whereas AEs disrupt the cooling area. This indicates that mesoscale eddies can modulate air-sea interaction processes.

Meanwhile, typhoons can significantly influence the intensity, size, and propagation speed of mesoscale eddies. Previous studies have found that typhoons lead to the dissipation of eddies, particularly AEs. In recent decades, there has been a notable intensification trend in typhoons east of Taiwan, accompanied by a corresponding increase in a positive sea level anomaly (SLA) in the ocean (Zhang *et al.*, 2020). Typhoon activity over CEs can lead to increase the intensity and expansion and may even trigger the formation of new CEs, thereby affecting the upper ocean structure and local circulation system (Sun *et al.*, 2014). Qualitative analysis of cyclonic eddy response to typhoons suggests that eddies on different sides of the typhoon track experience different impacts, with shorter-lived or smaller-radius eddies being more susceptible to the influence of typhoons (Yu *et al.*, 2019). Using altimeter data and numerical experiments, indicated that typhoons may alter the three-dimensional structure of CEs through the processes of typhoon-induced upward flow and dynamic adjustment, resulting in elliptical deformation and re-symmetrization of the eddy, ultimately strengthening the CE CEs in the upper ocean not only intensify sea surface cooling following typhoons, but also lead to a decrease of SLA, a deepening of the mixed-layer and a decrease in mixed-layer temperature. These eddies cause strong but shallow subsurface cooling, an increase in chlorophyll-a concentration within the eddy, and significant growth in EKE and available potential energy (APE) (Shang *et al.*, 2015; Liu *et al.*, 2017; Liu and Tang, 2018; Song *et al.*, 2018; Li *et al.*, 2021; Ma *et al.*, 2021).

Typhoons can also cause a reduction in pre-existing AEs, but the cooling is less significant. However, there is a noticeable cooling and increase in salinity in the subsurface layer, with rising of the 20°C isotherm, resulting in a decrease in heat content and kinetic energy, thereby weakening AEs (Lin *et al.*, 2005; Liu *et al.*, 2017; Huang and Wang, 2022). However, some researchers suggest that AEs may not weaken due to the strong cold suction effect from typhoons and may

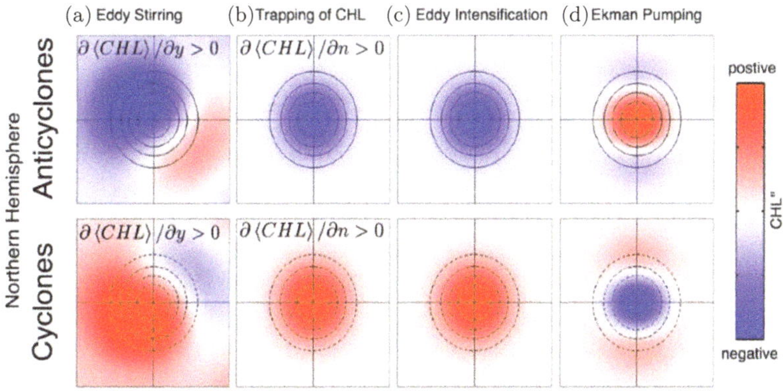

Fig. 2.5. Schematic for mechanisms of ocean eddies affecting surface chlorophyll distribution. (a) Eddy Stirring. (b) Eddy trapping. (c) Eddy enhancement. (d) Eddy-induced Ekman pumping.

even be strengthened by the typhoon-induced negative wind stress curl. This negative wind stress curl anomaly leads to the convergence and sinking of surface warm water in the AE region, increases the thickness of the mixed-layer, and consequently enhances AEs (Zhou *et al.*, 2019).

2.5 Ocean Eddy Roles in Marine Ecologic Environments

Mesoscale oceanic eddies not only affect the atmosphere but also have implications for biogeochemical and hydroacoustic processes within the ocean. The influence of mesoscale eddies on biogeochemistry extends beyond the horizontal and vertical transport of phytoplankton, encompassing their effects on the surrounding environment, thus indirectly impacting chlorophyll growth (Siegel *et al.*, 2011). The mechanisms of influence primarily include eddy stirring, trapping, enhancement, and eddy-induced Ekman pumping (Figure 2.5) (Gaube *et al.*, 2014). Among these mechanisms, eddy stirring and trapping mainly involve horizontal transport. Eddy stirring refers to the localized flow field rotation formed by eddies at their edges, driving horizontal movement of chlorophyll in the surrounding area (Siegel *et al.*, 2008, 2011). Trapping is due to the strong nonlinearity of eddies, causing water masses within the eddies to be entrapped and transported (Gaube *et al.*, 2014; McGillicuddy, 2016). Eddy enhancement and eddy-induced Ekman pumping involve vertical transport. Eddy enhancement refers to changes in isopycnals, leading to vertical transport during eddy intensification, resulting in an increase in chlorophyll concentration in CEs and a decrease in AEs (McGillicuddy *et al.*, 1998; Liu *et al.*, 2013; Gaube *et al.*, 2014; McGillicuddy, 2016; He *et al.*, 2019). Eddy-induced Ekman pumping occurs due to the interaction between the sea surface wind field and the flow field, generating vertical velocity, resulting in changes in chlorophyll concentration usually opposite to eddy enhancement (Gaube *et al.*, 2013, 2015).

For situations where chlorophyll concentration is higher in AEs Than that in CEs in oligotrophic subtropical circulation, previous investigators have proposed that this is due to eddy-induced Ekman pumping or convective mixing (Gaube *et al.*, 2013; He *et al.*, 2019). However, He *et al.* (2021) proposed a new mechanism: rather than changes in biomass, chlorophyll accumulation is primarily due to physiological responses.

In addition to studying eddy impact mechanisms on chlorophyll concentration, previous investigators have observed the peculiar phenomenon of increased chlorophyll concentration at the eddy edges. Xu *et al.* (2019) developed an automated identification algorithm for chlorophyll rings. They found high chlorophyll concentration rings around both AEs and CEs in the North Pacific (Figure 2.6), with the possible formation factors being horizontal advection and wind-flow interactions leading to eddy-induced Ekman pumping.

Horizontal advection dominates the formation of approximately two-thirds of the chlorophyll rings. Statistically analyzed the spatiotemporal characteristics of eddies and their impacts on chlorophyll distribution in the Bay of Bengal over 26 years, showing that such chlorophyll ring structures accounted for about 1% of the region, mainly forming during the enhancement and mature phases of eddies. The study also indicated that cyclonic (anticyclonic) eddies in the region led to approximately 11.15% (11.25%) increases (decreases) in chlorophyll, with the fastest changes occurring during the intensification and mature phases of eddies, where

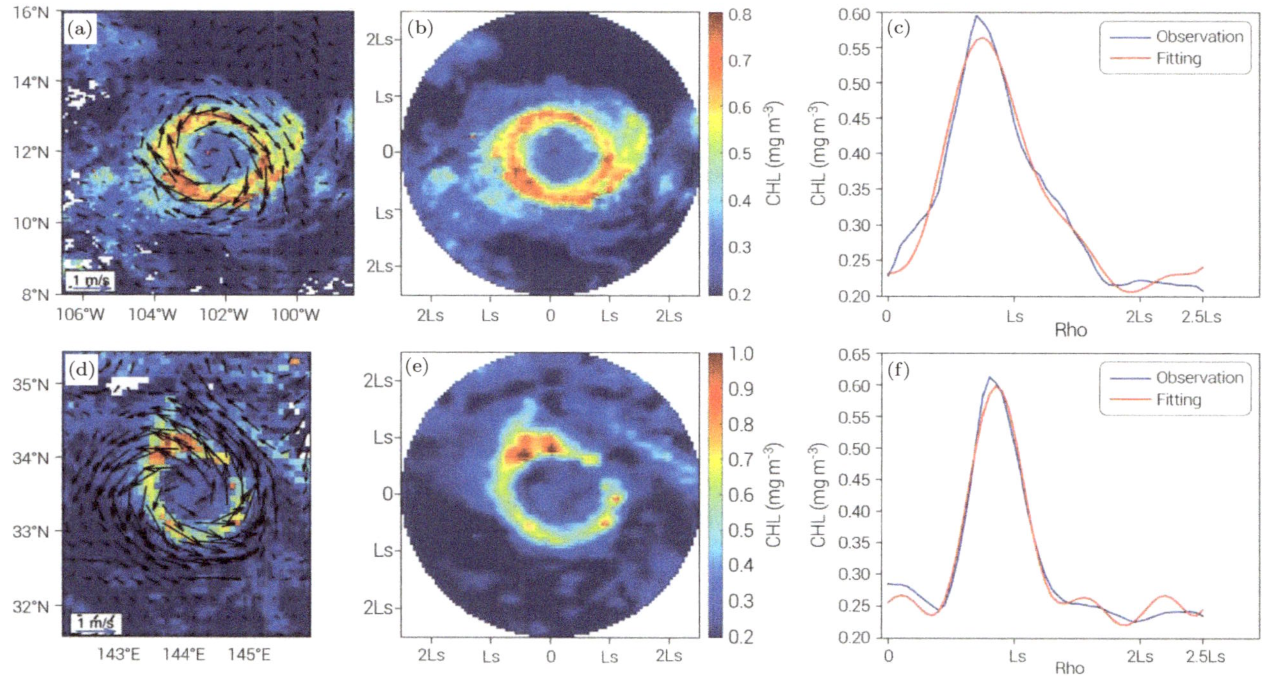

Fig. 2.6. (a) High chlorophyll concentration ring around an AE in the North Pacific on 31 December 2008. Vectors show the velocity field. (b) Conversion of chlorophyll region to polar coordinates. (c) Average chlorophyll distribution along the eddy radius. The peak average chlorophyll concentration is located near Ls, where Ls represents the eddy radius. Rho refers to the normalized distance from the eddy center. Panels (d-f) are the same as panels (a-c) but for a CE on 27 April 2005.

chlorophyll concentrations begin to return to their initial state during the early or late mature stages.

In hydro-acoustics, eddies are important oceanographic features. The study of the eddy effects on sound propagation originated in the 1970s, when large-scale oceanographic surveys and deep-sea hydroacoustic soundings were needed, and it was a major focus of attention due to its importance to naval operations. Temperature and salinity anomalies caused by mesoscale eddies can result in a large-scale inhomogeneous sound speed distribution. Although changes in the sound speed in seawater are generally small relative to itself, the resulting acoustic propagation characteristics can be significantly altered.

With the maturities of hydroacoustic modeling and ocean observation techniques, investigators have found that eddies are capable of causing significant changes in the main characteristics of the acoustic field, including shifts in the convergence zone, changes in the order of multipath arrivals, and horizontal refraction (Heathershaw *et al.*, 1991; Lawrence, 1983; Mellberg *et al.*, 1991). Most of the previous studies were based on parametric descriptions of mesoscale eddies. With the accumulation of oceanographic survey data and the increasing maturity of acoustic models, acoustic studies have gradually shifted to the real ocean environment. Xu *et al.* (2023) investigated the acoustic effects of a subsurface AE in the eastern waters of Taiwan Island. They found that the subsurface eddy can change the structural properties of the convergence zone, and the depth configuration of the source transmitter and receiver greatly impacts the acoustic properties. Chen *et al.* (2022) reported the acoustic effects of AEs in the SCS. They quantified the effects of topography and AEs in the acoustic convergence zones through the results of simultaneous eddy-acoustic field experiments, which also revealed the diversity and complexity of the acoustic field variability under the interaction of eddy environments and topographic changes.

2.6 Summaries

After separating from their parent currents, eddies become huge, movable kinetic energy reservoirs with more than 90% of the ocean total kinetic energy and warm or cold heat sources in the global ocean. This chapter specifies the recent research results of crucial roles that mesoscale eddies play in constituting the energy cascade chain of ocean circulation, and their modulation effects on submesoscale processes, taken near-inertial waves (NIWs) as an example, and small-scale processes, taken sea surface waves (SSWs) as example. On the other hand, as huge heat sources moving in the ocean upper layer, mesoscale eddies must have direct heat exchange with overlying atmosphere and in turn influences on weather and climate systems. Meanwhile, strong disturbances from the atmosphere such as tropical storms can generate ocean eddies and modulate existing eddies. In addition, stirring, trapping and Ekman pumping induced by mesoscale eddies directly affect marine ecological environments, including biogeochemistry, horizontal and vertical transports of phytoplankton and chlorophyll growth.

References

Adams, K. A., Hosegood, P., Taylor, J. R., Sallée, J. B., Bachman, S., Torres, R. and Stamper, M. (2017). Frontal circulation and submesoscale variability during the formation of a Southern Ocean mesoscale eddy, *J. Phys. Oceanogr.*, 47(7), pp. 1737–1753, doi:10.1175/JPO-D-16-0266.1.

Aiki, H., Zhai, X. and Greatbatch, R. J. (2016). Energetics of the global ocean: The role of mesoscale eddies, *Indo-Pacific Clim. Var. Pred.*, pp. 109–134, doi:10.1142/9789814696623_0004.

Alford, M. H., MacKinnon, J. A., Simmons, H. L. and Nash, J. D. (2016). Near-inertial internal gravity waves in the ocean, *Ann. Rev. Mar. Sci.*, 8(1), pp. 95–123, doi:10.1146/annurev-marine-010814-015746.

Aouf, L., Hauser, D., Chapron, B., Toffoli, A., Tourain, C. and Peureux, C. (2021). New directional wave satellite observations: Towards improved wave forecasts and climate description in Southern Ocean, *Geophys. Res. Lett.*, 48(5), pp. 1–10, doi:10.1029/2020GL091187.

Bunker, A. F. (1976). Computations of surface energy flux and annual air-sea interaction cycles of the North Atlantic Ocean, *Mon. Wea. Rev.*, 104(9), pp. 1122–1140, doi:10.1175/1520-0493(1976)_104< 1122:COSEFA>2.0.CO;2.

Businger, J. and Shaw, W. (1984). The response of the marine boundary layer to mesoscale variations in sea-surface temperature, *Dyn. Atmos. Oceans*, 8(3-4), pp. 267–281, doi:10.1016/0377-0265(84)_90012-5.

Bye, J. A. (1985). Large-scale momentum exchange in the coupled atmosphere-ocean, *Elsevier Oceanogr. Ser.*, 40, pp. 51–61, doi:10.1016/S0422-9894(08)_70702-5.

Callies, J. and Ferrari, R. (2013). Interpreting energy and tracer spectra of upper-ocean turbulence in the sub mesoscale range (1-200 km), *J. Phys. Oceanogr.*, 43(11), pp. 2456–2474, doi:10.1175/JPO-D-13-063.1.

Chelton, D. B., Schlax, M. G., Freilich, M. H. and Milliff, R. F. (2004). Satellite measurements reveal persistent small-scale features in ocean winds, *Science*, 303(5660), pp. 978–983, doi:10.1126/science.1091901.

Chelton, D. B. and Xie, S. P. (2010). Coupled ocean-atmosphere interaction at oceanic mesoscales, *Oceanography*, 23(4), pp. 52–69, doi:10.2307/24860862.

Chen, W., Zhang, Y., Liu, Y., Wu, Y., Zhang, Y. and Ren, K. (2022). Observation of a mesoscale warm eddy impacts acoustic propagation in the slope of the South China Sea, *Front. Mar. Sci.*, 9(2020), 1086799, doi:10.3389/fmars.2022.1086799.

Faller, A. J. (1966). Sources of energy for the ocean circulation and a theory of the mixed layer. In *Proc. Fifth US Congr. Appl. Mech.*, Minneapolis, Minnesota, *ASME*, 1966.

Feng, L., Liu, C., Köhl, A., Stammer, D. and Wang, F. (2021). Four types of baroclinic instability waves in the global oceans and the implications for the vertical structure of mesoscale eddies, *J. Geophys. Res. Oceans*, 126(3), e2020JC016966, doi:10.1029/2020JC016966.

Feng, L., Liu, C., Köhl, A. and Wang, F. (2022). Seasonality of four types of baroclinic instability in the global oceans, *J. Geophys. Res. Oceans*, 127(5), e2022JC018572, doi:10.1029/2022JC018572.

Frenger, I., Gruber, N., Knutti, R. and Münnich, M. (2013). Imprint of Southern Ocean eddies on winds, clouds and rainfall, *Nature Geosci.*, 6(8), pp. 608–612, doi:10.1038/ngeo1863.

Fu, L. L. (1981). Observations and models of inertial waves in the deep ocean, *Rev. Geophys.*, 19(1), pp. 141–170, doi:10.1029/RG019i001p00141.

Garratt, J. R. (1977). Review of drag coefficients over oceans and continents, *Mon. Wea. Rev.*, 105(7), pp. 915–929, doi:10.1175/1520-0493(1977)105<0915:RODCOO>2.0.CO;2.

Gaube, P., Chelton, D. B., Strutton, P. G. and Behrenfeld, M. J. (2013). Satellite observations of chlorophyll, phytoplankton biomass, and Ekman pumping in nonlinear mesoscale eddies, *J. Geophys. Res. Oceans*, 118(12), pp. 6349–6370, doi:10.1002/2013JC009027.

Gaube, P., McGillicuddy Jr, D. J., Chelton, D. B., Behrenfeld, M. J. and Strutton, P. G. (2014). Regional variations in the influence of mesoscale eddies on near surface chlorophyll, *J. Geophys. Res. Oceans*, 119(12), pp. 8195–8220, doi:10.1002/2014JC010111.

Gill, A. E., Green, J. S. A. and Simmons, A. J. (1974). Energy partition in the large-scale ocean circulation and the production of mid-ocean eddies, *Deep Sea Res. Oceanogr. Abst.*, 21(7), pp. 499–528, doi:10.1016/0011-7471(74)90010-2.

Hauser, D., Tourain, C., Hermozo, L., Alraddawi, D., Aouf, L., Chapron, B. and Tran, N. (2020). New observations from the SWIM radar on-board CFOSAT: Instrument validation and ocean wave measurement assessment, *IEEE Trans. Geosci. Rem. Sens.*, 59(1), pp. 5–26, doi:10.1109/TGRS.2020.2994372.

He, Q., Zhan, H., Xu, J., Cai, S., Zhan, W., Zhou, L. and Zha, G. (2019). Eddy-induced chlorophyll anomalies in the western South China Sea, *J. Geophys. Res. Oceans*, 124(12), pp. 9487–9506, doi:10.1029/2019JC015371.

He, Q., Zhan, H., Cai, S. and Zhan, W. (2021). Eddy-induced near-surface chlorophyll anomalies in the Subtropical Gyres: Biomass or physiology? *Geophys. Res. Let.*, 48(7), e2020GL091975, doi:10.1029/2020GL091975.

He, Q., Zhan, W., Cai, S., Du, Y., Chen, Z., Tang, S. and Zhan, H. (2023). Enhancing impacts of mesoscale eddies on Southern Ocean temperature variability and extremes, *Proc. Nat. Acad. Sci.*, 120(39), e2302292120, doi:10.1073/pnas.2302292120.

Heathershaw, A. D., Stretch, C. E. and Maskell, S. J. (1991). Coupled ocean-acoustic model studies of sound propagation through a front, *The J. Acous. Soc. Amer.*, 89(1), pp. 145–155, doi:10.1121/1.400520.

Ho, C. R., Zheng, Q., Zheng, Z. W., Kuo, N. J., Tai, C. K. and Su, F. C. (2009). Reply to comment by A. Wada *et al.* on "Importance of pre-existing oceanic conditions to upper ocean response induced by Super Typhoon Hai-Tang", *Geophys. Res. Lett.*, 36(9), L09604, doi:10.1029/2009GL037443.

Huang, R. X., Wang, W. and Liu, L. L. (2006). Decadal variability of wind-energy input to the world ocean, *Deep Sea Res. Part II: Top. Stu. Oceanogr.*, 53(1-2), pp. 31–41, doi:10.1016/j.dsr2.2005.11.001.

Huang, X. and Wang, C. (2022). Response of a mesoscale dipole eddy to the passage of a tropical cyclone: A case study using satellite observations and numerical modeling, *Rem. Sens.*, 14(12), 2865, doi:10.3390/rs14122865.

Jaimes, B. and Shay, L. K. (2010). Near-inertial wave wake of Hurricanes Katrina and Rita over mesoscale oceanic eddies, *J. Phys. Oceanogr.*, 40(6), pp. 1320–1337, doi:10.1175/2010JPO4309.1.

Ji, J., Ma, J., Dong, C., Chiang, J. C. and Chen, D. (2020). Regional dependence of atmospheric responses to oceanic eddies in the North Pacific Ocean, *Rem. Sens.*, 12(7), 1161, doi:10.3390/rs12071161.

Jia, Y., Calil, P. H. R., Chassignet, E. P., Metzger, E. J., Potemra, J. T., Richards, K. J. and Wallcraft, A. J. (2011). Generation of mesoscale eddies in the lee of the Hawaiian Islands, *J. Geophys. Res. Oceans*, 116(C11), C11009, doi:10.1029/2011JC007305.

Jing, Z., Fox-Kemper, B., Cao, H., Zheng, R. and Du, Y. (2021). Submesoscale fronts and their dynamical processes associated with symmetric instability in the northwest Pacific subtropical ocean, *J. Phys. Oceanogr.*, 51(1), pp. 83–100, doi:10.1175/JPO-D-20-0076.1.

Josey, S. A., Gulev, S. and Yu, L. (2013). Exchanges through the ocean surface, *Internat. Geophys.*, 103, pp. 115–140, doi:10.1016/B978-0-12-391851-2.00005-2.

Koseki, S. and Watanabe, M. (2010). Atmospheric boundary layer response to mesoscale SST anomalies in the Kuroshio Extension, *J. Clim.*, 23(10), pp. 2492–2507, doi:10.1175/2009JCLI2915.1.

Kunze, E. and Sanford, T. B. (1984). Observations of near-inertial waves in a front, *J. Phys. Oceanogr.*, 14(3), pp. 566–581, doi:10.1175/1520-0485(1984)_014<0566:OONIWI>2.0.CO;2.

Kunze, E. (1985a). Near-inertial wave propagation in geostrophic shear, *J. Phys. Oceanogr.*, 15(5), pp. 544–565, doi:10.1175/1520-0485(1985)_015<0544:NIWPIG>2.0.CO;2.

Kunze, E. (1985b), Corrigendum, *J. Phys. Oceanogr.*, 15(5), 1355.

Lawrence, M. W. (1983). Modeling of acoustic propagation across warm-core eddies, *The J. Acous. Soc. Amer.*, 73(2), pp. 474–485, doi:10.1121/1.388982.

Lee, D. K. and Niiler, P. P. (1998). The inertial chimney: The near-inertial energy drainage from the ocean surface to the deep layer, *J. Geophys. Res. Oceans*, 103(C4), pp. 7579–7591, doi:10.1029/97JC03200.

Lelong, M. P., Cuypers, Y. and Bouruet-Aubertot, P. (2020). Near-inertial energy propagation inside a Mediterranean anticyclonic eddy, *J. Phys. Oceanogr.*, 50(8), pp. 2271–2288, doi:10.1175/JPO-D-19-0211.1.

Li, X., Zhang, X., Fu, D. and Liao, S. (2021). Strengthening effect of super typhoon Rammasun (2014) on upwelling and cold eddies in the South China Sea, *J. Oceanol. Limnol.*, 39, pp. 403–419, doi:10.1007/s00343-020-9239-x.

Lin, I. I., Wu, C. C., Emanuel, K. A., Lee, I. H., Wu, C. R. and Pun, I. F. (2005). The interaction of Supertyphoon Maemi (2003) with a warm ocean eddy, *Mon. Wea. Rev.*, 133(9), pp. 2635–2649, doi:10.1175/MWR3005.1.

Liu, F., Tang, S. and Chen, C. (2013). Impact of nonlinear mesoscale eddy on phytoplankton distribution in the northern South China Sea, *J. Mar. Sys.*, 123, pp. 33–40, doi:10.1016/j.jmarsys.2013.04.005.

Liu, F. and Tang, S. (2018). Influence of the interaction between typhoons and oceanic mesoscale eddies on phytoplankton blooms, *J. Geophys. Res. Oceans*, 123(4), pp. 2785–2794, doi:10.1029/2017JC013225.

Liu, G., Chen, Z., Lu, H., Liu, Z., Zhang, Q., He, Q. and Cai, S. (2023). Energy transfer between mesoscale eddies and near-inertial waves from surface drifter observations, *Geophys. Res. Lett.*, 50(16), e2023GL104729, doi:10.1029/2023GL104729.

Liu, S. S., Sun, L., Wu, Q. and Yang, Y. J. (2017). The responses of cyclonic and anticyclonic eddies to typhoon forcing: The vertical temperature-salinity structure changes associated with the horizontal convergence/divergence, *J. Geophys. Res. Oceans*, 122(6), pp. 4974–4989, doi:10.1002/2017JC012814.

Lorenz, E. N. (1955). Available potential energy and the maintenance of the general circulation, *Tellus*, 7(2), pp. 157–167, doi:10.3402/tellusa.v7i2.8796.

Luo, S., Jing, Z. and Qi, Y. (2020). Submesoscale flows associated with convergent strain in an anticyclonic eddy of the Kuroshio extension: A high-resolution numerical study, *Ocean Sci. J.*, 55, pp. 249–264, doi:10.1007/s12601-020-0022-x.

Ma, J., Xu, H. and Dong, C. (2016). Seasonal variations in atmospheric responses to oceanic eddies in the Kuroshio Extension, *Tellus A: Dyn. Meteorol. Oceanogr.*, 68(1), p. 31563, doi:10.3402/tellusa.v68.31563.

Ma, J., Xu, H., Dong, C., Lin, P. and Liu, Y. (2015a). Atmospheric responses to oceanic eddies in the Kuroshio Extension region, *J. Geophys. Res. Atmos.*, 120(13), pp. 6313–6330, doi:10.1002/2014JD022930.

Ma, X., Chang, P., Saravanan, R., Montuoro, R., Hsieh, J. S., Wu, D. and Jing, Z. (2015b). Distant influence of Kuroshio eddies on North Pacific weather patterns? *Sci. Rep.*, 5(1), p. 17785, doi:10.1038/srep17785.

Ma, Z., Fei, J., Liu, L., Huang, X. and Li, Y. (2017). An investigation of the influences of mesoscale ocean eddies on tropical cyclone intensities, *Mon. Wea. Rev.*, 145(4), pp. 1181–1201, doi:10.1175/MWR-D-16-0253.1.

Ma, Z., Zhang, Z., Fei, J. and Wang, H. (2021). Imprints of tropical cyclones on structural characteristics of mesoscale oceanic eddies over the western North Pacific, *Geophys. Res. Lett.*, 48(10), e2021GL092601, doi:10.1029/2021GL092601.

Makin, V. K. (2005). A note on the drag of the sea surface at hurricane winds, *Boundary-layer Meteorol.*, 115, pp. 169–176, doi:10.1007/s10546-004-3647-x.

Matsuta, T. and Masumoto, Y. (2023). Energetics of the Antarctic Circumpolar Current. Part I: The Lorenz energy cycle and the vertical energy redistribution, *J. Phys. Oceanogr.*, 53(6), pp. 1467–1484, doi:10.1175/JPO-D-22-0133.1.

McGillicuddy Jr, D. J., Robinson, A. R., Siegel, D. A., Jannasch, H. W., Johnson, R., Dickey, T. D. and Knap, A. H. (1998). Influence of mesoscale eddies on new production in the Sargasso Sea, *Nature*, 394(6690), pp. 263–266, doi:10.1038/28367.

McGillicuddy Jr, D. J. (2016). Mechanisms of physical-biological-biogeochemical interaction at the oceanic mesoscale, *Ann. Rev. Mar. Sci.*, 8(1), pp. 125–159, doi:10.1146/annurev-marine-010814-015606.

McWilliams, J. C. (2016). Submesoscale currents in the ocean, *Proc. Royal Soc. A: Math., Phys. Engin. Sci.*, 472(2189), 20160117, doi:10.1098/rspa.2016.0117.

Mellberg, L. E., Johannessen, O. M., Connors, D. N., Botseas, G. and Browning, D. G. (1991). Acoustic propagation in the western Greenland Sea frontal zone, *The J. Acous. Soc. Amer.*, 89(5), pp. 2144–2156, doi:10.1121/1.400908.

Munk, W. and Wunsch, C. (1998). Abyssal recipes II: Energetics of tidal and wind mixing, *Deep Sea Res. Part I: Oceanogr. Res. Pap.*, 45(12), pp. 1977–2010, doi:10.1016/S0967-0637(98)_00070-3.

Müller, P. and Frankignoul, C. (1981). Direct atmospheric forcing of geostrophic eddies, *J. Phys. Oceanogr.*, 11(3), pp. 287–308, doi:10.1175/1520-0485(1981)011<0287:DAFOGE>2.0.CO;2.

Nonaka, M. and Xie, S. P. (2003). Covariations of sea surface temperature and wind over the Kuroshio and its extension: Evidence for ocean-to-atmosphere feedback, *J. Clim.*, 16(9), pp. 1404–1413, doi:10.1175/1520-0442(2003)16<1404:COSSTA>2.0.CO;2.

Olbers, D., Willebrand, J. and Eden, C. (2012). *Ocean Dynamics*, Springer Science and Business Media, Berlin.

Oort, A. H., Anderson, L. A. and Peixoto, J. P. (1994). Estimates of the energy cycle of the oceans, *J. Geophys. Res. Oceans*, 99(C4), pp. 7665–7688, doi:10.1029/93JC03556.

Pacanowski, R. C. (1987). Effect of equatorial currents on surface stress, *J. Phys. Oceanogr.*, 17(6), pp. 833–838, https://doi.org/10.1175/1520-0485(1987)017%3C0833:EOECOS%3E2.0.CO;2.

Park, K. A., Cornillon, P. and Codiga, D. L. (2006). Modification of surface winds near ocean fronts: Effects of Gulf Stream rings on scatterometer (QuikSCAT, NSCAT) wind observations, *J. Geophys. Res. Oceans*, 111(C3), C03021, doi:10.1029/2005JC003016.

Powell, M. D., Vickery, P. J. and Reinhold, T. A. (2003). Reduced drag coefficient for high wind speeds in tropical cyclones, *Nature*, 422(6929), pp. 279–283, doi:10.1038/nature01481.

Rai, S., Hecht, M., Maltrud, M. and Aliu, H. (2021). Scale of oceanic eddy killing by wind from global satellite observations, *Sci. Adv.*, 7(28), eabf4920, doi:10.1126/sciadv.abf4920.

Renault, L., McWilliams, J. C. and Penven, P. (2017). Modulation of the Agulhas Current retroflection and leakage by oceanic current interaction with the atmosphere in coupled simulations, *J. Phys. Oceanogr.*, 47(8), pp. 2077–2100, doi:10.1029/2018GL081211.

Renault, L., Marchesiello, P., Masson, S. and Mcwilliams, J. C. (2019). Remarkable control of western boundary currents by eddy killing, a mechanical air-sea coupling process, *Geophys. Res. Lett.*, 46(5), pp. 2743–2751, doi:10.1029/2018GL081211.

Romero, L., Hypolite, D. and McWilliams, J. C. (2020). Submesoscale current effects on surface waves, *Ocean Modelling*, 153, 101662, doi:10.1016/j.ocemod.2020.101662.

Scott, R. B. and Wang, F. (2005). Direct evidence of an oceanic inverse kinetic energy cascade from satellite altimetry, *J. Phys. Oceanogr.*, 35(9), pp. 1650–1666, doi:10.1175/JPO2771.1.

Scott, R. B. and Xu, Y. (2009). An update on the wind power input to the surface geostrophic flow of the World Ocean, *Deep Sea Res. Part I: Oceanogr. Res. Pap.*, 56(3), pp. 295–304, doi:10.1016/j.dsr.2008.09.010.

Shan, X., Jing, Z., Sun, B. and Wu, L. (2020). Impacts of ocean current-atmosphere interactions on mesoscale eddy energetics in the Kuroshio extension region, *Geosci. Lett.*, 7, pp. 1–8, doi:10.1186/s40562-020-00152-w.

Shang, X. D., Zhu, H. B., Chen, G. Y., Xu, C. and Yang, Q. (2015). Research on cold core eddy change and phytoplankton bloom induced by typhoons: Case studies in the South China Sea, *Adv. Meteor.*, 2015(1), 340432, doi:10.1155/2015/340432.

Siegel, D. A., Court, D. B., Menzies, D. W., Peterson, P., Maritorena, S. and Nelson, N. B. (2008). Satellite and in situ observations of the bio-optical signatures of two mesoscale eddies in the Sargasso Sea, *Deep Sea Res. Part II: Top. Stu. Oceanogr.*, 55(10–13), pp. 1218–1230, doi:10.1016/j.dsr2.2008.01.012.

Siegel, D. A., Peterson, P., McGillicuddy Jr, D. J., Maritorena, S. and Nelson, N. B. (2011). Bio-optical footprints created by mesoscale eddies in the Sargasso Sea, *Geophys. Res. Lett.*, 38(13), doi:10.1029/2011GL047660.

Small, R. D., deSzoeke, S. P., Xie, S. P., O'Neill, L., Seo, H., Song, Q. and Minobe, S. (2008). Air-sea interaction over ocean fronts and eddies, *Dyn. Atmos. Oceans*, 45(3-4), pp. 274–319, doi:10.1016/j.dynatmoce.j.dynatmoce.2008.01.0012008.01.001.

Small, R. J., Xie, S. P. and Wang, Y. (2003). Numerical simulation of atmospheric response to Pacific tropical instability waves,

J. Clim., 16(22), pp. 3723–3741, doi:10.1175/1520-0442(2003)016 <3723:NSOART>2.0.CO;2.

Song, D., Guo, L., Duan, Z. and Xiang, L. (2018). Impact of major typhoons in 2016 on sea surface features in the Northwestern Pacific, *Water*, 10(10), p. 1326, doi:10.3390/w10101326.

Song, F. and Zhou, T. (2013). FGOALS-s2 simulation of upper-level jet streams over East Asia: Mean state bias and synoptic-scale transient eddy activity, *Adv. Atmos. Sci.*, 30, pp. 739–753, doi:10.1007/s00376-012-2212-7.

Ssebandeke, J., von Storch, J. S. and Brüggemann, N. (2024). Sensitivity of the Lorenz energy cycle of the global ocean, *Ocean Dyn.*, 74(1), pp. 1–16, doi:10.1007/s10236-023-01568-6.

Stammer, D., Böning, C. and Dieterich, C. (2001). The role of variable wind forcing in generating eddy energy in the North Atlantic, *Progr. Oceanogr.*, 48(2-3), pp. 289–311, doi:10.1016/S0079-6611(01)00008-8.

Steinberg, J. M., Cole, S. T., Drushka, K. and Abernathey, R. P. (2022). Seasonality of the mesoscale inverse cascade as inferred from global scale-dependent eddy energy observations, *J. Phys. Oceanogr.*, 52(8), pp. 1677–1691, doi:10.1175/JPO-D-21-0269.1.

Sun, L., Li, Y. X., Yang, Y. J., Wu, Q., Chen, X. T., Li, Q. Y. and Xian, T. (2014). Effects of super typhoons on cyclonic ocean eddies in the western North Pacific: A satellite data-based evaluation between 2000 and 2008, *J. Geophys. Res. Oceans*, 119(9), pp. 5585–5598, doi:10.1002/2013JC009575.

Tan, K., Xie, L., Bai, P., Zheng, Q., Li, J., Xu, Y. and Li, M. (2023). Modulation effects of mesoscale eddies on sea surface wave fields in the South China Sea derived from a wave spectrometer onboard the China-France Ocean Satellite, *J. Geophys. Res. Oceans*, 128(1), e2021JC018088, doi:10.1029/2021JC018088.

Teng, F., Dong, C., Ji, J., Bethel, B. J., Pan, A. and Xu, C. (2021). Does the wind stress always damp an oceanic eddy? *Geosci. Lett.*, 8, pp. 1–6, doi:10.1186/s40562-021-00206-7.

Teng, F., Dong, C., Lim Kam Sian, K. T. C., Ji, J. and Zhu, W. (2023). Wind work on oceanic mesoscale eddies in the Northeast Tropical Pacific Ocean, *Front. Mar. Sci.*, 10, 1202875, doi:10.3389/fmars.2023.

Vecchi, G. A., Xie, S. P. and Fischer, A. S. (2004). Ocean-atmosphere covariability in the western Arabian Sea, *J. Clim.*, 17(6), pp. 1213–1224, doi:10.1175/1520-0442(2004)017<1213:OCITWA>2.0.CO;2.

Wallace, J. M., Smith, C. and Jiang, Q. (1990). Spatial patterns of atmosphere-ocean interaction in the northern winter, *J. Clim.*, 3(9), pp. 990–998, doi:10.1175/1520-0442(1990)003<0990:SPOAOI>2.0.CO;2.

Wang, G., Chen, D. and Su, J. (2008). Winter eddy genesis in the eastern South China Sea due to orographic wind jets, *J. Phys. Oceanogr.*, 38(3), pp. 726–732, doi:10.1175/2007JPO3868.1.

Wang, J., Bethel, B. J., Dong, C., Li, C. and Cao, Y. (2022). Numerical simulation and observational data analysis of mesoscale eddy effects on surface waves in the South China Sea, *Rem. Sens.*, 14(6), 1463, doi:10.3390/rs14061463.

Wang, R., Cheng, X., Xu, L. and Chen, J. (2020). Mesoscale eddy effects on the subduction of North Pacific eastern subtropical mode water, *J. Geophys. Res. Oceans*, 125(5), e2019JC015641, doi:10.1029/2019JC015641.

White, W. B. and Annis, J. L. (2003). Coupling of extratropical mesoscale eddies in the ocean to westerly winds in the atmospheric boundary layer, *J. Phys. Oceanogr.*, 33(5), 1095–1107, doi:10.1175/1520-0485(2003)_033<1095:COEMEI>2.0.CO;2.

Wu, J. (1980). Wind-stress coefficients over sea surface near neutral conditions - A revisit, *J. Phys. Oceanogr.*, 10(5), pp. 727–740, doi:10.1175/1520-0485(1980)_010<0727:WSCOSS>2.0.CO;2.

Wunsch, C. (1998). The work done by the wind on the oceanic general circulation, *J. Phys. Oceanogr.*, 28(11), pp. 2332–2340, doi:10.1175/1520-0485(1998)_028<2332:TWDBTW>2.0.CO;2.

Wunsch, C. and Ferrari, R. (2004). Vertical mixing, energy, and the general circulation of the oceans, *Annu. Rev. Fluid Mech.*, 36(1), pp. 281–314, doi:10.1146/annurev.fluid.36.050802.122121.

Xie, S. P. (2004). Satellite observations of cool ocean-atmosphere interaction, *Bull. Amer. Meteor. Soc.*, 85(2), pp. 195–208, doi:10.1175/BAMS-85-2-195.

Xu, A., Dong, C., Yu, F., Zhang, Y., Nan, F., Ren, Q. and Chen, Z. (2023). Effects of a subsurface eddy on acoustic propagation in the northwestern Pacific Ocean, *J. Mar. Sci. Engin.*, 11(9), 1785, doi:10.3390/jmse11091785.

Xu, G., Cheng, C., Yang, W., Xie, W., Kong, L., Hang, R. and Yang, J. (2019). Oceanic eddy identification using an AI scheme, *Rem. Sens.*, 11(11), pp. 1–12, doi:10.3390/rs11111349.

Xu, H., Wang, L. and He, J. (2008). Observed oceanic feedback to the atmosphere over the Kuroshio Extension during spring time and its possible mechanism, *Chin. Sci. Bull.*, 53(12), pp. 1905–1912, doi:10.1007/s11434-008-0138-3.

Xu, L., Li, P., Xie, S. P., Liu, Q., Liu, C. and Gao, W. (2016). Observing mesoscale eddy effects on mode-water subduction and transport in the North Pacific, *Nature Comm.*, 7(1), pp. 1–9, doi:10.1038/ncomms10505.

Yang, Q., Zhao, W., Liang, X., Dong, J. and Tian, J. (2017). Elevated mixing in the periphery of mesoscale eddies in the South China Sea, *J. Phys. Oceanogr.*, 47(4), pp. 895–907, doi:10.1175/JPO-D-16-0256.1.

Yu, F., Yang, Q., Chen, G. and Li, Q. (2019). The response of cyclonic eddies to typhoons based on satellite remote sensing data for 2001–2014 from the South China Sea, *Oceanologia*, 61(2), pp. 265–275, doi:10.1016/j.oceano.2018.11.005.

Yu, J., Lin, S., Jiang, Y. and Wang, Y. (2021). Modulation of typhoon-induced sea surface cooling by preexisting eddies in the South China Sea, *Water*, 13(5), 653, doi:10.3390/w13050653.

Yuan, Y. (1997). Representation of high frequency spectra of ocean waves and the basis for analyzing SAR images, *China J. Oceanol. Limnol.*, 28, pp. 1–5, doi:cir.nii.ac.jp/crid/1570854175812769536.

Zhai, X., Greatbatch, R. J. and Zhao, J. (2005). Enhanced vertical propagation of storm-induced near-inertial energy in an eddying ocean channel model, *Geophys. Res. Lett.*, 32(18), L18602, doi:10.1029/2005GL023643.

Zhang, Y., Zhang, Z., Chen, D., Qiu, B. and Wang, W. (2020). Strengthening of the Kuroshio current by intensifying tropical cyclones, *Science*, 368(6494), pp. 988–993, doi:10.1126/science.aax5758.

Zhang, Z., Liu, Y., Qiu, B., Luo, Y., Cai, W., Yuan, Q. and Tian, J. (2023). Submesoscale inverse energy cascade enhances Southern Ocean eddy heat transport, *Nature Comm.*, 14(1), 1335, doi:10.1038/s41467-023-36991-2.

Zhang, Z. and Qiu, B. (2020). Surface chlorophyll enhancement in mesoscale eddies by submesoscale spiral bands, *Geophys. Res. Lett.*, 47(14), e2020GL088820, doi:10.1029/2020GL088820.

Zhao, Z. K., Liu, C. X., Li, Q., Dai, G. F., Song, Q. T. and Lv, W. H. (2015). Typhoon air-sea drag coefficient in coastal regions, *J. Geophys. Res. Oceans*, 120(2), pp. 716–727, doi:10.1002/2014JC010283.

Zhong, Y., Bracco, A., Tian, J., Dong, J., Zhao, W. and Zhang, Z. (2017). Observed and simulated submesoscale vertical pump of an anticyclonic eddy in the South China Sea, *Sci. Rep.*, 7(1), 44011, doi:10.1038/srep44011.

Zhou, L., Chen, D., Lei, X., Wang, W., Wang, G. and Han, G. (2019). Progress and perspective on interactions between ocean and typhoon, *Chin. Sci. Bull.*, 64(1), pp. 60–72, doi:10.1360/N972018-00668.

Chapter 3

Generation, Development and Dissipation of Mesoscale Eddies

3.1 Introduction

The generation, development and dissipation mechanisms of ocean mesoscale eddies have been a research hot spot since they were discovered in the 1930s. Previous investigations have achieved a common consensus that mesoscale eddies are primarily generated through mechanisms, such as barotropic instability (Kang *et al.*, 2016; Liu *et al.*, 2017; Ji *et al.*, 2018), baroclinic instability (Gill *et al.*, 1974; Pedlosky, 1987; Itoh and Yasuda, 2010; Ji *et al.*, 2018), wind stress forcing (Chi *et al.*, 1998) and interactions between flow and topography (David *et al.*, 1972; Boyer and Davies, 1982; Heywood *et al.*, 1996). Liu *et al.* (2012) suggested that mesoscale eddies undergo three stages of evolution: "young", "mature" and "decaying". These stages collectively occupy approximately 1/5, 3/5, and 1/5 of the total eddy lifespan, respectively. During the evolution process, various parameters including size, energy and vorticity undergo an initial increase and stabilization, followed by a subsequent decline. As mesoscale processes, eddies act as a bridge between large- and small-scale processes. Large-scale processes, such as background ocean currents and the β-effect, influence the movement of eddies, affecting their propagation speeds and trajectories. Eddies also impact certain sub-mesoscale and small-scale processes, such as near-inertial energy transport, internal waves and sea surface wave distribution (see Part 5 for details). The generation mechanisms of mesoscale eddies can be theoretically summarized by factors related to nonlinear terms, baroclinic terms, vorticity input associated with wind stress curl, and changes in vorticity induced by the β-effect and other disturbance processes (Dong *et al.*, 2025).

3.2 Mechanisms of Eddy Generation

3.2.1 *General theories*

The generation of mesoscale eddies is inevitably accompanied by changes in vorticity. One of the most typical characteristics of eddies is the presence of a distinct vertical relative vorticity component. Therefore, discussing the generation mechanism of mesoscale eddies can be approached from the perspective of the vertical vorticity budget in the ocean. The equation expressing the budget of the vertical relative vorticity ζ can be formulated as follows:

$$\begin{aligned}\frac{\partial \zeta}{\partial t} &= -\nabla_h \times [(u \cdot \nabla)\boldsymbol{u}_h] - f(\nabla_h \cdot \boldsymbol{u}_h) - \beta v \\ &\quad + \frac{\nabla_h \rho \times \nabla_h p}{\rho^2} + \nabla_h \times \left[\frac{\partial}{\partial z}\left(A_z \frac{\partial \boldsymbol{u}_h}{\partial z}\right)\right] \\ &\quad + \nabla_h \times \left[\frac{\partial}{\partial x}\left(A_h \frac{\partial \boldsymbol{u}_h}{\partial x}\right) + \frac{\partial}{\partial y}\left(A_h \frac{\partial \boldsymbol{u}_h}{\partial y}\right)\right], \end{aligned} \tag{3.1}$$

where $\boldsymbol{u}$ represents the three-dimensional velocity vector, $\boldsymbol{u}_h$ is the horizontal velocity vector, f is the Coriolis parameter, β represents the variation of the Coriolis parameter f with latitude, v is the north–south velocity component, ρ is the seawater density, p is the seawater pressure, A_z and A_h are the vertical and horizontal eddy viscosity coefficients, ∇ denotes the three-dimensional gradient operator, and ∇_h denotes the horizontal gradient operator. Performing a vertical integration on the above equation yields the depth-integrated vertical relative vorticity budget control equation:

$$\frac{\partial}{\partial t}\int_{-H}^{0}\zeta dz = \underbrace{\int_{-H}^{0} -\nabla_h \times [(u\cdot\nabla)\boldsymbol{u}_h]dz}_{\text{I}} + \underbrace{\int_{-H}^{0}\frac{\nabla_h\rho\times\nabla_h p}{\rho^2}dz}_{\text{II}}$$
$$+\underbrace{\nabla_h\times\boldsymbol{\tau}_h}_{\text{III}} - \underbrace{\beta\int_{-H}^{0} v dz}_{\text{IV}} - \underbrace{\nabla_h\times\boldsymbol{\tau}_b}_{\text{V}} + \underbrace{f\boldsymbol{u}_h\cdot\nabla_h H}_{\text{VI}}$$
$$+\underbrace{\int_{-H}^{0}\nabla_h\times\left[\frac{\partial}{\partial x}\left(A_h\frac{\partial \boldsymbol{u}_h}{\partial x}\right)+\frac{\partial}{\partial y}\left(A_h\frac{\partial \boldsymbol{u}_h}{\partial y}\right)\right]dz}_{\text{VII}}. \tag{3.2}$$

The left side of Eq. (3.2) represents the changing rate of vorticity, and the various terms on the right side determine the change of vorticity. Term I is the nonlinear term, Term II is the baroclinic term, Term III is the vorticity input by the sea surface wind stress curl, Term IV is the vorticity change due to the β-effect, Term V is the vorticity change caused by the bottom frictional stress curl, Term VI is the vorticity change due to stretching and contraction of water columns caused by changes in seafloor topography, and Term VII is the vorticity change caused by lateral stresses.

3.2.2 *Barotropic instability*

The barotropic instability mechanism, i.e., horizontal current shear instability, primarily involves changing vorticity through the nonlinear term. For large-scale ocean circulation, when the horizontal shear is sufficiently strong (according to quasi-geostrophic theory, the horizontal gradient of potential vorticity of the flow field should change its sign in the horizontal direction), unstable perturbations gain energy from background kinetic energy and rapidly grow. The unstable waves, continuously enhanced by nonlinearity, eventually develop into mesoscale eddies. Baroclinic instability is considered one of the important mechanisms widely present in the generation of oceanic mesoscale eddies. It is a dominant factor regulating eddy variations in different regions, such as at the boundaries of strong currents in regions like western boundary currents (Kang *et al.*, 2016; Liu *et al.*, 2017; Ji *et al.*, 2018).

Using drifters and model data, Liu *et al.* (2017) discovered the asymmetry in the distribution of mesoscale eddies on both sides of the Kuroshio Current in the East China Sea. Cyclonic eddies (anticyclonic eddies) are mainly distributed in the west (east) of the Kuroshio, and the eddy radius is mostly smaller than the local first baroclinic Rossby deformation radius. The study indicates that the generation of such eddies is related to the horizontal velocity shear when the Kuroshio flows northeastward along the East China Shelf, and horizontal shear and vertical buoyancy flux are the primary mechanisms for the generation of eddies on both sides of the mainstream of the Kuroshio. Ji *et al.* (2018), based on satellite altimeter data, revealed two main mechanisms for eddy generation in the Kuroshio extension region. The first mechanism is dominated by baroclinic instability and horizontal shear, and the generated eddies are mostly distributed at the boundaries of the Kuroshio extension, generally with smaller radii and shorter lifetimes, with anticyclonic eddies mainly located to the south of the Kuroshio extension and cyclonic eddies to the north. Qiu *et al.* (2019) monitored a cyclonic eddy for three months and collected vertical profile data using autonomous underwater vehicles (AUV). Using an anisotropy analysis method, they revealed the interaction between eddy anisotropy and mean flow straining in baroclinic instability. The study found that this anisotropic partitioning helps eddies gain kinetic energy from the mean flow. The results indicate that changes in eddy shape and anisotropy lead to an increase in kinetic energy, a process that is particularly pronounced as the eddy moves eastward.

3.2.3 *Baroclinic instability*

This mechanism mainly modifies vorticity via baroclinic terms. For large-scale circulation with vertical shear, unstable disturbances can rapidly grow by gaining energy from the release of background potential energy (according to quasi-geostrophic theory, the horizontal gradient of potential vorticity in the flow field must change sign in the vertical direction), eventually evolving into eddies as nonlinearity intensifies. Baroclinic and barotropic instabilities are the two main types of instability mechanisms that generate mesoscale eddies. Numerous researchers contend that most oceanic eddies arise from baroclinic instability (Gill *et al.*, 1974; Pedlosky, 1977; Itoh and Yasuda, 2010; Ji *et al.*, 2018). The primary reason is that areas of active eddies in the actual ocean are often where baroclinic instability grows the fastest. Moreover, the spatial scale of eddies in the ocean exhibits significant latitude dependence, aligning with the deformation radius derived from baroclinic instability (Chelton *et al.*, 1998).

Qiu and Chen (2010) pointed out that within the subtropical counter-current region, the variation in larger mesoscale ocean eddies primarily arises from baroclinic instability induced by vertical shear between the subtropical counter-current and the North equatorial current. Wood *et al.* (1988) indicated that baroclinic instability provides energy and Reynolds stress for the growth of meanders in the Kuroshio extension jet stream, converting the eddy kinetic energy (EKE) into mean kinetic energy and strengthening the Kuroshio extension jet. The jet stream in the Kuroshio extension region exhibits a "meandering" path with strong baroclinic instability features, facilitating the generation of mesoscale eddies in this area. Tai and White (1990), based on satellite altimeter data, calculated the Reynolds stress in the Kuroshio extension region and used observational data to confirm that baroclinic instability enhances the Kuroshio extension jet stream. Stepanov *et al.* (2018), using an eddy-resolving ocean circulation numerical model, investigated the EKE budget (EKE sources and energy conversion rates) in the Sea of Okhotsk and found that wind stress and baroclinic instability play a leading role in the seasonal variation of mesoscale eddy generation on the western boundary of the Sea of Okhotsk during winter and spring, with baroclinic instability having a greater effect than barotropic instability in the eddy generation process in this region. Based on satellite altimeter data, Ji *et al.* (2018) revealed two main mechanisms for eddy generation in the Kuroshio extension region. (Figure 3.1) Apart from barotropic instability, the second mechanism is dominated by baroclinic processes leading to the "meandering" path of the Kuroshio extension, which generates eddies with larger spatial scales and longer lifetimes. The distribution of eddies is opposite to that produced by horizontal shear, specifically showing

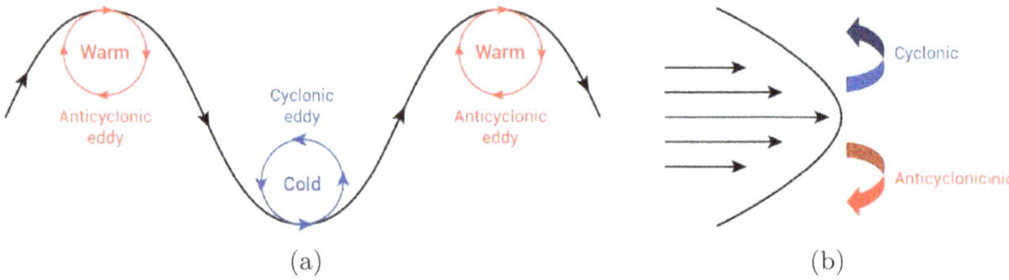

Fig. 3.1. Two instability mechanisms for the generation of mesoscale eddies: (a) baroclinic instability and (b) barotropic instability.
Source: Taken from Ji *et al.* (2018).

that anticyclonic (cyclonic) eddies are more distributed to the north (south) of the Kuroshio extension main axis.

3.2.4 *Wind stress forcing*

From the third component of the budget equation, it is evident that vorticity in the wind field over the ocean can input vorticity into the ocean through wind forcing, thereby generating mesoscale eddies. Numerical model results show that the action of wind fields is a key factor in generating warm and cold eddies in the South China Sea (SCS) basin (Chi *et al.*, 1998). The interaction between wind fields and topography is also an important mechanism for the rich eddy activity in the SCS. Satellite altimeter data reveal that islands alter wind stress vorticity, thus influencing eddy formation.

Offshore disturbances and the wind stress vorticity induced by the Ryukyu Islands restrict the generation of anticyclonic eddies on the east side of the Kuroshio Current, leading to a significantly lesser alignment of anticyclones with the Kuroshio changes compared to cyclonic eddies (Liu and Gan, 2012). In the Southern Hemisphere, the intensification of zonal wind stress leads to an increased isopycnal slope in the Southern Ocean, making the mean flow more prone to baroclinic instability. Owing to baroclinic instability, the potential energy of the mean flow is converted into eddy energy, resulting in many mesoscale eddies (Thompson and Sallée, 2012; Morrison *et al.*, 2015).

3.2.5 *Interaction between currents and topography*

From the last three items, it can be inferred that when topography exists, the resulting bottom friction stress, stretching and contraction of the water column, and the friction stress of the lateral boundary can change vorticity, thereby generating eddies. Among them, the island wake eddy is the most well known. The study of island wake phenomena has a long history. In classical fluid mechanics, a viscous boundary layer is generated at the fixed boundary under homogeneous and non-rotating conditions, resulting in local relative vorticity. According to Bernoulli's principle, when the flow passes an obstacle, the current field accelerates and the pressure decreases, while a reverse pressure gradient is rapidly generated downstream of the current field. This reverse pressure gradient induces secondary circulation and detaches from the fixed boundary to generate an eddy. When considering the rotation and stratification of the actual ocean, experiments, theories, and numerical simulations have shown that enhanced rotation suppresses eddy shedding (David *et al.*, 1972; Boyer and Davies, 1982; Heywood *et al.*, 1996). Boyer and Chen (1987) demonstrated that an increase in density stratification inhibits vertical motion through laboratory fluid experiments, resulting in the generation and shedding of recirculation eddies behind obstacles with a smaller Reynolds number (Re) compared to weak stratification. Island wakes can be classified into shallow- and deep-water island wakes. The criterion for this classification lies in whether the frictional force controlling the island wake is the bottom boundary layer friction stress or the boundary friction stress on both sides of the island (Tomczak, 1988). Dong and McWilliams (2007) pointed out that three types of instabilities coexist during deep-sea island wake phenomena, including centrifugal instability, barotropic instability and baroclinic instability. Through sensitivity experiments, they identified three dimensionless parameters: Reynolds number (Re), Rossby number (Ro) and Burger number (Bu), which together control the entire island wake process. The primary distinction between deep- and shallow-water island wakes lies in their oceanographic characteristics and the sources of vorticity. Vorticity can originate from three main sources: (1) lateral boundary stress, (2) bottom boundary stress, and (3) baroclinic current tilting. If lateral boundary stress is the dominant factor driving vorticity generation, the wake is classified as a deep-water island wake. Conversely, when bottom boundary stress is the main contributor, the wake is considered a shallow-water island wake (Pingree and Maddock, 1979; Signell and Geyer, 1991; Furukawa and Wolanski, 1998; Neill and Elliott, 2004). In shallow-water wakes, baroclinic processes play a crucial role in converting horizontal vorticity tilts into vertical vorticity (Smolarkiewicz and Rotunno, 1989). However, in deep-water wakes, the influence of bottom stress and vorticity tilting is negligible.

3.3 Mechanisms of Eddy Development and Dissipation

The evolution of ocean eddies encompasses multiple stages, each characterized by specific physical processes and mechanisms. The development and dissipation phases dominate most of the eddy lifecycle, representing crucial components in the overall eddy evolution process.

An eddy can be characterized by parameters such as radius, vorticity, kinetic energy, and deformation rate. EKE is defined as the average kinetic energy within the eddy boundary. Eddy deformation rate is defined as $(g_1^2 + g_2^2)^{1/2}$, where the shear deformation rate is $g_1 = \frac{\partial v}{\partial x} + \frac{\partial u}{\partial y}$ and the normal deformation rate is $g_2 = \frac{\partial u}{\partial x} - \frac{\partial v}{\partial y}$. These parameters evolve as the eddy develops. For instance, in the subtropical countercurrent region (15–28°N, 115°E–150°W), which is characterized by high EKE (Aoki and Imawaki, 1996), the evolution of characteristic parameters over time is evident for eddies with an average lifespan exceeding 20 weeks (Figure 3.2). During the first 1/5 of their lifespan (youth), the eddy size and energy increase. These parameters stabilize in the next 3/5 of the lifecycle (maturity) before undergoing a sharp decline in the final 1/5 (decay) of their lifespan, as observed by Liu *et al.* (2012). Interestingly, deformation rates follow a trend opposite to that of size and energy.

Hence, it can be found that the primary evolution process of eddies undergoes three stages, generation, development and dissipation, with certain stage-specific differences. Importantly, external influences may lead to variations in the eddy evolutionary patterns and processes.

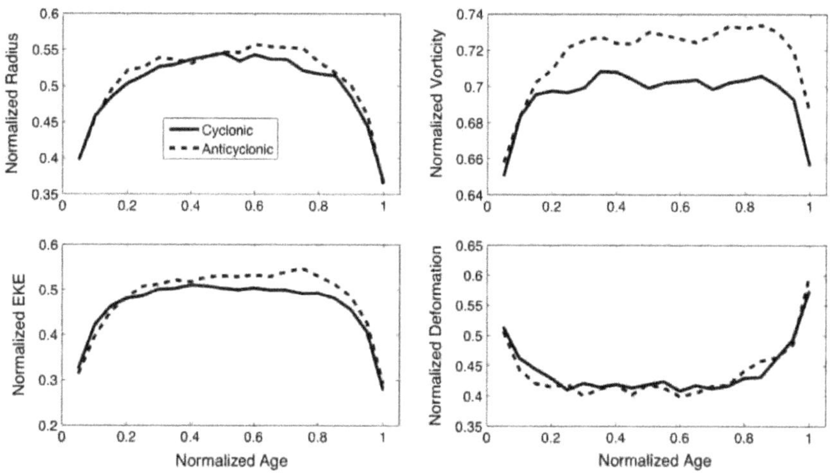

Fig. 3.2. Temporal evolution of characteristic parameters for eddies in the subtropical countercurrent region: radius (top left), vorticity (top right), eddy kinetic energy (bottom left), and deformation radius (bottom right). The lifecycles and parameters of eddies are normalized. Statistics are based on eddies with a lifespan greater than 20 weeks. Solid (dashed) lines correspond to cyclonic (anticyclonic) eddies, respectively.

Source: Taken from Liu *et al.* (2012).

3.3.1 *Background currents*

The background ocean currents (from geostrophic flow) and the β-effect play a crucial role in influencing the propagation speed and trajectory of eddies (McWilliams and Flierl, 1979). In global eddy movement statistics, eddies generally move westward. Cyclonic eddies display a poleward tendency, while anticyclonic eddies show an equatorward movement (Chelton *et al.*, 2011). The movement of eddies in regional seas varies due to the influence of background flow fields. As shown in the left panel of Figure 3.3, in the subtropical counter-current region, the westward velocity of eddies decreases with latitude at a rate of 0.5 cm s^{-1} per degree latitude. The right panel illustrates the variation in northward propagation speed with latitude at a rate of -0.1 cm s^{-1} per degree latitude. Eddies north of 21°N move equatorward, while those south of 21°N move poleward at a speed of 0–1 cm s^{-1}. The combination of equatorward movement in the northern latitude band and poleward movement in the southern

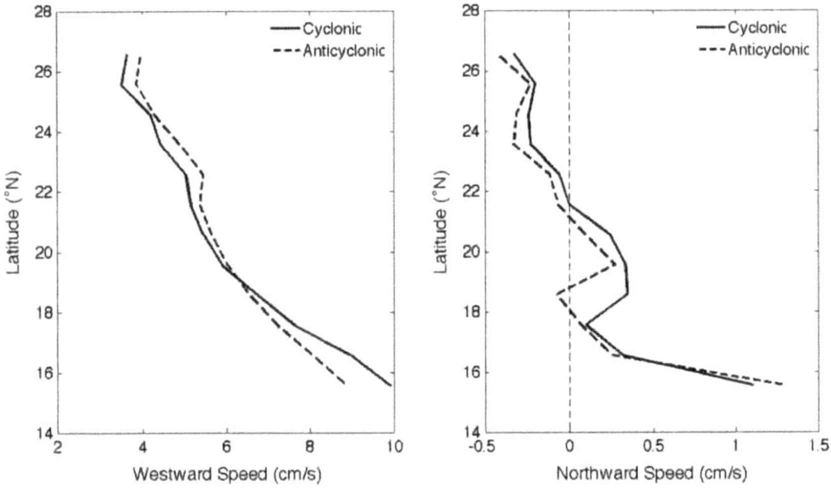

Fig. 3.3. Westward (northward) propagation speed (cm s^{-1}) of eddies in the subtropical countercurrent region. Solid (dashed) lines correspond to cyclonic (anticyclonic) eddies. Averaged values for eddies with a lifespan exceeding eight weeks are shown.

Source: Taken from Liu *et al.* (2012).

latitude band, as depicted in Figure 3.2, explains the characteristic of mature-stage eddies having a higher proportion in this region.

3.3.2 *Eddy–eddy interaction*

In the presence of complex ocean currents and vertical stratification, eddies are not always isolated entities. They undergo interactions, exhibiting phenomena such as the counter-rotating modulation of eddy flow fields and interactions like merging, pairing, or tearing with co-rotating eddies. Consequently, the interaction processes between eddies significantly influence their lifespan and energy distribution and constitute a fundamental element in the entire ocean layer (Sun and Oh, 2022; Lacassagne *et al.*, 2020). The interaction between mesoscale eddies leads to the exchange of energy and vorticity, often corresponding to the gradual decay of one eddy and the formation of a new merging eddy. These processes are important in understanding the mechanisms of eddy development and dissipation. As two eddies of the same polarity approach each other, complex

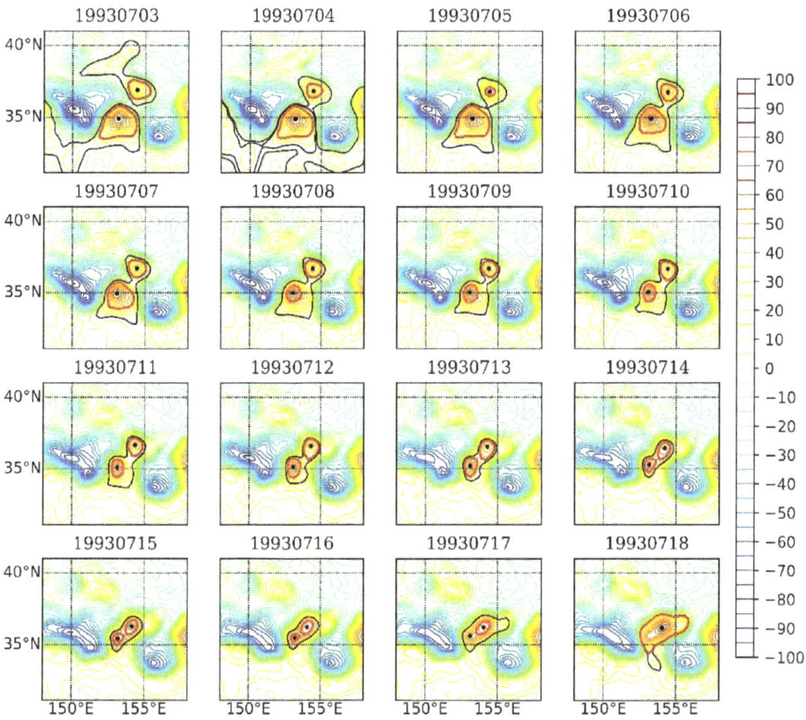

Fig. 3.4. Spatial and temporal distribution of SLA (color codes in cm) for mesoscale anticyclonic eddy merging event in the Northwest Pacific (148–157°E, 30–40°N) from 3 July to 18 July 1993. Black lines depict eddy contours, and red solid lines mark multi-core eddy structures.

Source: Taken from Tian *et al.* (2021).

behaviors and transitions arise, with a notable phenomenon being eddy merging (Marez *et al.*, 2020; Kuo and Chern, 2011).

Figure 3.4 shows a two mesoscale anticyclonic eddy merging event that occurred in the Northwest Pacific (148–157°E, 30–40°N) from 3 July to 18 July 1993. One can see that two anticyclonic eddies moved toward each other starting on 3 July 1993, forming a double-core eddy structure with shared contours in a week on 10 July, and eventually led to the formation of a single-core merging eddy with the outer sea level anomaly (SLA) closed contours as the boundary on 18 July. It implies that if eddies of the same polarity interact, their boundaries align, allowing them to coexist within the same closed SLA contours. Meanwhile, one can see that during the merging

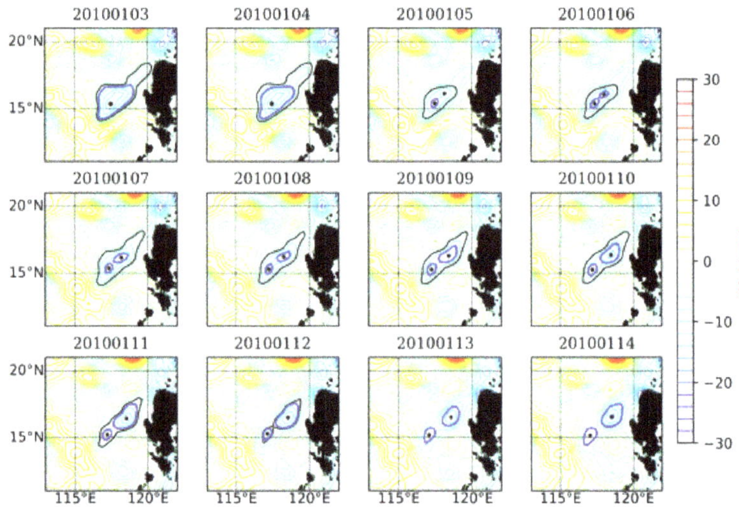

Fig. 3.5. Spatial and temporal distribution of SLA (in cm) for a splitting event of a mesoscale cyclonic eddy in the SCS (113–122°E, 11–21°N) from 3rd to 14th January 2010. Mononuclear eddies/multicore eddies and eddy groups are marked by blue lines and black lines, respectively. Eddy cores are marked by black dots.

Source: Taken from Tian *et al.* (2021).

process, the intensity of eddies (total surface area and central SLA) underwent complex fluctuations.

Figure 3.5 shows a mesoscale cyclonic eddy splitting event that occurred in the SCS (113–122°E, 11–21°N) from 3rd to 14th January 2010. One can see that the cyclonic eddy started to elongate northeastward before 3rd January, forming a double-core eddy structure with shared contours from 5 January to 11 January, and eventually led to the formation of two single-core splitting eddies after 12 January. Meanwhile, during the splitting event, the intensity of eddies (total surface area and central SLA) underwent a decay process. The above two cases reveal that merging and splitting events induced by eddy-eddy interaction play a crucial role in eddy development and dissipation processes.

3.3.3 *Viscosity-induced dissipation*

Early satellite observations in the 1970s had revealed that an ocean mesoscale eddy entering the continental shelf would quickly decay.

It manifests that the surface area of the eddy continuously expands, and the rotational angular velocity gradually decays with time. From a vorticity equation for a vertically stratified, horizontally homogeneous, viscous fluid, Zheng and Yuan (1989) derived analytical solutions, which indicate that viscous dissipation is a main mechanism for the surface area expansion and vorticity decay of the eddy.

3.4 Summary

The energy supplies from the ocean interior or exterior are necessary conditions for the growth and sustainable development of eddies. Meanwhile, eddies would finally dissipate if they lose energy supplies. This chapter focuses on general theories and research results of the generation, development and dissipation of ocean mesoscale eddies. Generation mechanisms include barotropic instability, baroclinic instability, wind stress forcing and interactions between flow and topography, and are formulated by a vertical relative vorticity budget equation. Development mechanisms involve three hypotheses: (1) For large-scale circulation with vertical shear, unstable disturbances such as eddies can rapidly grow by gaining energy from the release of background potential energy. (2) Interaction between eddy anisotropy and mean flow straining in baroclinic instability helps eddies gain kinetic energy from the mean flow. (3) Eddy and eddy merging leads to a combination of EKE and vorticity, corresponding to the formation of a new, larger eddy. Dissipation mechanisms involve two hypotheses: (1) Forced splitting of an eddy leads to loss of EKE and vorticity, corresponding to quick and sudden dissipation. (2) Viscous dissipation leads to the end of the eddy lifespan.

References

Aoki, S. and Imawaki, S. (1996). Eddy activities of the surface layer in the Western North Pacific detected by satellite altimeter and radiometer, *J. Oceanogr.*, 52, pp. 457–474, doi:10.1007/BF02239049.

Boyer, D. L. and Chen, R.-R. (1987). Laboratory simulation of mountain effects on large-scale atmospheric motion systems: The Rocky Mountains, *J. Atmos. Sci.*, 44, pp. 100–123, doi:10.1175/1520-0469(1987)0440100:LSOMEO.2.0.CO;2.

Boyer, D. and Davies, P. (1982). Flow past a circular cylinder on a beta plane, *Phil. Trans. Royal Soci. London Series A, Math. Phys. Sci.*, 306(1496), pp. 533–556, doi:10.1098/rsta.1982.0094.

Chelton, D. B., DeSzoeke, R. A., Schlax, M. G., El-Naggar, K., and Siwertz, N. (1998). Geographical variability of the first baroclinic Rossby radius of deformation, *J. Phys. Oceanogr.*, 28(3), pp. 433–460, doi:10.1175/1520-0485(1998)028<0433:GVOTFB>2.0.CO;2.

Chelton, D. B., Gaube, P., Schlax, M. G., Early, J. J., and Samelson, R. M. (2011). The influence of nonlinear mesoscale eddies on near-surface oceanic chlorophyll, *Science*, 334(6054), p. 328, doi:10.1126/science.1208897.

Chi, P., Chen, Y., and Lu, S. (1998). Wind-driven South China Sea deep basin warm-core/cool-core eddies, *J. Oceanogr.*, 54(4), pp. 347–360, doi:10.1007/BF02742619.

David, J., Walker, A., and Stewartson, K. (1972). The flow past a circular cylinder in a rotating frame, *ZAMP*, 23(5), pp. 745–752, doi:10.1007/BF01602641.

Dong, C. and McWilliams, J. C. (2007). A numerical study of island wakes in the Southern California Bight, *Cont. Shelf Res.*, 27, pp. 1233–1248, doi:10.1016/j.csr.2007.01.016.

Dong, C., You, Z., Dong, J., Ji, J., Sun, W., Xu, G., Lu, X., Xie, H., Teng, F., Liu, Y., Xu, A., Wang, Q., Xia, Q., Lin, X., Fu, M., Wang, J., Cao, Y., and Han, G. (2025). Oceanic mesoscale eddies, *Person. Comm.*, pp. 1–111, doi:10.34133/olar.0081.

Furukawa, K. and Wolanski, E. (1998). Shallow-water frictional effects in island wakes, *Estuar., Coast. Shelf Sci.*, 46, pp. 599–608, doi:10.1006/ecss.1997.0301.

Gill, A. E., Green, J. S. A., and Simmons, A. J. (1974). Energy partition in the large-scale ocean circulation and the production of mid-ocean eddies, *Deep Sea Res. Oceanogr. Abst.*, 21(7), pp. 499–528, doi:10.1016/0011-7471(74)90010-2.

Heywood, K. J., Stevens, D. P., and Bigg, G. R. (1996). Eddy formation behind the tropical island of Aldabra, *Deep Sea Res. Part I: Oceanogr. Res. Pap.*, 43(4), pp. 555–578, doi:10.1016/0967-0637(96)00097-0.

Itoh, S. and Yasuda, I. (2010). Characteristics of mesoscale eddies in the Kuroshio–Oyashio Extension region detected from the distribution of the sea surface height anomaly, *J. Phys. Oceanogr.*, 40, pp. 1018–1034, doi:10.1175/2009JPO4265.1.

Ji, J., Dong, C., Zhang, B., Liu, Y., Zou, B., King, G. P., Xu, G., and Chen, D. (2018). Oceanic eddy characteristics and generation mechanisms in the Kuroshio Extension region, *J. Geophys. Res. Oceans*, 123(11), pp. 8548–8567, doi:10.1029/2018JC014196.

Kang, D., Curchitser, E. N., and Rosati, A. (2016). Seasonal variability of the Gulf Stream kinetic energy, *J. Phys. Oceanogr.*, 46(4), pp. 1189–1207, doi:10.1175/JPO-D-15-0235.1.

Kuo, Y. C. and Chern, C. S. (2011). Numerical study on the interactions between a mesoscale eddy and a western boundary current, *J. Oceanogr.*, 67, pp. 263–272, doi:10.1007/s10872-011-0026-3.

Lacassagne, T. and Cagney, N. (2020). Vortex merging and splitting: A route to elastoinertial turbulence in Taylor-Couette flow, *Phys. Rev. Fluids*, 5, p. 113303, doi:10.1103/PhysRevFluids.5.113303.

Liu, S.-S., Sun, L., Wu, Q., and Yang, Y.-J. (2017). The responses of cyclonic and anticyclonic eddies to typhoon forcing: The vertical temperature-salinity structure changes associated with the horizontal convergence/divergence, *J. Geophys. Res. Oceans*, 122(6), pp. 4974–4989, doi:10.1002/2017JC012814.

Liu, Y., Dong, C., Guan, Y., Chen, D., McWilliams, J., and Nencioli, F. (2012). Eddy analysis in the subtropical zonal band of the North Pacific Ocean, *Deep Sea Res. Part I: Oceanogr. Res. Pap.*, 68(C11), pp. 54–67, doi:10.1016/j.dsr.2012.06.001.

Liu, Z. and Gan, J. (2012). Variability of the Kuroshio in the East China Sea derived from satellite altimetry data, *Deep Sea Res. Part I: Oceanogr. Res. Pap.*, 59, pp. 25–36, doi:10.1016/j.dsr.2011.10.008.

Marez, C. D., Carton, X., L'Hegaret, P., Meunier, T., and Stegner, A. (2020). Oceanic vortex mergers are not isolated but influenced by the β-effect and surrounding vortices, *Sci. Rep.*, 10, p. 2897, doi:10.1038/s41598-020-59800-y.

McWilliams, J. C. and Flierl, G. R. (1979). On the evolution of isolated non-linear vortices, *J. Phys. Oceanogr.*, 9, pp. 1155–1182, doi:10.1175/1520-0485(1979)009<1155:OTEOIN>2.0.CO;2.

Morrison, A. K., FröLicher, T. L., and Sarmiento, J. L. (2015). Upwelling in the Southern Ocean. *Phys. Today*, 68(1), pp. 27–32, doi:10.1063/PT.3.2654.

Neill, S. P. and Elliott, A. J. (2004). Observations and simulations of an unsteady island wake in the Firth of Forth, Scotland, *Ocean Dyn.*, 54, pp. 324–332, doi:10.1007/s10236-003-0084-1.

Pedlosky, J. (1977). On the radiation of meso-scale energy in the mid-ocean, *Deep Sea Res.*, 24(6), pp. 591–600, doi:10.1016/0146-6291(77)90529-X.

Pedlosky, J. (1987). *Geophysical Fluid Dynamics*, Springer-Verlag, New York, pp. 374–379, doi:10.1007/978-1-4612-4650-3.

Pingree, R. D. and Maddock, L. (1979). The tidal physics of headland flows and offshore tidal bank formation, *Mar. Geol.*, 32(3), pp. 269–289, doi:10.1016/0025-3227(79)90068-9.

Qiu, B. and Chen, S. (2010). Interannual variability of the North Pacific Subtropical Countercurrent and its associated mesoscale eddy field, *J. Phys. Oceanogr.,* 40(1), pp. 213–225, doi:10.1175/2009JPO4285.1.

Qiu, C. H., Mao, H. B., Wang, Y. H., Yu, J., Su, D., and Lian, S. (2019). An irregularly shaped warm eddy observed by Chinese underwater gliders, *J. Oceanogr.,* 75, pp. 139–148, doi:10.1007/s10872-018-0490-0.

Signell, R. P. and Geyer, W. R. (1991). Transient eddy formation around headlands, *J. Geophys. Res.,* 96, pp. 2561–2575, doi:10.1029/90JC02029.

Smolarkiewicz, P. K. and Rotunno, R. (1989). Low Froude number flow past three dimensional obstacles. Part I: Baroclinically generated lee vortices, *J. Atmos. Sci.*, 46, pp. 1154–1164, doi:10.1175/1520-0469(1989)046<1154:LFNFPT>2.0.CO;2.

Stepanov, D. V., Diansky, N. A., and Fomin, V. V. (2018). Eddy energy sources and mesoscale eddies in the Sea of Okhotsk, *Ocean Dyn.*, 68(7), pp. 825–845, doi:10.1007/s10236-018-1167-3.

Sun, W. Y. and Oh, T. J. (2022). Vortex Merger in Shallow Water Model. *Asia-Pacific J. Atmos. Sci.,* 58, pp. 533–547, doi:10.1007/s13143-022-00270-9.

Tai, C. K. and White, W. B. (1990). Eddy variability in the Kuroshio Extension as revealed by Geosat altimetry: Energy propagation away from the jet, Reynolds stress, and seasonal cycle, *J. Phys. Oceanogr.,* 20(11), pp. 1761–1777, doi:10.1175/1520-0485(1990)020<1761:EVITKE>2.0.CO;2.

Thompson, A. F. and Sallée, J.-B. (2012). Jets and topography: Jet transitions and the impact on transport in the Antarctic Circumpolar Current, *J. Phys. Oceanogr.*, 42(6), pp. 956–972, doi:10.1175/JPO-D-11-0135.1.

Tian, F., Li, Z., Yuan, Z., and Chen, G. (2021). EddyGraph: The tracking of mesoscale eddy splitting and merging events in the Northwest Pacific Ocean, *Rem. Sens.*, 13, p. 3435, doi:10.3390/rs13173435.

Tomczak, M. (1988). Island wakes in deep and shallow water, *J. Geophys. Res. Oceans,* 93(C5), pp. 5153–5154, doi:10.1029/JC093iC05p05153.

Wood, R. A. (1988). Unstable waves on oceanic fronts: Large amplitude behavior and mean flow generation, *J. Phys. Oceanogr.*, 18(5), pp. 775–787. doi:10.1175/1520-0485(1988)018<0775:UWOOFL>2.0.CO;2.

Zheng, Q. and Yuan, Y. (1989). Study on the analytical model of decay of mesoscale eddy on the continental shelf, *Sci. China Ser. B-Chem., Life Sci. Earth Sci.*, 32(9), p. 11.

Chapter 4

Detection and Observation Methods for Ocean Mesoscale Eddies

4.1 Introduction

The research on ocean mesoscale eddies initially relies on cruise observational data to explore their physical characteristics, hydrodynamic conditions and formation and dissipation processes. Over the past few decades, various methods, including observations, numerical simulations, laboratory simulations and theoretical analyses, have contributed to a foundational understanding of mesoscale eddies (Richardson, 1993; McWilliams and Flierl, 1979; Dickey *et al.*, 2008). Meanwhile, early studies were often limited to individual or idealized eddies. For example, Rovegno *et al.* (2009) investigated the thermohaline structure of two anticyclonic eddies in the Gulf of Alaska using cruise CTD observation data, and Ruiz *et al.* (2014) analyzed the water characteristics of a shallow mesoscale anticyclonic eddy (AE) captured by satellites using cruise observation data and revealed the mechanism of shallow AE formation. Similarly, Mandal *et al.* (2019) investigate CEs in the western Bay of Bengal using surface flow observations from a High-Frequency Radar (HFR). Taneda (1956, 1959) analyzed different development states of wake eddies in cylinder wake flows under different Reynolds numbers using laboratory experimental methods and then analyzed the characteristics of wake eddies under steady and unsteady conditions. Han *et al.* (2021) conducted laboratory experiments on island-induced wake eddies using the world largest French CORIOLIS rotating experimental platform and analyzed the dynamic mechanism of anticyclonic eddy decay.

Despite these advances, the uncertainty in eddy generation regarding location, time and scale, combined with the high cost of field observations, has resulted in limited observational data availability. Fortunately, recent developments in ocean satellite remote sensing technology and the continuous improvement of numerical simulation technology have led to significant progresses in mesoscale eddy research, moving beyond individual case studies to more comprehensive, global analyses (McGillicuddy *et al.*, 2007; McWilliams, 2008; Dong *et al.*, 2012; Nagai *et al.*, 2015; Lin *et al.*, 2015; Xu *et al.*, 2016; Lévy *et al.*, 2018). In particular, the launch of satellite altimeters has opened a new window for studying mesoscale eddies from a global perspective with continuous, all-weather observations of global sea surface height (SSH) data. Qiu (1999) studied the seasonal variation of mesoscale eddies in the North Pacific subtropical countercurrent region using satellite altimeter data. Hwang *et al.* (2004) tracked a CE and an AE in the subtropical countercurrent region based on geostrophic current data derived from satellite SSH data, studying the evolution of eddy characteristics during their propagations. Developed a mesoscale eddy dataset in the subtropical countercurrent region using satellite altimeter data and revealed that eddies in this region propagate westward at speeds of around 9.3 cm s^{-1} and dissipate near the Kuroshio east of Taiwan Island. Scharffenberg and Stammer (2010) revealed the transient eddy kinetic energy (EKE) characteristics of the Kuroshio Extension region based on satellite altimeter data from 2002 to 2005. Qiu and Chen (2005) found that the activity of mesoscale eddies in the Kuroshio Extension region is highly correlated with the stability of the Kuroshio Extension itself. Additionally, Chelton *et al.* (2007, 2011) analyzed various characteristics and movement features of global ocean mesoscale eddies based on satellite altimeter data, revealing that changes in SSH are mainly caused by mesoscale eddies.

Currently, the main observational datasets with global coverage used in mesoscale eddy research include: (1) SSH data (Chelton *et al.*, 2007; Ji *et al.*, 2017; Xu *et al.*, 2019b), (2) sea surface temperature (SST) data (Dong *et al.*, 2011a; Ji *et al.*, 2017), (3) global sea surface buoy data (Ji *et al.*, 2017; Dong *et al.*, 2011b), (4) satellite synthetic aperture radar (SAR) images (Karimova and Gade, 2014; Karimova,

2017), and (5) numerical products with global eddy resolution (Ji *et al.*, 2018) such as Hybrid Coordinate Ocean Model (HYCOM), OGCM for the Earth Simulator (OFES), and Estimating the Circulation & Climate of the Ocean (ECCO), as well as artificial intelligence algorithms (Xu *et al.*, 2019a).

As ocean observation technologies continue to diversify, researchers have begun to gradually understand the kinematic characteristics of eddies under geostrophic balance conditions. For instance, Codiga and Eriksen (1997) used Conductivity-Temperature-Depth (CTD) measurements to study the three-dimensional structure of eddies near the Cobb Seamount and discussed the relationship between the vertical displacement of eddies and geostrophic balance.

Timmermans *et al.* (2008) investigated eddies in the Canadian Basin using field observation data, indicating that anticyclonic eddies rotate faster than the geostrophic flow. Zhao *et al.* (2014) calculated the first baroclinic Rossby radius in the Arctic region using ice-tethered profiler data combined with climatological methods and observed eddies with scales comparable to the first baroclinic Rossby radius. These eddies, satisfying geostrophic balance, were found in the halocline layer and could be described by the Rankine vortex model with maximum azimuthal velocities between 0.05 and $0.4\,\mathrm{m\,s^{-1}}$. Scott *et al.* (2019) observed the three-dimensional structure of a unique Arctic anticyclonic cold eddy using shipboard acoustic Doppler current profiler (ADCP) data. By filtering out the slope current signal, they found that the eddy velocity field is symmetric, with a maximum azimuthal velocity of $10\,\mathrm{cm\,s^{-1}}$ and a Rossby number of approximately 0.4, consistent with the fact that observed geostrophic velocities are primarily controlled by the Coriolis component. Meunier *et al.* (2021) pointed out the feasibility of calculating the geostrophic velocity, relative vorticity and shear strain of eddies using the original glider flight distance as the horizontal coordinates. These studies primarily focused on the geostrophic velocity of individual oceanic eddies. However, to study regional and global curvature flow geostrophic characteristics, relying solely on cruise observation data is insufficient and often requires high-resolution satellite observation data or numerical model simulations. On the other hand, to study the flow of mesoscale eddies and fronts based on satellite altimeter data under the assumption of geostrophic balance is simple and feasible, for curved eddies and fronts, the non-geostrophic velocity components caused by the local curvature of streamlines cannot be ignored. Such geostrophic approximations are inaccurate for describing the structures of sub-mesoscale features with strong non-geostrophic characteristics (Chang *et al.*, 2013).

The inertial effect is highly important in western boundary currents. Direct field observations of ocean currents have shown significant differences in estimating sea level anomaly (SLA) in the Gulf Stream due to the introduction of centrifugal acceleration (Liu and Rossby, 1993). Furthermore, through comparative analysis with surface flow fields observed by drifters, after filtering out the Ekman flow signal, Maximenko and Niiler (2006) pointed out that the flow fields derived from sea level height overestimate (underestimate) the kinetic energy of cyclonic (anticyclonic) eddies. Therefore, studying the characteristics of geostrophic currents at the sea surface using satellite altimeter data also requires geostrophic corrections to the altimeter-derived flow fields.

The correction of geostrophic surface flow fields derived from altimeter data has long been a scientific challenge tackled by oceanographers. Uchida *et al.* (1998) first proposed that considering the non-geostrophic velocity component induced by the curvature of the Kuroshio could effectively improve the calculation of surface ocean currents from satellite altimetry (Topex/Poseidon). Maximenko and Niiler (2006) pointed out that the differences between the surface flow velocity and EKE (after filtering out the Ekman flow signal) measured by drifting buoys in the Gulf Stream and that evaluated from satellite altimeter data are mainly caused by centrifugal force. Recent studies have shown that correcting the bias in the surface flow velocity derived from satellite altimeter data due to curvature-induced effects can accurately assess the surface currents of strong mesoscale eddies in the Mozambique Channel (Penven *et al.*, 2014) and intense eddy rings in the Gulf of Mexico (Douglass and Richman, 2015). More recently, Ioannou *et al.* (2019) developed an iterative method to correct for non-geostrophic components of eddy currents in the Mediterranean using Archiving, Validation, and Interpretation of Satellite Oceanographic (AVISO) altimeter data. Correction studies on various types of geostrophic flows in the ocean have been conducted based on the latest version of AVISO altimeter data

(Cao *et al.*, 2022). Cao *et al.* (2023) further validated these corrections, providing a more accurate assessment of global geostrophic flows.

As above mentioned, the advancement of high-resolution ocean satellite remote sensing and the emergence of high-resolution ocean numerical models as well as reanalysis data have provided new opportunities for the great progress in ocean eddy science. Development of automated identification and detection of eddies using computer algorithms has become essential tasks. This chapter presents a comprehensive review of recent advances in survey technologies of mesoscale eddies, focusing on automated eddy identification and detection methods mainly used for satellite data. Meanwhile, cruise observation technologies used by typical eddy observation research programs are briefly introduced (Dong *et al.*, 2025).

4.2 Ocean Eddy Identification and Detection Methods

As the volume of ocean data continues to grow, methods for eddy detection have also become increasingly sophisticated. Based on the different types of data, eddy detection methods can be classified into Eulerian and Lagrangian approaches (Dong *et al.*, 2011a). Eulerian data refer to snapshot data at a particular moment in time or spatial field data, while Lagrangian data refers to the trajectory data of water masses or material particles.

4.2.1 *Ocean eddy identification and detection methods based on Eulerian data*

The detection methods of ocean eddies based on Eulerian data can be categorized into three main types: (1) Methods based on physical parameters. The Okubo-Weiss (OW) parameter method developed by Okubo (1970) and Weiss (1991) is the most widely applied so far. This method has been used in various regions, including the Mediterranean Sea (Isern-Fontanet *et al.*, 2003), the Peruvian Sea (Penven *et al.*, 2005), the Alaskan waters (Henson and Thomas, 2008), and for global eddy detection (Chelton *et al.*, 2007). Despite widespread application of the OW method, it has inherent flaws, notably the difficulty in determining the optimal threshold value for OW (Chelton *et al.*, 2011). Petersen *et al.* (2013) developed a new algorithm, R2, aimed at eliminating the dependency on threshold selection, which is based on the analogy between eddy and a given ideal Gaussian vortex. However, derivation of physical parameters can introduce noise and lead to increasing the false detection rate of eddies. Furthermore, physical criteria may also lead to failures in detecting eddies or underestimating their size (Ari Sadarjoen and Post, 2000; Chaigneau *et al.*, 2008; Nencioli *et al.*, 2010; Chelton *et al.*, 2011). (2) Methods based on geometric features of current field. Ari Sadarjoen and Post (2000) initially proposed the winding-angle (WA) method, which identifies eddies through closed curves. Nencioli *et al.* (2010) developed a new method based on the geometric features of current field vectors (VG). This method introduces four constraints corresponding to the definition and characteristics of the eddy velocity field. Points that satisfy all the constraints are defined as the eddy center. The four constraints are as follows: (i) For the velocity component v in the east-west direction at the eddy center, the signs of the values on both sides away from the center are opposite, and the magnitude linearly increases with the distance from the center; (ii) For the velocity component u in the north-south direction at the eddy center, the signs of the values on both sides away from the center are opposite, and the magnitude linearly increases with the distance from the center; (iii) The point of minimum velocity within the selected area is approximated as the eddy center; (iv) Near the approximate eddy center, the rotation direction of the velocity vectors must be consistent, meaning two adjacent velocity vectors must lie within the same quadrant or adjacent quadrants.

Once the eddy center is identified, the boundary of the eddy can be computed. After determining the eddy center, the stream function within the local range of the center is calculated, and the outermost closed streamline is identified as the boundary of the eddy. The average distance from points on the boundary to the eddy center is defined as the radius. To understand the temporal evolution of eddy characteristics, it is necessary to track the eddy, searching for the eddy at time $t + 1$ that is most similar and of the same polarity as the eddy at time t. The search range depends on the spatial and temporal resolution of the data, as well as the flow speed of the average background flow field. A better method is to estimate the size of the search range based on the average flow speed and the temporal resolution of the data.

Cyclonic and anticyclonic eddies can lead to localized sea surface elevation and depression, respectively, causing SSTs to decrease and increase, accordingly. The thermal wind velocity vector field (pseudo-velocity field) can be estimated from the SST field. Using the geometric features of the pseudo-velocity field facilitates the identification and detection of eddies, thereby obtaining characteristics such as the center, size, vorticity, movement speed and lifespan of the eddies (Dong *et al.*, 2011b). The main procedures include calculating the thermal wind speed, identifying the eddy centers within the thermal wind velocity field, calculating the size of eddies, and tracking the movement trajectory of eddies.

Over the past two decades, global remote sensing data has become increasingly abundant, with satellite observation data emerging as the most effective resource for mesoscale eddy research. The detection method based on the geometric features of current fields has been innovatively applied by Dong *et al.* (2022) for the global tracking and detection of mesoscale eddies, coupled temporally and spatially with multi-element data from the ocean, atmosphere and ecology. This has led to the compilation of basic characteristic information of global mesoscale eddies and their associated marine, biological and atmospheric observation data (near-global ocean mesoscale eddy atmospheric-ocean-biological interaction observational dataset, GOMEAD). (3) Hybrid methods combining physical parameters and geometric features. Chaigneau *et al.* (2008), Chelton *et al.* (2011) and Zhang *et al.* (2015) used local extremities of sea surface height anomalies (SSHA) as potential eddy centers, combining them with geometric features of closed streamlines surrounding the potential eddies. This approach mixes special physical parameter methods with geometric methods, considering both physical parameters and the geometric features of the flow. Le Vu *et al.* (2018) introduced a method that combines physical tracking the movement trajectory of eddies parameters, Local normalized angular momentum (LNAM), and the geometric features of streamlines to determine the centers and dynamic characteristics of eddies. This method can be applied across various spatial resolutions without the need for specific fine-tuning of parameters.

4.2.2 *Ocean eddy identification and detection methods based on Lagrangian data*

The Lagrangian tracking of drifting buoys is the important data used for ocean eddy detection. The main method is to identify the "loops" within the drifting tracking. Early studies manually identified these "loops" (Richardson, 1993; Shoosmith *et al.*, 2005; Fratantoni and Richardson, 2006). More than 230 acoustic calibration and ranging buoy data were collected over the period 1972–1989, and manual statistical work was performed from these buoy track data. However, manual identification is time-consuming and subject to human error or bias. To compensate for the shortcomings of manual methods, automated eddy detection methods using various mathematical methods have emerged (e.g., Glenn and Ebbesmeyer, 1993; Hamilton *et al.*, 1999; Boebel *et al.*, 2003; Lankhorst, 2006; Lilly and Gascard, 2006; Beron-Vera *et al.*, 2008; Griffa *et al.*, 2008; Lilly and Olhede, 2010). These detection methods can be divided into four categories: (1) Lagrangian Stochastic Model (LSM). Griffa *et al.* (2008) calculated the spin rate of a trajectory based on the two-dimensional first-order Lagrangian Stochastic Model (Veneziani *et al.*, 2005a, 2005b). When the spin rate exceeds a non-zero threshold, the part of the trajectory is identified as an eddy. Lankhorst (2006) used second-order autoregression (AR) to identify the part of the trajectory, where the two velocity components have similar frequencies, durations and persistence, as an eddy. The method is a one-dimensional second-order LSM-based method. (2) Elliptical pattern identification: Studies in this category include the work by Glenn and Ebbesmeyer (1993), Hamilton *et al.* (1999), Lilly and Gascard (2006), and Lilly and Olhede (2009). Using elliptical dispersion models, Glenn and Ebbesmeyer (1993), Hamilton *et al.* (1999), and Hamilton (2007) identify eddies from the paths of smoothed drifting buoys. Lilly and Gascard (2006) and Lilly and Olhede (2009) apply wavelet techniques to time-varying trajectories for ellipse detection. Lilly and Pérez-Brunius (2021) applied wavelet ridge analysis to extract displacement signals of coherence eddies from Lagrangian trajectories. (3) Lagrangian dynamic system tools: Beron-Vera *et al.* (2008) extracted Lagrangian coherent structures (LCS) from finite-time Lyapunov exponential fields. The structures represent the fluid domains with different advective properties and give the boundaries of the eddies. (4) Geometric methods: Boebel *et al.* (2003) determined the presence of eddies by calculating the curvature of the trajectory. When a drifting buoy at the surface moves with the eddy flow field, its trajectory forms a pendulum or "loop" shape (Dong *et al.*, 2011a). This pendulum-like trajectory occurs when the advective velocity exceeds the rotational velocity of the eddy. As the mean background velocity is removed, the drifting buoy trajectory follows a "loop" shape, which can be used to identify the presence of an eddy. Based on this principle, Dong *et al.* (2011a) developed a simple automated loop identification system (ALIS). Geometrically, a "loop" is defined as a closed, continuous curve, where the start and end points coincide, meaning that the drifting buoy eventually returns near its original position after some time. The following points must also be considered when carrying out eddy identification. (1) Since the observations of the drifting buoy trajectory are discrete data, when it returns to its "origin", one can only expect it to arrive at the "origin" within a certain threshold distance. (2) Even if a drifting buoy "returns" according to the definition of the threshold distance, it may remain in a cluster of points around the "origin" and not form a loop (rotation). Therefore, an angle of rotation must be calculated: only if the trajectory forms a complete and closed curve can the section of the trajectory be considered a "loop". (3) For higher frequency processes, such as inertial oscillations, drifting buoys form "loop" trajectories. For low-frequency processes, such as basin-scale circulation, drifting buoys can also develop "loop" trajectories. The Lagrangian eddy frequency lies between the intraseasonal and inertial scales, and a period criterion must be used to select only circulation within this period range. For identification, circulation with a period greater than the local inertial period or less than the intraseasonal scale can be selected (the maximum "loop" search time is 90 days). (4) To be conservative, two consecutive "loops" of the same polarity along the same drifting buoy track are only considered to be part of an eddy if at least two consecutive "loops" of the same polarity are detected and if the two "loops" are sufficiently close together (within advective distances). (5) If the background flow velocity (large-scale flow) exceeds the rotational velocity of the eddy, the trajectory of the drifting buoy will be pendulum-like rather than closed. Therefore, it is possible to "loop back" the trajectory identification after removing the mean background flow field. With these criteria, the ALIS method automatically identifies distinct "loops" in the trajectory of a drifting buoy. Eddies are then detected based on additional conditions such as specific periodicity and continuous polarity. For each "loop", the method provides eight key parameters: position, start and end times, period, polarity, intensity and eccentricity.

Building on ALIS algorithm (Dong *et al.*, 2011a), Liu *et al.* (2015) improved the distance threshold setting and integrated a self-intersecting method (Li *et al.*, 2021) to develop automated integrated loop identification system (AILIS), which enhances eddy identification on more complex trajectories, such as the trajectory of a "big loop" composed of a set of successive "small loops". The method can also extract partial eddy trajectories, such as the flow jacket (raindrop) structure found in an unclosed "loop".

4.2.3 *Lagrangian coherent eddy detection methods*

Lagrangian coherent eddy detection methods are based on the trajectories of fluid particles. They are characterized by objectivity (the detection results do not change with changes in the reference frame) and coherence (the water within the detected eddies remains highly compact, with minimal diffusion). Lagrangian eddy detection methods include Finite Time/Scale Lyapunov Exponent (FTLE/FSLE) based method, super coherent eddy (black hole eddy) detection method, Lagrangian Averaged Vorticity Deviation (LAVD) based method and spectral clustering eddy detection method (Hadjighasem *et al.*, 2017). Among these, the LAVD-based method Haller *et al.* (2016) has a higher rate of identifying coherent eddies, and the eddies detected by this method are referred to as rotationally coherent Lagrangian vortices. LAVD, as an objective parameter, is mainly used to describe the accumulated rotation angles of fluid particles in a fluid. As described by Haller *et al.* (2016), the calculation of LAVD first requires obtaining the Lagrangian trajectories of fluid particles, which can be obtained by time integration of the Eulerian velocity field, which satisfies the equation:

$$x = \dot{v}(x,\, t),\;\; t \in (t0,\, t1) \tag{4.1}$$

where $\boldsymbol{v}(x,t)$ is the Eulerian velocity field, x is the displacement parameter, t is the time parameter and $\dot{\boldsymbol{x}}$ denotes the derivative of displacement with respect to time. Integrating Eq. (4.1) yields the Lagrangian trajectories of fluid particles, and the collection of all Lagrangian trajectories of fluid particles is referred to as the flow map.

$$F_{t_0}^{t}(x0) \to x(t, t0, x0),\;\; t \in (t_0, t_1),\;\; x \in U. \tag{4.2}$$

The calculation of Lagrangian Averaged Vorticity Deviation (LAVD) is based on Lagrangian particle trajectories, and the equation is as follows,

$$\mathrm{LAVD}_{t_0}^{t_1}(x_0) = \int_{t_0}^{t_1} |\omega[x(s;\, x_0),\, s] - \bar{\omega}(s)|ds, \tag{4.3}$$

where $\omega(x(s;x_0),\, s)$ is the vorticity possessed by a fluid particle at position x at time s, and $\bar{\omega}(s)$ is the instantaneous mean vorticity within the spatial domain $U \cdot T = t_1 - t_0$ is defined as the Lagrangian time scale. Haller *et al.* (2016) defined rotationally coherent Lagrangian vortices as a group of nested LAVD material tubes (in three dimensions) or material lines (in two dimensions). Each tube or line represents an iso-surface or iso-line of LAVD, where fluid particles on the tube share the same average angular velocity. This results in a nested structure of LAVD tubes or lines, with decreasing LAVD values from the inside out. The rotational speed of the fluid particles corresponds to the LAVD magnitude, with local maxima representing the eddy centers.

To identify the boundary of an eddy, one can start from the eddy center and move outward, searching for the outermost closed LAVD surface or line. This boundary encloses all particles rotating around the eddy center with a specific angular velocity over the Lagrangian time scale T. Using this definition, rotationally coherent Lagrangian vortices can be detected. Since each material tube or line is composed of fluid particles, the LAVD eddy boundaries possess material properties, allowing them to act as material barriers. This characteristic makes LAVD eddies particularly promising for application in ocean material transport.

4.2.4 *Three-dimensional eddy detection methods*

An in-depth study of the three-dimensional structures of eddies is essential for deepening the understanding of the mechanisms behind their generation and decay. Such research can shed light on the significant role eddies play in regional and global ocean systems, as well as their influence on climate dynamics. Through in-situ observations, researchers can effectively delineate the three-dimensional structure of mesoscale eddies in the ocean. For instance, Johannessen *et al.* (1989) undertook a continuous 10-day observation of eddies along the Norwegian Sea coastline from February to March 1986. Using ADCP and CTD data, they analyzed the three-dimensional distributions of eddy flow velocity, temperature and salinity. Their analysis indicated that topographic effects, vorticity stretching, and baroclinic/barotropic instability are the main mechanisms driving mesoscale eddy formation in the region. This cold eddy persisted from the subsurface down to 1000 m. Complementing full-depth ADCP and CTD measurements (Zhang *et al.*, 2016), underwater gliders stand out as excellent instruments for observing three-dimensional structures of eddies (Wang *et al.*, 2020; Li *et al.*, 2019, 2020; Qiu *et al.*, 2019).

However, three-dimensional in-situ data remain scarce, driving the emergence of multi-data fusion as a new trend. This approach integrates satellite remote sensing data, numerical model outputs, or Argo data. Zhang *et al.* (2013) used global satellite altimeter and Argo data to derive a universal three-dimensional structure in a normalized stretching coordinate system through composite analysis. When combining satellite remote sensing data with Argo data for three-dimensional eddy detection, the identification of eddies can be directly conducted based on their temperature and salinity characteristics, followed by composite analysis (He *et al.*, 2018; Dai *et al.*, 2020; Zhan *et al.*, 2019). Typically, studies on the three-dimensional structures of eddies involve applying various two-dimensional eddy detection methods for each layer, followed by synthesizing the analysis of vorticity and investigating vertical temperature and salinity profiles (Dong *et al.*, 2017; Zhang *et al.*, 2023). Subsequent research has further delved into the three-dimensional eddy structure of subsurface strengthening (Zhang *et al.*, 2017).

4.2.5 *AI-based eddy detection methods*

Traditional automated methods for identifying ocean eddies face challenges including large detection errors, low operational efficiency and limited applicability to the global ocean. Constructing an automated method that is efficient, accurate and universally applicable for identifying ocean eddies remains a pressing and significant scientific challenge. The emergence of artificial intelligence (AI) methods has introduced new perspectives into the field of ocean eddy identification. Unlike traditional methods, which rely on extracting eddy features from data, AI methods adopt a paradigm shift by learning eddy features directly from data. This shift in approach bypasses the difficulties associated with accurately extracting the diverse features of eddies from data, thus substantially mitigating the limitations of traditional methods. Consequently, it holds considerable promise for achieving precise identification and tracking of mesoscale eddies in the global ocean. Ashkezari *et al.* (2016) used SLA data observed by satellite altimeters from January 1993 to September 2015 to apply support vector machine (SVM) and random forest (RF) models for detecting and predicting the lifecycles of daily eddies near the Hawaiian Islands and the North Pacific Subtropical Gyre region. The training process involved using 75% of the total samples, while the remaining 25% were used for testing to assess model accuracy and precision.

Compared to the local SLA extrema method proposed by Faghmous *et al.* (2015), the machine learning models showed lower errors in eddy identification. Notably, the random partitioning of training samples conferred an advantage by making the model independent of local ocean features, obviating the need for repeated parameter adjustments within specific marine regions. Franz *et al.* (2018) employed both convolutional neural network (CNN) and convolutional long- and short-term memory (ConvLSTM) models to detect and track mesoscale eddies in core regions around the Australian and South Equatorial Currents. The daily SLA data from 1993 to

2014 were randomly partitioned into training and validation datasets, except for March 1993 samples used for testing. Eddies identified by the OW parameter method served as labels during training, providing ground truth values, and were used to evaluate the model accuracy during testing. Lguensat *et al.* (2018) employed a CNN network to process daily SSH images from 1993 to 2011 in the South Atlantic Ocean, training a model named EddyNet for identifying and classifying mesoscale eddies. Subsequently, they applied the EddyNet model to SSH data in 2012 to detect cyclonic and anticyclonic mesoscale eddies, achieving an identification accuracy of over 88%. However, their study primarily focused on the occurrence locations and types of eddies without extracting other characteristic parameters.

Xu *et al.* (2019b) investigated the automated detection of eddies in the North Pacific Subtropical Countercurrent region using SLA data with a spatial resolution of 0.25° and a temporal resolution of 1 day. They employed the pyramid scene parsing network (PSPNet), comprising pyramid pooling modules and dilated convolutions to capture semantic information at different levels in various labels. The label data for the model were obtained from ocean eddy information detected using the VG algorithm. The results reveal that the PSPNet model exhibits comparable eddy detection capability to the VG algorithm, while detecting more small-scale eddies with higher accuracy. These findings underscore the high reliability of deep learning models in automated ocean eddy detection. Duo *et al.* (2019) studied ocean eddy identification in the South China Sea (SCS) using a deep learning model with a structure like the PSPNet model. Their model outperforms traditional algorithms for eddy identification. Xu *et al.* (2021) then investigated the performance of PSPNet, DeepLabV3+ and BiSeNet models in ocean eddy detection based on SLA data.

Compared to traditional methods, all three AI-based algorithms identify more eddies with the PSPNet algorithm detecting the largest number of ocean eddies. By incorporating spatial paths in the algorithm to avoid disruption of eddy edge information, the BiSeNet algorithm shows superior performance in detecting large-scale eddies. However, the DeepLabV3+ algorithm shows shorter lifespans for the detected eddies.

Santana *et al.* (2022) explored the influence of model complexity on eddy detection accuracy through a comparison of EddyNet, PSPNet, and the residual eddy detection network. Their findings revealed that deeper networks did not necessarily lead to improve performance. Thus, overly complex models were deemed unnecessary for ocean eddy detection. Saida *et al.* (2023) introduced a CNN incorporating an attention module, serial atrous spatial pyramid module, and feature-enhancing block for pixel-level segmentation using SLA images.

Experimental results showcased the consistent generalization of the model across diverse testing datasets, enabling precise segmentation of ocean eddies. Additionally, Zhao *et al.* (2023a) proposed the SymmetricNet, an end-to-end mesoscale eddy detection network leveraging multi-modal data fusion. This network seamlessly integrates various data types for ocean eddy detection, enriching contextual information while maintaining resolution.

Satellite altimeter observations serve as a primary data source for detecting ocean eddies. However, relying solely on data from a single sensor can significantly limit the available sea surface information, potentially impacting the reliability and accuracy of eddy identification. Recognizing the influence of ocean eddies on SST, Liu *et al.* (2021) fused SST and SLA features, using them as inputs for a deep learning model designed to detect ocean eddies. A comparison with results obtained solely using SLA as model input revealed improved eddy detection accuracy with the integration of multi-modal feature fusion. The inclusion of SST aided in distinguishing between normal and abnormal eddies, highlighting the effectiveness of multi-modal feature fusion in enhancing eddy detection model performance. Similarly, Zhao *et al.* (2023b) proposed a pyramid split attention eddy detection U-Net model for ocean eddy detection. Their model uses fused SST and SLA data as input, employing a pyramid split attention module to effectively capture spatial information at various scales while addressing large data volumes through group convolutions.

In addition to gridded altimetric data products, remote sensing imagery plays a crucial role in detecting ocean mesoscale eddies. Du *et al.* (2019) introduced the spatial pyramid pooling module into a CNN to develop a deep learning model named DeepEddy. This model equipped with principal component analysis filters can learn various eddy features across different levels. By applying the DeepEddy model to SAR images collected from 2005 to 2011 in the Kuroshio region, ocean eddies at different scales were successfully detected. The DeepEddy model exhibits a detection accuracy as high as $97.8 \pm 1\%$ and shows the ability to capture incomplete eddies in

SAR images, indicating the effective learning of underlying ocean eddy features by deep learning models. Drawing inspiration from traditional object detection networks, Xia *et al.* (2022a) introduced a context and edge association network, which uses a fused residual structure of Darknet-53 as the backbone for extracting multi-level features. Their network selectively incorporates features of different levels or sizes from hierarchical feature maps, showing a good performance in detecting multi-scale ocean sub-mesoscale eddies.

4.2.6 *Abnormal ocean eddy detection methods*

Abnormal mesoscale eddies can be categorized into two types: cyclonic warm eddies, characterized by a counterclockwise flow and a warmer core, and anticyclonic cold eddies, characterized by a clockwise flow and a colder core (Figure 4.1). These eddies deviate from the typical dynamics of conventional mesoscale eddies and are relatively rare. As a result, research on abnormal eddies has long been limited to isolated cases (Itoh and Yasuda, 2010; Yasuda *et al.*, 2000; Frenger *et al.*, 2015; Ji *et al.*, 2017). However, with the increasing availability of satellite observation data, these phenomena have gained wider recognition in recent years (An *et al.*, 2022; Sun *et al.*, 2022, 2023; Liu *et al.*, 2021; Ni *et al.*, 2021, 2023; Qi *et al.*, 2022; Moschos *et al.*, 2022; Sun *et al.*, 2021a, 2021b). Despite the growing attention, current studies still differ in their definition of abnormal eddies, particularly in how temperature anomalies within the eddies are delineated. Proposed two criteria for identifying abnormal eddies: (i) the difference between the average temperature inside the eddy (T_1) and the surrounding area (T_2) measured within a ring area 1.5 times the radius of the eddy. Cyclonic warm eddies require $T_1 - T_2 > T_0$, and anticyclonic cold eddies require $T_1 - T_2 < -T_0$, with T_0 set at 0.1°C; (ii) at least 60% of the grid points within the eddy must have temperatures warmer (for cyclonic warm eddies) or colder (for anticyclonic cold eddies) than that of the surrounding area. Ni *et al.* (2021), using satellite altimetry and microwave remote sensing data, applied only the first condition with T_0 set at 0°C. They found that surface cold core anticyclonic eddies (CAEs) and warm core cyclonic eddies (WCEs) are more common than previously thought, making up about 20% of the eddies globally. Their composite analysis from Argo float

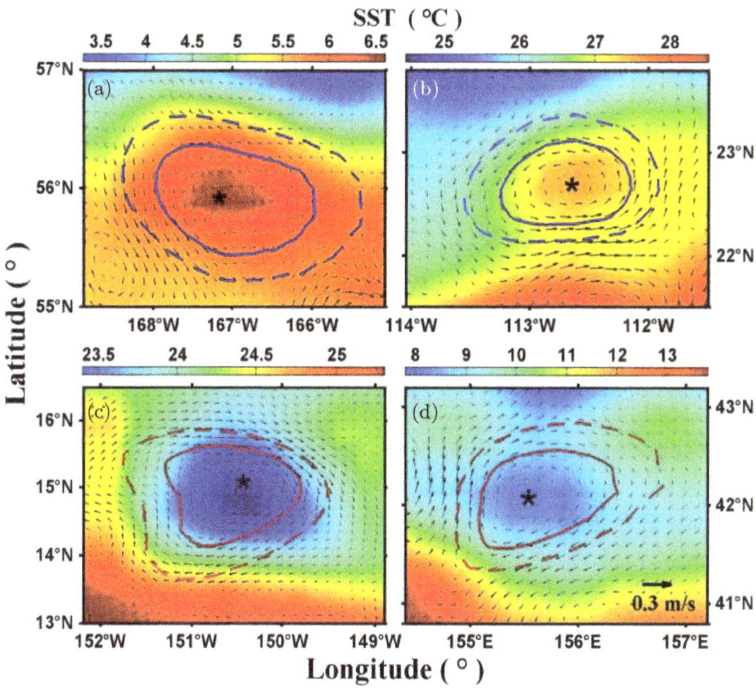

Fig. 4.1. Samples of warm-core eddies (a, b) and cold-core eddies (c, d) in the North Pacific. Shading indicates sea surface temperature (units: °C). Solid line contours represent the boundaries of the warm-core eddies and the cold-core eddies (CW-ACEs), which are also the inner boundaries of the background area. Dashed line contours indicate the outer boundaries of the background area. Black asterisks mark the centers of the CW-ACEs, and vectors show anomalies in sea surface geostrophic flow velocity (units: $\mathrm{m\,s^{-1}}$).

Source: Taken from Dong *et al.* (2025).

profiles indicates that the cold cores of CAEs and the warm cores of WCEs are typically confined to the upper ocean 50 m. Liu *et al.* (2021) developed a deep learning framework using 20 years of satellite SSH and SST data (1996–2015) to explore the global distribution of abnormal eddies. Their study found that abnormal eddies account for one-third of global eddies and are particularly active in equatorial currents and regions of high ocean instability. They also observed seasonal patterns in the Northern Hemisphere and found normal (abnormal) eddies are stronger in winter (summer), while the opposite is true in the Southern Hemisphere. Normal eddies show a greater annual amplitude than abnormal eddies, and the daily count of normal (abnormal) eddies increases (decreases) by 9.68 (11.80) per year.

4.3 Mesoscale Eddy Cruise Observation Projects

Mesoscale eddies are complex and variable ocean phenomena. High-resolution satellite remote sensing technologies and ocean numerical simulations have become Main methods for studying ocean eddies. However, cruise observations of ocean eddies have irreplaceable properties for the needs of ground truth data, even if still face a series of difficulties and challenges. Hereafter, practical experiences of five typical eddy cruise observation projects are briefly reviewed.

4.3.1 *E-FLUX*

To enhance the fundamental understanding of the coupling of physical, biological and biogeochemical processes occurring within mesoscale eddies, interdisciplinary observations of mesoscale eddies west of the Hawaiian Islands were conducted with the support of the US National Science Foundation (NSF) project E-FLUX. The E-FLUX field observation experiment was carried out in three phases from 4 November 2004 to 28 March 2005, with each phase lasting approximately three weeks. The same physical, bio-optical and chemical measurement methods were used during the three E-FLUX field observation experiments, including shipborne CTD observation, shipborne ADCP observation, drifting buoy observation and satellite observation.

The first voyage of E-FLUX (4–22 November 2004) primarily focused on CE Noah on the west side of the Big Island of Hawaii (Figure 4.2), with field observations commencing about three months after the formation of the eddy, roughly during the mature to decay phase. CTD and ADCP observations were conducted, and drifting buoys were deployed in the center of the eddy. Observations reveal that the isotherms, isopycnals, nutrient and chlorophyll-a isopleths within the CE are well-developed.

The third voyage of the E-FLUX expedition (10–28 March 2005) primarily observed the cyclonic eddy Opal (Figure 4.3), which was

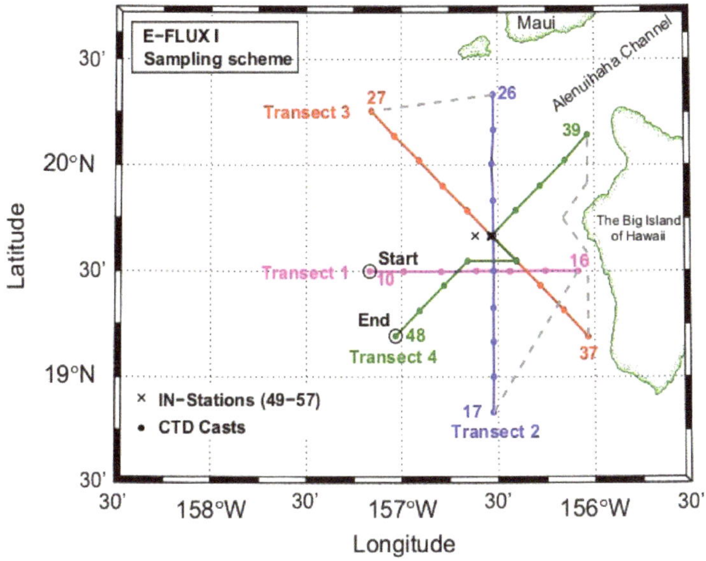

Fig. 4.2. Schematic of the first E-FLUX voyage.
Source: Taken from Dong *et al.* (2025).

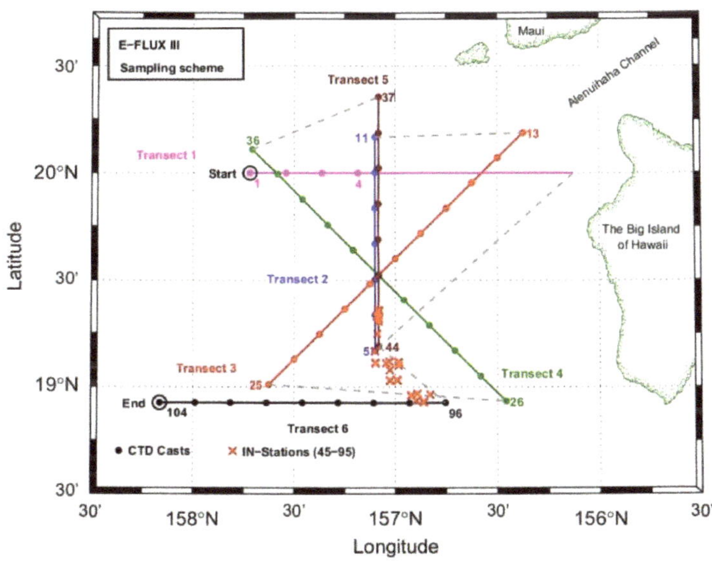

Fig. 4.3. Schematic of the third E-FLUX voyage.
Source: Taken from Dong *et al.* (2025).

in its mature stage. Like the cyclonic eddy Noah observed during the first voyage, the CTD and ADCP measurements show that the isotherms, isopycnals, nutrient and chlorophyll-a isopleths all exhibit upward bending towards the ocean surface.

Cruise observations of two eddies conducted through the E-FLUX program reveal a high correlation among physical, chemical and biological parameters, with changes in isopleths of nutrients and chlorophyll-a primarily induced by variations in isopycnals (Dickey *et al.*, 2008). The findings confirm that wind-driven first baroclinic CEs possess higher primary productivity and biomass, contributing minimally to the transport of particulate carbon and particulate nitrogen to the ocean's deeper layers (Benitez-Nelson *et al.*, 2007).

4.3.2 *EDDIES*

Supported by the US NSF, the Woods Hole Oceanographic Institution (WHOI) led the Eddies Dynamics, Mixing, Export, and Species Composition Program (EDDIES) from 2004 to 2005 (McGillicuddy *et al.*, 2007). This program conducted varying levels of sampling

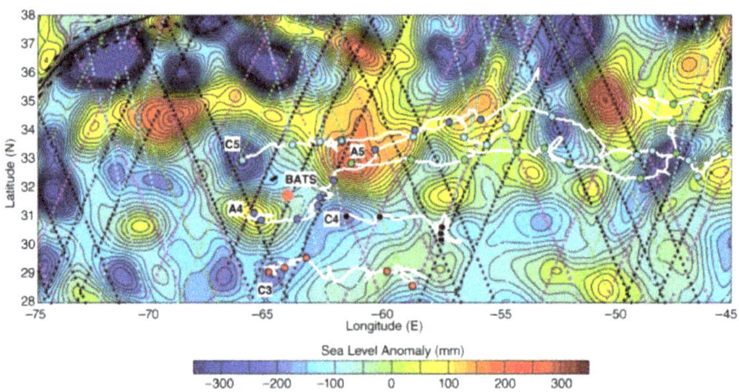

Fig. 4.4. Objective analysis of SLA for 17 June 2005, just before the first cruise of the EDDIES 2005 field season. The Gulf Stream mean path and meander envelope are indicated as solid and dashed black lines on the top left corner. Prior trajectories of eddies are indicated by white lines emanating from eddy centers, with dots at 30-day intervals. Satellite ground tracks are shown for the Jason (magenta), Topex 2 (green), Geosat Follow-on (black), and European Remote Sensing/Envisat (light blue) satellites.

Source: Taken from Dong *et al.* (2025).

across ten different eddies in the Sargasso Sea (Figure 4.4). The EDDIES experiment aimed to uncover the impacts of eddies and mixing on the structure of planktonic communities and biogeochemical cycles. The objectives include measuring the physiological response of phytoplankton to nutrient enhancement, evaluating changes in species composition induced by eddy disturbances, quantifying the impact of eddy disturbances on the upper ocean biogeochemical cycles, estimating the interaction between mixing processes within and below the mixed-layer along isopycnal surfaces and across them and integrating observed findings into a North Atlantic basin-scale eddy-resolving model to assess the impact of biogeochemical complexity on large-scale ocean material cycling.

The EDDIES was completed over two years, with cruise observations conducted during the summer months (June to September) when the mixed-layer depth is at its minimum. The initial phase involved identifying target eddies in the Sargasso Sea using advanced very high-resolution radiometer (AVHRR) SST and SeaWiFS ocean color data to determine the target sampling areas. Upon reaching the target eddy zones, a series of biogeochemical and hydrological observations were conducted, including towed observations, ADCP measurements and water sample collections (Figure 4.5).

The observations indicate that despite both cyclonic eddies and mode-water eddies experiencing phytoplankton growth, their ecological responses differ. Mode-water eddies can generate higher biomass and primary productivity in the deep ocean layers, maintained by the eddy/wind interaction, amplifying the effect of eddy-induced upwelling. Conversely, the eddy/wind interaction suppresses the eddy-induced upwelling within CEs (McGillicuddy *et al.*, 2007).

4.3.3 *LatMix*

The Scalable Lateral Mixing and Coherent Turbulence (LatMix) program of the WHOI was supported by both the Office of Naval Research (ONR) and the NSF. It is a study of mixing and stirring in the upper ocean on scales of 100 m to 10 km. Major field efforts occurred in June 2011 and in March 2012 in the region of the Gulf Stream off the coast of North Carolina. The program combines shipborne, autonomous, aerial observations and numerical simulations to understand ocean mixing across scales, particularly the energy

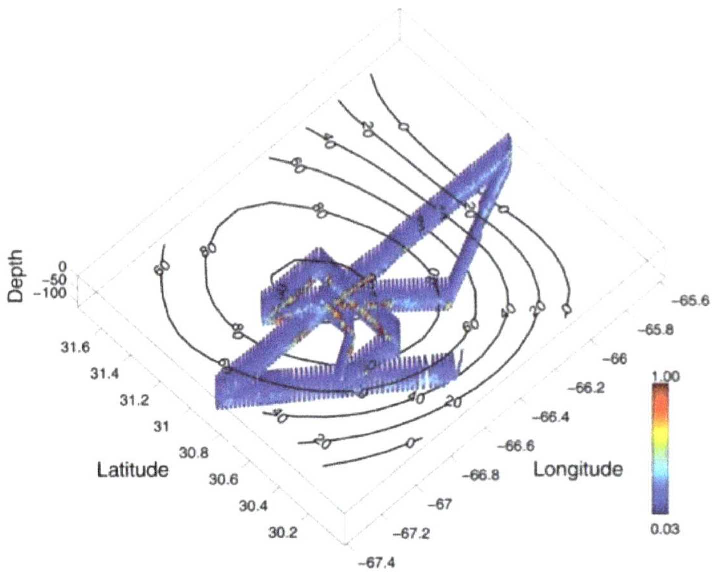

Fig. 4.5. Three-dimensional distribution of chlorophyll-a fluorescence inside the anticyclonic eddy A4 measured by the EDDIES cruise observation program in summer 2005.

Source: Taken from Dong *et al.* (2025).

cascade and mixing from mesoscale to sub-mesoscale. The initial scientific vision of LatMix identified four potential processes inducing mixing at the sub-mesoscale. Two hypotheses were related to internal wave processes, while the other two were associated with mesoscale and sub-mesoscale phenomena, specifically mesoscale flow stretching, frontogenesis and mixed-layer instabilities, which were key mechanisms for sub-mesoscale mixing. Building on this framework, LatMix conducted comprehensive multi-vessel observations in 2011 and 2012. In addition to CTD observations of temperature and salinity and ADCP measurements of current fields, the program employed gliders and tracer particles for more refined observations of ocean dynamics. To clarify the role of mesoscale current fields, the observations in 2011 focused on mesoscale eddies (Shcherbina *et al.*, 2015), selecting areas of weak and moderate stretching for detailed examination (Figure 4.6). The analysis results of these observations indicate that the energy cascade induced by mesoscale current fields

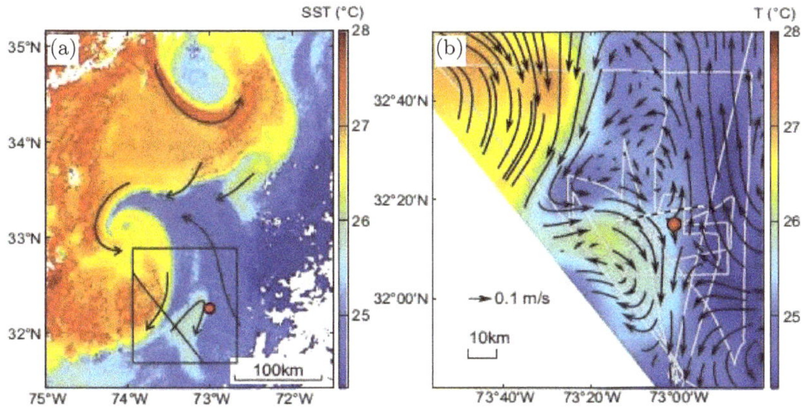

Fig. 4.6. Study area before the dye release on 13 June 2011. (a) AVHRR sea surface temperature image. Black arrows show the subjective interpretation of the surface flow patterns. (b) Sea surface temperature and velocity field, based on the reconnaissance survey conducted by three research vessels (R/Vs) from 10 to 12 June 2011. White lines show the R/V tracks. The survey area is shown as black box in (a). Red dots represent the flow stagnation points targeted by the study.

Source: Taken from Dong *et al.* (2025).

is a significant mechanism for the observed sub-mesoscale mixing (Shcherbina *et al.*, 2015).

To study mixing processes in the Gulf Stream front (Figure 4.7), the 2012 LatMix observations in the Gulf Stream region measured the physical characteristics of sub-mesoscale flows between mesoscale eddies and small-scale and three-dimensional turbulent motions. Observations of tracer particles reveal enhanced cross-frontal mixing at the Gulf Stream front. Moreover, numerical simulations show that rapid cross-frontal mixing occurred through shear dissipation, driven by frontal instabilities and vertical mixing. This provides direct evidence of the role of sub-mesoscale fronts in generating horizontal mixing, which is crucial for determining the horizontal structure of the ocean mixed layer. These observations further suggest that shear dissipation at the Gulf Stream front along the North Atlantic can provide significant freshwater flux to account for the freshwater deficit in the subtropical mode water budget and offer nutrient fluxes controlling primary productivity in the subtropical circulation (Wenegrat *et al.*, 2020).

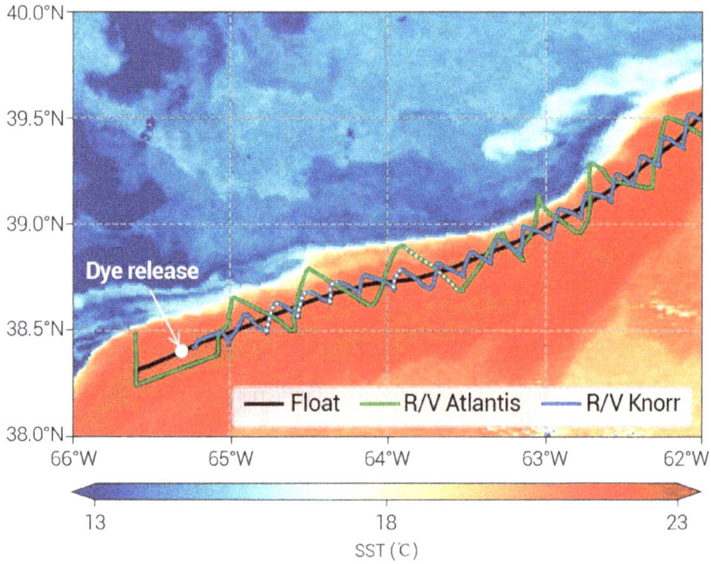

Fig. 4.7. Overview of the LatMix Dye Release Experiment along the Gulf Stream Front. Trajectories of Lagrangian drifters (black), R/V Atlantis (green), and R/V Knorr (blue) from Day 64 to Day 68, overlaid with satellite SST image on Day 71.

Source: Taken from Dong *et al.* (2025).

4.3.4 *OSMOSIS*

The Ocean Surface Mixing-Ocean Sub-mesoscale Interaction Study (OSMOSIS) project was a collaboration between scientists at various universities of the United Kingdom in the North Atlantic Ocean from September 2012 to September 2013. The project aims to observe the annual cycle of turbulence motions in the ocean surface boundary layer, spanning the transition from small-scale to sub-mesoscale and mesoscale processes. The OSMOSIS dataset provides a unique opportunity to address variability of the mixed-layer instability characteristics throughout the year. The observation program is broadly divided into two observational components (Figure 4.8). One consists of an array of nine mooring systems. The outer four systems form a rectangular mooring array with a spacing of about 15 km, and the inner five are composed of a rectangle formed by four moorings and one central mooring, with spacings of about 3 km. The moorings

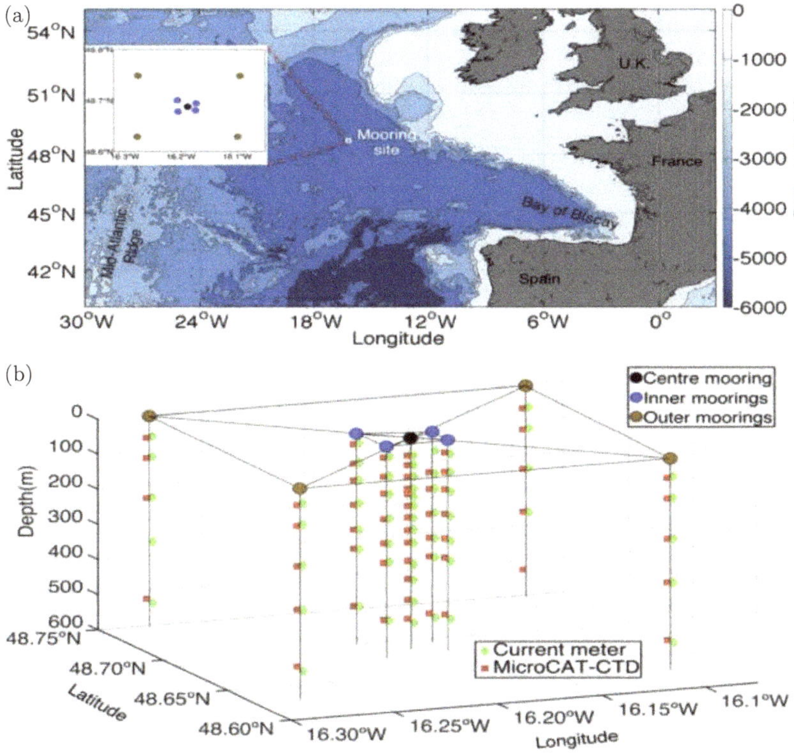

Fig. 4.8. (a) Location of the OSMOSIS mooring array. The European continent is shaded in grey, with color filling showing changes in water depth in meter. An inset map shows a zoomed view of the mooring sites. The mooring array consists of a central mooring (in black), four inner moorings (in blue), and four outer moorings (in dark yellow). (b) Vertical section of the OSMOSIS array. Current meters are shown as green diamonds, MicroCAT-CTDs are shown as red squares. *Source*: Taken from Dong *et al.* (2025).

are equipped with CTDs and current meters to observe the temperature, salinity and current velocities in the upper ocean layers in the Northeast Atlantic Ocean throughout the year from September 2012 to September 2013. The other observational component involves glider observations, which provide detailed structures of the temperature and salinity in the area around the moorings throughout the mooring observation period and can be used to calculate ocean dissipation rate (Thompson *et al.*, 2016; Yu *et al.*, 2019).

The year-long OSMOSIS mooring observations enable the evaluation of sub-mesoscale vertical transport processes against the backdrop of mesoscale motions. The findings suggest that sub-mesoscale motions play a continuous role in the re-stratification of the upper ocean throughout the year, with the strongest effects observed during winter and spring (Yu *et al.*, 2019). Persistent frontogenesis induced by mesoscale strain fields typically occurs when sub-mesoscale horizontal buoyancy gradients are enhanced. During this process, sub-mesoscale vertical velocities and buoyancy fluxes increase significantly, showing the role of mesoscale frontogenesis in enhancing sub-mesoscale turbulence (Yu *et al.*, 2019). Moreover, energy cascade analyses between mesoscale and sub-mesoscale motions based on OSMOSIS data indicate a two-phase energy transfer: from sub-mesoscale to mesoscale during winter and from mesoscale to sub-mesoscale during spring. The winter inverse cascade of kinetic energy is attributed to baroclinic instability within the mixed layer, while the spring forward energy cascade is associated with mesoscale frontogenesis (Garabato *et al.*, 2022).

The OSMOSIS data also provide rare observational evidence of sub-mesoscale eddies generated in the open ocean by baroclinic instability. The eddies are caused by mesoscale stirring, filamentation and subsequent frontal instabilities. Meanwhile horizontal shear or barotropic instability is one mechanism for generating such eddies. Conclusions drawn from linear theory combined with the OSMOSIS data suggest that mixed-layer or sub-mesoscale baroclinic instability is a more plausible explanation for the observed sub-mesoscale eddies. These results have significant implications for the ocean energy budget and the parameterization of ocean circulation models (Buckingham *et al.*, 2017).

4.3.5 *S-MEE*

The South China Sea Mesoscale Eddy Experiment (S-MEE) project was carried out by Ocean University of China with joint supports by Chinese government funding agencies from 2013 to 2014. The project aims to investigate the 3D structure and generation and dissipation processes of ocean eddies in the northern SCS. As part of S-MEE, two bottom-anchored subsurface mooring arrays were deployed along the historical pathway of mesoscale eddies in the region (Figure 4.9). The first array, cross-shaped and consisting of 10 moorings, was deployed

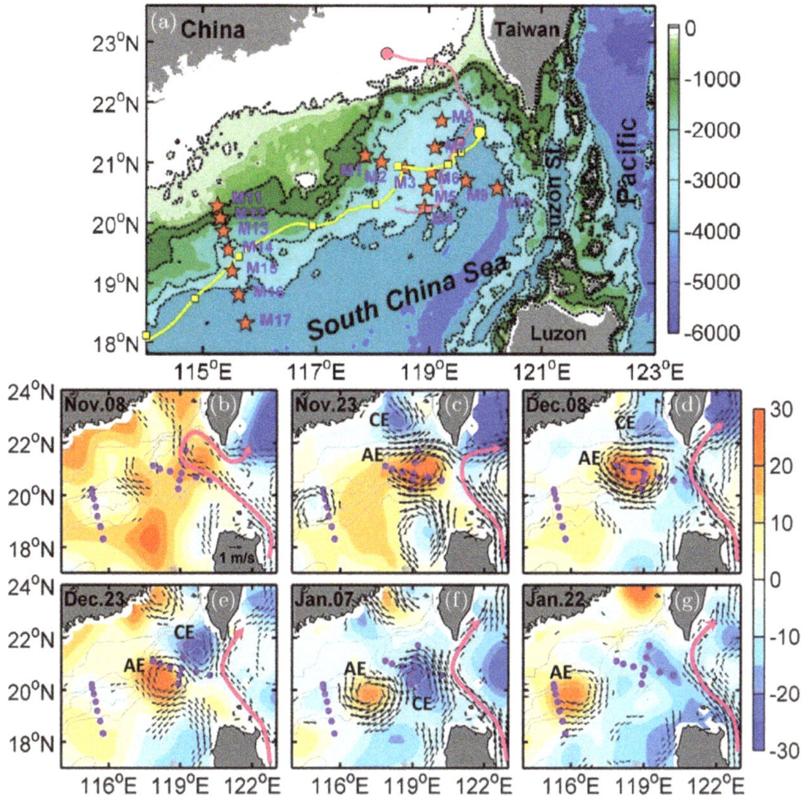

Fig. 4.9. (a) Locations of the S-MEE mooring arrays (red stars). The color shading shows bathymetry with black dashed lines indicating the isobaths of 100, 500, 1000, 2000 and 3000 m, respectively. Yellow and magenta lines denote the trajectories of AE and CE in (b–g), respectively. The dot on the trajectory indicates the generation site of eddy, and the squares indicate eddy's locations every 15 days after it was generated. Bathymetry data is downloaded from www.ngdc.noaa.gov/mgg/global/. (b–g) Maps of altimeter SLA (shading, in cm) and surface absolute geostrophic velocities (arrows, those with magnitudes smaller than $0.3\,\mathrm{m\,s^{-1}}$ are not shown) during the period of AE and CE (from 8 November 2013 to 22 January 2014). Purple dots indicate the mooring locations. Magenta arrows roughly denote axes of the Kuroshio current path.

Source: Taken from Zhang *et al.* (2016).

west of the Luzon Strait in the generation region of eddies in late October 2013 and successfully recovered in early June 2014. Before the experiment, mooring M1 at the western end of this array had been deployed and recovered twice and provided continuous measurements for a total length of long than 2 years. The second array, consisting of 7 moorings along a line normal to the local isobaths, was deployed in the dissipation region of eddies in mid-January 2013 and successfully recovered in mid-June 2014. All the moorings were equipped with ADCPs, recording current meters (RCMs), CTDs and temperature chains to monitor horizontal current velocity, pressure and temperature/salinity (T/S) in the whole water column.

To identify and track the oceanic eddies, AVISO merged satellite altimeter sea level anomaly (SLA) and absolute surface geostrophic velocity data during the experiment period (October 2013–June 2014) are used. Total 5 distinct ocean eddies were observed to cross the mooring arrays. Among these eddies, one anticyclonic and cyclonic eddy pair was fully captured by the first mooring array from mid-November to early January (Figure 4.9(b–e)).

The S-MEE for the first time captured full-depth 3D structures of an anticyclonic and cyclonic eddy pair, which are characterized by a distinct vertical tilt of their axes. Meanwhile, the eddy evolution at an upstream versus downstream location and eddy energy budget analysis reveal that generation of sub-mesoscale motions most likely constitutes the dominant dissipation mechanism for the observed eddies.

4.4 Summary

This chapter focuses on detection and observation methods for ocean mesoscale eddies. At present, information traction of the global ocean distribution and long-term variability of mesoscale eddies mainly relies on satellite remote sensing data and numerical model reanalysis data products. Therefore, various computer-based automated eddy identification and detection methods have been developed in recent years. This chapter briefly reviews advantages and disadvantages of broadly used eddy identification and detection methods, including the methods based on Eulerian data, methods based on Lagrangian data, Lagrangian coherent eddy detection methods, three-dimensional eddy detection methods, AI-based eddy detection methods and abnormal ocean eddy detection methods.

On the other hand, the cruise observation is an irreplaceable method to collect three-dimensional, high resolution ground truth data for ocean mesoscale eddy studies. Thus, five typical cruise observation projects during ten years between 2004 are briefly reviewed.

Among them, two projects were conducted in the Pacific Ocean, i.e., the E-FLUX field observation experiment supported by the US and carried out in the water area west of the Hawaiian Islands from 2004 to 2005, and the South China Sea Mesoscale Eddy Experiment (S-MEE) project supported by China carried out in the northeastern SCS from 2013 to 2014. The other three projects were conducted in the Atlantic Ocean, i.e., the Eddies Dynamics, Mixing, Export, and Species Composition Program (EDDIES) supported by the US and carried out in the Sargasso Sea from 2004 to 2005, the Scalable Lateral Mixing and Coherent Turbulence (LatMix) program supported by the US and carried out in the Gulf Stream region off the coast of North Carolina from 2011 to 2012 and the Ocean Surface Mixing-Ocean Sub-mesoscale Interaction Study (OSMOSIS) project supported by the UK and carried on in the North Atlantic Ocean from 2012 to 2013. All the five projects achieved success and shared the practical experiences and results with communities.

References

An, M., Liu, J., Liu, J., Sun, W., Yang, J., Tan, W. and Dong, C. (2022). Comparative analysis of four types of mesoscale eddies in the north pacific subtropical countercurrent region–part I spatial characteristics, *Fron. Mar. Sci.*, 9, 1004300, doi:10.3389/fmars.2022.1004300.

Ashkezari, M. D., Hill, C. N., Follett, C. N., Forget, G. and Follows, M. J. (2016). Oceanic eddy detection and lifetime forecast using machine learning methods, *Geophys. Res. Lett.*, 43(23), pp. 12–234, doi:10.1002/2016GL071269.

Benitez-Nelson, C. R., Bidigare, R. R., Dickey, T. D., Landry, M. R., Leonard, C. L., Brown, S. L. and Yang, E. J. (2007). Mesoscale eddies drive increased silica export in the subtropical Pacific Ocean, *Science*, 316(5827), pp. 1017–1021, doi:10.1126/science.1136221.

Beron-Vera, F. J., Olascoaga, M. J. and Goni, G. J. (2008). Oceanic mesoscale eddies as revealed by Lagrangian coherent structures, *Geophys. Res. Lett.*, 35(12), L12603, doi:10.1029/2008GL033957.

Boebel, O., Lutjeharms, J., Schmid, C., Zenk, W., Rossby, T. and Barron, C. (2003). The Cape Cauldron: A regime of turbulent inter-ocean exchange, *Deep Sea Res. Part II: Top. Stu. Oceanogr.*, 50(1), pp. 57–86, doi:10.1016/S0967-0645(02)_00379-X.

Buckingham, C. E., Khaleel, Z., Lazar, A., Martin, A. P., Allen, J. T., Naveira Garabato, A. C. and Vic, C. (2017). Testing Munk's hypothesis for submesoscale eddy generation using observations in the

North Atlantic, *J. Geophys. Res. Oceans*, 122(8), pp. 6725–6745, doi:10.1002/2017JC012910.

Cao, Y., Dong, C., Qiu, Z., Bethel, B. J., Shi, H., Lü, H. and Cheng, Y. (2022). Corrections of mesoscale eddies and Kuroshio extension surface velocities derived from satellite altimeters, *Rem. Sens.*, 15(1), 184, doi:10.3390/rs15010184.

Cao, Y., Dong, C., Stegner, A., Bethel, B. J., Li, C., Dong, J. and Yang, J. (2023). Global sea surface cyclogeostrophic currents derived from satellite altimetry data, *J. Geophys. Res. Oceans*, 128(1), e2022JC019357, doi:10.1029/2022JC019357.

Chaigneau, A., Gizolme, A. and Grados, C. (2008). Mesoscale eddies off Peru in altimeter records: Identification algorithms and eddy spatio-temporal patterns, *Progr. Oceanogr.*, 79(2–4), pp. 106–119, doi:10.1016/j.pocean.2008.10.013.

Chang, M. H., Tang, T. Y., Ho, C. R. and Chao, S. Y. (2013). Kuroshio-induced wake in the lee of Green Island off Taiwan, *J. Geophys. Res. Oceans*, 118(3), pp. 1508–1519, doi:10.1002/jgrc.20151.

Chelton, D. B., Schlax, M. G., Samelson, R. M. and de Szoeke, R. A. (2007). Global observations of large oceanic eddies, *Geophys. Res. Lett.*, 34(15), pp. 87–101, doi:10.1029/2007GL030812.

Chelton, D. B., Gaube, P., Schlax, M. G., Early, J. J. and Samelson, R. M. (2011). The influence of nonlinear mesoscale eddies on near-surface oceanic chlorophyll, *Science*, 334(6054), pp. 328–332, doi:10.1126/science.1208897.

Codiga, D. L. and Eriksen, C. C. (1997). Observations of low-frequency circulation and amplified subinertial tidal currents at Cobb Seamount, *J. Geophys. Res. Oceans*, 102(C10), pp. 22993–23007, doi:10.1029/97JC01451.

Dai, J., Wang, H., Zhang, W., An, Y. and Zhang, R. (2020). Observed spatiotemporal variation of three-dimensional structure and heat/salt transport of anticyclonic mesoscale eddy in Northwest Pacific, *J. Oceanol. Limnol.*, 38(6), pp. 1654–1675, doi:10.1007/s00343-019-9148-z.

Dickey, T. D., Nencioli, F., Kuwahara, V. S., Leonard, C., Black, W., Rii, Y. M. and Zhang, Q. (2008). Physical and bio-optical observations of oceanic cyclones west of the island of Hawaii, *Deep Sea Res. Part II: Top. Stu. Oceanogr.*, 55(10–13), pp. 1195–1217, doi:10.1016/j.dsr2.2008.01.006.

Dong, C., Lin, X., Liu, Y., Nencioli, F., Chao, Y., Guan, Y. and McWilliams, J. C. (2012). Three-dimensional oceanic eddy analysis in the Southern California Bight from a numerical product, *J. Geophys. Res. Oceans*, 117(C7), C00H14, doi:10.1029/2011JC007354.

Dong, C., Liu, L., Nencioli, F., Bethel, B. J., Liu, Y., Xu, G. and Zou, B. (2022). The near-global ocean mesoscale eddy atmospheric-oceanic-biological interaction observational dataset, *Sci. Data*, 9(1), 436, doi:10.1038/s41597-022-01550-9.

Dong, C., Nencioli, F., Liu, Y. and McWilliams, J. C. (2011a). An automated approach to detect oceanic eddies from satellite remotely sensed sea surface temperature data, *IEEE Geosci. Rem. Sens. Lett.*, 8(6), pp. 1055–1059, doi:10.1109/LGRS.2011.2155029.

Dong, C., Liu, Y., Lumpkin, R., Lankhorst, M., Chen, D., McWilliams, J. C. and Guan, Y. (2011b). A scheme to identify loops from trajectories of oceanic surface drifters: An application in the Kuroshio Extension region, *J. Atmos. Ocean. Tech.*, 28(9), pp. 1167–1176, doi:10.1175/JTECH-D-10-05028.1.

Dong, D., Brandt, P., Chang, P., Schütte, F., Yang, X., Yan, J. and Zeng, J. (2017). Mesoscale eddies in the northwestern Pacific Ocean: Three-dimensional eddy structures and heat/salt transports, *J. Geophys. Res. Oceans*, 122(12), pp. 9795–9813, doi:10.1002/2017JC013303.

Dong, C., You, Z., Dong, J., Ji, J., Sun, W., Xu, G. and Han, G. (2025). Oceanic mesoscale eddies, *Ocean-Land-Atmos. Res.*, 4, pp. 1–111, doi:10.34133/olar.0081.

Douglass, E. M. and Richman, J. G. (2015). Analysis of ageostrophy in strong surface eddies in the Atlantic Ocean, *J. Geophys. Res. Oceans*, 120(3), pp. 1490–1507, doi:10.1002/2014JC010350.

Du, Y., Song, W., He, Q., Huang, D., Liotta, A. and Su, C. (2019). Deep learning with multi-scale feature fusion in remote sensing for automatic oceanic eddy detection, *Inform. Fusion*, 49, pp. 89–99, doi:10.1016/j.inffus.2018.09.006.

Duo, Z., Wang, W. and Wang, H. (2019). Oceanic mesoscale eddy detection method based on deep learning, *Rem. Sens.*, 11(16), pp. 1–14, doi:10.3390/rs11161921.

Faghmous, J. H., Frenger, I., Yao, Y., Warmka, R., Lindell, A. and Kumar, V. (2015). A daily global mesoscale ocean eddy dataset from satellite altimetry, *Sci. data*, 2(1), pp. 1–16, doi:10.1038/sdata.2015.28.

Franz, K., Roscher, R., Milioto, A., Wenzel, S. and Kusche, J. (2018). Ocean eddy identification and tracking using neural networks, In *IGARSS 2018-2018 IEEE Intern. Geosci. Rem. Sens. Symp.*, pp. 6887–6890, IEEE, doi:10.1109/IGARSS.2018.8519261.

Fratantoni, D. M. and Richardson, P. L. (2006). The evolution and demise of North Brazil Current rings, *J. Phys. Oceanogr.*, 36(7), pp. 1241–1264, doi:10.1175/JPO2907.1.

Frenger, I., Münnich, M., Gruber, N. and Knutti, R. (2015). Southern Ocean eddy phenomenology, *J. Geophys. Res. Oceans*, 120(11), pp. 7413–7449, doi:10.1002/2015JC011047.

Garabato, A. C. N., Yu, X., Callies, J., Barkan, R., Polzin, K. L., Frajka-Williams, E. E. and Griffies, S. M. (2022). Kinetic energy transfers between mesoscale and submesoscale motions in the open ocean's upper layers, *J. Phys. Oceanogr.*, 52(1), pp. 75–97, doi:10.1175/JPO-D-21-0099.1.

Glenn, S. M. and Ebbesmeyer, C. C. (1993). Drifting buoy observations of a loop current anticyclonic eddy, *J. Geophys. Res. Oceans*, 98(C11), pp. 20105–20119, doi:10.1029/93JC02078.

Griffa, A., Lumpkin, R. and Veneziani, M. (2008). Cyclonic and anticyclonic motion in the upper ocean, *Geophys. Res. Lett.*, 35(1), L01608, doi:10.1029/2007GL032100.

Hadjighasem, A., Farazmand, M., Blazevski, D., Froyland, G. and Haller, G. (2017). A critical comparison of Lagrangian methods for coherent structure detection, *Chaos: An Interdis. J. Nonl. Sci.*, 27(5), 053104, doi:10.1063/1.4982720.

Haller, G., Hadjighasem, A., Farazmand, M. and Huhn, F. (2016). Defining coherent vortices objectively from the vorticity, *J. Fluid Mech.*, 795, pp. 136–173, doi:10.1017/jfm.2016.151.

Hamilton, P. (2007). Eddy statistics from Lagrangian drifters and hydrography for the northern Gulf of Mexico slope, *J. Geophys. Res.*, 112, C09002, doi:10.1029/2006JC003988.

Hamilton, P., Fargion, G. S. and Biggs, D. C. (1999). Loop Current eddy paths in the western Gulf of Mexico, *J. Phys. Oceanogr.*, 29(6), pp. 1180–1207, https://doi.org/10.1175/1520-0485(1999)029%3C1180:LCEPIT%3E2.0.CO;2.

Han, G., Dong, C., Yang, J., Sommeria, J., Stegner, A. and Caldeira, R. (2021). Strain evolution and instability of an anticyclonic eddy from a laboratory experiment, *Fron. Mar. Sci.*, 8, 645531, doi:10.3389/fmars.2021.645531.

He, Q., Zhan, H., Cai, S., He, Y., Huang, G. and Zhan, W. (2018). A new assessment of mesoscale eddies in the South China Sea: Surface features, three-dimensional structures, and thermohaline transports, *J. Geophys. Res. Oceans*, 123(7), pp. 4906–4929, doi:10.1029/2018JC014054.

Henson, S. A. and Thomas, A. C. (2008). A census of oceanic anticyclonic eddies in the Gulf of Alaska, *Deep Sea Res. Part I: Oceanogr. Res. Pap.*, 55(2), pp. 163–176, doi:10.1016/j.dsr.2007.11.005.

Hwang, C., Wu, C. R. and Kao, R. (2004). TOPEX/Poseidon observations of mesoscale eddies over the Subtropical Countercurrent: Kinematic characteristics of an anticyclonic eddy and a cyclonic eddy, *J. Geophys. Res. Oceans*, 109(C8), C08013, doi:10.1029/2003JC002026.

Ioannou, A., Stegner, A., Tuel, A., Le Vu, B., Dumas, F. and Speich, S. (2019). Cyclostrophic corrections of AVISO/DUACS surface velocities and its application to mesoscale eddies in the Mediterranean Sea, *J. Geophys. Res. Oceans*, 124(12), pp. 8913–8932, doi:10.1029/2019JC015031.

Isern-Fontanet, J., García-Ladona, E. and Font, J. (2003). Identification of marine eddies from altimetric maps, *J. Atmos. Ocean. Tech.*, 20(5), pp. 772–778, doi:10.1175/1520-0426(2003)20<772:IOMEFA>2.0.CO;2.

Itoh, S. and Yasuda, I. (2010). Characteristics of mesoscale eddies in the Kuroshio–Oyashio Extension region detected from the distribution of the sea surface height anomaly, *J. Phys. Oceanogr.*, 40(5), pp. 1018–1034, doi:10.1175/2009JPO4265.1.

Ji, J., Dong, C., Zhang, B. and Liu, Y. (2017). An oceanic eddy statistical comparison using multiple observational data in the Kuroshio Extension region, *Acta Oceanol. Sin.*, 36, pp. 1–7, doi:10.1007/s13131-016-0882-1.

Ji, J., Dong, C., Zhang, B., Liu, Y., Zou, B., King, G. P. and Chen, D. (2018). Oceanic eddy characteristics and generation mechanisms in the Kuroshio Extension region, *J. Geophys. Res. Oceans*, 123(11), pp. 8548–8567, doi:10.1029/2018JC014196.

Johannessen, J. A., Sandven, S., Lygre, K., Svendsen, E. and Johannessen, O. M. (1989). Three-dimensional structure of mesoscale eddies in the Norwegian Coastal Current, *J. Phys. Oceanogr.*, 19(1), pp. 3–19, doi:10.1175/1520-0485(1989)019<0003:TDSOME>2.0.CO;2.

Karimova, S. (2017). An approach to automated spiral eddy detection in SAR images, In *2017 IEEE Intern. Geosci. Rem. Sens. Symp. (IGARSS)*, pp. 743–746, IEEE, doi:10.1109/IGARSS.2017.8127059.

Karimova, S. S. and Gade, M. (2014). Eddies in the Red Sea as seen by satellite SAR imagery, In *Rem. Sens. African Seas*, pp. 357–378, doi:10.1007/978-94-017-8008-7_18.

Lankhorst, M. (2006). A self-contained identification scheme for eddies in drifter and float trajectories, *J. Atmos. Ocean. Tech.*, 23(11), pp. 1583–1592, doi:10.1175/JTECH1931.1.

Le Vu, B., Stegner, A. and Arsouze, T. (2018). Angular momentum eddy detection and tracking algorithm (AMEDA) and its application to coastal eddy formation, *J. Atmos. Ocean. Tech.*, 35(4), pp. 739–762, doi:10.1175/JTECH-D-17-0010.1.

Lévy, M., Franks, P. J. and Smith, K. S. (2018). The role of submesoscale currents in structuring marine ecosystems, *Nature Comm.*, 9(1), pp. 1–16, doi:10.1038/s41467-018-07059-3.

Lguensat, R., Sun, M., Fablet, R. and Mason, E. (2018). EddyNet: A deep neural network for pixel-wise classification of oceanic eddies, *IGARSS 2018-2018 IEEE Intern. Geosci. Rem. Sens. Symp.*, IEEE, pp. 1764–1767, doi:10.1109/IGARSS.2018.8518411.

Li, S., Wang, S., Zhang, F. and Wang, Y. (2019). Constructing the three-dimensional structure of an anticyclonic eddy in the South China Sea using multiple underwater gliders, *J. Atmos. Ocean. Tech.*, 36(12), pp. 2449–2470, doi:10.1175/JTECH-D-19-0006.1.

Li, S., Zhang, F., Wang, S., Wang, Y. and Yang, S. (2020). Constructing the three-dimensional structure of an anticyclonic eddy with the optimal configuration of an underwater glider network, *Appl. Ocean Res.*, 95, 101893, doi:10.1016/j.apor.2019.101893.

Li, X., Zhang, X., Fu, D. and Liao, S. (2021). Strengthening effect of super typhoon Rammasun (2014) on upwelling and cold eddies in the South China Sea, *J. Oceanol. Limnol.*, 39, pp. 403–419, doi:10.1007/s00343-020-9239-x.

Lilly, J. M. and Pérez-Brunius, P. (2021). Extracting statistically significant eddy signals from large Lagrangian datasets using wavelet ridge analysis, with application to the Gulf of Mexico, *Nonl. Proc. Geophys.*, 28(2), pp. 181–212, doi:10.5194/npg-28-181-2021.

Lilly, J. M. and Olhede, S. C. (2009). Bivariate instantaneous frequency and bandwidth, *IEEE Trans. Sig. Proc.*, 58(2), pp. 591–603, doi:10.1109/TSP.2009.2031729.

Lilly, J. M. and Gascard, J. C. (2006). Wavelet ridge diagnosis of time-varying elliptical signals with application to an oceanic eddy, *Nonl. Proc. Geophys.*, 13(5), pp. 467–483, doi:10.5194/npg-13-467-2006.

Lin, X., Dong, C., Chen, D., Liu, Y., Yang, J., Zou, B. and Guan, Y. (2015). Three-dimensional properties of mesoscale eddies in the South China Sea based on eddy-resolving model output, *Deep Sea Res. Part I: Oceanogr. Res. Pap.*, 99, pp. 46–64, doi:10.1016/j.dsr.2015.01.007.

Liu, M. and Rossby, T. (1993). Observations of the velocity and vorticity structure of Gulf Stream meanders, *J. Phys. Oceanogr.*, 23(2), pp. 329–345, doi:10.1175/1520-0485(1993)023<0329:OOTVAV>2.0.CO;2.

Liu, Y., Zheng, Q. and Li, X. (2021). Characteristics of global ocean abnormal mesoscale eddies derived from the fusion of sea surface height and temperature data by deep learning, *Geophys. Res. Lett.*, 48(17), e2021GL094772, doi:10.1029/2021GL094772.

Liu, Z., Du, Y. and Xu, K. (2015). An improved scheme of identifying loops using Lagrangian drifters, In *2015 2nd IEEE Intern. Conf. Spatial Data Mining Geogr. Knowl. Serv. (ICSDM)*, pp. 125–128, IEEE, doi:10.1109/ICSDM.2015.7298038.

Mandal, S., Sil, S., Pramanik, S., KS, A. and Jena, B. K. (2019). Characteristics and evolution of a coastal mesoscale eddy in the Western Bay of Bengal monitored by high-frequency radars, *Dyn. Atmos. Oceans*, 88, 101107, doi:10.1016/j.dynatmoce.2019.101107.

Maximenko, N. and Niiler, P. (2006). Mean surface circulation of the global ocean inferred from satellite altimeter and drifter data, In *Proc. Symp. 15 years Progr. Radar Alt., ESA Spec. Publ.*, ESA SP, 614.

McGillicuddy Jr, D. J., Anderson, L. A., Bates, N. R., Bibby, T., Buesseler, K. O., Carlson, C. A. and Steinberg, D. K. (2007). Eddy/wind interactions stimulate extraordinary mid-ocean plankton blooms, *Science*, 316(5827), pp. 1021–1026, doi:10.1126/science.1136256.

McWilliams, J. C. (2008). The nature and consequences of oceanic eddies, *Geophys. Monogr. Ser.*, 177, pp. 5–15, doi:10.1029/177GM03.

McWilliams, J. C. and Flierl, G. R. (1979). On the evolution of isolated, nonlinear vortices, *J. Phys. Oceanogr.*, 9(6), pp. 1155–1182, doi:10.1175/1520-0485(1979)009<1155:OTEOIN>2.0.CO;2.

Meunier, T., Pallás Sanz, E., de Marez, C., Pérez, J., Tenreiro, M., Ruiz Angulo, A. and Bower, A. (2021). The dynamical structure of a warm core ring as inferred from glider observations and along-track altimetry, *Rem. Sens.*, 13(13), 2456, doi:10.3390/rs13132456.

Moschos, E., Barboni, A. and Stegner, A. (2022). Why do inverse eddy surface temperature anomalies emerge? The case of the Mediterranean Sea, *Rem. Sens.*, 14(15), 3807, doi:10.3390/rs14153807.

Nagai, T., Gruber, N., Frenzel, H., Lachkar, Z., McWilliams, J. C. and Plattner, G. K. (2015). Dominant role of eddies and filaments in the offshore transport of carbon and nutrients in the California Current System, *J. Geophys. Res. Oceans*, 120(8), pp. 5318–5341, doi:10.1002/2015JC010889.

Nencioli, F., Dong, C., Dickey, T., Washburn, L. and McWilliams, J. C. (2010). A vector geometry–based eddy detection algorithm and its application to a high-resolution numerical model product and high-frequency radar surface velocities in the Southern California Bight, *J. Atmos. Ocean. Tech.*, 27(3), pp. 564–579, doi:10.1175/2009JTECHO725.1.

Ni, Q., Zhai, X., Jiang, X. and Chen, D. (2021). Abundant cold anticyclonic eddies and warm cyclonic eddies in the global ocean, *J. Phys. Oceanogr.*, 51(9), pp. 2793–2806, doi:10.1175/JPO-D-21-0010.1.

Ni, Q., Zhai, X., Yang, Z. and Chen, D. (2023). Generation of cold anticyclonic eddies and warm cyclonic eddies in the tropical oceans, *J. Phys. Oceanogr.*, 53(6), pp. 1485–1498, doi:10.1175/JPO-D-22-0197.1.

Okubo, A. (1970). Horizontal dispersion of floatable particles in the vicinity of velocity singularities such as convergences, *Deep Sea Res. Oceanogr. Abst.*, 17(3), pp. 445–454, doi:10.1016/0011-7471(70)90059-8.

Penven, P., Echevin, V., Pasapera, J., Colas, F. and Tam, J. (2005). Average circulation, seasonal cycle, and mesoscale dynamics of the Peru Current System: A modeling approach, *J. Geophys. Res. Oceans*, 110(C10), C10021, doi:10.1029/2005JC002945.

Penven, P., Halo, I., Pous, S. and Marié, L. (2014). Cyclogeostrophic balance in the Mozambique Channel, *J. Geophys. Res. Oceans*, 119(2), pp. 1054–1067, doi:10.1002/2013JC009528.

Petersen, M. R., Williams, S. J., Maltrud, M. E., Hecht, M. W. and Hamann, B. (2013). A three-dimensional eddy census of a high-resolution global ocean simulation, *J. Geophys. Res. Oceans*, 118(4), pp. 1759–1774, doi:10.1002/jgrc.20155.

Qi, Y., Mao, H., Du, Y., Li, X., Yang, Z., Xu, K. and Xing, H. (2022). A lens-shaped, cold-core anticyclonic surface eddy in the northern South China Sea, *Fron. Mar. Sci.*, 9, 976273, doi:10.3389/fmars.2022.976273.

Qiu, B. (1999). Seasonal eddy field modulation of the North Pacific Subtropical Countercurrent: TOPEX/Poseidon observations and theory, *J. Phys. Oceanogr.*, 29(10), pp. 2471–2486, doi:10.1175/1520-0485(1999)029<2471:SEFMOT>2.0.CO;2.

Qiu, B. and Chen, S. (2005). Variability of the Kuroshio Extension jet, recirculation gyre, and mesoscale eddies on decadal time scales, *J. Phys. Oceanogr.*, 35(11), pp. 2090–2103, doi:10.1175/JPO2807.1.

Qiu, C., Mao, H., Wang, Y., Su, D. and Lian, S. (2019). An irregularly shaped warm eddy observed by Chinese underwater gliders, *J. Oceanogr.*, 75, pp. 139–148, doi:10.1007/s10872-018-0490-0.

Richardson, P. L. (1993). A census of eddies observed in North Atlantic SOFAR float data, *Progr. Oceanogr.*, 31(1), pp. 1–50, doi:10.1016/0079-6611(93)90022-6.

Rovegno, P. S., Edwards, C. A. and Bruland, K. W. (2009). Observations of a Kenai eddy and a Sitka eddy in the northern Gulf of Alaska, *J. Geophys. Res. Oceans*, 114(C11), C11012, doi:10.1029/2009JC005451.

Ruiz, S., Pelegrí, J. L., Emelianov, M., Pascual, A. and Mason, E. (2014). Geostrophic and ageostrophic circulation of a shallow anticyclonic eddy off Cape Bojador, *J. Geophys. Res. Oceans*, 119(2), pp. 1257–1270, doi:10.1002/2013JC009169.

Sadarjoen, I. A. and Post, F. H. (2000). Detection, quantification, and tracking of vortices using streamline geometry, *Comp. Graph.*, 24(3), pp. 333–341, doi:10.1016/S0097-8493(00)00029-7.

Saida, S. J., Sahoo, S. P. and Ari, S. (2023). Deep convolution neural network based semantic segmentation for ocean eddy detection, *Exp. Sys. Appl.*, 219, 119646, doi:10.1016/j.eswa.2023.119646.

Santana, O. J., Hernández-Sosa, D. and Smith, R. N. (2022). Oceanic mesoscale eddy detection and convolutional neural network complexity, *Intern. J. Appl. Earth Observ. Geoinfor.*, 113, 102973, doi:10.1016/j.jag.2022.102973.

Scharffenberg, M. G. and Stammer, D. (2010). Seasonal variations of the large-scale geostrophic flow field and eddy kinetic energy inferred from the TOPEX/Poseidon and Jason-1 tandem mission data, *J. Geophys. Res. Oceans*, 115(C2), C02008, doi:10.1029/2008JC005242.

Shcherbina, A. Y., Sundermeyer, M. A., Kunze, E., D'Asaro, E., Badin, G., Birch, D. and Ledwell, J. R. (2015). The LatMix summer campaign: Submesoscale stirring in the upper ocean, *Bull. Amer. Meteorol. Soc.*, 96(8), pp. 1257–1279, doi:10.1175/BAMS-D-14-00015.1.

Shoosmith, D. R., Richardson, P. L., Bower, A. S. and Rossby, H. T. (2005). Discrete eddies in the northern North Atlantic as observed by looping RAFOS floats, *Deep Sea Res. Part II: Top. Stu. Oceanogr.*, 52(3-4), pp. 627–650, doi:10.1016/j.dsr2.2004.12.011.

Scott, R. M., Pickart, R. S., Lin, P., Münchow, A., Li, M., Stockwell, D. A. and Brearley, J. A. (2019). Three-dimensional structure of a cold-core Arctic eddy interacting with the Chukchi Slope Current, *J. Geophys. Res. Oceans*, 124(11), pp. 8375–8391, doi:10.1029/2019JC015523.

Sun, W., An, M., Liu, J., Liu, J., Yang, J., Tan, W. and Dong, C. (2023). Comparative analysis of four types of mesoscale eddies in the North Pacific Subtropical Countercurrent region-part II seasonal variation, *Fron. Mar. Sci.*, 10, 1121731, doi:10.3389/fmars.2023.1121731.

Sun, W. Y. and Oh, T. J. (2022). Vortex merger in shallow water model, *Asia-Pacific J. Atmos. Sci.*, 58(4), pp. 533–547, doi:10.1007/s13143-022-00270-9.

Sun, W., Liu, Y., Chen, G., Tan, W., Lin, X., Guan, Y. and Dong, C. (2021a). Three-dimensional properties of mesoscale cyclonic warm-core and anticyclonic cold-core eddies in the South China Sea, *Acta Oceanol. Sin.*, 40, pp. 17–29, doi:10.1007/s13131-010-0003-5.

Sun, Z., Hu, J., Chen, Z., Zhu, J., Yang, L., Chen, X. and Wu, X. (2021b). A strong Kuroshio intrusion into the South China Sea and its accompanying cold-core anticyclonic eddy in winter 2020–2021, *Rem. Sens.*, 13(14), p. 2645, doi:10.3390/rs13142645.

Taneda, S. (1956). Experimental investigation of the wakes behind cylinders and plates at low Reynolds numbers, *J. Phys. Soc. Japan*, 11(3), pp. 302–307, doi:10.1143/JPSJ.11.302.

Taneda, S. (1959). Downstream development of the wakes behind cylinders, *J. Phys. Soc. Japan*, 14(6), pp. 843–848, doi:10.1143/JPSJ.14.843.

Thompson, A. F., Lazar, A., Buckingham, C., Naveira Garabato, A. C., Damerell, G. M. and Heywood, K. J. (2016). Open-ocean submesoscale motions: A full seasonal cycle of mixed layer instabilities from gliders, *J. Phys. Oceanogr.*, 46(4), pp. 1285–1307, doi:10.1175/JPO-D-15-0170.1.

Timmermans, M. L., Toole, J., Proshutinsky, A., Krishfield, R. and Plueddemann, A. (2008). Eddies in the Canada Basin, Arctic Ocean, observed from ice-tethered profilers, *J. Phys. Oceanogr.*, 38(1), pp. 133–145, doi:10.1175/2007JPO3782.1.

Uchida, H., Imawaki, S. and Hu, J. (1998). Comparison of Kuroshio surface velocities derived from satellite altimeter and drifting buoy data, *J. Oceanogr.*, 54, pp. 115–122, doi:10.1007/BF02744385.

Veneziani, M., Griffa, A., Garraffo, Z. D. and Chassignet, E. P. (2005a). Lagrangian spin parameter and coherent structures from trajectories released in a high-resolution ocean model, *J. Mar. Res.*, 63, pp. 753–788, doi:10.1357/0022240054663187.

Veneziani, M., Griffa, A., Reynolds, A. M., Garraffo, Z. D. and Chassignet, E. P. (2005b). Parameterizations of Lagrangian spin statistics and particle dispersion in the presence of coherent vortices, *J. Mar. Res.*, 63, pp. 1057–1083, doi:10.1357/002224005775247571.

Wang, Y. H., Shen, X. R., Yang, S. Q. and Gao, Z. K. (2020). Three-dimensional dynamic analysis of observed mesoscale eddy in the South China Sea based on complex network theory, *Europhys. Lett.*, 128(6), p. 60005, doi:10.1209/0295-5075/128/60005.

Weiss, J. (1991). The dynamics of enstrophy transfer in two-dimensional hydrodynamics, *Phys. D: Nonl. Phen.*, 48(2-3), pp. 273–294, doi:10.1016/0167-2789(91)_90088-Q.

Wenegrat, J. O., Thomas, L. N., Sundermeyer, M. A., Taylor, J. R., D'Asaro, E. A., Klymak, J. M. and Lee, C. M. (2020). Enhanced mixing across the gyre boundary at the Gulf Stream front, *Proc. Nat. Aca. Sci.*, 117(30), pp. 17607–17614, doi:10.1073/pnas.2005558117.

Xia, L., Chen, G., Chen, X., Ge, L. and Huang, B. (2022a). Submesoscale oceanic eddy detection in SAR images using context and edge association network, *Fron. Mar. Sci.*, 9, 1023624, doi:10.3389/fmars.2022.1023624.

Xu, L., Li, P., Xie, S. P., Liu, Q., Liu, C. and Gao, W. (2016). Observing mesoscale eddy effects on mode-water subduction and transport in the North Pacific, *Nature Comm.*, 7(1), pp. 1–9, doi:10.1038/ncomms 10505.

Xu, G., Cheng, C., Yang, W., Xie, W., Kong, L., Hang, R. and Yang, J. (2019a). Oceanic eddy identification using an AI scheme, *Rem. Sens.*, 11(11), pp. 1–12, doi:10.3390/rs11111349.

Xu, G., Dong, C., Liu, Y., Gaube, P., and Yang, J. (2019b). Chlorophyll rings around ocean eddies in the North Pacific, *Sci. Rep.*, 9(1), pp. 1–8, doi:10.1038/s41598-018-38457-8.

Xu, G., Xie, W., Dong, C. and Gao, X. (2021). Application of three deep learning schemes into oceanic eddy detection, *Fron. Mar. Sci.*, 8, 672334, doi:10.3389/fmars.2021.672334.

Yasuda, I., Ito, S. I., Shimizu, Y., Ichikawa, K., Ueda, K. I., Honma, T. and Koizumi, K. (2000). Cold-core anticyclonic eddies south of the Bussol

Strait in the northwestern subarctic Pacific, *J. Phys. Oceanogr.*, 30(6), pp. 1137–1157, doi:10.1175/1520-0485(2000)030<1137:CCAESO>2.0.CO;2.

Yu, X., Naveira Garabato, A. C., Martin, A. P., Buckingham, C. E., Brannigan, L. and Su, Z. (2019). An annual cycle of submesoscale vertical flow and restratification in the upper ocean, *J. Phys. Oceanogr.*, 49(6), pp. 1439–1461, doi:10.1175/JPO-D-18-0253.1.

Zhan, P., Krokos, G., Guo, D. and Hoteit, I. (2019). Three-dimensional signature of the Red Sea eddies and eddy-induced transport, *Geophys. Res. Lett.*, 46(4), pp. 2167–2177, doi:10.1029/2018GL081387.

Zhang, C., Li, H., Liu, S., Shao, L., Zhao, Z. and Liu, H. (2015). Automatic detection of oceanic eddies in reanalyzed SST images and its application in the East China Sea, *Sci. China Earth Sci.*, 58(12), pp. 2249–2259, doi:10.1007/s11430-015-5101-y.

Zhang, Z., Tian, J., Qiu, B., Zhao, W., Chang, P., Wu, D. and Wan, X. (2016). Observed 3D structure, generation, and dissipation of oceanic mesoscale eddies in the South China Sea, *Sci. Rep.*, 6(1), 24349, doi:10.1038/srep24349.

Zhang, Z., Zhang, Y., Wang, W. and Huang, R. X. (2013). Universal structure of mesoscale eddies in the ocean, *Geophys. Res. Lett.*, 40(14), pp. 3677–3681, doi:10.1002/grl.50736.

Zhang, Z., Zhang, Y. and Wang, W. (2017). Three-compartment structure of subsurface-intensified mesoscale eddies in the ocean, *J. Geophys. Res. Oceans*, 122(3), pp. 1653–1664, doi:10.1002/2016JC012376.

Zhang, Z., Zheng, Y. and Li, H. (2023). Imprints of tropical cyclone on three-dimensional structural characteristics of mesoscale oceanic eddies, *Fron. Earth Sci.*, 10, 1057798, doi:10.3389/feart.2022.1057798.

Zhao, N., Huang, B., Yang, J., Radenkovic, M. and Chen, G. (2023a). Oceanic eddy identification using pyramid split attention U-Net with remote sensing imagery, *IEEE Geosci. Rem. Sens. Lett.*, 20, pp. 1–5, doi:10.1109/LGRS.2023.3243902.

Zhao, Y., Fan, Z., Li, H., Zhang, R., Xiang, W., Wang, S. and Zhong, G. (2023b). SymmetricNet: end-to-end mesoscale eddy detection with multi-modal data fusion, *Fron. Mar. Sci.*, 10, 1174818, doi:10.3389/fmars.2023.1174818.

Zhao, M., Timmermans, M. L., Cole, S., Krishfield, R., Proshutinsky, A. and Toole, J. (2014). Characterizing the eddy field in the Arctic Ocean halocline, *J. Geophys. Res. Oceans*, 119(12), pp. 8800–8817, doi:10.1002/2014JC010488.

Part 2

Generation Mechanisms of Eddies in the South China Sea

Chapter 5

Local Basin Processes for Eddy Genesis

5.1 Introduction

The South China Sea (SCS) is characterized by energetic eddies of high-frequency occurrence, long life and regular distribution patterns (Zheng *et al.*, 2017; He *et al.*, 2018). The eddy diameters range from 100 to 300 km, the lifetimes vary from several days to several months, and the eddy disturbances may reach down to the bottom layer (Zhang *et al.*, 2016). Early investigations treated eddies in the SCS as a single process (Li *et al.*, 1998; Wang *et al.*, 2003; Xie *et al.*, 2011). Thus, their generation was individually attributed to various mechanisms, such as the Kuroshio intrusion (Metzger and Hurlburt, 2001), local winds (Wang *et al.*, 2008), current instability (Hu *et al.*, 2001), bottom topography effects (Lin *et al.*, 2015) and disturbance from the Pacific (Zheng *et al.*, 2011; Hu *et al.*, 2012).

In recent years, investigators have found that eddies in the SCS behaved as grouped phenomena (Xie and Zheng, 2017; Xie *et al.*, 2018). Nan *et al.* (2011) observed three long-lived anticyclonic eddies (AEs) appearing as an eddy train along 18°N in the northern SCS in summer 2007 and suggested the frontal instability of the local current as their generation mechanism. Zheng *et al.* (2014) found that there is internal coherence among alternatingly distributed anticyclonic and cyclonic eddies (CEs) in the SCS and suggested existence of two-dimensional (2D) standing wave modes in the SCS deep basin. This is the first time to treat all eddies in the SCS basin as a whole process.

Xie *et al.* (2018) developed the theory of the baroclinic Rossby normal mode (RNM) to analyze grouped moving eddies in the SCS deep basin. Their results give a new insight into the SCS eddy dynamics. The inherent dynamic connection among eddies and the importance of the basin geometry in eddy generation are revealed. Later studies revealed effects of other basin structures on genesis of mesoscale eddies (Mensah and Ohshima, 2020; Lin *et al.*, 2023a, 2023b; Zhao *et al.*, 2023; Yuan *et al.*, 2024).

This chapter focuses on effects of the SCS basin processes to eddy generation and variation. The RNM phenomena and theories are introduced first. Then, the theoretical RNM solutions and grouped eddies are compared. In addition, wind forcing, current instability and topographic effects are also discussed.

5.2 Rossby Normal Mode Theories

5.2.1 *RNMs in ocean and atmosphere*

The RNMs are intrinsic oscillation properties of an enclosed fluid body when considering the rotation effects of the earth (Pedlosky, 1987). Their spatial structures and phase speeds are determined by the resonant characteristics of the fluid body rather than by forcing mechanism (Ahlquist, 1985). The RNMs in the atmosphere have been evidenced by observations since 1960s (Eliasen and Machenhauer, 1969; Madden and Julian, 1972; Salby, 1981; Wu *et al.*, 1994; Day and Mitchell, 2010; Sassi *et al.*, 2012).

In the case of the ocean, Longuet-Higgins (1965) first derived the eigenfunctions of the barotropic Rossby waves in a rectangular basin, which shows a 2D chessboard patterns with alternating cyclonic and anticyclonic circulation. Theories of free RNMs in a circular, equatorial or polar basin and wind-forced modes in a rectangular basin were developed in 1960–1970s (Pedlosky, 1965; Rattray and Charnell, 1966; Flierl, 1977; LeBlond and Mysak, 1978). Later, Pedlosky and Spall (1999) calculated the barotropic RNMs in a basin with a meridional barrier. LaCasce (2000) studied the wind-forced baroclinic RNMs in a square basin. LaCasce and Pedlosky (2002) showed that the modes are robust even to changes in basin geometry. Graef (2016) developed theories of free and forced RNMs in

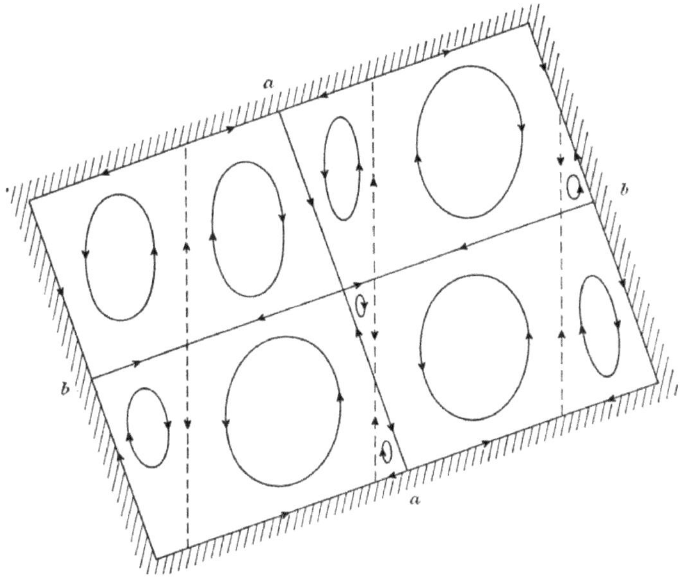

Fig. 5.1. Form of the streamlines for RNMs in a rectangular basin: Mode (2, 2). *Source*: Taken from Longuet-Higgins *et al.* (1965).

a semi-enclosed rectangular gulf. Xie *et al.* (2018) developed the baroclinic RNMs in the enclosed basin, which has been applied in the Kuril Basin of the Sea of Okhotsk by Mensah and Ohshima (2020) (Figure 5.1).

5.2.2 *RNMs in an enclosed ocean basin*

According to Pedlosky (1987), the small-amplitude oscillations in an enclosed ocean basin with a uniform depth and homogeneous fluid can be governed by the linear potential vorticity equation on the β-plane in the form of

$$\frac{\partial}{\partial t}\left\{\nabla^2\psi - \frac{1}{R^2}\psi\right\} + \beta\frac{\partial\psi}{\partial x} = 0, \tag{5.1}$$

where ψ is the stream function, i.e., the sea level undulation ζ, R is the Rossby deformation radius, β is the northward gradient of the Coriolis parameter f. Assuming the 2D oscillation in the form of

$\psi(x, y, t) = e^{-i\sigma t}\Phi(x, y)$, Eq. (5.1) becomes

$$\nabla^2\Phi + \frac{i\beta}{\sigma}\frac{\partial\Phi}{\partial x} - \frac{1}{R_i^2}\Phi = 0, \tag{5.2}$$

where σ is the frequency. To remove the first x-deviation from Eq. (5.2), Φ is transformed to be $\Phi(x, y) = e^{\frac{-i\beta}{2\sigma}x}\varphi(x, y)$. The boundary condition $\varphi(x, y) = \Phi(x, y) = 0$ implies that φ are spatially identical to the finite class of clamped-membrane modes.

Consider the rectangular ocean basin with a zonal length $L(0 \leq x \leq L)$, a meridional width $l(0 \leq y \leq l)$ and a uniform depth. Any of the set

$$\varphi(x, y) = \varphi_{mn} = \sin\left(\frac{m\pi}{L}x\right)\sin\left(\frac{n\pi}{l}y\right),$$
$$m = 1, 2, 3\ldots, \quad n = 1, 2, 3\ldots \tag{5.3}$$

satisfies Eq. (5.2) and the boundary condition $\Phi = 0$ at $x = 0, L$ and $y = 0, l$, provided that σ equals the corresponding eigenvalue

$$\sigma_{mn} = \frac{\beta}{2\left[\left(\frac{m\pi}{L}\right)^2 + \left(\frac{n\pi}{l}\right)^2 + \left(\frac{1}{R}\right)^2\right]^{1/2}}. \tag{5.4}$$

In the stratified ocean, the eigen frequency is the same as that shown in Eq. (5.4) except that R should be the baroclinic deformation radius R_i, $i = 1, 2, 3\ldots$. Thus, the corresponding eigenvalue period is

$$T_{mn} = \frac{4\pi}{\beta}\left[\left(\frac{m\pi}{L}\right)^2 + \left(\frac{n\pi}{l}\right)^2 + \left(\frac{1}{R_i}\right)^2\right]^{\frac{1}{2}}. \tag{5.5}$$

The three-dimensional (3D) stream function ψ should be

$$\begin{aligned}\psi(x, y, z, t) &= \phi(z)\psi(x, y, t) = \sum_{m,n,i=1}^{\infty} A_{mn}^i\phi^i(z)\psi_{mn}(x, y, t)\\ &= \sum_{m,n,i=1}^{\infty} A_{mn}^i\phi^i(z)\cos\left(\frac{\beta}{2\sigma_{mn}}x + \sigma_{mn}t\right)\\ &\quad\times\sin\left(\frac{m\pi}{L}x\right)\sin\left(\frac{n\pi}{l}y\right),\end{aligned} \tag{5.6}$$

where A^i_{mn} is the amplitude independent of x, y and t, and $\phi^i(z)$ is the ith vertical mode of the stream function.

From Eq. (5.6), one can see that the solution consists of two wave components: westward traveling waves, $\cos\left(\frac{\beta}{2\sigma_{mn}}x + \sigma_{mn}t\right)$, and 2D $(x - y)$ standing waves, $\sin\left(\frac{m\pi}{L}x\right)\sin\left(\frac{n\pi}{l}y\right)$. Within a short observation time interval, the 2D sea level patterns of the RNMs should show very much like the standing waves because of their long periods (>3 months). Within a long observation time interval, the 2D sea level patterns of the RNMs should vary with the time and possibly generate complex structures due to wave-wave interaction. Each horizontal mode has fixed and moving nodes separating cells of motion. Each cell alternatively decreases and then increases in the size as the moving node traveling waves approach and then pass a fixed node. The phase speed of westward propagation is

$$C_{mn} = -\frac{2\sigma^2_{mn}}{\beta} = -\frac{\beta}{2\left[\left(\frac{m\pi}{L}\right)^2 + \left(\frac{n\pi}{l}\right)^2 + \left(\frac{1}{R_i}\right)^2\right]}. \tag{5.7}$$

5.3 RNMs in the SCS Deep Basin

5.3.1 *Theoretical solutions*

As shown in Figure 5.3, the SCS deep basin is surrounded by lands or broad continental shelves on the north, west and south sides and a generally north–south running island chain consisting of Taiwan Island and Philippine Islands on the east side. Below the depth of 500 m, the SCS deep basin has only one gap to connect the western North Pacific (WNP), i.e., the Luzon Strait located between the Taiwan Island and the Luzon Island, of which the width is about 400 km, less than 10% of the total length of the basin perimeter. Below 3000 m, the SCS deep basin becomes a fully enclosed one with a maximum depth deeper than 5500 m, lying southwest–northeastwards at the central and eastern SCS.

For simplicity without losing physics, we use the rectangular ocean basin with the uniform depth to model the SCS deep basin as shown by the box enclosed by red dashed lines in Figure 5.2. The model basin covers an area ranging from 11°to 19°N and from 110°to 120°E

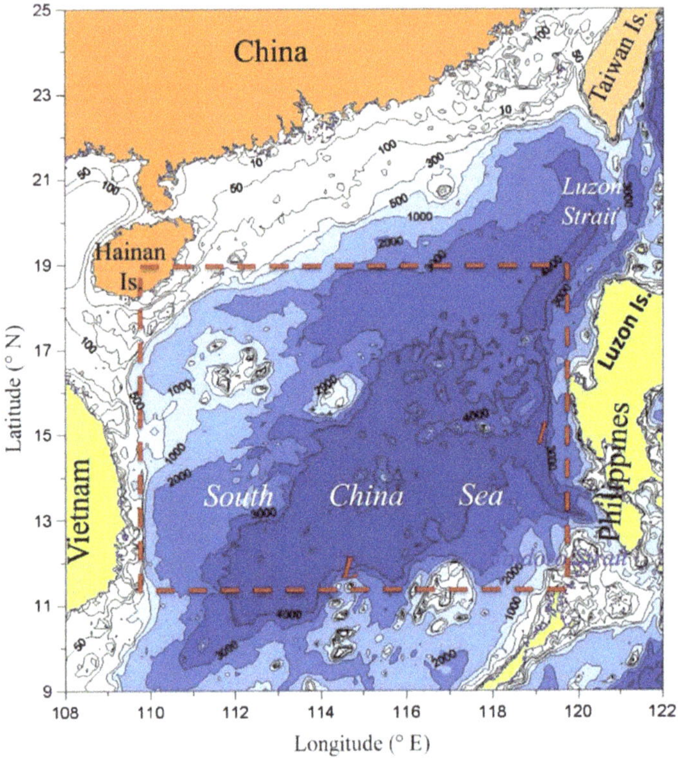

Fig. 5.2. Map of the SCS deep basin: red rectangle represents a model ocean basin used for calculation of RNMs. Numerals on isobaths are in m. The ETOPO2 data are used for topography.
Source: Taken from Xie *et al.* (2018).

with the central latitude at 15°N. The rectangular ocean basin covers 90% of the area of the SCS deep basin. Vertically, the SCS deep basin is characterized by strong, seasonally variable stratification. According to cruise observations, the first baroclinic mode R_1 in the deep SCS basin varies within a range from 40 to 100 km (Cai *et al.*, 2008; Xie *et al.*, 2016).

Based on the above parameters of the model SCS deep basin, i.e., $L = 1100$ km, $l = 880$ km, central latitude is 15°N, and R_0 and R_1 are taken as 4000 and 60 km, and the RNM solutions (5.6), the 2D sea level distribution patterns and corresponding parameters of the barotropic and baroclinic RNMs can be calculated. The normalized sea level patterns of the first nine barotropic modes ζ^0_{mn} and baroclinic modes ζ^1_{mn} $(m, n = 1, 2, 3)$ are shown in Figure 5.3.

From Figure 5.3, one can see that all normal modes consist of alternatingly distributed elevation and depression centers.

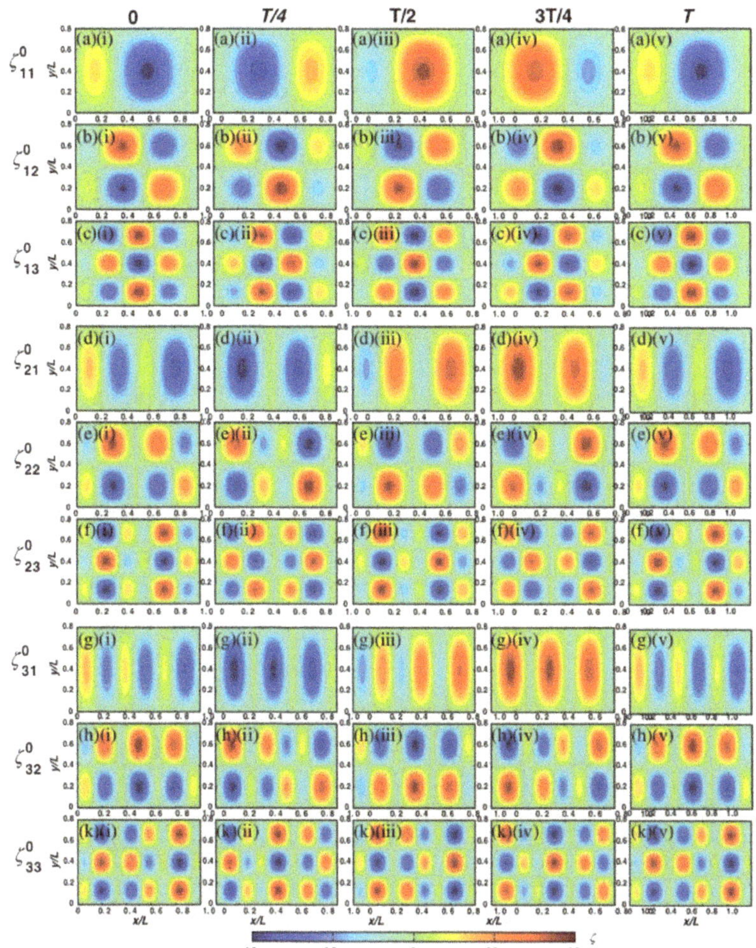

Fig. 5.3. Normalized 2D sea level distribution patterns of first nine barotropic normal modes $(a - k, \zeta^0_{mn})$ and first nine baroclinic modes $(l - f, \zeta^1_{mn})$ of model SCS deep basin. Color codes for normalized sea level ζ are dimensionless with a maximum value of 1.

Source: Taken from *Xie et al.* (2018).

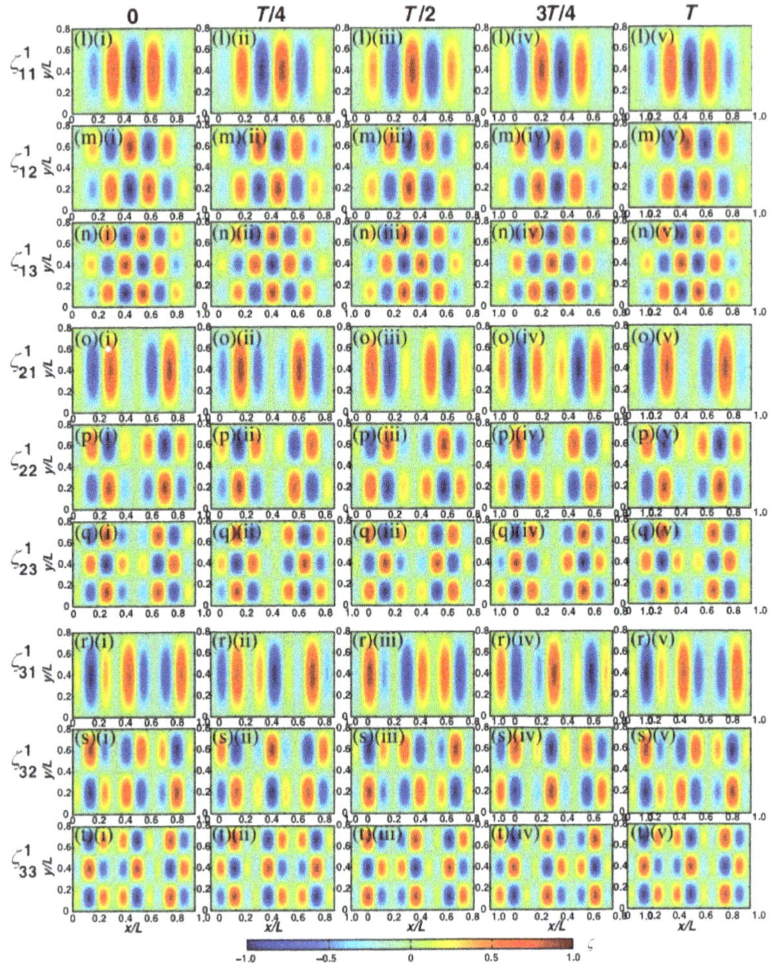

Fig. 5.3. (*Continued*).

The centers have different sizes and different amplitudes at different locations. The elevation and depression centers simultaneously move westward, and their sizes and amplitudes vary periodically with the time. For example, for mode ζ_{11}^0, the depressing cyclone is centered at $x = 0.6$, $y = 0.4$ with the maximum amplitude $\zeta_{\max} = -1$ and zonal diameter $d_1 = 0.5L$ at $t = 0$. It moves to $x = 0.4$ at $t = T/4$ (T is the period), with $\zeta_{\max}$ and d_1 decreasing to -0.9 and 0.4, respectively. It further moves westward to $x = 0.2$ at $t = T/2$, and $\zeta_{\max}$

and d_1 dramatically decrease to -0.2 and 0.3. At $t = 3T/4$, a new depressing center appears at $x = 0.8$ on the eastern side of the basin. It moves westward with increasing amplitude and size, and recovers to the same location, amplitude and size as that at $t = 0$ at $t = T$. Then, westward propagation with decreasing amplitude and size will repeat the above process in later periods. These variable processes cause complex 2D sea level patterns varying periodically with time.

Comparing to the barotropic modes, the baroclinic modes with the same horizontal mode number have more elevation and depression centers in the zonal direction. For example, the mode ζ_{11}^0 has one elevation center and one depression centre at a certain time, while ζ_{11}^1 has three elevation centers and three depression centers at the same time. The sizes of the baroclinic cyclones and anticyclones are of $0.3 - 0.5$ of the barotropic ones. The normal modes contain different wavelength components, periods and westward moving speeds. For example, for baroclinic mode ζ_{11}^1, the zonal wavelength of the carrier wave λ_{11}^1 is 364 km, the period is 114 days and the westward moving speed is -3.7 cm s^{-1}. For a higher mode ζ_{12}^1, the wavelength λ_{12}^1 decreases to 342 km, the period increases to 121 days and the moving speed decrease to -2.8 cm s^{-1}. Generally, the higher normal modes have the shorter wavelengths, the longer periods and the slower westward moving speeds. The periods of the barotropic modes increase 3 times from 30 days of the lowest mode λ_{11}^0 to 91 days of mode λ_{33}^0, and the corresponding westward moving speeds decrease one order from -52.5 to -5.9 cm s^{-1}. For baroclinic modes, the periods of the first nine modes are distributed in a range from 114 to 143 days. Their westward moving speeds are within a range of -3.7 to -2.2 cm s^{-1}. The zonal wavelengths of the carrier wave components λ_{mn}^1 are within a range of 364 to 291 km. For the baroclinic and barotropic modes with the same mode numbers m and n, the wavelengths λ_m and λ_n of the standing wave components are the same as the numbers shown in columns 3–4 and rows 2–10 and 11–19, while the baroclinic zonal wavelengths of the carrier wave components λ_{mn}^1 are one half less than that of the barotropic modes λ_{mn}^0 as shown in column 1 and rows 2–10 and 11–19. For each mode, the zonal wavelengths of the standing wave components λ_m shown in column 3 are at least twice larger than that of the carrier wave components λ_{mn} shown in column 2. Thus, the zonal wavelengths of the sea level patterns in Figure 5.4 are mainly dominated by the wavelength of the

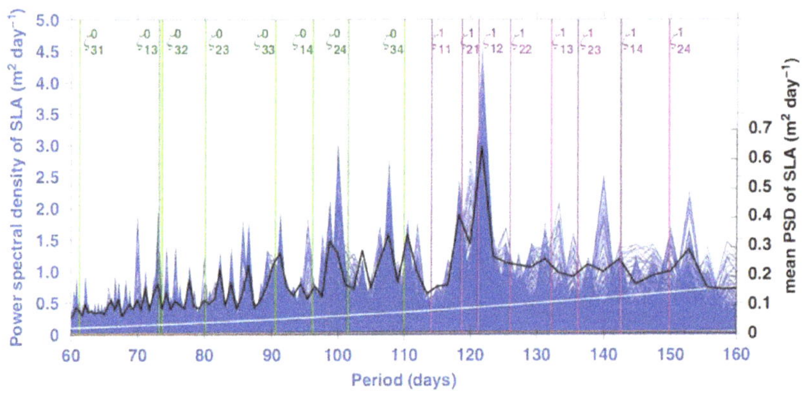

Fig. 5.4. The Fourier period power density spectra of SLA time series from 1 January 1993 to 31 December 2015 in the SCS deep basin. Black line shows the mean power spectrum of all 1280 grids. White line represents 95% confident level. Vertical green and magenta lines represent the periods of barotropic modes ζ^0_{mn} and baroclinic modes ζ^1_{mn}, respectively.

Source: Taken from Xie *et al.* (2018).

carrier wave λ_{mn}. For example, for mode ζ^1_{11}, the zonal wavelength between two elevation/depression centers is about 360 km and close to the zonal wavelength λ^1_{11} of the carrier wave.

5.3.2 *Comparison to satellite observations*

The sea level anomaly (SLA) images from 1 January 1993 to 31 December 2015 were used to analyze the RNMs in the SCS by Xie *et al.* (2018). In the time domain, the periods of the RNMs compare well with that of the SLA variation in the SCS deep basin. As shown in Figure 5.4, the discrete distribution of the power spectral density (PSD) peaks implies that the mechanisms that induce the SLA variations should have discrete periods. The periods of the RNMs just have such discrete distribution patterns. The relatively high barotropic normal modes cover the peaks with periods shorter than 110 days, and the low baroclinic modes cover the peaks with periods longer than 114 days. The correlation coefficient (R^2) of the peak periods of SLA with that of the barotropic and baroclinic modes reaches 0.98 with a standard deviation of 2.3 days. These results indicate that (1) the RNMs indeed exist in the SCS deep basin and

(2) they are dominant components to be responsible for the SLA variation within a period range from 60 to 160 days.

In the space domain, the RNMs compare well with the 2D SLA patterns in the SCS deep basin. As shown in Figure 5.5, the distribution of the SLA on 17 August 2015 is featured by alternating positive/negative centers in the zonal direction, like mode ζ_{11}^{1}. The alternating SLA patterns maintained on 25 August 2015, but the SLA centers moved a little westward with respect to that on 17 August. Similarities of mode ζ_{11}^{1} to SLA patterns are 6/6 and 6/6 on the two days. Westward movement of the SLA centers is consistent with the theoretically predicted westward propagation of the RNMs. For modes ζ_{12}^{1} and ζ_{13}^{1}, positive and negative centers appear as 2D chessboard-like distribution patterns. On the other hand, the SLA

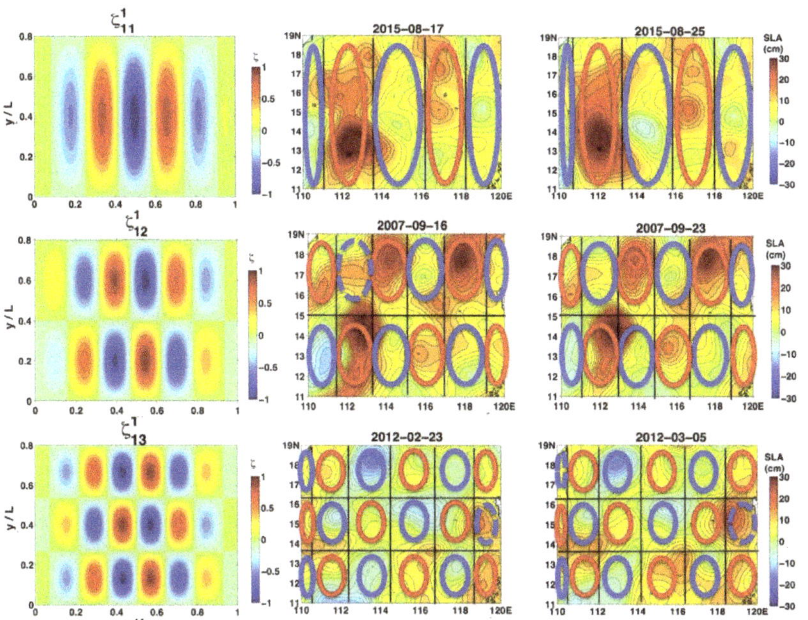

Fig. 5.5. Comparisons of 2D sea level patterns of the first three baroclinic normal modes ζ_{11}^{1}, ζ_{12}^{1}, and ζ_{13}^{1} to the satellite SLA patterns on 17 and 25 August 2015, 16 and 23 September 2007 and 23 February and 5 March 2012 in the SCS deep basin from 11°N to 19°N and from 110°E to 120°E. Color codes for SLA are in cm.

Source: Taken from Xie *et al.* (2018).

images show 2-row chessboard-like patterns on 16 and 23 September 2007 and that on 23 February and 5 March 2012 show 3-row patterns, respectively. Similarities of mode ζ_{12}^1 to SLA patterns are 11/12 and 12/12 on 16 and 23 September 2007, and that of ζ_{13}^1 to SLA patterns are 17/18 and 16/18 on 23 February and 5 March 2012. The high similarities and westward movement of the SLA centers provide evidence for the existence of RNMs in the SCS deep basin.

Table 5.1 lists comparisons of wavelengths and westward speeds of RNMs to that derived from 2D SLA patterns in Figure 5.5. One can see that the zonal wavelengths of SLA patterns decrease from 447 km for mode ζ_{11}^1 on 17 and 25 August 2015 to 341 km for mode ζ_{13}^1 on 23 February and 5 March 2012, and the corresponding meridional wavelengths decrease from 924 to 638 km. The decreasing trend of wavelengths is consistent with the theoretical results of the RNMs. The standard deviations of these two datasets are 69 and 82 km for the zonal and meridional wavelengths, respectively. The westward moving speeds of the chessboard-like SLA patterns are -5.0 cm s^{-1} from 17 August to 25 August 2015 for ζ_{11}^1, -3.5 cm s^{-1} from 16 September to 23 September 2007 for ζ_{12}^1 and -2.4 cm s^{-1} from 23 February to 5 March 2012 for ζ_{13}^1. They are close to the theoretical results of -3.7, -3.3, and -2.8 cm s^{-1}. These westward speeds are close to the eddy westward moving speeds around 3 cm s^{-1} derived from eddy resolving numerical models by Lin *et al.* (2015).

Table 5.1. Comparisons of wavelengths and westward speeds of RNMs to that derived from 2D SLA patterns in Figure 5.5.

		Zonal wavelength (km)		Meridional wavelength (km)		Westward speed (cm s^{-1})	
Mode	Occurrence time	SLA	Mode	SLA	Model	SLA	Mode
ζ_{11}^1	17 and 25 Aug. 2015	447 ± 41	364	N/A	1760	-5.0 ± 1.4	-3.7
ζ_{12}^1	16 and 23 Sept. 2009	385 ± 45	342	924 ± 62	880	-3.5 ± 2.2	-3.3
ζ_{13}^1	23 Feb. and 5 March 2012	341 ± 60	314	638 ± 31	587	-2.4 ± 0.7	-2.8
SD*		69		82		1.0	

Note: *Standard deviation.

The RNM theory gives satisfactory explanations for both 2D SLA patterns of the grouped eddy distribution, the westward movement and associated complex temporal variation of the eddy patterns. Zhao *et al.* (2023) further separated mesoscale eddy fields analogous to RNMs from the satellite observation of the SCS using the multi-scale window transform (MWT) method. They found the oscillating RNM mode with period of 6 months and wavelength of 250 km, named as RNM6, dominates the eddy group with the lifetime of 3 months.

5.3.3 *Resonance of the Pacific disturbance with RNMs in the SCS deep basin*

The occurrence and sustenance of the RNMs in the SCS require initial disturbances and power supplies. The energetic disturbances originating from the WNP have a high possibility to penetrate the Luzon Strait and enter the SCS (Zheng *et al.*, 2011; Hu *et al.*, 2012; Xie *et al.*, 2016). Xie *et al.* (2018) clarified the resonant relation of the RNMs in the SCS deep basin to the impinging long waves facing the Luzon Strait at 19°–21.2°N, 124°–130°E in the WNP. Zhao *et al.* (2023) found the transition between penetration and leaping of the Kuroshio at the Luzon Strait is a key process to the development of RNM6 (see Section 6.3 for details).

5.4 Instability of Background Flow

Diagnosis of the mesoscale eddy energy budget reveals that barotropic (BT) and baroclinic (BC) instabilities, wind work, advection and pressure work are essential ingredients of the eddy energy sources and sinks in the SCS, and their contributions vary from region to region (Zhao *et al.*, 2022). The BT and BC instabilities of background flow are two important dynamic terms in the energy budget of mesoscale eddies. In the SCS, the BC and BT are essential for eddy generation and variation, but their contributions vary from region to region (Zhang *et al.*, 2013; Zhao *et al.*, 2022).

Using linear instability theory, Chen *et al.* (2012) showed that the BC is critical to the seasonal variation of eddy kinetic energy (EKE) in the SCS. Based on a high-resolution ocean general circulation model, Yang *et al.* (2013) suggested that 60% of the mesoscale eddy energy in the SCS was attributed to the release of large-scale APE via the BC. Wang *et al.* (2020) found that strong baroclinic instability associated with the Kuroshio intrusion contributes to the mesoscale eddy variability in the northern SCS. In the southwestern SCS, the regional mesoscale eddy energy is mainly generated by the BC (Zhao *et al.*, 2022). Energy budget analysis by Ding *et al.* (2021) reveals that both the BT and BC increase the eddy kinetic energy (EKE) in fall off southeast of Vietnam, whereas only the BT contributes to the EKE in west of the Luzon Strait in winter. After the onset of monsoon in June, the BC increases sharply, which allows the eddy to intensify to attain its maximum energy near Xisha Islands (Chu *et al.*, 2014).

5.5 Topography Effects

In the ocean, topography is not only one of the factors leading to disappearance of mesoscale eddies, but also the source of mesoscale/sub-mesoscale eddies. Wang *et al.* (2008) propose that orographic wind jets associated with the northeast winter monsoon and the gaps in the mountainous island chain along the eastern boundary of the SCS can spin up CEs and AEs. Shu *et al.* (2022) found that the interaction of a CE and AE pair with seamount topography contribute to the generation of eddies. Owing to the potential vorticity conservation, the westward-propagating tilted interface across the eddy pair squeezes the deep-water column, thereby giving rise to negative vorticity west of the seamount. The strong front between the eddy pair induces a northward deep flow, thereby generating a strong horizontal velocity shear because of lateral friction and enhanced negative vorticity.

5.6 Winds and Atmospheric Forcing

Chen *et al.* (2019) investigated mesoscale eddy generation mechanisms in the eastern SCS (ESCS) using $1\frac{1}{2}$-layer reduced-gravity model. They found that as the primary forcing for eddy genesis, the high-frequency winds in the interior SCS explain about 54% of mesoscale eddies generated in the ESCS, while that from the WNP explain about 40%.

Using in situ observations and high-resolution reanalysis products in the northern SCS, Liu and Jing (2024) investigate the underlying generation mechanism of the intra-thermocline eddies (ITEs) with the potential vorticity (PV) budget analysis. The results suggested that wintertime surface buoyancy loss rather than frictional forcing is a crucial favorable condition for the ITE formation, highlighting the crucial role of atmospheric diabatic forcing.

5.7 Eddy Dissipation Processes

5.7.1 *Topographic and stratification controls*

The bottom topography of the SCS plays a pivotal role in eddy dissipation, with several mechanisms contributing to energy loss as eddies interact with topographic features. Previous studies have highlighted that rough bottom topography significantly enhances energy dissipation, especially in the NSCS. This occurs through increased viscous dissipation and bottom frictional drag near the ocean floor (Yang *et al.*, 2022). The interaction of eddies with the topography leads to the generation of internal lee waves, which account for a substantial portion of the enhanced viscous dissipation (Yang *et al.*, 2019; Hu *et al.*, 2000). Non-wave-related form drag caused by the topography also plays a significant role in dissipating EKE (Yang *et al.*, 2022). It enhances energy dissipation by creating resistance to the flow. This resistance is not associated with wave generation but rather with the direct interaction between the eddy and the topographic features. The presence of island chains, such as the Paracel (Xisha) Islands, acts as a focal point for eddy dissipation (Yang *et al.*, 2019). This study reported that westward-propagating eddies interact with these islands, their kinetic energy decreases sharply. The energy cascade occurs through the generation of sub-mesoscale currents and lee waves, ultimately leading to enhanced turbulent mixing and dissipation.

5.7.2 *Vertical mixing*

Vertical mixing processes are intimately linked to eddy dissipation, influencing the distribution and fate of eddy energy. Energy cascade of mesoscale eddies dominates dissipation processes (Yang *et al.*, 2019). Strong winds and internal tides contribute to enhancement of vertical mixing within mesoscale eddies in the SCS, which is critical for dissipating eddy energy by redistributing it vertically and facilitating interactions with deeper layers (Zhang *et al.*, 2016). AEs can influence vertical mixing in different ways depending on their amplitude. Smaller-amplitude AEs can suppress downward mixing, leading to cooling in the subsurface. Larger-amplitude AEs, on the other hand, can induce strong vertical warm-water subsidence, resulting in warming in the subsurface (Song *et al.*, 2025).

5.7.3 *Seasonal and interannual variability*

Seasonal and interannual variability also plays a role in eddy dissipation processes. Winter eddies persist longer due to stronger stratification, while summer eddies dissipate rapidly through wind-driven mixing (Long *et al.*, 2021). Seasonal changes in surface heat flux affect the thermal stratification of the water column, influencing the efficiency of mixing and dissipation mechanisms. Surface cooling in winter can lead to a cold, salty core in the upper part of AEs, while surface warming in spring can create a seasonal thermocline that caps the winter-mixed water inside the eddy. Interannually, El Niño event enhances basin-scale gyres that disrupt eddy coherence, whereas La Niña event favors prolonged eddy lifetimes through sustained Kuroshio forcing (Chu *et al.*, 2017).

5.8 Summary

As a semi-enclosed marginal sea, the geometry and location of the SCS basin regulate the generation and variation of mesoscale eddies due to the intrinsic features of the RNMs. The theoretical solutions of RNMs of a rectangular ocean basin with a uniform depth are derived. The complex and variable chessboard-like SLA patterns in the basin, consisting of alternatingly distributed CEs and AEs, can be explained by the RNMs with different tempo-spatial scales. The periods, 2D patterns and the propagation speeds of the modes are evidenced by the spectra peak periods, the grouped eddy patterns, and the westward movement of the eddy patterns on SLA time series images in the SCS deep basin. The intrinsic basin effects give a new insight into the SCS eddy dynamics in view of eddy groups. Besides the RNMs, the recent research results on the roles of instability of background flow, topography effects, vertical mixing, winds and atmospheric forcing played in generating and dissipating of mesoscale eddies in the SCS are also summarized.

References

Ahlquist, J. E. (1985). Climatology of normal mode Rossby waves, *J. Atmos. Sci.*, 42(19), pp. 2059–2068, doi:10.1175/1520-0469(1985)042<2059:CONMRW>2.0.CO;2.

Cai, S., Long, X., Wu, R. and Wang, S. (2008). Geographical and monthly variability of the first baroclinic Rossby radius of deformation in the South China Sea, *J. Mar. Sys.*, 74(1), pp. 711–720, doi:10.1016/j.jmarsys.2007.12.008.

Chen, G., Gan, J., Xie, Q., Chu, X., Wang, D. and Hou, Y. (2012). Eddy heat and salt transports in the South China Sea and their seasonal modulations, *J. Geophys. Res. Oceans*, 117(C5), doi:10.1029/2011JC007724.

Chen, J., Cheng, X. and Chen, X. (2019). Eddy generation mechanism in the eastern South China Sea, *Acta Oceanol. Sin.*, 38, pp. 20–28, doi:10.1007/s13131-019-1409-3.

Chu, X., Dong, C. and Qi, Y. (2017). The influence of ENSO on an oceanic eddy pair in the South China Sea, *J. Geophys. Res. Oceans*, 122(3), pp. 1643–1652, doi:10.1002/2016JC012642.

Chu, X., Xue, H., Qi, Y., Chen, G., Mao, Q., Wang, D. and Chai, F. (2014). An exceptional anticyclonic eddy in the South China Sea in 2010, *J. Geophys. Res. Oceans*, 119(2), pp. 881–897, doi:10.1002/2013JC009314.

Day, K. A. and Mitchell, N. J. (2010). The 16-day wave in the Arctic and Antarctic mesosphere and lower thermosphere, *Atmos. Chem. Phys.*, 10(3), pp. 1461–1472, doi:10.5194/acp-10-1461-2010.

Ding, R., Xuan, J., Zhang, T., Zhou, L., Zhou, F., Meng, Q. and Kang, I. S. (2021). Eddy-induced heat transport in the South China Sea, *J. Phys. Oceanogr.*, 51(7), pp. 2329–2349, doi:10.1175/JPO-D-20-0206.1.

Eliasen, E. and Machenhauer, B. (1969). On the observed large-scale atmospheric wave motions, *Tellus*, 21(2), pp. 149–165, doi:10.3402/tellusa.v21i2.10069.

Flierl, G. R. (1977). Simple applications of McWilliams' "A note on a consistent quasi-geostrophic model in a multiply connected domain", *Dyn. Atmos. Oceans*, 1(5), pp. 443–454, doi:10.1016/0377-0265(77)90003-3.

Graef, F. (2016). Free and forced Rossby normal modes in a rectangular gulf of arbitrary orientation, *Dyn. Atmos. Oceans*, 75, pp. 46–57, doi:10.1016/j.dynatmoce.2016.05.005.

He, Q., Zhan, H., Cai, S., He, Y., Huang, G. and Zhan, W. (2018). A new assessment of mesoscale eddies in the South China Sea: Surface features, three-dimensional structures, and thermohaline transports, *J. Geophys. Res. Oceans*, 123(7), pp. 4906–4929, doi:10.1029/2018JC014054.

Hu, J., Kawamura, H., Hong, H., Kobashi, F. and Wang, D. (2001). 3∼6 months variation of sea surface height in the South China Sea and its adjacent ocean, *J. Phys. Oceanogr.*, 57(1), pp. 69–78.

Hu, J., Kawamura, H., Hong, H. and Qi, Y. (2000). A review on the currents in the South China Sea: seasonal circulation, South China Sea warm current and Kuroshio intrusion, *J. Oceanogr.*, 56, pp. 607–624, doi:10.1023/A:1011117531252.

Hu, J., Zheng, Q., Sun, Z. and Tai, C.-K. (2012). Penetration of nonlinear Rossby eddies into South China Sea evidenced by cruise data, *J. Geophys. Res. Oceans*, 117(C3), p. C03010, doi:10.1029/2011JC007525.

LaCasce, J. H. (2000). Baroclinic Rossby waves in a square basin, *J. Phys. Oceanogr.*, 30(12), pp. 3161–3178, doi:10.1175/1520-0485(2000)030<3161:BRWIAS>2.0.CO;2.

LaCasce, J. H. and Pedlosky, J. (2002). Baroclinic Rossby waves in irregular basins, *J. Phys. Oceanogr.*, 32(10), pp. 2828–2847, doi:10.1175/1520-0485(2002)032<2828:BRWIIB>2.0.CO;2.

Li, L., Nowlin Jr, W. D. and Su, J. (1998). Anticyclonic rings from the Kuroshio in the South China Sea, *Deep Sea Res. Part I: Oceanogr. Res. Pap.*, 45(9), pp. 1469–1482, doi:10.1016/S0967-0637(98)00026-0.

Lin, X., Dong, C., Chen, D., Liu, Y., Yang, J., Zou, B. and Guan, Y. (2015). Three-dimensional properties of mesoscale eddies in the South China Sea based on eddy-resolving model output, *Deep Sea Res. Part I: Oceanogr. Res. Pap.*, 99, pp. 46–64, doi:10.1016/j.dsr.2015.01.007.

Lin, Z., Xie, L., Huang, R. and Bai, P. (2023a). Seasonal variation of mesoscale eddies in the Andaman Sea, *Haiyang Xuebao*, 45(4), pp. 1–16.

Lin, Z., Zhang, S., Zhang, Z., Yu, X. and Gao, Y. (2023b). The Rossby normal mode as a physical linkage in a machine learning forecast model for the SST and SSH of South China Sea deep basin, *J. Geophys. Res. Oceans*, 128(9), e2023JC019851, doi:10.1029/2023JC019851.

Liu, Y., and Jing, Z. (2024). Intrathermocline eddy with lens-shaped low potential vorticity and diabatic forcing mechanism in the South China Sea, *J. Phys. Oceanogr.*, 54(3), pp. 929–948, doi:10.1175/JPO-D-23-0149.1.

Long, Y., Zhu, X. H., Guo, X., Ji, F. and Li, Z. (2021). Variations of the Kuroshio in the Luzon Strait revealed by EOF analysis of repeated XBT data and sea-level anomalies, *J. Geophys. Res. Oceans*, 126(7), e2020JC016849, doi:10.1029/2020JC016849.

Longuet-Higgins, M. S. (1965). Planetary waves on a rotating sphere, II, *Proc. Royal Soc. London. Series A, Math. Phys. Sci.*, 284(1396), pp. 40–68, doi:10.1098/rspa.1965.0051.

Madden, R. A. and Julian, P. R. (1972). Further evidence of global-scale 5-day pressure waves, *J. Atmos. Sci.*, 29(8), pp. 1464–1469, doi:10.1175/1520-0469(1972)029<1464:FEOGSD>2.0.CO;2.

Mensah, V. and Ohshima, K. I. (2020). Variabilities of the sea surface height in the Kuril Basin of the Sea of Okhotsk: Coherent shelf-trapped mode and Rossby normal modes, *J. Phys. Oceanogr.*, 50(8), pp. 2289–2313, doi:10.1175/JPO-D-19-0216.1.

Metzger, E. J., and Hurlburt, H. E. (2001). The nondeterministic nature of Kuroshio penetration and eddy shedding in the South China Sea, *J. Phys. Oceanogr.*, 31(7), pp. 1712–1732, doi:10.1175/1520-0485(2001)031<1712:TNNOKP>2.0.CO;2.

Nan, F., Xue, H., Xiu, P., Chai, F., Shi, M. and Guo, P. (2011). Oceanic eddy formation and propagation southwest of Taiwan, *J. Geophys. Res. Oceans*, 116(C12), p. C05002, doi:10.1029/2011JC007386.

Pedlosky, J. (1965). A study of the time dependent ocean circulation, *J. Atmos. Sci.*, 22(3), pp. 267–272, doi:10.1175/1520-0469(1965)022<0267:ASOTTD>2.0.CO;2.

Pedlosky, J. (1987). *Geophysical Fluid Dynamics*, Springer-Verlag, New York, pp. 374–379, doi:10.1007/978-1-4612-4650-3.

Pedlosky, J. and Spall, M. A. (1999). Rossby normal modes in basins with barriers, *J. Phys. Oceanogr.*, 29(9), pp. 2332–2349, doi:10.1175/1520-0485(1999)029<2332:RNMIBW>2.0.CO;2.

Rattray, M. and Charnell, R. L. (1966). Quasigeostrophic free oscillations in enclosed basins, *J. Mar. Res.*, 24(1), pp. 82–101.

Salby, M. L. (1981). Rossby normal modes in nonuniform background configurations. Part I: Simple nonuniformities, *J. Atmos. Sci.*, 38(9), pp. 1803–1826, doi:10.1175/1520-0469(1981)038<1803:RNMINB>2.0.CO;2.

Sassi, F., Garcia, R. R., and Hoppel, K. W. (2012). Large-scale Rossby normal modes during some recent northern hemisphere winters, *J. Atmos. Sci.*, 69(3), pp. 820–839, doi:10.1175/JAS-D-11-0103.1.

Song, X., Sun, R., He, S., Zhang, H., Gu, Y., Li, P. and Song, J. (2025). Impact of anticyclonic mesoscale eddies on the vertical structures of marine heatwaves in the South China Sea, *Rem. Sens.*, 17(3), 370, doi:10.3390/rs17030370.

Shu, Y., Wang, J., Xue, H., Huang, R. X., Chen, J., Wang, D. and Wang, W. (2022). Deep-current intraseasonal variability interpreted as topographic Rossby waves and deep eddies in the Xisha Islands of the South China Sea, *J. Phys. Oceanogr.*, 52(7), pp. 1415–1430, doi:10.1175/JPO-D-21-0147.1.

Wang, G., Su, J. and Chu, P. C. (2003). Mesoscale eddies in the South China Sea observed with altimeter data, *Geophys. Res. Lett.*, 30(21), 2121, doi:10.1029/2003GL018532.

Wang, G., Chen, D. and Su, J. (2008). Winter eddy genesis in the eastern South China Sea due to orographic wind jets, *J. Phys. Oceanogr.*, 38(3), pp. 726–732, doi:10.1175/2007JPO3868.1.

Wang, Q., Zeng, L., Chen, J., He, Y., Zhou, W. and Wang, D. (2020). The linkage of Kuroshio intrusion and mesoscale eddy variability in the northern South China Sea: Subsurface speed maximum, *Geophys. Res. Lett.*, 47(11), e2020GL087034, doi:10.1029/2020GL087034.

Wu, D. L., Hays, P. B. and Skinner, W. R. (1994). Observations of the 5-day wave in the mesosphere and lower thermosphere, *Geophys. Res. Lett.*, 21(24), pp. 2733–2736, doi:10.1029/94GL02660.

Xie, L., Tian, J., Zhang, S., Zhang, Y. and Yang, Q. (2011). An anticyclonic eddy in the intermediate layer of the Luzon Strait in Autumn 2005, *J. Oceanogr.*, 67, pp. 37–46, doi:10.1007/s10872-011-0004-9.

Xie, L., Zheng, Q., Tian, J., Zhang, S., Feng, Y. and Yi, X. (2016). Cruise observation of Rossby waves with finite wavelengths propagating from the Pacific to the South China Sea, *J. Phys. Oceanogr.*, 46(10), pp. 2897–2913, doi:10.1175/JPO-D-16-0071.1.

Xie, L., Zheng, Q., Zhang, S., Hu, J., Li, M., Li, J. and Xu, Y. (2018). Rossby wave normal modes in the South China Sea deep basin evidenced by satellite altimetry, *Int. J. Rem. Sens.*, 39(2), pp. 399–417, doi:10.1080/01431161.2017.1384591.

Yang, H., Wu, L., Liu, H. and Yu, Y. (2013). Eddy energy sources and sinks in the South China Sea, *J. Geophys. Res. Oceans*, 118(9), pp. 4716–4726, doi:10.1002/jgrc.20362.

Yang, Q., Nikurashin, M., Sasaki, H., Sun, H. and Tian, J. (2019). Dissipation of mesoscale eddies and its contribution to mixing in the northern South China Sea, *Sci. Rep.*, 9, 556, doi:10.1038/s41598-018-36610-x.

Yang, Z., Jing, Z. and Zhai, X. (2022). Effect of small-scale topography on eddy dissipation in the northern South China Sea, *J. Phys. Oceanogr.*, 52(10), pp. 2397–2416, doi:10.1175/JPO-D-21-0208.1.

Yuan, X., Wang, Q., Hu, D., Feng, J., Wang, F. and Tian, Z. (2024). Intraseasonal variability of full-depth currents in the Philippine Sea, *J. Geophys. Res. Oceans*, 129(5), e2023JC019827, doi:10.1029/2023JC019827.

Zhao, Y., Yang, Y., Mao, L., and Zhang, Y. (2022). On the genesis of the South China Sea mesoscale eddies, *J. Mar. Sci. Eng.*, 10(2), pp. 188, doi:10.3390/jmse10020188.

Zhao, Y., Mao, J. and Liang, X. (2023). Characteristics and attributions of a Rossby normal mode in the South China Sea, *Haiyang Xuebao*, 45(10), pp. 31–41.

Zhang, Z., Tian, J., Qiu, B., Zhao, W., Chang, P., Wu, D. and Wan, X. (2016). Observed 3D structure, generation, and dissipation of oceanic mesoscale eddies in the South China Sea, *Sci. Rep.*, 6, 24349, doi:10.1038/srep24349.

Zhang, Z., Zhao, W., Tian, J., and Liang, X. (2013). A mesoscale eddy pair southwest of Taiwan and its influence on deep circulation, *J. Geophys. Res. Oceans*, 118(12), pp. 6479–6494, doi:10.1002/2013JC008994.

Zheng, Q., Hu, J., Zhu, B., Feng, Y., Jo, Y.-H., Sun, Z., Zhu, J., Lin, H., Li, J. and Xu, Y. (2014). Standing wave modes observed in the South China Sea deep basin, *J. Geophys. Res. Oceans*, 119(7), pp. 4185–4199, doi:10.1002/2014JC009957.

Zheng, Q., Tai, C.-K., Hu, J., Lin, H., Zhang, R., Su, F.-C. and Yang, X. (2011). Satellite altimeter observations of nonlinear Rossby eddy–Kuroshio interaction at the Luzon Strait, *J. Oceanogr.*, 67, pp. 365–376, doi:10.1007/s10872-011-0035-2.

Chapter 6

Disturbance Processes from the Pacific for Eddy Genesis in the SCS

6.1 Introduction

Ocean mesoscale eddies are one of the typical and common dynamic processes in the open ocean and the marginal seas. Studying the characteristics and evolution processes of eddies can provide further insights into the transport of heat or plankton and be used in the planning and organization of ocean research. The South China Sea (SCS), a marginal sea in the western North Pacific (WNP), is connected to the Pacific through a major channel, the Luzon Strait. On the east side of the Luzon Strait, there is the Kuroshio, a strong western boundary current flowing from south to north (Figure 6.1). The direct exchange of seawater between the WNP and the SCS is generally blocked by the strong Kuroshio in the most cases. However, the SCS is still disturbed by the WNP through activities of energetic mesoscale eddies as shown in Figure 6.2. These eddies are usually generated by disturbances from ocean currents and wind patterns in the Pacific and propagate westward across the Kuroshio and the Luzon Strait into the SCS (Wu and Chiang, 2007; Zheng *et al.*, 2007; Sheu *et al.*, 2010; Hu *et al.* 2012). They can interact with local currents and topography to form big eddies. These big eddies may persist for some time and impact on ocean and atmospheric dynamics in the region.

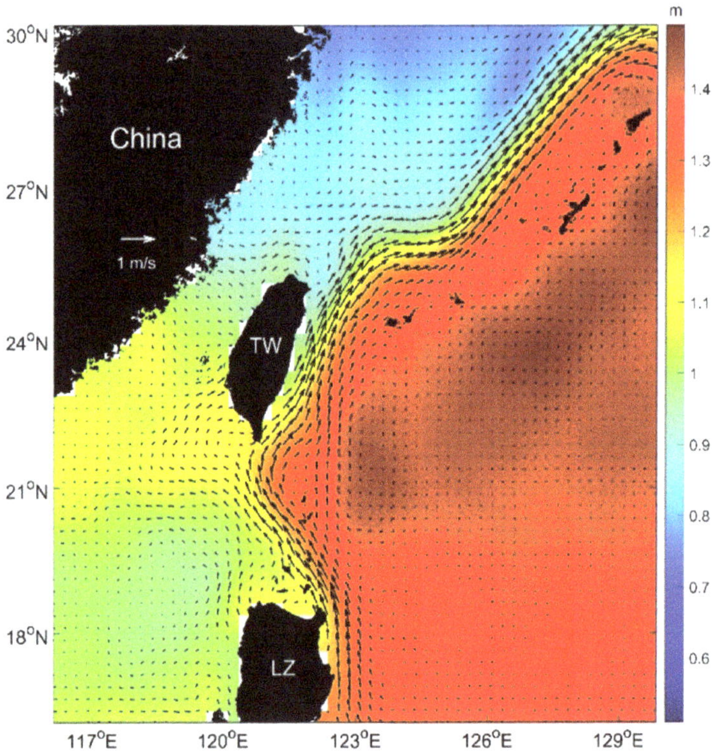

Fig. 6.1. Long-term averages of the absolute dynamic height (color codes in m) and the derived geostrophic current (in m s^{-1}) in the western boundary area of the WNP. TW: Taiwan Island. CMEMS Global Ocean Gridded L4 Reprocessed Product (SEALEVEL_GLO_PHY_L4_MY_008_047)) from 1993 to 2019. LZ: Luzon Island.

Besides mesoscale eddies passing through the Luzon Strait from the WNP to the SCS, the eddies in the SCS may be generated by the shedding of the Kuroshio looping path passing through the Luzon Strait. Hu *et al.* (2000) reviewed previous literature and proposed four types of the Kuroshio intrusion into the SCS. They are direct branch, loop, extend, and ring. Nan *et al.* (2011, 2015) summarized it into three types: looping path, leaping path, and leaking path, as shown in Figure 6.3. The leaking path allows eddies to penetrate directly from the WNP through the Luzon Strait to the SCS, while the looping path provides opportunities for the loop shedding to form eddies in the SCS (see Chapter 7 for details).

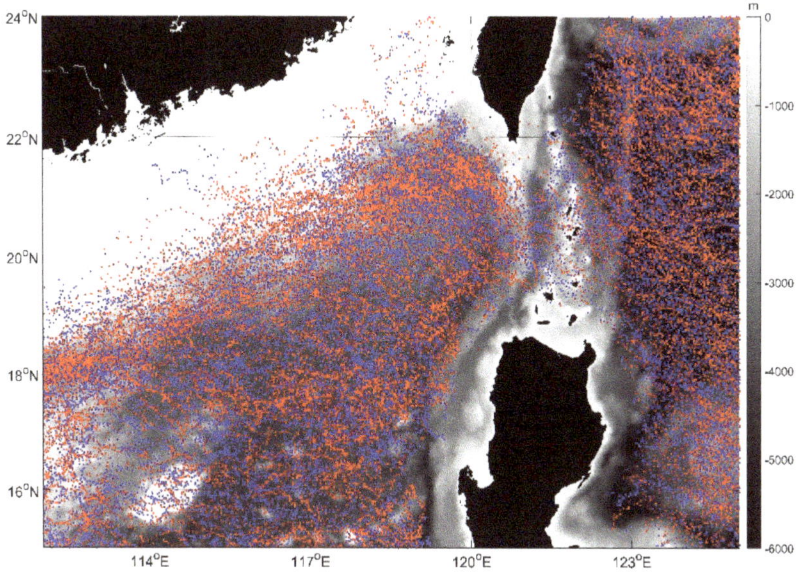

Fig. 6.2. Distribution of eddies from 15°N to 24°N and from 112°E to 125°E in the SCS and the WNP during 1993–2016. Blue and red dots represent cyclonic eddies (CEs) and anticyclonic eddies (AEs), respectively.

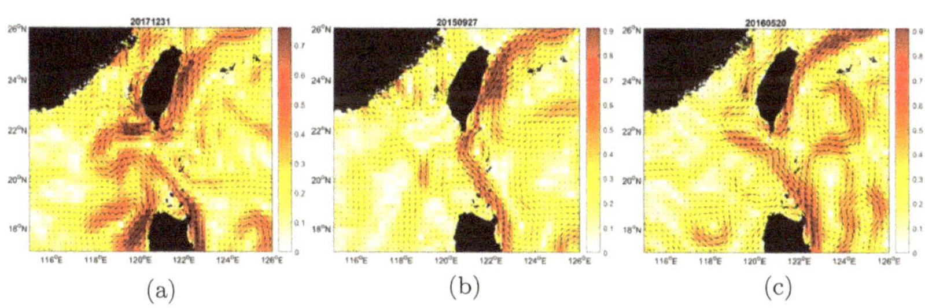

Fig. 6.3. Examples of three Kuroshio paths in the Luzon Strait shown in sea surface geostrophic current (m s^{-1}) field: (a) the looping path on 31 December 2017, (b) the leaping path on 27 September 2015 and (c) the leaking path on 20 May 2016.

This chapter specifies the research results of the roles of disturbance processes from the Pacific played in eddy genesis in the SCS. The processes include westward propagating eddies (Section 6.2), wave-wave resonance (Section 6.3), wind stress forcing (Section 6.4) and climate change (Section 6.5).

6.2 Westward Propagating Eddies

6.2.1 *Statistics of eddies approaching the western boundary*

Using the ocean eddy-tracking data set produced by the satellite SSH-based eddy identification algorithm downloaded from http://cioss.coas.oregonstate.edu/eddies (Chelton *et al.*, 2011), statistical distributions of trajectories of eddies approaching western boundary (E_{WB}) approaching the eddy effect zone (EEZ), which is defined as a belt 250 km from Kuroshio mean axis, from October 1992 to April 2012 were obtained as shown in Figures 6.4 and 6.5 (Cheng *et al.*, 2017). The mean axis of the Kuroshio maximum velocity at the sea surface (solid line in Figure 6.4(a)) was derived from average satellite altimeter absolute dynamic topography (ADT) fields from October 1992 to April 2012. The mean width of the Kuroshio is approximately 100 km (Liang *et al.*, 2003).

Total 315 eddies with the radius larger than 50 km including 138 AEs and 177 CEs were detected from 19.5 years of satellite altimeter SSH data, with more than 95% of eddies being generated in west of the Mariana Islands. Eddy trajectories show that eddies

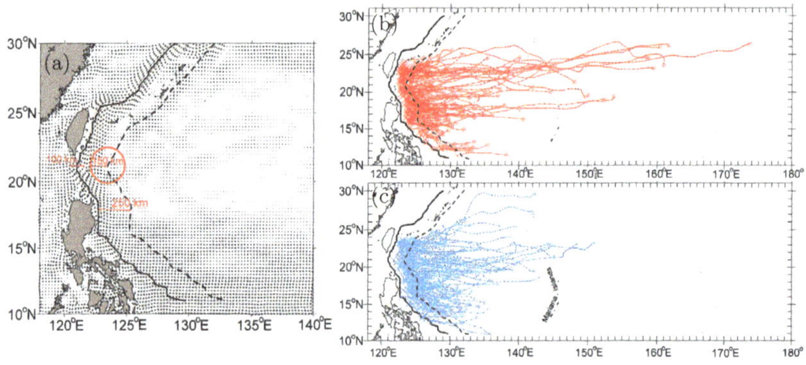

Fig. 6.4. (a) Mean geostrophic current field of the Northwest Pacific from October 1992 to April 2012. Solid line and dashed line denote the mean axis of the Kuroshio and the EEZ of 250 km from the mean axis, respectively. Red circle with a diameter of 150 km represents an E_{WB}. (b) and (c) Maps of trajectories of AEs and CEs. Circles and crosses on trajectories represent locations of eddy generation and termination, respectively.

Source: Taken from Cheng *et al.* (2017).

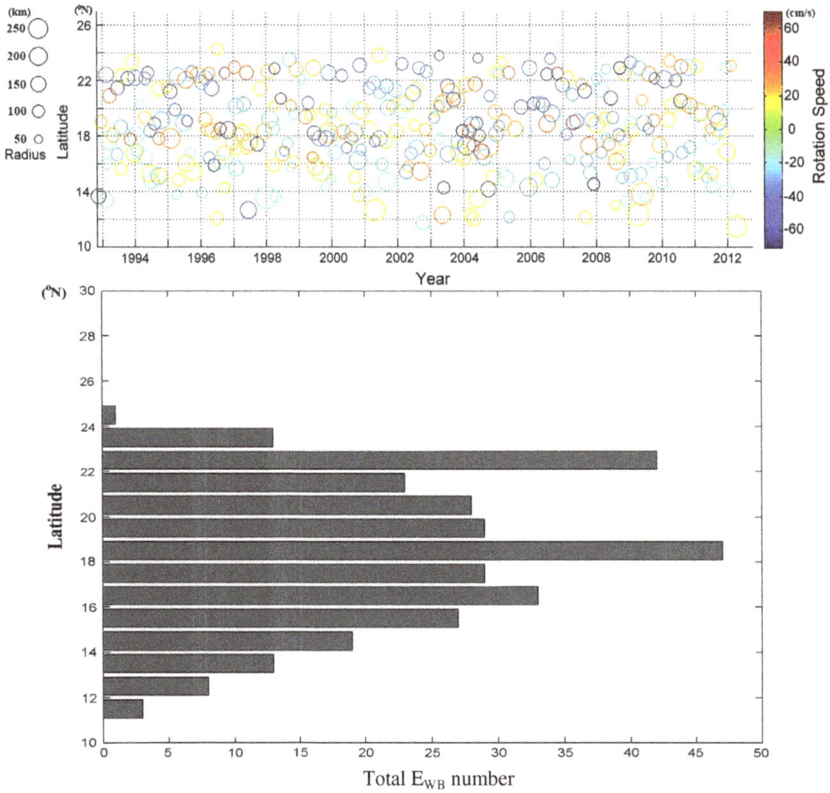

Fig. 6.5. (a) Spatiotemporal distribution of E_{WB}. Circle sizes on left denote the eddy radii. Color codes on right denote the maximum rotational speed within eddies. Positive (negative) speed values represent AEs (CEs). (b) Latitudinal distribution histogram of E_{WB} from October 1992 to April 2012.

Source: Taken from Cheng *et al.* (2017).

frequently intrude into the Kuroshio region at two latitude bands 18°–19°N and 22°–23°N (also see Figure 6.5(b)), with periods of 146 ± 62 and 165 ± 46 days, respectively. The interaction time within the two active bands is (33 ± 10) days at 18°–19°N and (45 ± 17) days at 22°–23°N. The two eddy-concentrated bands are associated with the northern and southern Subtropical Countercurrents (STCCs), which provide a key energy source for eddy generation. In addition, when westward-propagating eddies approach the Ryukyu Islands, the southwestward recirculation flow east of the island chain as well as topographic effects cause some eddies to head southwestward to east of Taiwan and intrude into the Kuroshio region at 22°–23°N. Therefore, the STCCs play a key role in inducing the eddies to frequently intrude into the Kuroshio region at 18°–19°N and 22°–23°N. In addition, the Ryukyu Islands are responsible for focusing eddies within 22°–23°N.

Figure 6.5(a) shows the spatiotemporal distributions of E_{WB}. One can see that in the time domain census statistics shows generally continuous distribution with an annual mean of 16 eddies with the radius larger than 50 km. The mean sea surface radius of eddies is 130 km with that of more than 80% of eddies larger than 100 km and the maximum radius exceeding 150 km (Chelton *et al.*, 2011). The sea surface rotational speeds of higher than 90% of eddies are within a range of ±20 cm s^{-1} and the maximum rotational speed of approximately 30 cm s^{-1}. In the space domain, Figure 6.5(b) shows that eddies are concentrated within the western boundary between 14° and 24°N with the latitudinal double-peaked distribution at 18°–19°N and 22°–23°N.

6.2.2 *Eddies directly crossing the Luzon Strait*

Eddies generated in the WNP propagate westward towards the western boundary with the phase velocity of long baroclinic Rossby waves and a possibility colliding with the Kuroshio flow. Using sea level data from satellite altimeters and momentum analysis, Zheng *et al.* (2011). found that eddies with a radius greater than 150 km are sufficiently strong to significantly alter the Kuroshio path. For eddies with a radius of less than 100 km, the Kuroshio exerts a dominant influence.

When eddies interact with the Kuroshio, the Kuroshio path can be significantly altered. These alterations include meandering, where the Kuroshio path becomes curved; cut-offs, where eddies disrupt the Kuroshio, and branching, where the Kuroshio splits into multiple streams. The schematic diagram of these three paths is shown in Figure 6.6 and examples from satellite altimeter data are shown in Figure 6.7. For eddy direct penetration, the Luzon Strait is like the cut-off path. Figure 6.7(b) shows the Kuroshio cut-off event east of the Luzon Island on 21 December 1994. One can see that a CE northeast of the Luzon Island cut off the Kuroshio upstream. Meanwhile an AE southeast of the Taiwan Island maintained the Kuroshio

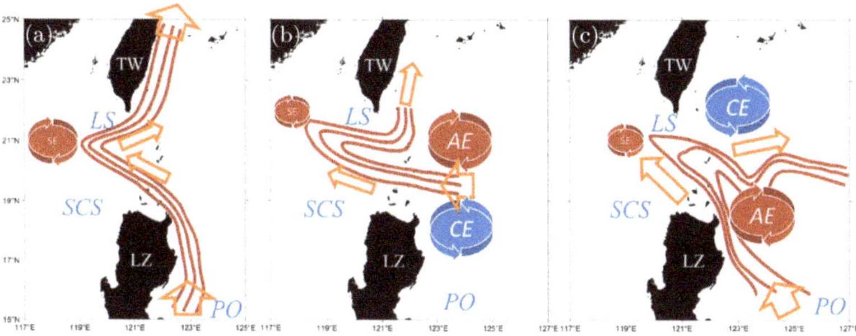

Fig. 6.6. Schematic diagram of the three types of paths that the Kuroshio near the Luzon Strait forced by big eddies from the Pacific. (a) Loop current pattern. (b) Jet flow pattern. (c) Bifurcating pattern. Streamlines represent the modified Kuroshio path (see Chapter 7 for definitions). AE: Anticyclonic eddy. CE: Cyclonic eddy. LS: Luzon Strait. LZ: Luzon Island. PO: Pacific Ocean. SCS: South China Sea. SE: Shed eddy. TW: Taiwan Island.

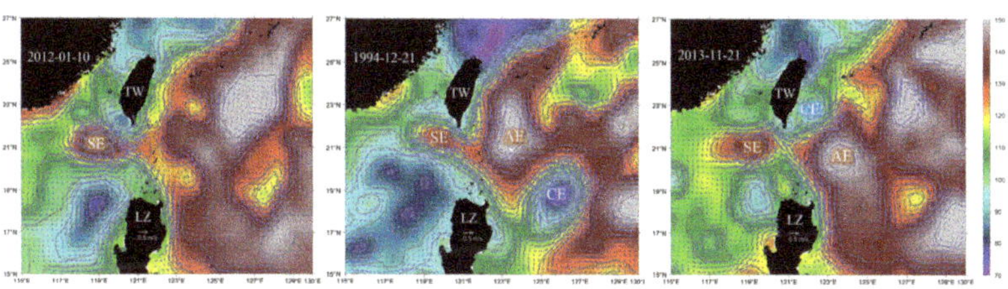

(a) (b) (c)

Fig. 6.7. Examples of three modified Kuroshio paths forced by interacting with big eddies from the Pacific shown on satellite altimeter SLH images on (a) 10 January 2012, (b) 21 December 1994 and (c) 21 November 2013. Arrows represent geostrophic current velocity vectors. Color codes of SLH are in centimeter.

downstream and at the same time separated a northwestward jet flow intruding into the SCS through the Luzon Strait. The peak sea level height (SLH) of the jet was 130 cm, and then an AE with a peak SLH of 120 cm was generated in the southwest of the Taiwan Island. Using temperature and salinity data from Conductivity Temperature Depth (CTD) and current velocity data from Acoustic Doppler Current Profiler (ADCP), Hu *et al.* (2012) found that the jet flow in the Luzon Strait contains high-temperature and high-salinity water masses at the depths of 100 to 250 m, indicating that properties of the jet flow water are close to that of the WNP water, implying that the jet flow and AE southwest of the Taiwan Island originate from the Pacific. The dynamic analysis indicates that these eddies are generated by baroclinic instability, pass through the Luzon Strait and inject anomalous vorticity into the SCS (Lu and Liu, 2013) with a penetration speed of about 0.6 m s^{-1} (Hu *et al.*, 2012). Their vertical structure, which typically extends to the depth of 1500 m, enables substantial heat and salt redistribution with winter sea surface temperature anomaly up to 2°C (Lin *et al.*, 2022).

Using satellite altimeter data, Trott *et al.* (2021) found most westward-propagating eddies occur during boreal fall and winter when the basin-scale circulation favors their propagation. A key condition for eddies to propagate westward across the Luzon Strait occurs when the Kuroshio exhibits a "leaking path". This is where the Kuroshio bifurcates, with one branch entering the northern SCS (NSCS) and the other continuing northward along the east of the Taiwan Island. In this "leaking path" scenario, an AE can generate in the east of the Kuroshio. The eddy once forms in the east of the Kuroshio, is then advected into the SCS. Nan *et al.* (2011) mentioned that the leakage path appears the most frequently, at 68.2%, and mostly in the second half of the year, the loop path and jump path account for 16.4% and 15.4%, respectively. This means that there is more than a 50% chance that a vortex will form in the Pacific and pass directly through the Luzon Strait into the SCS.

In addition, the strength and path of the Kuroshio have a significant impact on the eddy propagation through the Luzon Strait. When the Kuroshio loops westward into the SCS, the potential vorticity across the current is weak, allowing eddies to propagate freely (Nan *et al.*, 2013; Wu *et al.*, 2017). Conversely, when the Kuroshio leaps northward, bypassing the Luzon Strait, the potential vorticity strengthens, blocking eddies and forcing them to follow the northward path. Therefore, it is also possible that the eddy directly passes through the Luzon Strait into the SCS when the looping path occurs.

6.2.3 *Eddy-Kuroshio interaction*

6.2.3.1 *Modification of Kuroshio path*

As shown in Figures 6.6(a) and 6.7(a), eddies in the SCS can be shed from the Kuroshio looping path. Sun *et al.* (2021) found that strong Kuroshio intrusion into the SCS may generate a larger AE, verse visa.

Satellite altimeter data and model results have also confirmed the eddy detachment from the Kuroshio loop current. Analyzed multi-decadal data, Wu *et al.* (2017) suggested that the looping path of Kuroshio is correlated with weakened Kuroshio upstream transport. Meanwhile, looping path generated eddies alter SCS stratification and salinity distribution. This phenomenon is also observed in the subsurface layer. Wang *et al.* (2023) reported a subsurface AE with salinity exceeding 34.7, distinct from the SCS water, confirming its Pacific origin. The subsurface AE exhibited a lens-shaped core between 20–200 m depth, with a double thermocline structure and well-mixed water, distinct from typical SCS water. Sea surface signals of AE include a sea-level anomaly of 32 cm and a diameter >200 km. As the AE developed, it detached from the Kuroshio loop current and migrated southwestward into the SCS. The formation of the subsurface AE is caused by baroclinic instability during the Kuroshio loop current intrusion, driven by seasonal subsurface speed maxima in the Luzon Strait. Winter temperature cooling and reduced Kuroshio transport enhance intrusion and eddy shedding (Qi *et al.*, 2022; Wang *et al.*, 2023). This process highlights how Kuroshio looping paths directly shed eddies into the SCS basin.

6.2.3.2 *Kuroshio downstream cut-off*

Recent studies indicated that strong mesoscale eddies from the Pacific cut off the Kuroshio east of the Taiwan Island, enhance the Kuroshio intrusion into the Luzon Strait, create a looping path and increase the eddy kinetic energy in the NSCS. Kuroshio cut-off events east of Taiwan Island occur when westward-propagating mesoscale CEs interact with the Kuroshio. These interactions disrupt the northward flow, leading to a temporary cessation or redirection of the Kuroshio main path (KMP).

Zheng *et al.* (2019) reported that a KMP cut-off event east of the Taiwan Island occurred during fall-winter 2013–2014, which enhanced the eddy kinetic energy in the NSCS. The study analyzed satellite altimeter and mooring observations to document a rare KMP cut-off event caused by a sizeable CE impingement from the Pacific with the western boundary current near the Taiwan Island. The upper ocean volume transport of KMP was reduced by over 82%, dropping from 17×10^6 m^3 s^{-1} in September 2013 to 3×10^6 m^3 s^{-1}

in November 2013. The cut-off persisted until mid-January 2014, with gradual recovery observed as the CE dissipated. Time-series images reveal the westward movement of a strong CE, which deformed and eventually split into smaller eddies after colliding with the Kuroshio near the Taiwan Island. The interaction of CE with the Kuroshio caused significant path deformation, leading to a complete cut-off at the northern Luzon Strait by 1 December 2013. The data from moored instruments deployed in the east of Taiwan Island confirm the drastic reduction of the Kuroshio volume transport during the CE impingement. Figure 6.8 shows how Kuroshio disruptions intensify the mesoscale activities in the downstream region.

During the cut-off event, the Kuroshio branched into two streams: (i) a northward branch that turned eastward, and (ii) a northwestward branch that flowed into the SCS, intensifying pre-existing AE west of the Luzon Strait. These changes significantly altered regional circulation patterns. The Kuroshio intrusion into the SCS led to a substantial increase in the eddy kinetic energy (EKE) in the

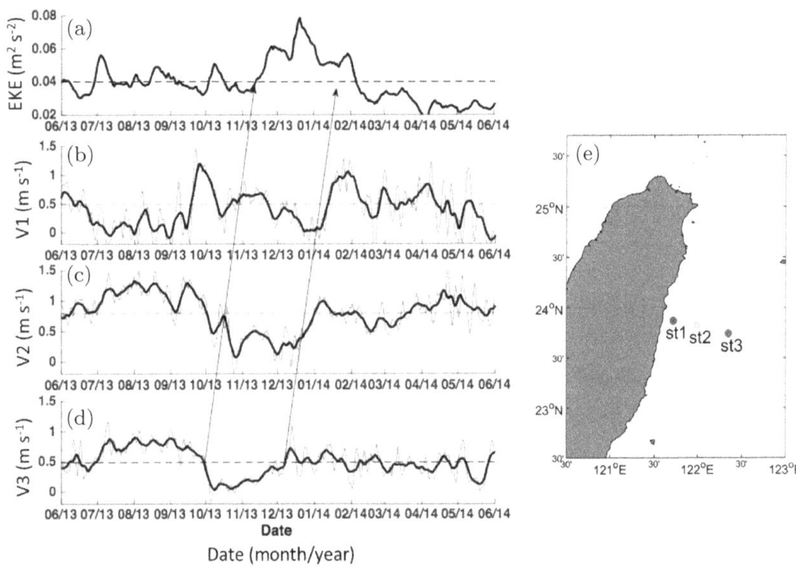

Fig. 6.8. (a) EKE in NSCS derived from satellite altimeter data, (b)–(d) current velocities at the depth of 50 m derived from three ADCP moorings deployed in the east of Taiwan Island, and (e) the locations of three ADCP moorings along a section at around 23.75°N from 121.7°E to 123.0°E.

Source: Redrawn from Zheng *et al.* (2019).

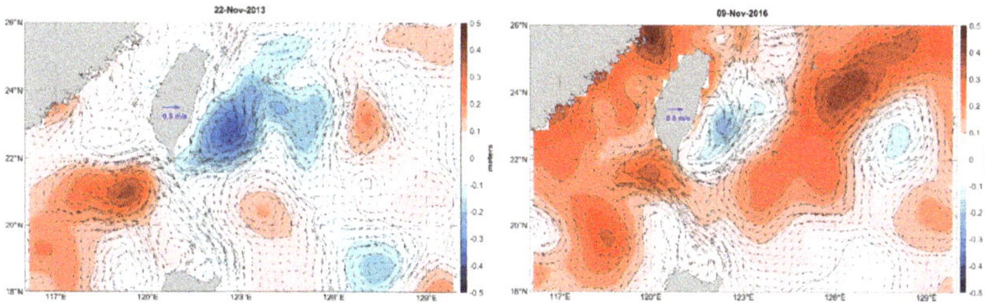

(a) (b)

Fig. 6.9. SLA (color codes in m) images with absolute geostrophic current velocities (arrows in m s^{-1}) derived from satellite altimetry on (a) 22 November 2013 and (b) 9 November 2016.

NSCS. The EKE peaked at twice its annual mean during November 2013–February 2014, indicating intensified mesoscale activities due to Kuroshio disruptions. Zheng *et al.* (2019) introduced a physical criterion based on total momentum to explain how eddy collisions influence KMP dynamics. Figure 6.7 shows two cases showing the changes in absolute geostrophic current and sea level anomaly (SLA), when CEs impinge the Kuroshio east of the Taiwan Island. One case occurred in November 2013, and another in November 2016. It is obvious that CE in the east of the Taiwan Island forces the Kuroshio to become very chaotic, causing the upstream of the Kuroshio from the east coast of the Luzon Island to flow into the Luzon Strait, and the EKE in the NSCS increases significantly (Figure 6.9).

6.3 Wave-Wave Resonance

The energetic disturbances originating from the WNP have a high possibility to penetrate the Luzon Strait and enter the SCS (Zheng *et al.*, 2011; Hu *et al.*, 2012; Xie *et al.*, 2016). Xie and Zheng (2017) clarified the resonant relation of the Rossby normal modes (RNMs) in the SCS deep basin to the impinging long waves facing the Luzon Strait at 19°–21.2°N, 124°–130°E in the WNP. They found the amplification factor for the periods of RNMs to that of the Pacific wave components are all above 0.9 (Figure 6.10), indicating that the incoming waves from the WNP contributes to resonate with the RNMs in the SCS deep basin. Zhao *et al.* (2023) found the transition between penetration and leaping of the Kuroshio at the Luzon Strait is a key process to the development of RNM6. The CEs and AEs

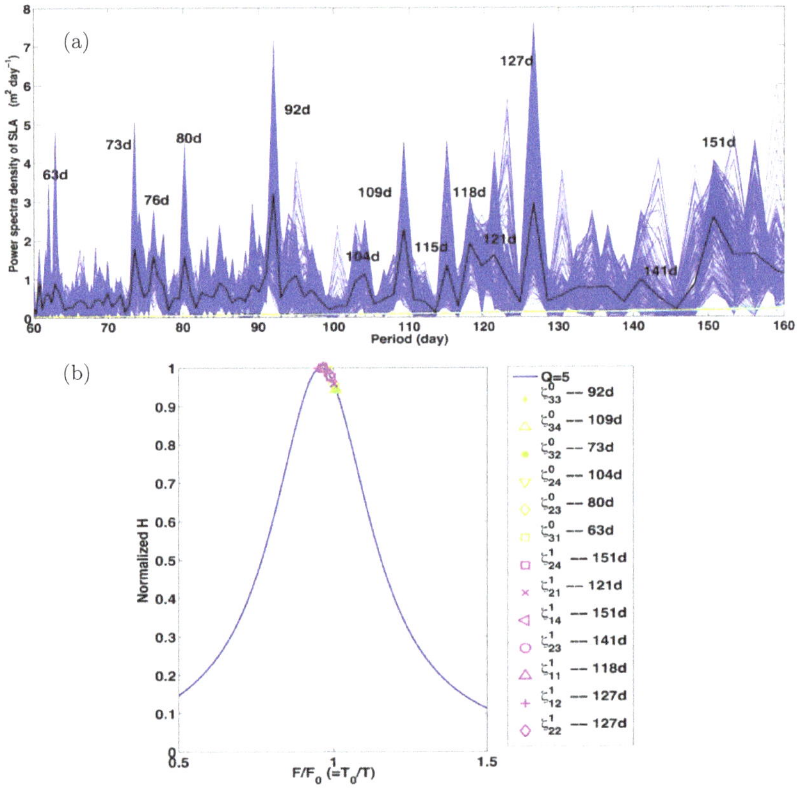

Fig. 6.10. (a) The Fourier period power density spectra of the long waves derived from satellite altimeter SLA data at 216 grids at 19°–21.2°N, 124°–130°E in the WNP from 1 January 1993 to 31 December 2016. (b) Resonant relations of the RNMs with peak period components of the incoming long waves. Curve represents a normalized theoretical resonant relation.

Source: Taken from Xie *et al.* (2018).

west of the Luzon Strait, and hence the RNMs are easily generated as the leaking and looping branches occur.

6.4 Wind Stress Forcing

Wind forcing also play a significant role in forming the looping path of the Kuroshio. The Kuroshio intrusion the into the SCS is influenced by wind-driven Ekman transport (Metzger and Hurlbust, 2001;

Nan *et al.*, 2011). The winter northeasterly winds enhance the westward flow of Kuroshio surface water into the Luzon Strait, causing the Kuroshio to undergo anticyclonic bending and subsequently intrude into the SCS (Farris and Wimbush, 1996). Jia and Chassignet (2011) found that when the integrated Ekman transport exceeds 2×10^{12} m^3, the Kuroshio loop expands and eddy shedding events occur. Metzger and Hurlburt (2001) suggested that stronger negative wind stress curl in the northern Luzon Strait increases Ekman pumping, deepens the thermocline and enhances the westward Kuroshio intrusion. Wu and Hsin (2012) conducted a series of numerical simulations to examine the wind forcing mechanism and suggested that the northeast wind in winter causes the Ekman drift to the northwest, which strengthens the Kuroshio in the upstream of the Luzon Strait and intensifies the Kuroshio intrusion into the SCS, and wind stress curl off the southwest coast of Taiwan Island is the main cause of the Kuroshio intrusion. Zhong *et al.* (2021) suggested more intrusion occurring in winter due to northeasterly monsoon. Winter monsoons enhance positive wind stress curl, which can influence the Kuroshio path by modifying vorticity fluxes and frictional stresses within the Luzon Strait.

6.5 Climate Change

6.5.1 *Effects of PDO*

Based on the studies of Zheng *et al.* (2019) and Lin *et al.* (2022), Tseng *et al.* (2025) examines the impact of Kuroshio cut-off events east of the Taiwan Island on vorticity dynamics and the looping path formation in the Luzon Strait, with a particular focus on climate variability during different Pacific Decadal Oscillation (PDO) regimes. They identified 12 significant Kuroshio cut-off events between 1993 and 2021 caused by mesoscale CEs impinging on the Kuroshio main path east of the Taiwan Island. These cut-off events result in intensified loop currents west of the Luzon Strait and increased negative vorticity in the NSCS, as shown in Figure 6.11. The phenomenon is more pronounced during the negative PDO regime (1998–2013), with negative vorticity increasing by 79% compared to a 34% increase during the positive PDO regime (Figure 6.12). During the negative

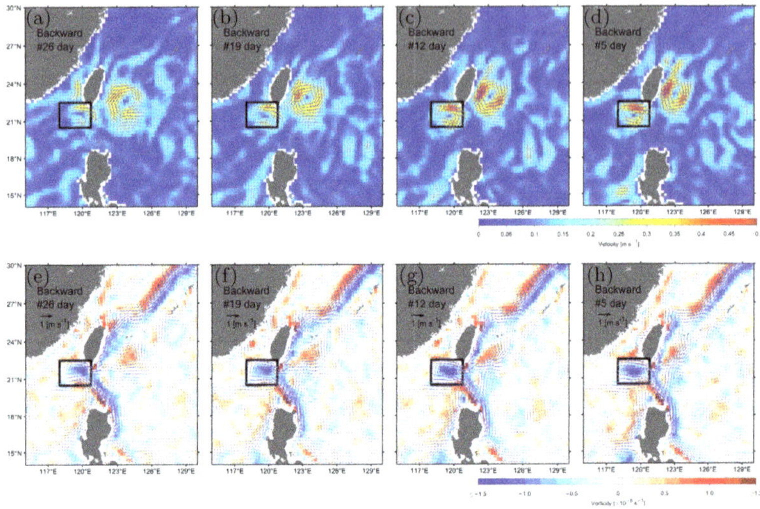

Fig. 6.11. The process of the Kuroshio cut-off event east of the Taiwan Island. (a–d) Variations of current velocity fields (color codes in m s^{-1}). (e–h) Variation of vorticity fields (color codes in 10^{-6} s^{-1}). Notation "#n day" represents a specific segment of the n-th backward day sequentially. The zero-th day is defined as the end of day for interaction between the eddy and the Kuroshio.
Source: Redrawn from Tseng *et al.* (2025).

PDO regime, Kuroshio velocity weakened east of Taiwan Island but strengthened east of Luzon Island. This disparity is linked to changes in sea level differences across the Luzon Strait, driving stronger loop currents into the NSCS. Negative PDO phases align with La Niña conditions, which weaken westerly winds and drive southward shifts in the North Equatorial Current bifurcation (NECB), concentrating flow near the Luzon Island. This research contributes valuable knowledge about mesoscale processes, climate-ocean interactions, and their impacts on circulation patterns in the SCS. It also highlights key drivers behind the variability in ocean currents under changing climatic conditions.

6.5.2 *Effects of ENSO*

The previous investigations have fond that the strength of the Kuroshio affects its path as it passes through the Luzon Strait (Wu and Hsin, 2012). If the Kuroshio is stronger, the occurrence probabilities of its looping path or leaking path may decrease, so the

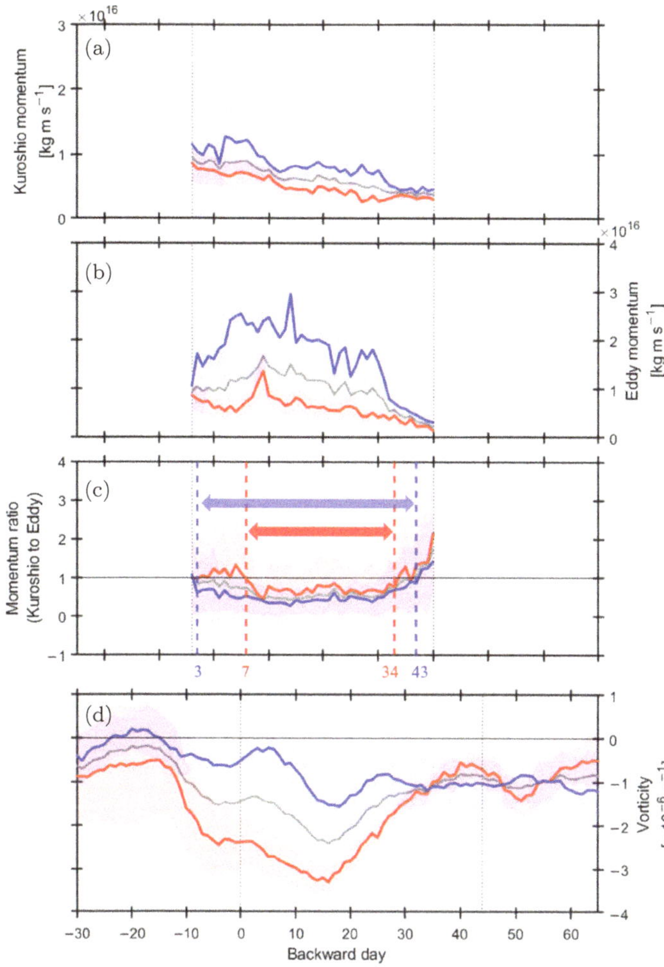

Fig. 6.12. Backward composite time series of (a) the Kuroshio momentum, (b) CE momentum, (c) the momentum ratio of the Kuroshio to CE and (d) the mean relative vorticity in NSCS. Red, blue, and gray lines (shading areas) are the composite data (90% confidence intervals of the composite data) of periods of negative PDO (1998–2013), positive PDO (1993–1997 and 2014–2021), and the whole (1993–2021), respectively. Two red (blue) dash lines in (c) indicate the period when the momentum ratio is below one during period of negative PDO (positive PDO).

Source: Redrawn from Tseng *et al.* (2025).

probability of the SCS eddy being formed by the Pacific disturbances may also decrease. The variability of the Kuroshio strength may arise from the bifurcation of the North Equatorial Current (NEC), which shifts latitudinally in response to ENSO phases (Kim *et al.*, 2004). During a La Niña event, the warming of the western Pacific warm pool strengthens the northeast trade winds, and the northerly winds in the Philippines Sea push the bifurcation of the NEC to shift southward. (Qiu and Chen, 2010; Wang *et al.*, 2020; Huang *et al.*, 2022). Thus, the strong trade winds prevail during the La Niña event and enhance the upstream of the Kuroshio (England *et al.*, 2014). Conversely, the bifurcation of the NEC shifts northward during a El Niño event and thus weakens the upstream of the Kuroshio. The northward bifurcation shifts weaken the Kuroshio mainstream (Tozuka *et al.*, 2002) and enhance Kuroshio intrusion into the SCS and eddy generation (Qu *et al.*, 2004). During the La Niña event, the NEC bifurcation point shifts southward. This shift causes more seawater to flow northward into the Kuroshio, increasing its velocity and volume transport (Yaremchuk and Qu, 2004). The strength of the Kuroshio is maintained through geostrophic balance, with higher sea levels in the interior Pacific and lower sea levels in marginal seas. The sea level decrease south of the NEC increases the sea surface height across the NEC, strengthening it. This contributes to the intensification of the Kuroshio during the La Niña event (Wang *et al.*, 2024).

6.6 Summaries

The South China Sea, as a marginal sea of the Pacific Ocean, experiences significant influence from Pacific-origin mesoscale disturbances through multiple mechanisms and pathways. This chapter has examined the complex interactions of these ocean systems, revealing the key processes that govern regional circulation patterns and water mass properties.

The Pacific eddies influence the SCS through two principal pathways: direct penetration through the Luzon Strait and shedding from the Kuroshio intrusion. The westward-propagating eddies, driven by baroclinic instability, interact with the Kuroshio and significantly alter its path through meandering, cut-offs or branching mechanisms.

These interactions create conditions favorable for eddy penetration into the SCS.

The wave-wave resonance of the Rossby normal modes (RNMs) in the SCS deep basin to the impinging long waves from the Pacific has been found. It is indicated that the incoming waves from the Pacific contribute to genesis and growth of the RNMs in the SCS deep basin.

Wind forcing represents another significant external factor, particularly through generation of Ekman transport mechanisms. Winter northeasterly monsoon enhances the westward flow of Kuroshio surface water into the Luzon Strait, promoting anticyclonic bending and SCS intrusion. These wind-driven dynamics highlight the atmospheric-oceanic coupling that influences SCS circulation patterns.

The broader climate system, particularly ENSO phases, modulates these interactions through the effect on the NEC bifurcation. During El Niño events, the northward shift of the NEC bifurcation point weakens the Kuroshio mainstream, favoring increased intrusion into the SCS and subsequent eddy generation. Conversely, La Niña conditions strengthen the Kuroshio through southward NEC bifurcation shifts, potentially limiting intrusion events. This teleconnection reveals how basin-scale climate oscillations manifest in regional oceanographic processes.

Future research directions should focus on quantifying how changing climate conditions might alter interactions between the Pacific and the SCS. As monsoon patterns shift and ENSO characteristics evolve under climate change scenarios, the frequency and intensity of eddy penetration events may change accordingly.

References

Chelton, D. B., Schlax, M. G. and Samelson, R. M. (2011). Global observations of nonlinear mesoscale eddies, *Prog. Oceanogr.*, 91, pp. 167–216, doi:10.1016/j.pocean.2011.01.002.

Cheng, Y.-H., Ho, C.-R., Zheng, Q., Qiu, B., Hu, J. and Kuo, N.-J. (2017). Statistical features of eddies approaching the Kuroshio east of Taiwan Island and Luzon Island, *J. Oceanogr.*, 73, pp. 427–438, doi:10.1007/s10872-017-0411-7.

England, M. H., McGregor, S., Spence, P., Meehl, G. A., Timmermann, A., Cai, W., Gupta, A. S., McPhaden, M. J., Purich, A. and Santoso, A. (2014). Recent intensification of wind-driven circulation in the Pacific and the ongoing warming hiatus, *Nat. Clim. Change*, 4(3), pp. 222–227, doi:10.1038/nclimate2106.

Farris, A. and Wimbush, M. (1996). Wind-induced intrusion into the South China Sea, *J. Oceanogr.*, 52, pp. 771–784, doi:10.1007/BF02239465.

Hu, J., Kawamura, H., Hong, H. and Qi, Y. (2000). A review on the currents in the South China Sea: seasonal circulation, South China Sea warm current and Kuroshio intrusion, *J. Oceanogr.*, 56, pp. 607–624, doi:10.1023/A:1011117531252.

Hu, J., Zheng, Q., Sun, Z. and Tai, C. K. (2012). Penetration of nonlinear Rossby eddies into South China Sea evidenced by cruise data, *J. Geophys. Res. Oceans*, 117, C03010, doi:10.1029/2011JC007525.

Huang, Y., Zhang, L., Wang, F., Wang, F. and Hu, D. (2022). Interannual variations of the North Equatorial Current/Undercurrent from mooring array observations, *Front. Mar. Sci.*, 9, 979442, doi:10.3389/fmars.2022.979442.

Jia, Y. and Chassignet, E. P. (2011). Seasonal variation of eddy shedding from the Kuroshio intrusion in the Luzon Strait, *J. Oceanogr.*, 67, pp. 601–611, doi:10.1007/s10872-011-0060-1.

Kim, Y. Y., Qu, T., Jensen, T., Miyama, T., Mitsudera, H., Kang, H. W. and Ishida, A. (2004). Seasonal and interannual variations of the North Equatorial Current bifurcation in a high-resolution OGCM, *J. Geophys. Res. Oceans*, 109, C03040, doi:10.1029/2003JC002013.

Liang, W. D., Tang, T. Y., Yang, Y. J., Ko, M. T. and Chuang, W. S. (2003). Upper-ocean currents around Taiwan, *Deep-Sea Res. II*, 50, pp. 1085–1105, doi:10.1016/S0967-0645(03)00011-0.

Lin, J. Y., Zheng, Z. W., Zheng, Q., Wu, D. R., Gopalakrishnan, G., Ho, C. R., Pan, J., Lin, Y. C. and Xie, L. L. (2022). Satellite observed new mechanism of Kuroshio intrusion into the northern South China Sea, *Int. J. Appl. Earth Obs. Geoinf.*, 115, 103119, doi:10.1016/j.jag.2022.103119.

Lu, J. and Liu, Q. (2013). Gap-leaping Kuroshio and blocking westward-propagating Rossby wave and eddy in the Luzon Strait, *J. Geophys. Res. Oceans*, 118(3), pp. 1170–1181, doi:10.1002/jgrc.20116.

Metzger, E. J. and Hurlburt, H. E. (2001). The nondeterministic nature of Kuroshio penetration and eddy shedding in the South China Sea, *J. Phys. Oceanogr.*, 31, pp. 1712–1732, doi:10.1175/1520-0485(2001)031<1712:TNNOKP>2.0.CO;2.

Nan, F., Xue, H. and Yu, F. (2015). Kuroshio intrusion into the South China Sea: A review, *Prog. Oceanogr.*, 137, pp. 314–333, doi:10.1016/j.pocean.2014.05.012.

Nan, F., Xue, H., Chai, F., Shi, L., Shi, M. and Guo, P. (2011). Identification of different types of Kuroshio intrusion into the South China Sea, *Ocean Dyn.*, 61, pp. 1291–1304, doi:10.1007/s10236-011-0426-3.

Nan, F., Xue, H., Chai, F., Wang, D., Yu, F., Shi, M., Guo, P. and Xiu, P. (2013). Weakening of the Kuroshio intrusion into the South China Sea over the past two decades, *J. Clim.*, 26(20), pp. 8097–8110, doi:10.1175/JCLI-D-12-00315.1.

Qi, Y., Mao, H., Du, Y., Li, X., Yang, Z., Xu, K., Yang, Y., Zhong, W., Zhong, F., Yu, L. and Xing, H. (2022). A lens-shaped, cold-core anticyclonic surface eddy in the northern South China Sea, *Front. Mar. Sci.*, 9, 976273, doi:10.3389/fmars.2022.976273.

Qiu, B. and Chen, S. (2010). Interannual-to-decadal variability in the bifurcation of the North Equatorial Current off the Philippines, *J. Phys. Oceanogr.*, 40, pp. 2525–2538, doi:10.1175/2010JPO4462.1.

Qu, T., Kim, Y. Y., Yaremchuk, M., Tozuka, T., Ishida, A. and Yamagata, T. (2004). Can Luzon Strait transport play a role in conveying the impact of ENSO to the South China Sea? *J. Clim.*, 17, pp. 3644–3657, doi:10.1175/1520-0442(2004)017<3644:CLSTPA>2.0.CO;2.

Sheu, W. J., Wu, C. R. and Oey, L. Y. (2010). Blocking and westward passage of eddies in the Luzon Strait, *Deep-Sea Res. II: Top. Stud. Oceanogr.*, 57, pp. 1783–1791, doi:10.1016/j.dsr2.2010.04.004.

Sun, Z., Hu, J., Chen, Z., Zhu, J., Yang, L., Chen, X. and Wu, X. (2021). A strong Kuroshio intrusion into the South China Sea and its accompanying cold-core anticyclonic eddy in winter 2020–2021, *Rem. Sens.*, 13, 2645, doi:10.3390/rs13142645.

Tozuka, T., Kagimoto, T., Masumoto, Y. and Yamagata, T. (2002). Simulated multiscale variations in the western tropical Pacific: The Mindanao Dome revisited, *J. Phys. Oceanogr.*, 32, pp. 1338–1359, doi:10.1175/1520-0485(2002)032<1338:SMVITW>2.0.CO;2.

Trott, C. B., Metzger, E. J. and Yu, Z. (2021). Investigating mesoscale eddy characteristics in the Luzon Strait region using altimetry, *Ocean Dyn.*, 71, pp. 679–698, doi:10.1007/s10236-021-01460-1.

Tseng, Y. H., Ho, C. R., Zheng, Q. and Zheng, Z. W. (2025). Loop current west of the Luzon strait triggered by Kuroshio cut-off events east of Taiwan Island, *J. Geophys. Res. Oceans*, 130(2), e2024JC021572, doi:10.1029/2024JC021572.

Wang, X., Du, Y., Zhang, Y., Wang, T., Wang, M. and Jing, Z. (2023). Subsurface anticyclonic eddy transited from Kuroshio shedding eddy in the northern South China Sea, *J. Phys. Oceanogr.*, 53, pp. 841–861, doi:10.1175/JPO-D-22-0106.1.

Wang, X., Tong, B., Wang, D. and Yang, L. (2020). Variations of the North Equatorial Current bifurcation and the SSH in the western

Pacific associated with El Niño flavors, *J. Geophys. Res. Oceans*, 125(1), e2019JC015733, doi:10.1029/2019JC015733.
Wang, Y. L., Jin, F. F., Wu, C. R. and Qiu, B. (2024). Northwestern Pacific Oceanic circulation shaped by ENSO, *Sci. Rep.*, 14, 11684, doi:10.1038/s41598-024-62361-z.
Wu, C. R. and Chiang, T. L. (2007). Mesoscale eddies in the northern South China Sea, *Deep-Sea Res. II: Top. Stud. Oceanogr.*, 54, pp. 1575–1588, doi:10.1016/j.dsr2.2007.05.008.
Wu, C. R. and Hsin, Y. C. (2012). The forcing mechanism leading to the Kuroshio intrusion into the South China Sea, *J. Geophys. Res. Oceans*, 117, C07015, doi:10.1029/2012JC007968.
Wu, C. R., Wang, Y. L., Lin, Y. F. and Chao, S. Y. (2017). Intrusion of the Kuroshio into the south and East China Seas, *Sci. Rep.*, 7, 7895, doi:10.1038/s41598-017-08206-4.
Xie, L. and Zheng, Q. (2017). New insights to the South China Sea: Rossby normal modes, *Acta Oceanol. Sin.*, 36(7), pp. 1–3.
Xie, L., Zheng, Q., Tian, J., Zhang, S., Feng, Y. and Yi, X. (2016). Cruise observation of Rossby waves with finite wavelengths propagating from the Pacific to the South China Sea, *J. Phys. Oceanogr.*, 46(10), pp. 2897–2913, doi:10.1175/JPO-D-16-0071.1.
Xie, L., Zheng, Q., Zhang, S., Hu, J., Li, M., Li, J. and Xu, Y. (2018). The Rossby normal modes in the South China Sea deep basin evidenced by satellite altimetry, *Int. J. Remote Sens.*, 39, pp. 399–417.
Yaremchuk, M. and Qu, T. (2004). Seasonal variability of the large-scale currents near the coast of the Philippines, *J. Phys. Oceanogr.*, 34, pp. 844–855, doi:10.1175/1520-0485(2004)034<0844:SVOTLC>2.0.CO;2.
Zhao, R., Zhu, X.-H., Zhang, C., Zheng, H., Zhu, Z.-N., Ren, Q., Liu, Y., Nan, F. and Yu, F. (2023). Summer anticyclonic eddies carrying Kuroshio waters observed by a large CPIES array west of the Luzon Strait, *J. Phys. Oceanogr.*, 53, pp. 341–359, doi:10.1175/JPO-D-22-0019.1.
Zheng, Z. W., Ho, C. R. and Kuo, N. J. (2007). Mechanism of weakening of west Luzon eddy during La Nina years, *Geophys. Res. Lett.*, 34(11), L11604, doi:10.1029/2007GL030058.
Zheng, Q., Ho, C. R., Xie, L. and Li, M. (2019). A case study of a Kuroshio main path cut-off event and impacts on the South China Sea in fall-winter 2013–2014, *Acta Oceanol. Sin.*, 38, pp. 12–19, doi:10.1007/s13131-019-1411-9.

Zheng, Q., Tai, C. K., Hu, J., Lin, H., Zhang, R. H., Su, F. C. and Yang, X. (2011). Satellite altimeter observations of nonlinear Rossby eddy–Kuroshio interaction at the Luzon Strait, *J. Oceanogr.*, 67, pp. 365–376, doi:10.1007/s10872-011-0035-2.

Zhong, Y., Zhou, M., Waniek, J. J., Zhou, L. and Zhang, Z. (2021). Seasonal variation of the surface Kuroshio intrusion into the South China Sea evidenced by satellite geostrophic streamlines, *J. Phys. Oceanogr.*, 51(8), pp. 2705–2718, doi:10.1175/JPO-D-20-0242.1.

Chapter 7

Intermediary Role of the Kuroshio Current in Eddy Generation in the South China Sea

7.1 Introduction

The previous investigators have paid attention to the Kuroshio's role in eddy generation in the SCS. From the hydrographic survey data, Wang and Chern (1987) and Li *et al.* (1998) found connections of water mass properties of anticyclonic eddies (AEs) in the northeastern SCS to that of the Kuroshio. After the satellite altimeter sea surface height (SSH) data are available, distribution statistics of mesoscale eddies in the SCS reveals that the northeastern SCS west of the Luzon Strait near the Kuroshio intrusion region is the most concentrated region of mesoscale eddies (Wang *et al.*, 2003; Ho *et al.*, 2004; Chen *et al.*, 2011).

For dynamic mechanisms inducing the Kuroshio intrusion and eddy shedding, early investigations attributed them to the monsoon wind forcing (Wang and Chern, 1987; Farris and Wimbush, 1996; Centurioni *et al.*, 2004; Jia and Chassignet, 2011). Meanwhile, Li *et al.* (1993) suggested that variations of the Kuroshio and the whole circulation system of the Pacific have an important influence on the Kuroshio current entering the SCS via Luzon Strait, based on numerical modeling results (Qu, 2000; Metzger and Hurlburt, 2001).

As a marginal sea along the western boundary of the Pacific, the SCS connects the Pacific through the Luzon Strait with three-dimensional (3D) scales of $O(400)$ km (north-south width) by $O(200)$ km (west-east length) by $O(2000)$ m (depth). Thus, it is an easy way the westward deflected Kuroshio path and westward propagating eddies from the Pacific enter the SCS as predicted by Li *et al.* (1993). Caruso *et al.* (2006) found that the Kuroshio path near the Luzon Strait takes various patterns. Using time series of satellite altimeter absolute dynamic topography (ADT) images surrounding the Luzon Strait, Zheng *et al.* (2011) investigated a case in that forced by a couple of energetic eddies from the Pacific, the Kuroshio intruded the SCS and shed an AE in June–July 2004. Nan *et al.* (2011a) classified the Kuroshio intrusion patterns into three types: looping path, leaping path, and leaking path based on ADT images and the surface geostrophic currents near the Luzon Strait. From the cruise conductivity-temperature-depth (CTD) data measured in January 2010, Hu *et al.* (2012) found an AE with a diameter of about 150 km west of the Luzon Strait, of which the high-temperature and high-salinity water within the subsurface layer core originated from the Kuroshio, implying that it was a Kuroshio-intrusion-shed eddy.

Later, interaction of the Kuroshio current in the EEZ with the E_{WB} propagating westward from the Pacific has attracted more and more attentions of the research community (Lien *et al.*, 2014; Jan *et al.*, 2015; Tsai *et al.*, 2015; Andres *et al.*, 2017). Zhang *et al.* (2017) examined three events for the Kuroshio intruded into the SCS in the pattern of loop current and shed AEs in winter and autumn of 2012 and autumn 2013. They found that T–S diagrams of shed eddies are close to the Kuroshio curve between 100 and 200 m, implying that the Kuroshio is the major water supplier of shed eddies. Zheng *et al.* (2019) examined a Kuroshio cut-off event caused by collision of an energetic cyclonic eddy (CE) from the Pacific with the western boundary and its impacts on the SCS in fall–winter 2013–2014. The event forced the Kuroshio bifurcated into the SCS deep basin, shed an AE, and raised the local peak eddy kinetic energy (EKE) up to 100%. McMahon *et al.* (2021) focused on the problems regarding the Kuroshio loop current shedding eddies through the synergetic use of moored and satellite data and data-assimilative model products. Recently, Zhao *et al.* (2023) reported an AE generated north of Luzon Island propagating northwestward into the SCS in summer. Summarizing the above results, this chapter aims to clarify the roles of the Kuroshio and the Pacific play in modification of the Kuroshio path, Kuroshio intrusion into the SCS, and shedding eddies.

7.2 Properties of the Kuroshio

The Kuroshio current is the western boundary current (WBC) of the North Pacific Ocean. The word of Kuroshio originates from Japanese word "黑(kuro)潮(shio)", which means the black tides owing to its high-temperature and high-salinity water originating from the North Equatorial Current in dark blue color. The Kuroshio forms in east of the Philippines (Nitani, 1972) and travels northward along the east coast of Luzon Island. Its average width is of $O(000)$ km, average depth is of $O(1000)$ m, average velocity is of $O(0.7)$ m s^{-1} and average transport of 21 Sv (1 Sv = 10^6 m^3 s^{-1}) (Qiu, 2001; Centurioni *et al.*, 2004; Yaremchuk and Qu, 2004; Ho *et al.*, 2020; McMahon *et al.*, 2021).

Figure 7.1(a) shows the yearly mean Kuroshio axis and path derived from the yearly averaged absolute geostrophic velocity from 1993 to 2017 and Argos drifting buoy velocity from 1979 to 2016. One can see that after leaving the Philippines, the Kuroshio traverses

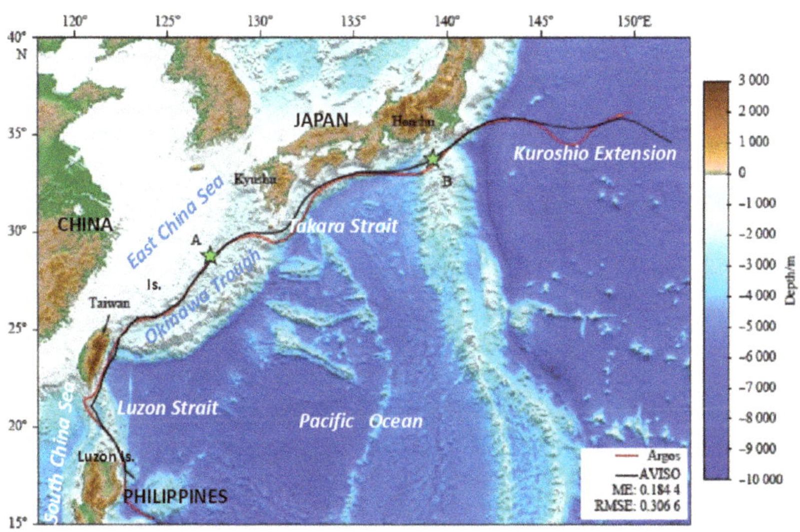

Fig. 7.1. The Kuroshio axis and path with bathymetry in m derived from the yearly averaged absolute geostrophic velocity from 1993 to 2017 (dark blue line) and Argos drifting buoy velocity from 1979 to 2016 (red line). Mean error (ME) and root mean square error (RMSE) between the two lines were calculated at green stars A and B located at 28.75°N, 127.25°E and 33.75°N, 139.25°E.

Source: Taken and labeled from Zhuang *et al.* (2020).

the Luzon Strait and flows further northward along the east coast of Taiwan Island. Then it turns northeastward into the Okinawa Trough in the East China Sea, returns to the Pacific through the Tokara Strait and continues its long journey along the continental slope south of Japanese Chain Islands till injecting into the Kuroshio Extension (Zhuang *et al.*, 2020).

The behavior of the Kuroshio Current near the Luzon Strait is of great significance for generation of eddies in the SCS. The Luzon Strait sandwiched by the Taiwan Island in north and the Luzon Island in south is a big gap of the western boundary of the North Pacific as shown in Figure 6.1. 3D scales of the strait are the north-south width of $O(400)$ km, the west-east length of $O(200)$ km and the average depth of $O(2000)$ m, connecting the North Pacific to east and the SCS to west. Previous investigators have documented that the Kuroshio path near the Luzon Strait deflects with the seasonal variation. Farris and Wimbush (1996) attributed the phenomenon to the northeast monsoon if the local wind-stress exceeds a critical value. Centurioni *et al.* (2004) analyzed the seasonal variation of drifter trajectories near the Luzon Strait and found most drifters crossed the Luzon Strait into the SCS from October to March and attributed the phenomenon to the Ekman effect during the prevalent northeast monsoon season. Figure 7.2 shows the Kuroshio path patterns near the Luzon Strait in four seasons derived from the satellite altimeter surface geostrophic current data (Ho *et al.*, 2020). One can see that the Kuroshio is strengthened with its flow path along the eastern boundary of the Luzon Strait in spring and summer and weakened with deflected path intruding toward the Luzon Strait in autumn and winter.

Owing to inhibition of 3D topography of the Luzon Strait, the Ekman effect during the prevalent northeast monsoon season, collision, and blockage of westward propagating eddies from the Pacific, the Kuroshio path near the Luzon Strait takes various patterns (Caruso *et al.*, 2006). Based on the research results of the modified Kuroshio path near the Luzon Strait and its role in generation of mesoscale eddies in the SCS, this chapter proposes three patterns for the Kuroshio intrusion into the SCS: loop current (LOC), jet flow (JEF) and bifurcating current (BIC) as shown in Figure 7.3.

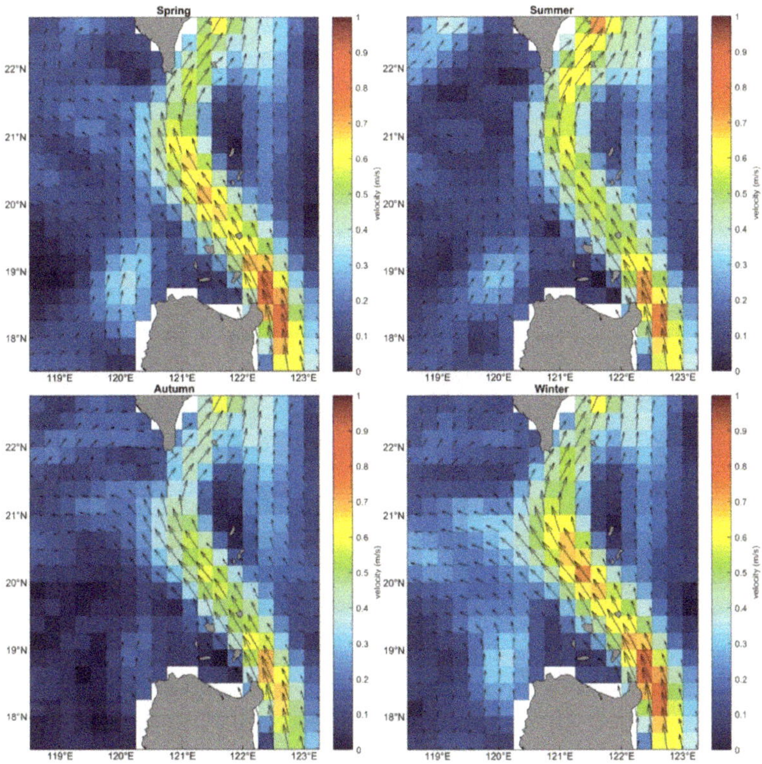

Fig. 7.2. Seasonal swing of the Kuroshio path near the Luzon Strait derived from the satellite altimeter surface geostrophic current data.
Source: Taken from Ho *et al.* (2020).

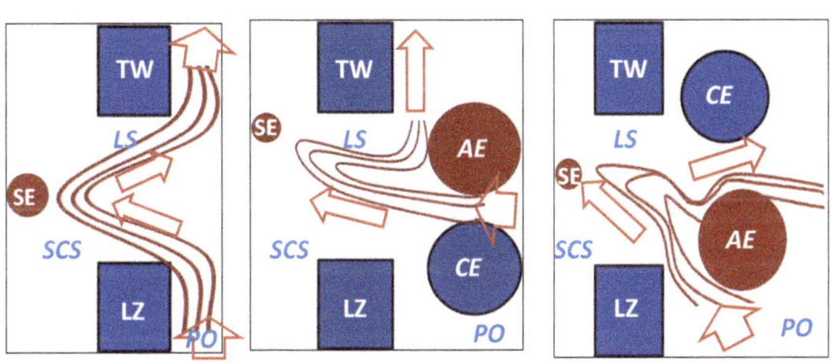

(a) (b) (c)

Fig. 7.3. Schematics of three patterns of the modified Kuroshio path near the Luzon Strait. (a) Loop current pattern. (b) Jet flow pattern. (c) Bifurcating pattern. Streamlines represent the local modified Kuroshio path. AE: Anticyclonic eddy. CE: Cyclonic eddy. LS: Luzon Strait. LZ: Luzon Island. PO: Pacific Ocean. SCS: South China Sea. SE: Shed eddy. TW: Taiwan Island.

7.3 Eddy Shedding in the SCS by Modified Kuroshio Path

7.3.1 *Modified Kuroshio path by Luzon Strait: Loop current pattern*

As shown in Figure 7.3(a), the loop current pattern (LOC) is featured by that the Kuroshio intrudes the SCS through the Luzon Strait, sheds an AE, and then loops back to the Pacific. Previous investigators have figured out that the LOC mainly occurs in fall–winter season and attributed to the Ekman effect of strong northeast monsoon winds and inhibition of 3D topography of the Luzon Strait (Li *et al.*, 1989, 1998; Caruso *et al.*, 2006; Nan *et al.*, 2011a, 2011b). In fact, forcing of energetic eddies from the Pacific is also an important mechanism for formation of the LOC as shown in the next subsection.

Nan *et al.* (2011a) classified the Kuroshio intrusion patterns into three types: looping path, leaping path, and leaking path based on the images of mean ADT and the surface geostrophic currents near the Luzon Strait as shown in Figure 7.4. Comparing Figure 6.4 to Figure 6.3, one can see that their looping path and leaping path are like the LOC defined in this chapter, and leaking path like the BIC.

Suppose that the entering flow and the exiting flow of the LOC have the symmetric structure with the Gaussian-type of velocity distribution, the vorticities of the entering and the exiting flows can be estimated as (Qiu *et al.*, 2009; Qiu and Chen, 2010)

$$\zeta_{lp\mp} = \mp \frac{2U_{wst}}{L_\sigma}, \tag{7.1}$$

where U_{wst} is the peak value of westward velocity and L_σ is the e-fold width of the flow. It implies that the looping action enforces the Kuroshio to lose the vorticity of $\left(-\frac{4U_{wst}}{L_\sigma}\right)$ to the SCS, i.e., to shed an anticyclonic eddy with the initial vorticity of

$$\zeta_e = -\frac{4U_{wst}}{L_\sigma}. \tag{7.2}$$

Zhang *et al.* (2013) gave an empirical expression for the sea surface dynamic height of mesoscale eddy in the form of

$$\eta = \eta_0 \left(1 - \frac{r_n^2}{2}\right) e^{-\frac{r_n^2}{2}}, \tag{7.3}$$

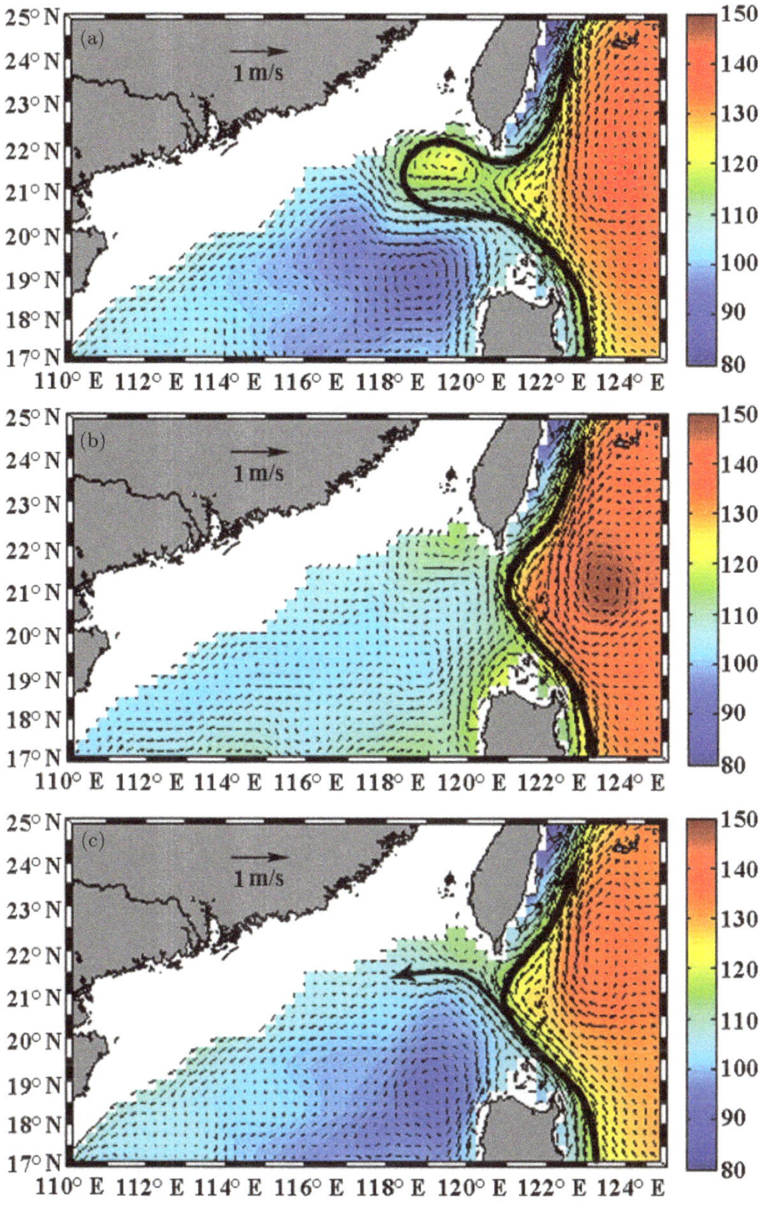

Fig. 7.4. Images of mean ADT (in cm) and surface geostrophic currents near the Luzon Strait. (a–c) were defined as three types of Kuroshio intrusion: looping path, leaping path, and the leaking path by Nan *et al.* (2011a). Black line represents the Kuroshio path axis.

Source: Taken from Nan *et al.* (2011a).

where η_0 is the peak value of the sea surface dynamic height of mesoscale eddy, $r_n\left(=\frac{r}{R}\right)$ is the normalized radius vector, in which R is the eddy radius. From Eq. (7.3), the relative vorticity of the eddy shear flow can be derived as

$$\zeta = \frac{g\eta_0}{fR^2}\left(\frac{r_n^4}{2} - 4r_n^2 + 4\right)e^{-\frac{r_n^2}{2}}, \tag{7.4}$$

where g is the gravitational acceleration and f is the Coriolis parameter.

In comparison of Eqs. (7.2) and (7.4), the initial radius of shed eddy can be estimated as

$$R_{se}^2 = \frac{g\eta_0 L_\sigma}{4fU_{wst}}. \tag{7.5}$$

If we reasonably choose $\eta_0 = 1.5$ m (see Figure 7.5), $L_\sigma = 50$ km (see Figure 6.4(a)), $f = 3.5 \times 10^{-5}$ s^{-1}(20°N), and $U_{wst} = 0.7$ m s^{-1}

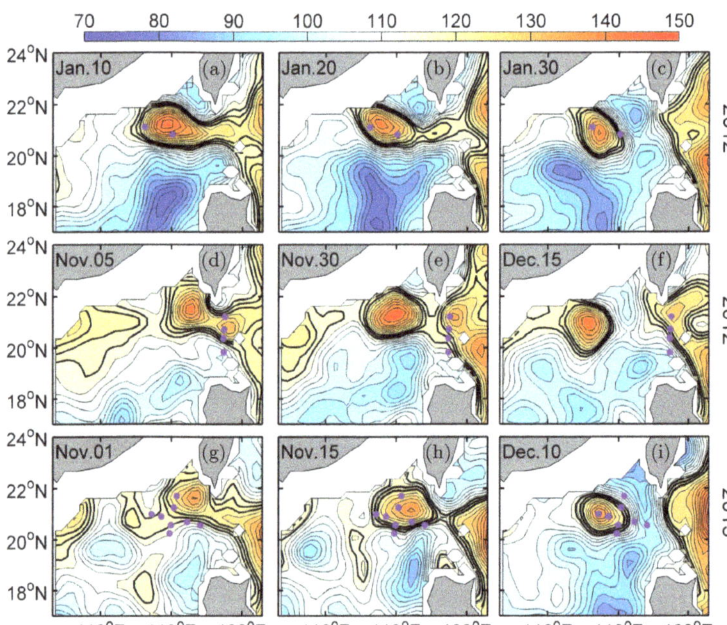

Fig. 7.5. Satellite altimeter SSH images of the LOC and SEs in the northeastern SCS in winter and fall of 2012 and fall of 2013. Black thick contours of 117, 120, and 123 cm are used to denote the LOC. Regions shallower than 100 m are masked by white shading. Purple dots denote mooring locations. The date (month and day) is shown on top-left corner of each image. Color codes of SSH are in cm. *Source*: Taken from Zhang *et al.* (2017).

(Ho *et al.*, 2020), the estimate of R_{se} is 86 km. The possible fate of shed eddy, i.e., growing, sustaining, or dumping, is determined by the background vorticity field.

Figure 7.5 shows satellite altimeter SSH images of the LOC and SEs in the northeastern SCS in winter and fall of 2012 and fall of 2013 (Zhang *et al.*, 2017). From the images, we measure the average peak SSH of the SEs $\eta_0 = 1.5$ m and the average sea surface radius, $R_{se} = 95$ km, which is close to 86 km of R_{se} estimated by Eq. (7.5).

Figure 7.6 shows T–S diagrams of water masses of the northeastern SCS, the Kuroshio and shed eddies shown in Figures 7.5(a–c) and g-i (Zhang *et al.*, 2017). One can see that above about 230 m,

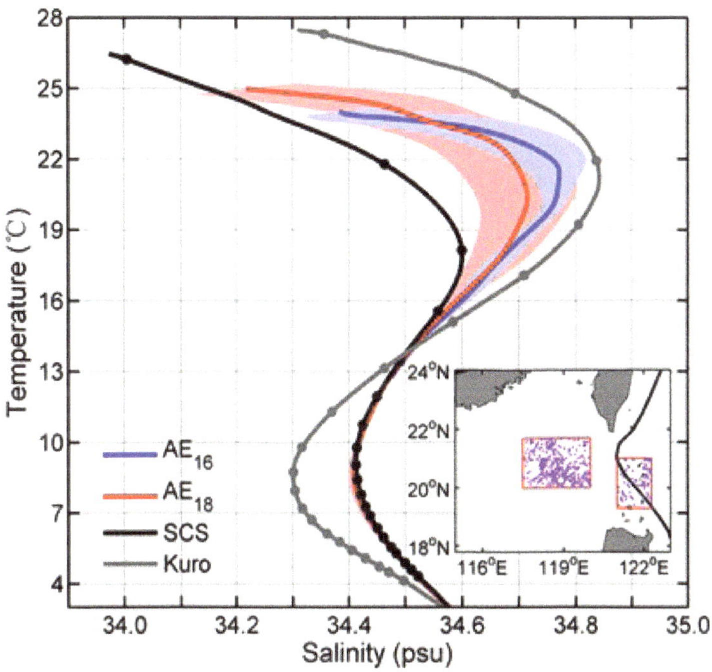

Fig. 7.6. Mean T–S diagrams of water masses of the northeastern SCS (black line) and the Kuroshio (gray line) computed from historical Argo T–S profiles within the red boxes of the inset figure, and eddies shown in Figures 7.5(a–c) (AE_{16}; blue line) and g–i (AE_{18}; red line) obtained from the moored CTD data. Shadings denote the standard deviations of the T–S. Dots on black and gray curves denote depths of 50, 100, 150, and 200 m, and so on. Purple dots in the inset figure represent Argo profile locations, and the black curve denotes the time-mean Kuroshio axis.

Source: Taken from Zhang *et al.* (2017).

T–S diagrams of shed eddies are sandwiched by that of the SCS and the Kuroshio, and close to the Kuroshio curve between 100 and 200 m, implying that the Kuroshio is the major water supplier of shed eddies. Below about 230 m, shed eddies share the same curve with the SCS.

7.3.2 *Modified Kuroshio path by eddies from the Pacific I: Jet flow pattern*

As shown in Figure 7.3(b), the jet flow (JEF) pattern is featured by that a couple of energetic AE and CE from the Pacific modifies the Kuroshio path and functions as a three-way valve. CE in south shuts the upstream Kuroshio current down and AE in north releases the downstream Kuroshio current and the JEF into the northern continental slope of SCS through the Luzon Strait. The JEF sheds an AE on the slope. Previous investigators have figured out that the JEF may occur in any season (Zheng *et al.*, 2011; Hu *et al.*, 2012).

Figure 7.7 shows satellite altimeter ADT images of a winter case of the JEF, showing the dynamic conditions for occurrence of the

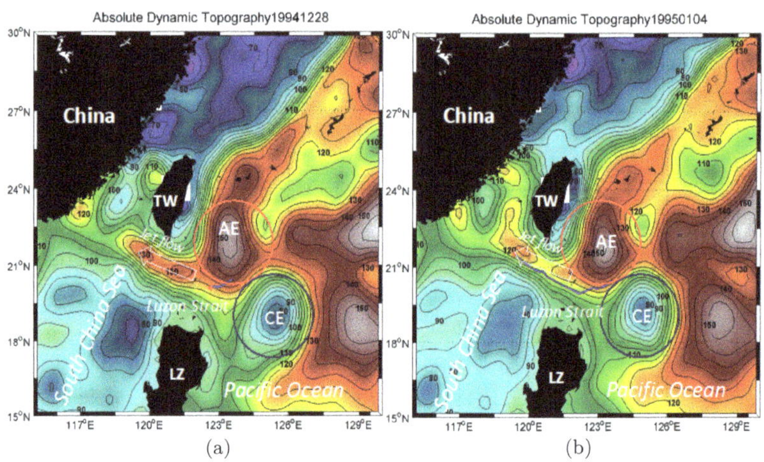

Fig. 7.7. Satellite altimeter ADT (in cm) images of a winter case of the JEF events on (a) 28 December 1994 and (b) 4 January 1995. AE: Anticyclonic eddy. CE: Cyclonic eddy. LZ: Luzon Island. TW: Taiwan Island.

Source: Courtesy of Dr. Yu-Hsin Cheng, NTOU.

JEF event on 28 December 1994 and 4 January 1995. One can see that during the event, the normal Kuroshio path was destroyed by the couple of energetic AE and CE from the Pacific. The horizontal length scale of eddies reaches $O(400)$ km the same as the N-S width of the Luzon Strait. CE with a peak SSH of 80 cm southeast of the Luzon Strait cuts the upstream Kuroshio current. AE with a peak SSH of 150 cm northeast of the Luzon Strait maintains the downstream Kuroshio current and meanwhile bifurcates a northwestward jet flow intruding the northern continental slope of SCS through the Luzon Strait. The jet flow has the peak SSH of 130 cm and the shed eddy southwest of the Taiwan Island has the initial peak SSH of 120 cm.

Revisited the conductivity-temperature-depth (CTD) profiler data, acoustic Doppler current profiler (ADCP) data, simultaneous satellite altimeter ADT data and Argo float profiling data measured during the JEF event in January 2010 (Hu *et al.*, 2012), one can see that the jet flow in the Luzon Strait had a subsurface high-salinity layer from 100 to 250 m with the highest salinity higher than 34.8. Meanwhile, an SE with a radius of about 75 km centered at around 21°N and 118°E west of the Luzon Strait also had a subsurface high-salinity water prism with a salinity higher than 34.8. The T–S diagram of the SE is close to that of the Northwest Pacific water. These results indicate the Pacific origin of the jet flow and SE waters.

It is of interest that as CE of the couple of eddies is much weaker than AE, the JEF may evolve into the LOC. Figure 7.8 shows time series of ADT images surrounding the Luzon Strait in June–July 2004 (Zheng *et al.*, 2011). One can see the evolution process that a couple of AE and CE modified the Kuroshio path and formed the LOC rather than the JEF. The case is featured by a seriously asymmetric couple of eddies, strong AE of a horizontal scale of $O(400)$ km and weak CE of $O(200)$ km. Furthermore, from the images, choosing $\eta_0 = 1.3$ m, $L_\sigma = 40$ km, $f = 3.5 \times 10^{-5}$ s^{-1}, and $U_{wst} = 0.7$ m s^{-1}, Eq. (7.5) predicts the initial radius of SE, R_{se} of 72 km, which is close to 65 km directly measured from the image of 2004/06/30, further confirming the LOC formation.

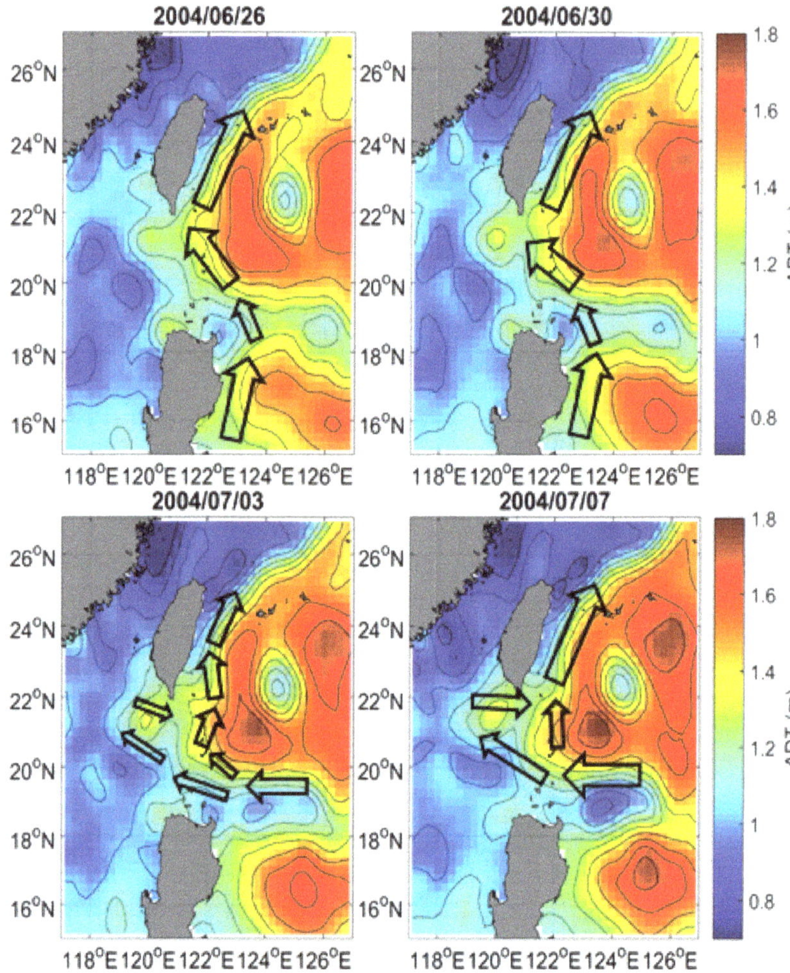

Fig. 7.8. Time series of ADT images surrounding the Luzon Strait in June–July 2004. Arrows show modified Kuroshio path. Color codes of SSH is in m.
Source: Taken from Zheng *et al.* (2011).

7.3.3 *Modified Kuroshio path by eddies from the Pacific II: Bifurcating current pattern*

Figure 7.3(c) shows schematics of the bifurcating current (BIC) pattern. Comparing Figure 7.3(c) with Figure 7.3(b), one can see that the physics of JEF and BIC is the same, i.e., a couple of energetic

AE and CE from the Pacific modifies the Kuroshio path and functions as a three-way valve. The only difference from the two is the upside-down positions of AE and CE. CE in north shuts the Kuroshio current down and AE in south detours the Kuroshio current and releases a bifurcating current into the SCS deep basin through the Luzon Strait. The BIC sheds an AE in the SCS deep basin.

Figure 7.9 shows an example of the BIC observed in November–December 2013 (Zheng *et al.*, 2019). From time series ADT images,

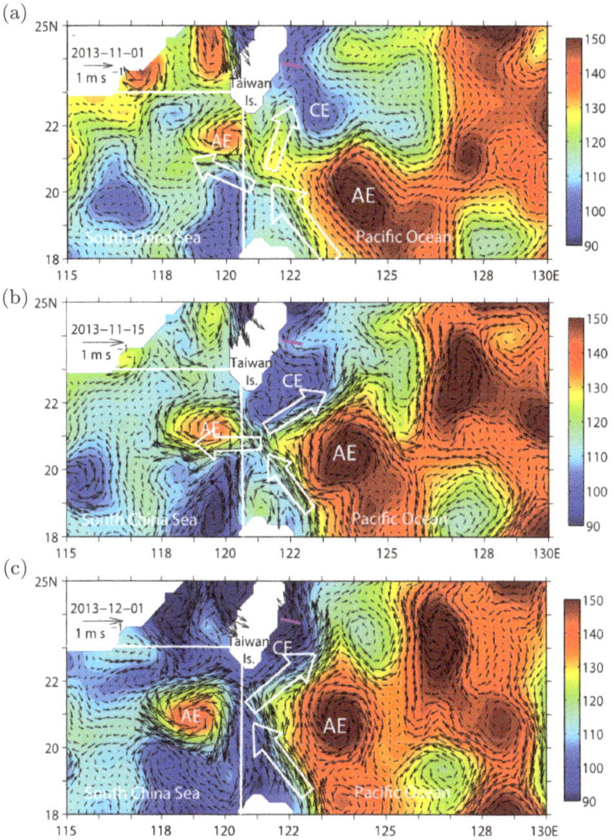

Fig. 7.9. Time series ADT images surrounding the Luzon Strait during the period of November–December 2013. Small black arrows represent the geostrophic current vectors and large white arrows represent the Kuroshio path, respectively. Area enclosed by white lines on the lower left side is used for calculating the EKE. Color codes are in cm.

Source: Taken from Zheng *et al.* (2019).

one can see that a couple of energetic AE and CE with the horizontal scale of about $O(300)$ km modified the Kuroshio path. CE in north cut-off the Kuroshio current and AE in south detoured the Kuroshio current northeastward and released a bifurcating current into the SCS deep basin through the Luzon Strait. The BIC shed an AE in the SCS, of which the horizontal scale (radius) increased from $O(50)$ km on 1 November to $O(100)$ km on 1 December. Meanwhile, the maximum tangential velocity increased from $O(0.6)$ to $O(1)$ m s^{-1}.

To estimate impact of the BIC-shed AE on the SCS, Zheng *et al.* (2019) calculated the EKE in the test area from 18° to 23°N and from

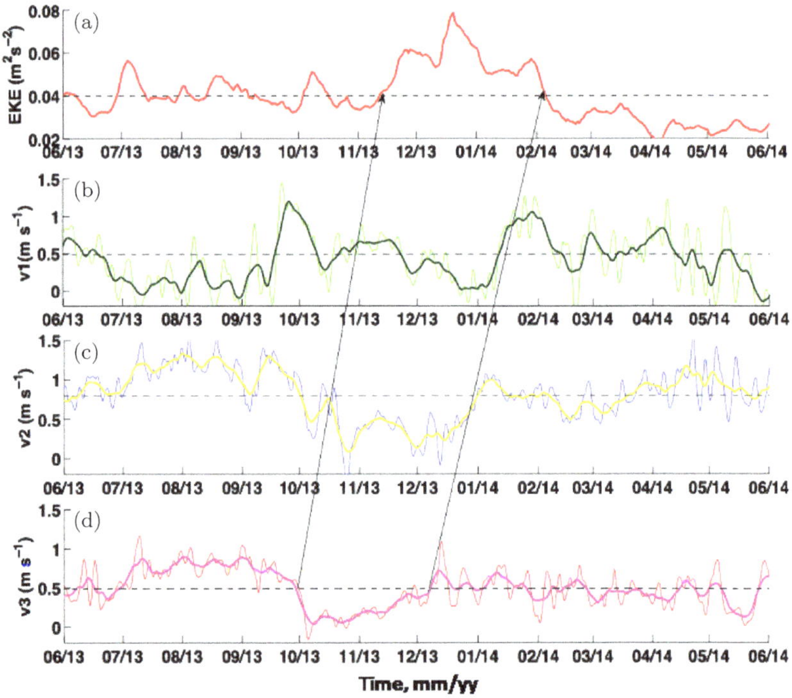

Fig. 7.10. Comparison of time series of (a) EKE in the test area with the northward velocity components (b) v1, (c) v2 and (d) v3 measured at 50 m at three mooring stations S1 (121.72°E, 23.87°N), S2 (121.99°E, 23.82°N) and S3 (122.35°E, 23.74°N) from 1 June 2013 to 31 May 2014. Thin and thick lines represent the weekly mean and three-point boxcar running mean, respectively. Two rays show the period for the Kuroshio path cut-off by CE.

Source: Taken from Zheng *et al.* (2019).

115° to 120.5°E, i.e., the white box marked in Figure 7.9, defined as

$$EKE = \frac{1}{2}\left(u_{ij}'^2 + v_{ij}'^2\right) \tag{7.6}$$

where (u_{ij}', v_{ij}') is the geostrophic velocity anomaly at grid i, j computed with respect to a 20-year (1993–2012) mean (downloaded from https://www.aviso.altimetry.fr/en/data/products/sea-surface-height-products/global/msla-uv.html). The grid size is 1°/4 by 1°/4. Figure 7.10 shows comparison of the time series of calculated EKE in the test area with the northward velocity components measured at 50 m at three mooring stations S1 (121.72°E, 23.87°N), S2 (121.99°E, 23.82°N) and S3 (122.35°E, 23.74°N) deployed across the mean Kuroshio path east of the Taiwan Island from 1 June 2013 to 31 May 2014. One can see that during the period for the event of Kuroshio path cut-off by CE lasting for about 2 months from 1 October to 1 December at S3 and from 15 October to 31 December at S1 and S2, the northward velocities v2 and v3 at two offshore stations S2 and S3 decreased 80–100%. At station S1 nearest the shore, v1 decreased below the annual mean in late November 2013, and lasted till January 2014. Correspondingly the peak EKE in the test area of SCS increased 100% with a time delay of one month with respect to the time for the Kuroshio cut-off at S3, implying that the BIC transported the massive EKE from the Pacific eddies to the SCS.

7.4 Interaction of the Kuroshio with Eddies from the Pacific

From the analysis results in Section 7.3, one can see that in most cases of the Kuroshio intrusion into the SCS, energetic eddies from the Pacific play a key role of controller behind the phenomena. Thus, this section examines interaction of the Kuroshio with the E_{WB} approaching the EEZ from the Pacific.

7.4.1 *Statistics of the E_{WB}*

Using the ocean eddy-tracking data set produced by the satellite SSH-based eddy identification algorithm (Chelton *et al.*, 2011), Cheng *et al.* (2017) examined statistical distributions of trajectories of eddies approaching western boundary (E_{WB}) approaching the eddy effect zone (EEZ), which is defined as a belt 250 km from Kuroshio mean axis, from October 1992 to April 2012.

They found that eddy trajectories show that eddies frequently intrude into the Kuroshio region at two latitude bands 18°–19°N and 22°–23°N, with periods of 146 ± 62 and 165 ± 46 days, respectively. These westward propagating eddies from the Pacific comprise significant disturbance sources to modify the Kuroshio path, force the Kuroshio intrusion into the SCS and directly generate mesoscale eddies in the SCS (see Section 6.2 for details).

7.4.2 *Dynamic criteria for the Kuroshio-E_{WB} interaction*

Physically, interaction of the Kuroshio with impinging eddies is a problem of collision of two moving water bodies in the same medium. Therefore, the total momentum of individual water body is a suitable parameter to characterize its role during collision (Zheng *et al.*, 2019). The amplitude of total momentum vector, carried by a segment of the Kuroshio with the same length as the impinging eddy diameter $2R$, can be calculated in a Cartesian coordinate system as (Zheng *et al.*, 2011)

$$M_k = \int_0^{L_k} \int_0^{2R} \int_o^{D_k} \rho(x, y, z) v_k(x, y, z) dx dy dz, \tag{7.7}$$

where integral limits $L_k, 2R$, and D_k represent the width, the length, and the depth of the Kuroshio, $\rho(x, y, z)$ and $v_k(x, y, z)$ are the water density and the velocity vector amplitude, respectively. Using the intermediate value theorem for integral calculation and assuming a homogeneous ocean (a constant density ρ) yield

$$M_k = 2\rho R L_k D_k V_k, \tag{7.8}$$

where V_k is the intermediate value of mean northward velocity vector amplitude.

For the eddy, the total momentum is a vector sum of the horizontal motion momentum and the rotational momentum. Using the same approach, the total momentum amplitude can be calculated in a cylindrical coordinate system as

$$M_e = \sqrt{M_{\text{horizontal}}^2 + M_{\text{rotation}}^2}$$
$$= \sqrt{(\pi\rho R^2 D_e V_e)^2 + \left(\int_0^{D_e}\int_0^{2\pi}\int_0^{R} \rho v_r(r,\theta,z) r dr d\theta dz\right)^2}$$
$$= \pi\rho R^2 D_e (V_e^2 + V_r^2)^{1/2}, \tag{7.9}$$

where R is the eddy radius, D_e is the eddy depth, V_e is the mean horizontal moving velocity amplitude, and V_r is the intermediate value of mean tangential velocity amplitude.

From Eqs. (7.8) and (7.9), the ratio of the total momentum amplitude of the Kuroshio to that of the eddy is defined as

$$\Gamma = \frac{2}{\pi}\left(\frac{D_k}{D_e}\right)\left(\frac{L_k}{R}\right)\frac{V_k}{(V_e^2 + V_r^2)^{1/2}}. \tag{7.10}$$

One can see that parameter Γ depends on not only the ratio of the eddy and the Kuroshio velocities, but also that of the horizontal and vertical scales.

Figure 7.11 shows graphic expressions of Γ vs. R in summer and winter, respectively. Parameters are taken as follows: D_k as 500 m in winter and 1000 m in summer (Zheng *et al.*, 2011), D_e as 2000 m

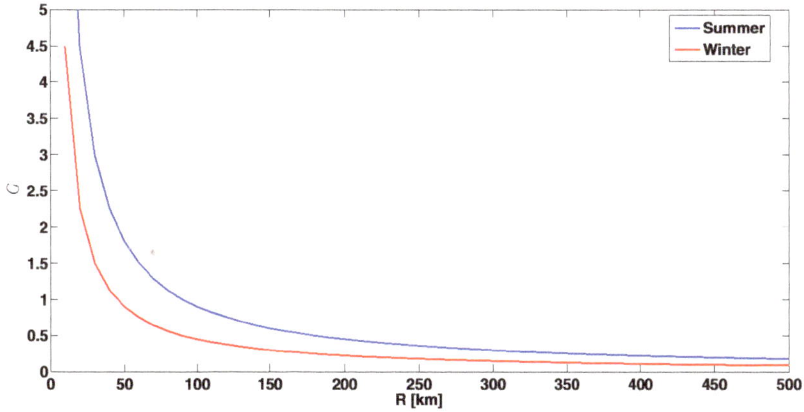

Fig. 7.11. Ratio of the total momentum amplitude of the Kuroshio to that of E_{WB} Γ vs. the eddy radius R.

Source: Taken from Zheng *et al.* (2019).

(Xie *et al.*, 2016; Zhang *et al.*, 2016), L_k as 100 km (Section 7.4.1), V_k as 0.4 m s^{-1} (Lien *et al.*, 2014; Tsai *et al.*, 2015), V_e as 0.1 m s^{-1} and V_r as 0.1 m s^{-1} (Zhang *et al.*, 2016; Zheng *et al.*, 2017). One can see that parameter Γ decreases quickly with the eddy radius. For R is 50 km in winter and 90 km in summer, the critical value of $\Gamma = 1$, implying that the total momentum of the Kuroshio balances that of the E_{WB}. Beyond the critical point, the Kuroshio plays a dominant role in the interaction for smaller R, while the E_{WB} plays a dominant role and enables to modify even to cut the Kuroshio path off for larger R, if they collide.

Note that the radii of all eddies shown in Figure 7.12(a) are larger than 50 km and that of 80% larger than 100 km. The annual mean number of E_{WB} is 16. Therefore, in the winter half a year, the mean number of E_{WB} stronger than the Kuroshio is 8, and in the summer half a year, that is 6. Namely, the annual mean number of strong E_{WB} is 14, i.e., the occurrence possibility of the Kuroshio path being modified by the E_{WB} is 1.2 times per month. Further consider that the Kuroshio-E_{WB} interaction durations reach (33 ± 10) days at 18°–19°N and (45 ± 17) days at 22°–23°N, implying that impact of the E_{WB} from the Pacific on the Kuroshio near the Luzon Strait sustains all year round.

7.5 Novel Mechanisms of the Kuroshio Intrusion

Lin *et al.* (2022) proposed a novel mechanism to explain the Kuroshio intrusion into the NSCS through the Luzon Strait based on satellite observations spanning 1993–2018. The impingement of CEs with the downstream Kuroshio current reduces northward inertial advection and triggers the Kuroshio intrusion in the Luzon Strait. This mechanism introduces a backward feedback process, where downstream changes in the Kuroshio influence upstream dynamics at the Luzon Strait. They analyzed satellite altimeter data and used the outputs from HYCOM/NCODA numerical simulations to conduct dynamic diagnosis of the physical processes driving the looping path of the Kuroshio by the cut-off events east of the Taiwan Island using the vorticity equation. The main processes include: planetary vorticity stretching, which squeezes the planetary vorticity due to eddy

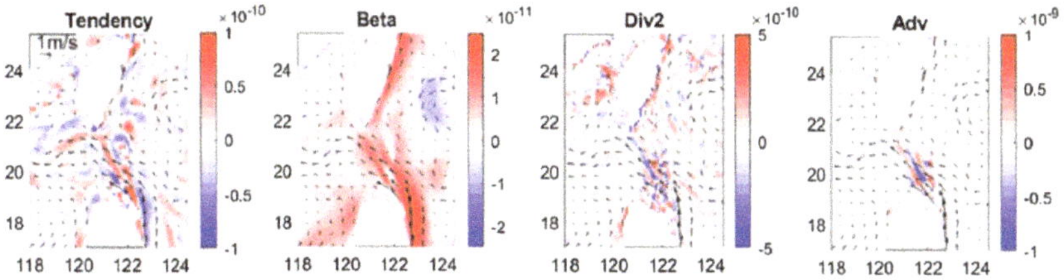

Fig. 7.12. Major terms (color shadings) in the vorticity equation, tendency ($\partial\zeta/\partial t$), ADV ($u\partial\zeta/\partial x + \nu\partial\zeta/\partial y$), beta ($\beta\nu$), and Div2 ($f(\partial u/\partial x + \partial v/\partial y)$) during the period of generation of Kuroshio intrusion in the Luzon Strait in response to the eddy impingement east of the Taiwan Island on 24 November 2000. Arrows denote background currents corresponding to certain dates. Because the relative vorticity ζ is usually much smaller than planetary vorticity f, term Div2 is the only stretching term shown in this analysis.

Source: Redrawn from Lin *et al.* (2022).

impingement and forces the Kuroshio to move westward into the SCS; advection of negative vorticity, whose interaction with CEs modifies the flow patterns both downstream and upstream, and the β effect contribution, which, although weaker overall, dominates on the left flank of the Kuroshio by reducing flow divergence and enhancing the intrusion. Figure 7.12 shows the major terms in the vorticity equation This new mechanism emphasizes how downstream features east of the Taiwan Island can critically modify upstream dynamics at the Luzon Strait. The findings refine existing theories on Kuroshio intrusion mechanisms, complementing factors like wind forcing, pressure gradients, β effect, potential vorticity conservation and eddy activities.

7.6 Summaries

This chapter summaries the research results on the Kuroshio's role in eddy generation in the SCS. Key points are concluded as follows.

(1) As a big gap of the western boundary of the North Pacific with a north-south width of $O(400)$ km provides a water channel for the western boundary current, the Kuroshio, to intrude the SCS.

The Kuroshio intrusion patterns are categorized into three patterns: loop current (LOC), jet flow (JEF) and bifurcating current (BIC). All three patterns shed mesoscale eddies in the SCS. The LOC is driven by two forcings: the northeast monsoon winds and modification of the Kuroshio by a couple of size-asymmetric eddies east of the Luzon Strait, a strong anticyclonic E_{WB} with the horizontal scale larger than $O(400)$ km in north and a weak cyclonic E_{WB} in south. The JEF and BIC patterns are driven by a couple of size-symmetric eddies east of the Luzon Strait, but the positions of anticyclonic E_{WB} and cyclonic E_{WB} switch each other.

(2) Energetic eddies from the Pacific approaching the EEZ, i.e., the western boundary from 13°N to 24°N within 250 km from the Kuroshio mean axis, play a key role of controller behind the phenomena of the Kuroshio intrusion into the SCS.

(3) Parameter Γ, the ratio of the total momentum of the Kuroshio to that of the eddy, is used as the dynamic criterion to determine the possible occurrence of modification of the Kuroshio by E_{WB}. Calculation results indicate that for eddy radius R is 50 km in winter and 90 km in summer, the critical value of $\Gamma = 1$, implying that the total momentum of the Kuroshio balances that of the E_{WB}. Beyond the critical point, the Kuroshio plays a dominant role in the interaction for smaller R, while the E_{WB} plays a dominant role and enables to modify even to cut the Kuroshio path off for larger R, if they collide. On the other hand, the occurrence possibility of the Kuroshio path being modified by the E_{WB} is 1.2 times per month and interaction durations reach (33±10) days at 18°–19°N and (45±17) days at 22°–23°N, implying that impact of the E_{WB} from the Pacific on the Kuroshio in the EEZ and meanwhile on eddy generation in the SCS sustains all year round.

(4) The above results can be used to support the following conclusions. The Kuroshio is the western boundary current of the North Pacific, therefore, its behavior is certainly controlled by the dynamic processes from the Pacific, such as westward-propagating energetic eddies and wave motions. Therefore, the Kuroshio is not an independent third party between the SCS and the Pacific, but an attachment of the Pacific. The intrusion into the SCS is not an active action of the Kuroshio but in most cases

forced by energetic eddies from the Pacific and seasonally driven by the northeast monsoon. Namely, in most cases, the Kuroshio path modification, intrusion into the SCS, and shedding eddies should be attributed to impacts of E_{WB} originating from the Pacific. Thus, the Kuroshio plays an intermediary role in eddy shedding in the SCS, which transfers the EKE of the Pacific to the SCS and provides high-temperature and high-salinity water to the eddy water mass.

(5) Novel mechanisms include feedback processes, where CEs impinging on the downstream Kuroshio trigger upstream intrusion into the Luzon Strait. These findings reveal how localized mesoscale disturbances can propagate effects throughout the connected ocean systems.

References

Andres, M., Mensah, V., Jan, S., Chang, M.-H., Yang, Y.-J., Lee, C. M., Ma, B. and Sanford, T. B. (2017). Downstream evolution of the Kuroshio's time-varying transport and velocity structure, *J. Geophys. Res. Oceans*, 122, pp. 3519–3542, doi:10.1002/2016JC012519.

Caruso, M. J., Gawwarkiewicz, G. G. and Beardsley, R. C. (2006). Interannual variability of the Kuroshio intrusion in the South China Sea, *J. Oceanogr.*, 62, pp. 559–575, doi:10.1007/s10872-006-0076-0.

Centurioni, L. R., Niiler, P. P. and Lee, D.-K. (2004). Observations of inflow of Philippine Sea surface water into the South China Sea through the Luzon Strait, *J. Phys. Oceanogr.*, 34, pp. 113–121, doi:10.1175/1520-0485(2004)034<0113:OOIOPS>2.0.CO;2.

Chelton, D. B., Schlax, M. G. and Samelson, R. M. (2011). Global observations of nonlinear mesoscale eddies, *Progr. Oceanogr.*, 91, pp. 167–216, doi:10.1016/j.pocean.2011.01.002.

Chen, G., Hou, Y. and Chu, X. (2011). Mesoscale eddies in the South China Sea: Mean properties, spatiotemporal variability, and impact on thermohaline structure, *J. Geophys. Res.*, 116, C06018, doi:10.1029/2011JD016244.

Cheng, Y.-H., Ho, C.-R., Zheng, Q., Qiu, B., Hu, J. and Kuo, N.-J. (2017). Statistical features of eddies approaching the Kuroshio east of Taiwan Island and Luzon Island, *J. Oceanogr.*, 73, pp. 427–438, doi:10.1007/s10872-017-0411-7.

Farris, A. and Wimbush, M. (1996). Wind-induced Kuroshio intrusion into the South China Sea, *J. Oceanogr.*, 52, pp. 771–784, doi:10.1007/BF02239465.

Ho, C.-R., Zheng, Q., Kuo, N.-J., Tsai, C.-H. and Huang, N. E. (2004). Observation of the Kuroshio intrusion region in the South China Sea from AVHRR data, *Int. J. Remote Sens.*, 25, pp. 4583–4591, doi:10.1080/0143116042000192376.

Ho, C.-R., Zheng, Z.-W., Hsu, P.-C. and Lu, C.-Y. (2020). Kuroshio variation and its impact on the South China Sea, Chapter 8 in *Regional Oceanography in the South China Sea*, (eds.) Hu, J. *et al.*, World Scientific, London, pp. 189–215, doi:10.1142/9789811206917_0008.

Hu, J., Zheng, Q. Sun, Z. and Tai C.-K. (2012). Penetration of nonlinear Rossby eddies into South China Sea evidenced by cruise data, *J. Geophys. Res.*, 117, C03010, doi:10.1029/2011JC007525.

Jan, S., Yang, Y. J., Wang, J., Mensah, V., Kuo, T.-H., Chiou, M.-D., Chern, C.-S., Chang, M.-H. and Chien, H. (2015). Large variability of the Kuroshio at 23.758°N east of Taiwan, *J. Geophys. Res. Oceans*, 120, pp. 1825–1840, doi:10.1002/2014JC010614.

Jia, Y. and Chassignet, E. P. (2011). Seasonal variation of eddy shedding from the Kuroshio intrusion in the Luzon Strait, *J. Oceanogr.*, 67, pp. 601–611, doi:10.1007/s10872-011-0060-1.

Li, R.-F., Zeng, Q.-C., Gan, Z.-J. and Wang, W.-Z. (1993). Numerical simulation of the South China Sea Warm Current and the Taiwan Strait flow in winter, *Progr. Natl. Sci.*, 3, pp. 123–129.

Li, L., Nowlin, W. D. Jr. and Su, J. (1998). Anticyclonic rings from the Kuroshio in the South China Sea, *Deep-Sea Res. I*, 45, pp. 1469–1482, doi:10.1016/S0967-0637(98)00026-0.

Li, L. and Wu, B. Y. (1989). A Kuroshio loop in South China Sea?–On circulations of the north-eastern South China Sea, *J. Oceanogr. Taiwan Str.*, 8(1), pp. 89–95.

Lien, R.-C., Ma, B., Cheng, Y.-H., Ho, C.-R., Qiu, B., Lee, C. M. and Chang, M.-H. (2014). Modulation of Kuroshio transport by mesoscale eddies at the Luzon Strait entrance, *J. Geophys. Res. Oceans*, 119, pp. 2129–2142, doi:10.1002/2013JC009548.

Lin, J. Y., Zheng, Z. W., Zheng, Q., Wu, D. R., Gopalakrishnan, G., Ho, C. R., Pan, J., Lin, Y. C. and Xie, L. L. (2022). Satellite observed new mechanism of Kuroshio intrusion into the northern South China Sea, *Int. J. Appl. Earth Obs. Geoinf.*, 115, 103119, doi:10.1016/j.jag.2022.103119.

McMahon, C. W., Kuehl, J. J. and Sheremet, V. A. (2021). Dynamics of gap-leaping western boundary currents with throughflow forcing, *J. Phys. Oceanogr.*, 51, pp. 2243–2256, doi:10.1175/JPO-D-20-0216.1.

Metzger, E. J. and Hurlburt, H. E. (2001). The nondeterministic nature of Kuroshio penetration and eddy shedding in the South China Sea, *J. Phys. Oceanogr.*, 31(7), pp. 1712–1732, doi:10.1175/1520-0485(2001)031<1712:TNNOKP>2.0.CO;2.

Nan, F., Xue, H., Chai, F., Shi, L., Shi, M. and Guo, P. (2011a). Identification of different types of Kuroshio intrusion into the South China Sea, *Ocean Dyn.*, doi:10.1007/s10236-011-0426-3.

Nan, F., Xue, H., Xiu, P., Chai, F., Shi, M. and Guo, P. (2011b). Oceanic eddy formation and propagation southwest of Taiwan, *J. Geophys. Res.*, 116, C12045, doi:10.1029/2011JC007386.

Nitani, H. (1972). *Beginning of the Kuroshio, in Kuroshio, Its Physical Aspects*, eds. Stommel, H. and Yoshida, K., Univ. of Tokyo Press, Tokyo, Japan, pp. 129–163.

Qiu, B. (2001). Kuroshio and Oyashio currents, in *Encyclopedia of Ocean Sciences*, eds. Steele, J. H. and Turekian, K. K., Elsevier, New York, pp. 1413–1425, doi:10.1006/rwos.2001.0350.

Qiu, B. and Chen, S. (2010). Interannual variability of the North Pacific Subtropical Countercurrent and its associated mesoscale eddy field, *J. Phys. Oceanogr.*, 40, pp. 213–225, doi:10.1175/2009JPO4285.1.

Qiu, B., Chen, S. and Kessler, W. S. (2009). Source of the 70-day mesoscale eddy variability in the Coral Sea and the North Fiji Basin, *J. Phys. Oceanogr.*, 39, pp. 404–420, doi:10.1175/2008JPO3988.1.

Qu, T. (2000). Upper-layer circulation in the South China Sea, *J. Phys. Oceanogr.*, 30, pp. 1450–1460, doi:10.1175/1520-0485(2000)030 <1450:ULCITS>2.0.CO;2.

Tsai, C.-J., Andres, M., Jan, S., Mensah, V., Sanford, T. B., Lien, R.-C. and Lee, C. M. (2015). Eddy-Kuroshio interaction processes revealed by mooring observations off Taiwan and Luzon, *Geophys. Res. Lett.*, 42, pp. 8098–8105, doi:10.1002/2015GL065814.

Wang, J. and Chern, C-S. (1987). The warm-core eddy in the northern South China Sea, I. Preliminary observations on the warm-core eddy, *Acta Oceanogr. Taiwan*, 18, pp. 92–103.

Wang, G., Su, J. and Chu, P. C. (2003). Mesoscale eddies in the South China Sea observed with altimeter data, *Geophys. Res. Lett.*, 30(21), pp. 1221–1226, doi:10.1029/2003GL018532.

Xie, L., Zheng, Q., Tian, J., Zhang, S., Feng, Y. and Yi, X. (2016). Cruise observation of Rossby waves with finite wavelengths propagating from the Pacific to the South China Sea, *J. Phys. Oceanogr.*, 46(10), pp. 2897–2913, doi:10.1175/JPO-D-16-0071.1.

Yaremchuk, M. and Qu, T. (2004). Seasonal variability of the large-scale currents near the coast of the Philippines, *J. Phys. Oceanogr.*, 34, pp. 844–855, doi:10.1175/1520-0485(2004)034<0844:SVOTLC>2.0.CO;2.

Zhao, R., Zhu, X.-H., Zhang, C., Zheng, H., Zhu, Z.-N., Ren, Q., Liu, Y., Nan, F. and Yu, F. (2023). Summer anticyclonic eddies carrying Kuroshio waters observed by a large CPIES array west of the Luzon Strait, *J. Phys. Oceanogr.*, 53, pp. 341–359, doi:10.1175/JPO-D-22-0019.1.

Zhang, Z., Zhang, Y., Wang, W. and Huang, R. X. (2013). Universal structure of mesoscale eddies in the ocean, *Geophys. Res. Lett.*, 40, pp. 3677–3681, doi:10.1002/grl.50736.

Zhang, Z., Tian, J., Qiu, B., Zhao, W., Chang, P., Wu, D. and Wan, X. (2016). Observed 3D structure, generation, and dissipation of oceanic mesoscale eddies in the South China Sea, *Sci. Rep.*, 6, 24349, doi:10.1038/srep24349.

Zhang, Z., Zhao, W., Qiu B. and Tian, J. (2017). Anticyclonic eddy shedding from Kuroshio loop and the accompanying cyclonic eddy in the northeastern South China Sea, *J. Phys. Oceanogr.*, 47, pp. 1243–1259, doi:10.1175/JPO-D-16-0185.1.

Zheng Q., Ho, C.-R., Xie, L. and Li M. (2019). A case study of a Kuroshio main path cut-off event and impacts on the South China Sea in fall-winter 2013-2014, *Acta Oceanol. Sin.*, 38(4), pp. 1–8, doi:10.1007/s13131-019-1411-9.

Zheng, Q., Tai, C.-K., Hu, J., Lin, H., Zhang, R., Su, F.-C. and Yang, X. (2011). Satellite altimeter observations of nonlinear Rossby eddy-Kuroshio interaction at the Luzon Strait, *J. Oceanogr.*, 67, pp. 365–376, doi:10.1007/s10872-011-0035-2.

Zheng, Q., Xie, L., Zheng, Z.-W. and Hu, J. (2017). Progress in research of mesoscale eddies in the South China Sea, *Adv. Mar. Sci.*, 35, pp. 131–158. (in Chinese with English abstract).

Zhuang Z., Hui, Z., Yang, G., Zhao, X. and Yuan, Y. (2020). Principal-component estimates of the Kuroshio Current axis and path based on the mathematical verification between satellite altimeter and drifting buoy data, *Acta Oceanol. Sin.*, 39(1), pp. 14–24, doi:10.1007/s13131-019-1523-2.

Chapter 8

Typhoon Forced Ocean Mesoscale Eddies

8.1 Introduction

Typhoon, an alias of hurricane in the western North Pacific (WNP), is featured by strong winds ($\geq$64 knots or 33 m s^{-1}) and a low-pressure center. As a rapidly rotating storm system, typhoon injects a huge amount of momentum and energy episodically into the ocean. Thus, the ocean inexorably has strong response to this powerful atmospheric forcing. Previous investigators have reported typhoon impacts on the behavior of ocean eddies. For example, Sun *et al.* (2014) examined the effects of super typhoons on the pre-typhoon ocean cyclonic eddies (CEs) in the WNP using satellite data and Argo floats. Cheng *et al.* (2018) investigated mesoscale CEs induced or strengthened by slow-moving super typhoons using satellite SSH data. An important progress obtained recently is that Lin *et al.* (2025) provided direct evidence that confirms the linkage between tropical cyclone (TC) passage and the subsequent generation of CEs.

The South China Sea (SCS) occupies the southwestern WNP with dense typhoon tracks. Meanwhile it is a sea area with broad distribution of massive and energetic eddies (ref. Chapters 1 and 11). Zhang *et al.* (2022) examined statistical distribution features of mesoscale eddies entering the northern continental shelf of SCS (NCSS). The further analysis of their results reveals that the seasonal distribution of eddy numbers has a close linkage with the seasonal variations of typhoon forcing index (TFI) (Wang, 2020). Linear regression correlation coefficients between the two reach as high as 0.91 for CEs and 0.53 for anticyclonic eddies (AEs). For the intruding shelf type of CEs, the typhoon forcing is a dominant one among typhoon, Kuroshio intrusion and monsoon forcing. The correlation coefficient reaches 0.82, which is 2.3 and 4.6 times higher than that of other two (ref. Chapter 12). This chapter focuses on recent research results and evidence, which help to confirm that the typhoon forcing is one of generation mechanisms for mesoscale eddies in the SCS.

8.2 Statistics of Typhoon Tracks in the WNP

Figure 8.1 shows statistics of tropic cyclone (TC) tracks in the WNP (0°–50°N, 105°–180°E) derived from the best-track data

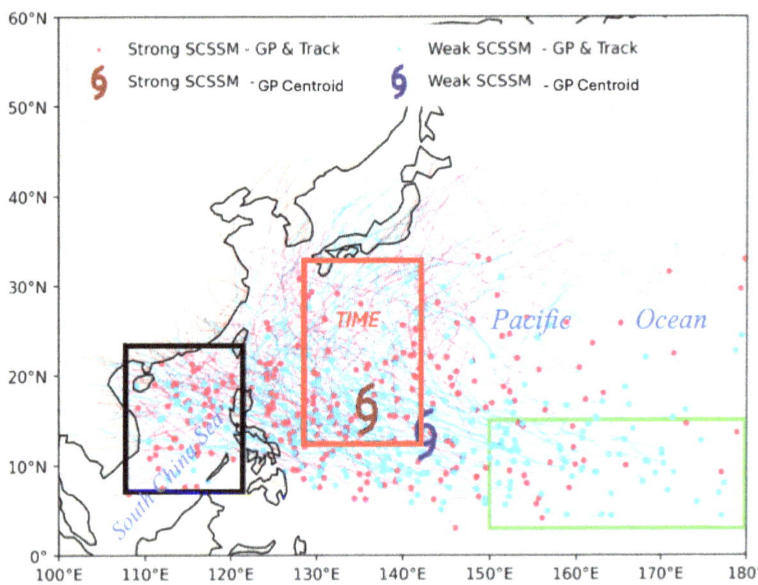

Fig. 8.1. Genesis positions (GPs) and activity ranges of TCs over the WNP in the TC season from 1979 to 2020. Curves represent the TC tracks, dots the GP, and cyclone symbols the centroid of horizontal distribution of TC genesis positions. Pink and light blue colors distinguish strong and weak SCS summer monsoon (SCSSM) years. Black, red (TIME) and green boxes represent the SCS, the study area for typhoon forced CEs shown in Figure 8.2 and far zone of CT genesis, respectively.

Source: Taken and redrawn from Liu *et al.* (2023).

obtained from the Joint Typhoon Warning Center (JTWC), archived in the International Best Track Archive for Climate Stewardship (IBTrACS) v04r00 (Knapp *et al.*, 2010). The CTs (including typhoons) that occur in the TC season from May to December (Wang *et al.*, 2007) and have maximum wind speeds of at least 25 knots (13 m s^{-1}) during the period from 1979 to 2020 are counted (Liu *et al.*, 2023).

From Figure 8.1, one can see that the SCS sits at the westmost end of a belt between 10°N and 20°N with the densest TC tracks not only originating from the Pacific but also from locally generated TCs. According to statistics of TC track records from 1981 to 2010, average 11 TCs per year passed over the SCS (Zheng *et al.*, 2015; Wang, 2020). Chapters 1 and 11 of this book show that the SCS is also featured by broad distribution of massive and energetic ocean eddies. Thus, it is of significance to clarify the dynamic relations between the two. Hereafter, the discussion will focus on the causality relation, i.e., the typhoon serves as an external forcing to generate ocean eddies.

8.3 Direct Evidence for Typhoon Forced CEs

8.3.1 *Satellite altimeter observations*

To find the direct evidence for typhoon-induced mesoscale cyclonic eddies (TIME), Lin *et al.* (2025) identified 69 potential candidates using global mesoscale eddy trajectory atlas and JTWC typhoon data from 1995 to 2018. The search area covering a range of 13°–32°N and 128°–145°E in the WNP as shown as the red box (TIME) in Figure 8.1. The new atlas is used to detect the appearance and eradication of mesoscale eddies during individual TC passages. The atlas contains the data and information on eddy position, amplitude, speed, effective radius, sea surface height (SSH) contours and speed profile derived from multi-satellite altimetry merged data including merged sea surface height anomaly (SSHA) and absolute geostrophic current (AGC) with a spatial resolution of 0.25° by 0.25° (Mason *et al.*, 2014). The atlas is distributed by the Archiving, Validation, and Interpretation of Satellite Data in Oceanographic (AVISO), Centre National d'Etudes Spatiales (CNES) of France

(AVISO, 2008; Pegliasco *et al.*, 2022). The daily SSHA is used to systematically characterize the instantaneous state of the upper ocean under the influence of TC passages, including the evolution process, strengths, sizes and kinetic energies of eddies (Lin *et al.*, 2025).

To identify potential TIME events, Lin *et al.* (2025) selected 69 TCs with intensities stronger than Category 1 on the Saffir-Simpson hurricane wind scale to match with the global mesoscale eddy trajectory atlas from 1995 to 2018. To find the most representative candidates from selected 69 TCs for dynamic reconstruction, a series of selection procedures are performed: (1) Initiation of the CE must strictly coincide with TC passage within a 1° square area. (2) The sea surface characteristic in neutral condition must last for more than 5 consecutive days before a CE is generated. (3) The spin-up CEs must have an amplitude greater than 7.5 cm (Wang *et al.*, 2003). (4) The negative SSH anomaly (stronger than 7.5 cm) must last for more than 14 consecutive days following the TC passage. (5) The CEs must appear in the open ocean rather than in shelf or coastal regions. Only events meeting all these criteria simultaneously are considered as representative TIME candidates for further analysis. During the study period, five events corresponding to typhoons Rosie (1997), Haitang (2005), Nida (2009), Ma-On (2011) and Nuri (2014) are identified as possible TIME events as listed in Table 8.1.

To obtain reliable and determinative conclusions, three TIME candidates with the largest SSHA amplitudes associated with

Table 8.1. Information of the five TIME candidates.

TC name	CE genesis date	CE position lat. lon.	CE SSHA amp diff*, cm	TC category	Transit speed**, m s^{-1}
Rosie	22 Jul 1997	17.5°N 132.6°E	22.6	3	3.06
Haitang	16 Jul 2005	20.7°N 127.7°E	8	5	5.75
Nida	29 Nov 2009	19.3°N 139.4°E	22.3	4	1.52
Ma-On	17 Jul 2011	26.0°N 133.5°E	11.5	2	5.47
Nuri	5 Nov 2014	26.4°N 138.8°E	8.3	1	5.68

Notes: *CE amplitude difference is defined as the max SSHA difference within 5 days before and after TC passage.
**Transit speed is defined as the TC speed averaged over interval from entering to leaving CE genesis position.

typhoons Rosie, Nida and Ma-On (Table 8.1) are selected for the further analysis as shown in column A in Figure 8.2. One can see the common features of the three representative cases as follows. (1) Typhoon tows a negative SSHA wake on the left side against the typhoon track. The negative wake has a horizontal scale within a range from 200 to 500 km, and spatially variable SSHA within a range from 0 to -0.7 m. (2) The variable negative SSHA wake may be enhanced and develop to strong CEs at certain locations. The dynamic conditions will be re-constructed with numerical models in the next section. (3) Examining the SSHA imagery features, one can see that typhoon also tows a positive SSHA wake on the right side against the typhoon track. The positive wake has the same horizontal scale as that of negative one, but a little small range of variable SSHA from 0 to 0.5 m. The variable positive SSHA wake may also be enhanced and develop to strong AEs at locations corresponding to CEs. In the other words, typhoon behaves like a ship propeller, which generate the wakes on both sides of navigating track.

8.3.2 *Numerical simulations*

The causal relationship between TCs and the resulting CEs was systematically examined through a series of semi-idealized experiments (Lin *et al.*, 2025). Three TIME candidates corresponding to typhoons Rosie (1997), Nida (2009) and Ma-On (2011) were reconstructed using the Regional Ocean Modeling System (ROMS) through standard experiments $\text{Rosie}_{\text{STD}}$, Nida_{STD} and $\text{Ma-On}_{\text{STD}}$ (all wind forcing (MERRA-2) and current ocean conditions (HYCOM/NCODA)), respectively. The simulated results are shown in Column B in Figure 8.2. One can see that satellite-observed SSHA and AGC patterns in Column A and outputs from standard runs in Column B coincide very well. Take typhoon Nida as an example. As the strongest typhoon in 2009, Nida formed near 6°N on 21 November 2009 and headed northwestward. On 25 November, it rapidly intensified into a Category 5 typhoon, then slightly weakened to Category 4. By 27 November, Nida re- intensified to Category 5 and moved slowly around 19°N, 138°E. Subsequently, Nida lingered for nearly 2 days, and a strong CE formed beneath its path. Figure 8.3 shows the model-simulated evolution of the CE generated during the passage of typhoon Nida. One can see that the model-simulated characteristics

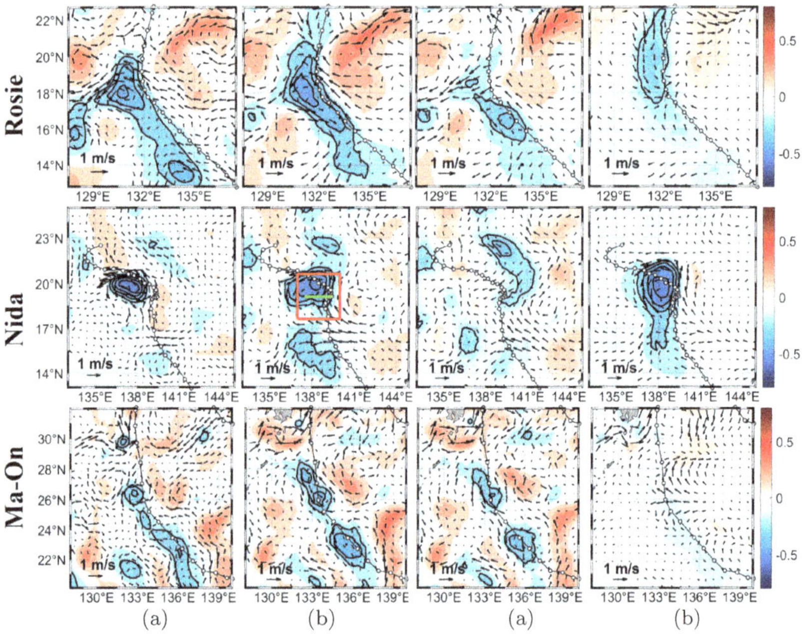

Fig. 8.2. Column (a): satellite-observed SSHA (m) images with absolute geostrophic current (AGCs) (ms^{-1}). Column (b): outputs from standard runs of numerical simulation experiments. Columns (c) and (d): outputs from idealized experiments EXP_{NOTC} and EXP_{TCONLY}. Upper, middle and bottom rows show the mature stage of CEs corresponding to typhoons Rosie, Nida and Ma-On. Color shadings represent the SSHAs, while black contours indicate the possible CE centers with negative SSHAs. Black lines with hollow circles show the typhoon tracks. Green line in red box in Column B for typhoon Nida marks the transect location for the energy analysis.

Source: Taken from Lin *et al.* (2025).

and strengths of the negative SSHAs and current responses show considerable consistency with satellite observations, once the CE reaches its mature stage in the geostrophic balance status. Meanwhile, sustaining of the CE after typhoon passage indicates its ocean wake nature.

Moreover, in addition to standard experiments, two more semi-idealized experiments EXP_{NOTC} (excluded TC wind forcing ($4° \times 4°$ area at the TC center) and current ocean conditions (HYCOM/NCODA)) and EXP_{TCONLY} (all wind forcing (MERRA-2) and climatic ocean conditions (Global Ocean Physics Reanalysis))

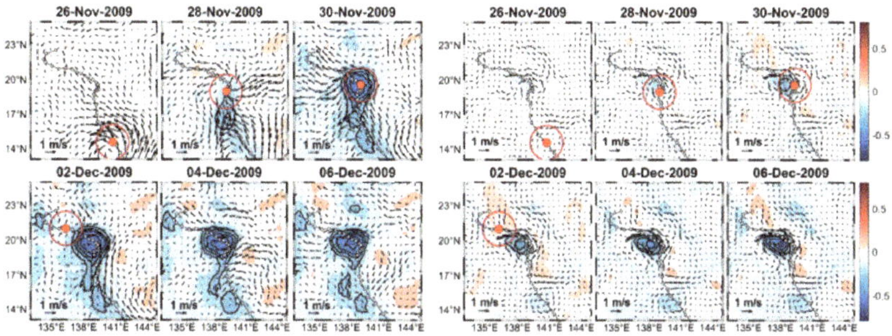

(a) (b)

Fig. 8.3. Progression of CE under typhoon Nida (2009). (a) ROMS model-simulated evolution of SSHAs and upper-layer currents. Arrows depict depth-averaged (0–50 m) current velocities (m s^{-1}) during the typhoon passage. (b) Satellite observations of SSHAs and AGCs. Color shading represents SSHAs (m) relative to 24 November 2009. Black contours represent SSHA levels, ranging from −0.20 m with 0.10 m intervals. Black lines show the typhoon tracks. Red dots with circles mark the typhoon center positions on each day (00 UTC). Arrows denote AGCs (m s^{-1}).

Source: Taken from Lin *et al.* (2025).

were designed and conducted to systematically examine the causal relationship between TCs and the resulting CEs. The results from these semi-idealized experiments are shown in Columns C and D of Figure 8.2, respectively. A realistic TC-induced CE process can be obtained for detailed analysis by cross-comparing CEs from all experiments (see Figure 8.2). First, the comparison reveals that the CEs during typhoons Rosie and Ma- On were primarily triggered by ocean dynamic conditions. This is evident from the fact that the characteristics of CEs for these two events were reproduced even without the contribution of TC wind forcing (Column C in Figure 8.2). During typhoon Rosie, wind forcing (as shown in Column D) contributed to some variations in SSHA, but the ocean dynamic environments also played a significant role. Therefore, these two cases do not represent ideal TIME events. In contrast, in the case of Nida, typhoon wind forcing plays a dominant role in the generation of CE (Figure 8.2). In addition, $\text{Nida}_{\text{TCONLY}}$ exhibits the patterns like Nida_{STD}, indicating that the ocean forcing contributes less to CE generation compared to TC wind forcing. This suggests that the wind forcing associated with typhoon Nida serves as a primary trigger for a realistic TIME, with minimal influences from ocean conditions, such as preexisting eddies, currents, jets, upwelling and stratification.

Compared to the other two TC cases, therefore, Nida provides a more favorable environment for an in-depth exploration of the dynamic process of TIME. Figure 8.4 shows the evolution of the model-simulated horizontal velocity and temperature fields from Nida_{STD} at selected depths of 0 (top-level), 40 (mid-level), and 80 m (bottom-level) during the passage of typhoon Nida. These depths are chosen to represent the characteristics of the mixed layer (ML), the ML-thermocline interface, and the upper thermocline. Driven by TC Nida, a complete three-dimensional (3D) CE structure forms and extends throughout the water column gradually as shown in Figure 8.4(a–c).

The simulation results in Figure 8.4(a–c) show that the TIME was initially generated out of nothing. The ocean current response to typhoon Nida can be divided into three major components. (1) Flow divergence: the momentum response primarily involves current divergence in the ML around the TC footprint, resulting in net water transport outward from the TC center (top-level of Figure 8.4(a) and (b)). At these times, the structure of a CE is not evident in the current field. Additionally, the current divergence in the ML generates significant vertical shear across the ML base and the top of the seasonal thermocline. This vertical current shear drives the vertical

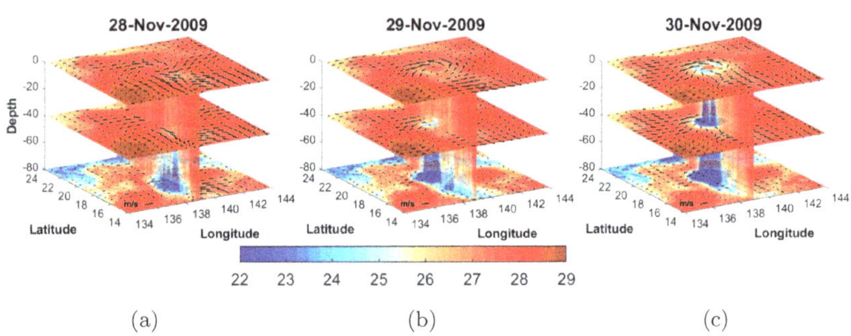

Fig. 8.4. 3D evolution images of current and temperature responses to typhoon Nida derived from standard experiment Nida_{STD}. Three sections show the current velocity (vectors in m s^{-1}) and temperature (color shading in °C) fields at selected depths of (a) 0 (top-level), (b) 40 (mid-level) and (c) 80 m (bottom-level). The current scale is marked in the lower left corner of the bottom-level.

Source: Taken from Lin *et al.* (2025).

mixing process (Price, 1981; Shay *et al.*, 2000), inducing substantial cooling (bottom- level of Figure 8.4(a)). (2) Ekman pumping: the spatial variability of the wind at the sea surface results in varying Ekman transports. To conserve mass, this spatial variability generates vertical velocities at the top of the Ekman layer. Consequently, the current divergence in the ML tied to positive wind stress curl (WSC) leads to upwelling of cooler water from the lower layers. (3) Geostrophic balance: on 29 November, a cyclonic current pattern associated with geostrophic balance began to develop. Concurrently, the SSH amplitude tended to decrease drastically (Figure 8.5), indicating a stable barotropic response due to surface divergence in the ML combined with a baroclinic response resulting from the uplift of colder, higher-density subsurface water (Geisler, 1970; Ginis and Sutyrin, 1995). With the difference in SSH, a pressure gradient force was generated to establish a geostrophic flow. It is evident that the surface current transforms from being divergent (as shown in top-level of Figure 8.4(a)) to exhibiting a rotating eddy (top-level of Figure 8.4(c)). It is a perfect physics solution of the dynamic problem of TIME in the form of 3D numerical expression.

Figure 8.5 shows temporal variation of the SSH along a zonal section across the CE center at 18.7°N, 139.0°E derived from simulation experiment Nida_{STD}. One can see that the typhoon forced CE was initially generated on 28 November 2009, reached its peak phase on 30 November and sustained till December 10 and afterward. Meanwhile, a high SSH signal was concurrent on the east side against CE.

To better understand the dynamic linkage between TC-induced currents and baroclinicity during the CE generation, the kinetic energy (KE) and the available potential energy (APE) quantities are estimated. APE in the ocean is defined as the difference between the potential energy in the current state and that in the reference state (Lorenz, 1955; Huang, 1998; Luecke *et al.*, 2017; Rudzin and Chen, 2022). For instance, the formation of a CE due to TC passage will result in the increase of APE, because it represents an anomaly relative to the long-term climatological reference state. Additionally, the input of KE into the ocean from TC stirring causes isopycnals to tilt, increasing APE relative to the stable reference state (indicating conversion of KE to APE). KE and APE are computed at each grid point and averaged along the zonal transect through the CE center in the case of typhoon Nida (green line in middle Column B of

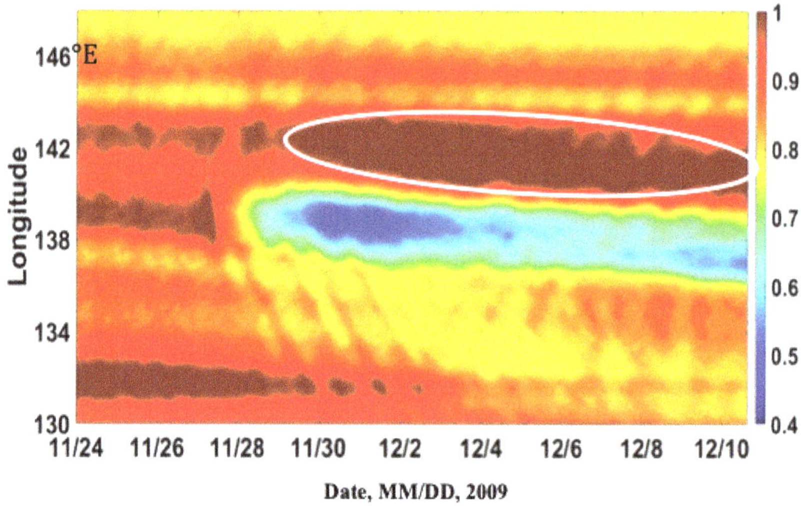

Fig. 8.5. Temporal variation of the SSH (in m) along a zonal section across the CE center at 18.7° N, 139.0° E derived from simulation experiment Nida_{STD}. White circle marks a high SSH signal concurrent with CE.

Source: Taken from Lin *et al.* (2025).

Figure 8.2) for the time-varying outputs, following the methods of Rudzin and Chen (2022):

$$KE = \frac{1}{2}\left(\sqrt{u^2 + v^2}\right)^2, \tag{8.1}$$

and

$$APE = -\frac{1}{2}\frac{g}{\rho_0}\frac{(\rho_{\text{in situ}} - \bar{\rho})^2}{\partial\bar{\rho}/\partial z}, \tag{8.2}$$

where u and v are the zonal and meridional current velocities (m s^{-1}), respectively, g is the gravitational acceleration ρ_0 is the reference density of seawater, $\rho_{\text{in situ}}$ is the instantaneous density calculated from the modeled temperature and salinity at each grid point, and ρ is the average density calculated from the temperature and salinity at each grid point using the monthly climatology from the Global Ocean Physics Reanalysis provided by the E.U. Copernicus Marine Environment Monitoring Service (CMEMS).

Figure 8.6 shows the time series of KE and APE across the CE of Nida at depths of 20, 40, 60, 80, 100 and 200 m. These depth

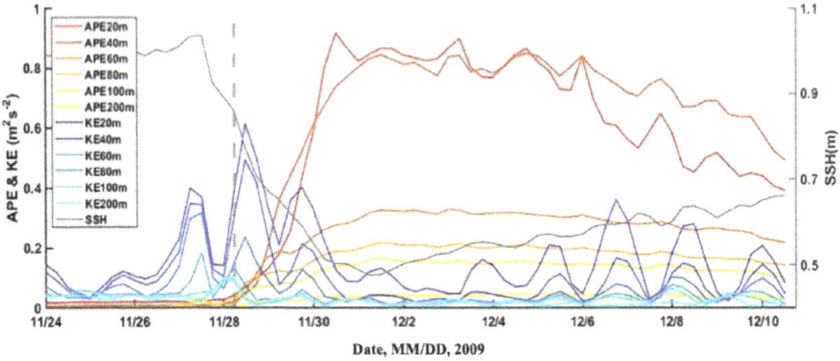

Fig. 8.6. Accumulation and transition processes of the averaged APE and KE at depths of 20, 40, 60, 80, 100 and 200 m during the passage of typhoon Nida derived from simulation experiment Nida$_{\mathrm{STD}}$ along the transect across the CE center as marked by green line in red box in Column B of Figure 8.2 for typhoon Nida. Curves in warm colors, cold colors and gray show temporal variations of APE, KE and SSH, respectively. Vertical gray dash line marks the noon time on 28 November when typhoon center passed over the CE location.

Source: Taken from Lin *et al.* (2025).

levels are selected based on the velocity fields, where the greatest velocity gradients occur. One can see that KE and APE are near zero before the TC passage at all depths. As typhoon Nida approached the location, where the CE formed, the currents started to respond, and KE rapidly increased at all depths and peaked as the typhoon center passed through at noon time on 28 November. This increase of KE is attributed to the typhoon wind forcing, which imparts momentum into the ML and creates uniform ML currents. KE peaked near-simultaneously occurring at all depths on 28 November as shown in Figure 8.6.

In the earlier progression of TIME by typhoon Nida, the Ekman pumping was triggered by a positive WSC induced by Nida. Later, the upwelling of deeper cold and denser waters due to Ekman pumping resulted in the inclination of the thermocline and an increase in baroclinity. According to Eq. (8.2), this process eventually causes the increase of APE. In Figure 8.6, KE at different depths rapidly and nearly simultaneously increased from 27 November to 29 November. Subsequently, KE begun to decrease and partially to convert into APE. This process aligns with the described TC wind forcing, which drives upwelling, inclines the thermocline and increases APE through enhanced baroclinicity.

After the KE–APE transitions, APE reached its maximum on 30 November just before the TC center left. At that time, KE is converted into APE, leading to the decrease of KE, which reached the lowest value. However, APE peaked at slightly different times at various depths, i.e., it showed a more pronounced peak at shallow depths compared to the deep water. This difference is attributed to the temperature gradient created by cold-water upwelling, which is the strongest near the sea surface and decreases with depth. By 30 November, KE had minimized, and turbulent mixing from wind forcing also diminished with the leaving TC center. Concurrently, APE at all depths reached maxima, indicating the mature stage of the CE with the strongest baroclinicity. The period from 28–30 November could be defined as the spin-up stage of the TIME event. After the TC center left on 1 December, APE at all depths gradually decreased and stabilized as wind-induced currents subsided. Additionally, the peaks of APE propagated from the surface to deep layers, reflecting the fact that wind forcing is concentrated in the upper ocean layers.

Figure 8.6 shows a significant phenomenon that the curves of temporal variations of KE occurred with strong fluctuations after typhoon passed over. The fluctuations as a dominant KE component have a period near local inertial period of 1.46 days, implying that they are the internal near-inertial waves (NIWs) induced by typhoon and propagate as ocean wake waves. The typhoon forced CE was concurrent with NIWs, implying its typhoon wake nature.

In summary, the mechanisms for the generation and evolution processes of the typhoon forced CE can be summarized as follows. During the typhoon passage, the momentum induced by typhoon stirs within the ML and gradually increases the KE. Ekman pumping is triggered by the positive WSC associated with the typhoon. The KE inputs into the ocean via Ekman pumping lead to inclination of isopycnals and increase in baroclinity, which in turn raise APE. The continuous conversion of KE into APE occurs during the initial stage of CE development from 28 to 30 November as shown in Figures 8.4 and 8.6. As the CE matures, it reaches a geostrophic balanced state characterized by the currents generated by positive WSC. Meanwhile, the upwelling of the cooler and denser subsurface water and the upper ocean transport directed away from the typhoon track all contribute to the establishment of geostrophic balance as shown in Figure 8.4 (Shay, 2009). Finally, once the typhoon leaves the CE location, APE gradually decreases back to zero at all depths, and the CE circulation weakens and eventually disappears (Rudzin and Chen, 2022).

8.4 Satellite Altimeter Observed Typhoon Forced AEs

Examining satellite altimeter images shown in Figure 8.2 and numerical simulation results shown in Figure 8.5 reveals that typhoon not only tows a negative SSHA wake on the left side against its track, but also a positive SSHA wake on the right side against the track concurrently. The positive wake has the same horizontal scale as that of negative one, but a little small range of variable SSHA from 0 to 0.5 m. The variable positive SSHA wake may also be enhanced and develop to strong AEs at locations corresponding to CEs. This section includes the analysis results in four cases of satellite altimeter along-track sea level anomaly (SLA) measured across typhoon centers in the SCS. The purposes are to clarify the occurrence of typhoon forced AEs and explore their dynamic features.

8.4.1 *Satellite altimeter measured along-track SLA across typhoon centers*

Satellite altimeter along-track SLA data used here were measured near simultaneously across the centers of four typhoons (tropic depression), TD28W, Nanmadol, Fengshen and Chanchu in the SCS deep basin deeper than 3000 m from 2000 to 2010. The data downloaded from website of AVISO are computed with respect to a seven-year mean. Meanwhilc, the corrections for dry troposphere, wet troposphere, ionosphere and tide, as well as multi-mission cross-calibration were applied to the data AVISO, 2008). The width of satellite altimeter antenna nadir footprint, i.e., horizontal resolution of satellite track on the sea surface, is about 10 km. The SLA accuracy is 4 cm and the temporal resolution is 1 s.

Figure 8.7(a–d) show the along-track SLA measured by Topex/Poseidon (T/P) (French–US altimeter satellite), Jason-1 (French–US altimeter satellite) and Envisat (European Remote

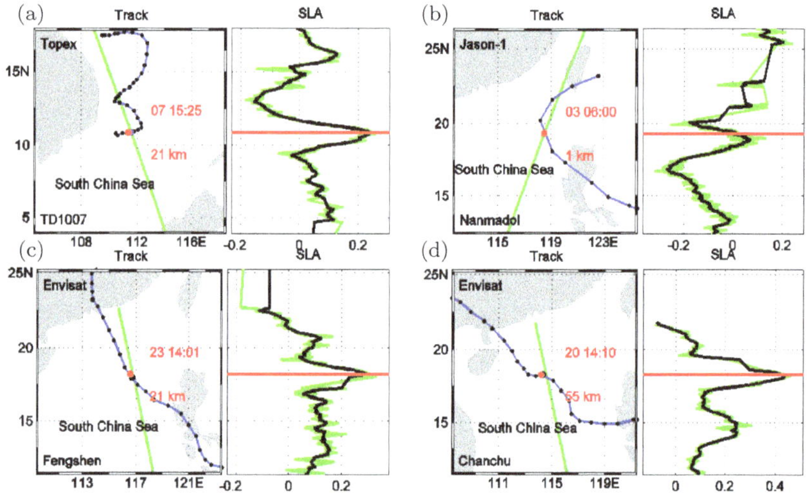

Fig. 8.7. Four cases of along-track SLA measurements (in m) near simultaneously across the centers of (a) tropic depression TD1007 from T/P on 7 October 2000, (b) typhoon Nanmadol from Jason-1 on 3 December 2004, (c) typhoon Fengshen from Envisat-2 on 23 June 2008 and (d) typhoon Chanchu from Envisat-2 on 20 July 2010, respectively. Numbers in red are: date UTC time and distance from the satellite ground track to the typhoon center. In left panel, green lines represent satellite tracks, blue lines represent typhoon tracks with red dots showing typhoon center locations at the time for satellites passed over. In right panel, green curves and solid lines represent raw and smoothed along-track SLA measurements within an interval of 20 min. Red lines mark the locations of typhoon centers.

Source: Taken from Zheng *et al.* (2015).

Sensing satellite) and the above four typhoon tracks: tropic depression TD1007 on 7 October 2000, typhoons Nanmadol on 3 December 2004, Fengshen on 23 June 2008 and Chanchu on 20 July 2010. The time intervals are shorter than 3 hours between times for typhoon center passed over and satellite ground track just or near scanned the typhoon centers. The information and parameters interpreted from the figures are listed in Table 8.2.

From Figure 8.7(a–d), one can see that in all the four cases, typhoon-forced sea level signatures appear as isolated, positive pulses on the right sides against the typhoon tracks (across the track in the case of Nanmadol). The average horizontal scale (characteristic half width) of the SLA pulse is (115 ± 12) km and the average

Table 8.2. Information and parameters in the four cases of satellite altimeter tracks near simultaneously scanning the typhoon centers.

Satellite cycle/ pass	Typhoon name	UTC Time mm/dd/yyyy hh:mm	Typhoon center N lat. E log.	Water depth m	Dist. to T-center km	Char. half width km	Peak ampl. cm
T/P 297/38	TD1007	10/07/2000 15:25	10.9° 111.5°	3396	21	99 ± 5*	40 ± 2*
Jason-1 107/51	Nanmadol	12/03/2004 06:00	19.2° 118.7°	3599	1	110 ± 5	33 ± 2
Envisat-2 69/798	Fengshen	06/23/2008 14:01	18.0° 116.4°	3906	21	124 ± 5	25 ± 2
Envisat-2 91/421	Chanchu	07/20/2010 14:10	18.3° 114.5°	3582	55	128 ± 5	38 ± 2

Note: $*\frac{1}{2}$ resolution.

peak amplitude is (34 ± 6) cm (Table 8.2). For the further analysis, function

$$\eta = \eta_p \text{sech}^2\left(\frac{y}{\Delta}\right), \tag{8.3}$$

where η_p is a peak value of SLA pulse, y-axis is along the satellite track with an origin at the symmetric point of the pulse, Δ is the characteristic half width, is used to fit the typhoon-forced along-track SLA pulses as shown in Figure 8.8. The correlation coefficients are as high as 0.97, 0.99, 0.94 and 0.93 for the four typhoons. Meanwhile, the characteristic half widths are determined as 99, 110, 124 and 128 km (Table 8.2), respectively (Zheng *et al.*, 2015).

8.4.2 *Derivation of AE dynamic parameters*

Based on the principles of geophysical fluid dynamics (Pedlosky, 1987), an isolated, positive SLA pulse forced by typhoon represents an ocean AE in the northern hemisphere. From along-track SLA curves in Figure 8.7(a–d), one can see that AEs are approximately co-centered with the typhoon centers (T-centers). However, the satellite tracks may not cross the T-center exactly. Thus, the dynamic parameters of AEs need to be calculated by the along-track SLA data. Figure 8.9 shows geometry of along-track SLA pulse and the AE. The circle represents the eddy boundary at a plane of SLA $\eta_1 = \eta_p \text{sech}^2(1) \approx 0.42\eta_p$ Parameters Δ_0 and s represent the characteristic half width of AE and the distance from the eddy center to

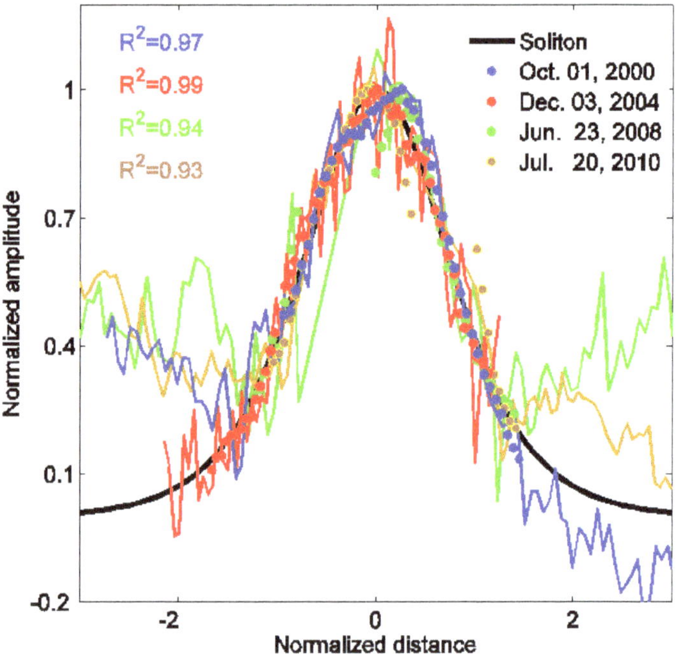

Fig. 8.8. Comparison of single soliton model (black solid curve) to normalized typhoon-forced sea level signatures (raw data: color curves, smoothed: dotted lines) in four cases shown in Figure 8.7(a–d).

Source: Taken from Zheng *et al.* (2015).

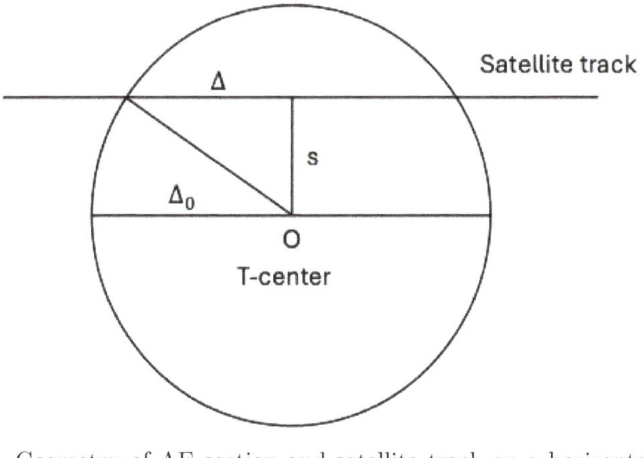

Fig. 8.9. Geometry of AE section and satellite track on a horizontal plane at SLA $\approx 0.42\eta_p$.

the satellite track. The geometry gives an approximate relation

$$\Delta_0 = (\Delta^2 + s^2)^{1/2}. \tag{8.4}$$

On the other hand, based on Eq. (8.3), the SLA amplitude of AE can be derived from

$$\eta_0 = \eta_p \text{sech}^{-2}\left(\frac{s}{\Delta_0}\right). \tag{8.5}$$

Furthermore, the tangential velocity of AE can be calculated using the geostrophic relation (Podlosky, 1987)

$$u = -\frac{g}{f}\frac{\partial \eta}{\partial y}, \tag{8.6}$$

where g is the gravitational acceleration (= 9.8 m s^{-2}), f is the Coriolis parameter (= $2\Omega \sin\phi$, where ϕ is the latitude), η is the sea level height, and y is the distance along the y-axis of the Cardician coordinate system positive northward. In the study case, an approximate form of Eq. (8.6) is used

$$u \approx \frac{g}{f}\frac{[1 - \text{sech}^2(1)]\eta_0}{\Delta_0}, \tag{8.7}$$

where $(1 - \text{sech}^2(1)) \approx 0.58$.

Using Eqs. (8.4), (8.5) and (8.7), dynamic parameters of typhoon forced AEs in the SCS are derived from the satellite along-track data listed in Table 8.2. The calculated results and dynamic parameters of typhoons are listed in Table 8.3. One can see that in the four cases of typhoons, satellite altimeters near simultaneously scanned AEs, which are featured by co-existence with typhoons (Li, 2020). Their mean characteristic half width or horizontal scale, SLA amplitude and tangential velocity are (119 ± 15) km, (36.2 ± 7.4) cm and (0.47 ± 0.23) m s^{-1}, respectively. These characteristic scales are within statistical mean ranges of dynamic parameters of mesoscale eddies in the SCS (see Chapter 11), and evidence that typhoons are important generators for mesoscale eddies in the SCS.

Table 8.3. Dynamic parameters of four typhoons and concurrent AEs in the SCS derived from the satellite along-track data from 2000 to 2010.

	Typhoon parameters				Eddy parameters		
Typhoon name	Highest wind speed m s^{-1}	Transit speed m s^{-1}	Lowest P. hP_a	f 10^{-5} s^{-1}	Δ_0 km	η_0 cm	u m s^{-1}
TD1007	15.6	2.1	998	2.75	101	41.8	0.85
Nanmadol	46.9	8.5	935	4.78	110	33	0.36
Fengshen	46.9	4.2	945	4.49	126	25.7	0.26
Chanchu	35.8	2.8	970	4.56	139	44.4	0.40
Mean					(119 ± 15)	(36.2 ± 7.4)	(0.47 ± 0.23)

8.4.3 *Statistical correlationship of eddy numbers with typhoons*

Chapter 12 of this book analyzes the statistical correlationship of eddy numbers in the northern continental shelf of the SCS (NCSS) with typhoon forcing index (TFI). The results show that the linear regression correlation coefficients (R^2) of CE and AE numbers to TFI_s reach as high as 0.91 and 0.53, respectively, implying a close causality relationship between the two.

8.5 Summary

This chapter specifies the role of typhoon played in generation of mesoscale eddies in the SCS. Satellite altimeter SSH images for three representative typhoons Rosie (1997), Nida (2009) and Ma-On (2011) near the SCS in the WNP, and along-track SLA data measured near simultaneously with passages of typhoon centers for four typhoons TD1007 (2000). Nanmadol (2004), Fengshen (2008) and Chanchu (2010) in the SCS are interpreted. The results indicate that typhoon generates the belt-like wakes in the ocean as long as 500–1000 km, which consist of a negative and a positive SSH branches aligned on the left and the right sides of the typhoon track, respectively. At certain locations of the two branches, the SSH anomalies develop to CEs and AEs. Time series satellite SSH images show that the maximum amplitudes of eddies are strengthened by −22.6, −22.3 and −11.8 cm for CEs and about 20 cm for AEs within 5 days before and after typhoon passages in the WNP, and about 20 cm for AEs within one day in the SCS. Some of eddies are generated from initial disturbances, and some out of nothing.

Numerical simulation experiment results verify that being a sole external forcing, the typhoon can generate an ocean CE out of nothing through effects of flow divergence, Ekman pumping and geostrophic balance. The SSH at the CE center is strengthened to 1 m from near zero within 2 days. The simulation results also verify that the typhoon ocean wakes have two branch structures: the negative SSH branch against the left side of typhoon track and the positive SSH branch against the right side.

References

AVISO (2008). *SSALTO/DUACS User Handbook: (M)SLA and (M)ADT Near-Real Time and Delayed Time Products, CLS-DOS-NT-06-034.* Toulouse, France.

Cheng, Y.-T., Tseng, R.-S., Chang, Y.-C. and Chen, J.-M. (2018). Detecting cyclonic eddies induced by global super typhoons by using satellite altimetry, *J. Photogr. Rem. Sens.*, 23, pp. 15–24.

Geisler, J. E. (1970). Linear theory of the response of a two layer ocean to a moving hurricane, *Geophys. Astrophys. Fluid Dyn.*, 1, pp. 249–272, doi:10.1080/03091927009365774.

Ginis, I. and Sutyrin, G. (1995). Hurricane-generated depth-averaged currents and sea surface elevation, *J. Phys. Oceanogr.*, 25, pp. 1218–1242, doi:10.1175/1520-0485(1995)025<1218:HGDACA>2.0.CO;2.

Huang, R. X. (1998). Mixing and available potential energy in a Boussinesq ocean, *J. Phys. Oceanogr.*, 28, pp. 669–678, doi:10.1175/1520-0485(1998)028<0669:MAAPEI>2.0.CO;2.

Knapp, K. R., Kruk, M. C., Levinson, D. H., Diamond, H. J. and Neumann, C. J. (2010). The international best track archive for climate stewardship (IBTrACS): Unifying tropical cyclone data, *Bull. Am. Meteorol. Soc.*, 91, pp. 363–376, doi:10.1175/2009BAMS2755.1.

Li, J. (2020). Continental shelf waves in the northern South China Sea, Chapter 13 in J. Hu *et al.* (eds.), *Regional Oceanography in the South China Sea,* World Scientific, Singapore, pp. 355–378, doi:10.1142/9789811206917_0013.

Lin, J.-Y., Ho, H., Gopalakrishnan, G., Zheng, Z.-W., Tseng, R.-S., Pan, J., Ho, C.-R. and Zheng, Q. (2025). Typhoon induced mesoscale cyclonic eddy a long neglected linkage between atmosphere ocean and climate, *npj Clim. Atmos. Sci.*, (2025) 8.64, doi:10.1038/s41612-025-00946-9.

Liu, S., Xu, J., Tu, S., Zheng, M. and Chen, Z. (2023). Relationship between South China Sea summer monsoon and Western North Pacific tropical cyclones linkages with the interaction of Indo-Pacific Pattern, *Atmosphere*, 14, pp. 645, doi:10.3390/atmos14040645.

Lorenz, E. N. (1955). Available potential energy and the maintenance of the general circulation, *Tellus*, 7, pp. 157–167, doi:10.1111/j.2153-3490.1955.tb01148.x.

Luecke, C. A. Arbic, B. K., Bassette, S. L., Richman, J. G., Shriver, J. F., Alford, M. H., Smedstad, O. M., Timko, P. G., Trossman, D. S. and Wallcraft, A. J. (2017). The global mesoscale eddy available potential energy field in models and observations, *J. Geophys. Res. Oceans*, 122, pp. 9126–9143, doi:10.1002/2017JC013136.

Mason, E., Pascual, A. and McWilliams, J. C. (2014). A new sea surface height–based code for oceanic mesoscale eddy tracking, *J. Atmos. Ocean. Technol.*, 31, pp. 1181–1188, doi:10.1175/JTECH-D-14-00019.1.

Pedlosky, J. (1987). *Geophysical Fluid Dynamics*, 2nd ed., Springer, New York, pp. 45–50, doi:10.1007/978-1-4612-4650-3.

Podlosky, J. (1987). *Geophysical Fluid Dynamics*, 2nd edn., Springer, New York, pp. 86–216.

Pegliasco, C. Delepoulle, A., Mason, E., Morrow, R., Faugère, Y. and Dibarboure, G. (2022). META3.1exp: A new global mesoscale eddy trajectory atlas derived from altimetry, *Earth Syst. Sci. Data*, 14, pp. 1087–1107, doi:10.5194/essd-14-1087-2022.

Price, J. F. (1981). Upper ocean response to a hurricane, *J. Phys. Oceanogr.*, 11, pp. 153–175.

Rudzin, J. E. and Chen, S. (2022). On the dynamics of the eradication of a warm core mesoscale eddy after the passage of Hurricane Irma (2017), *Dyn. Atmos. Oceans*, 100, 101334, doi:10.1016/j.dynatmoce.2022.101334.

Shay, L. K. (2009). Upper ocean structure: Responses to strong atmospheric forcing events, in J. H. Steele (ed.), *Encyclopedia of Ocean Sciences (Second Edition)*, Academic Press, Oxford, pp. 192–210, doi:10.1016/B978-012374473-9.00628-7.

Shay, L. K., Goni, G. J. and Black, P. G. (2000). Effects of a warm oceanic feature on Hurricane Opal, *Mon. Weather Rev.*, 128, pp. 1366–1383, doi:10.1175/1520-0493(2000)128<1366:EOAWOF>2.0.CO;2.

Sun, L., Li, Y.-X., Yang, Y.-J., Wu, Q., Chen, X.-T., Li, Q.-Y., Li, Y.-B. and Xian, T. (2014). Effects of super typhoons on cyclonic ocean eddies in the western North Pacific: A satellite data-based evaluation between 2000 and 2008, *J. Geophys. Res. Oceans*, 119, pp. 5585–5598, doi:10.1002/2013JC009575.

Wang, G., Su, J. and Chu, P. C. (2003). Mesoscale eddies in the South China Sea observed with altimeter data, *Geophys. Res. Lett.*, 30(21), 2121, doi:10.1029/2003GL018532.

Wang, G., Su, J., Ding, Y. and Chen, D. (2007). Tropical cyclone genesis over the South China Sea, *J. Mar. Syst.*, 68, pp. 318–326, doi:10.1016/j.jmarsys.2006.12.002.

Wang, Y. (2020). Composite of typhoon induced sea surface temperature and chlorophyll-a responses in the South China Sea, *J. Geophys. Res. Oceans*, 125, e2020JC016243, doi:10.1029/2020JC016243.

Zhang, T., Li, J., Xie, L. and Zheng, Q. (2022). Statistical analysis of mesoscale eddies entering the continental shelf of the northern South China Sea, *J. Mar. Sci. Eng.*, 10, 206, doi:10.3390/jmse10020206.

Zheng, Q., Zhu, B., Li, J., Sun, Z., Xu, Y. and Hu, J. (2015). Growth and dissipation of typhoon-forced solitary continental shelf waves in the northern South China Sea, *Clim Dyn.*, 45, pp. 853–865, doi:10.1007/s00382-014-2318-y.

Part 3

Cruise Observations of Eddies in the South China Sea

Chapter 9

Observational Studies of Mesoscale Eddies in the South China Sea

9.1 Introduction

The South China Sea (SCS), the largest marginal sea in the Northwest Pacific, is characterized by frequent and complex mesoscale eddy activities, which play a crucial role in regional circulation, material transport and ecological dynamics. Over the past two decades, advancements in marine observation technologies have led to significant progress in understanding the spatiotemporal distribution and dynamic mechanisms of mesoscale eddies in the SCS, revealing key features of these phenomena. This chapter synthesizes *in situ* data to present research findings on mesoscale eddies in the SCS, with a focus on 11 typical mesoscale eddies (Figure 9.1).

9.2 Characteristics of Typical Mesoscale Eddies in the SCS

9.2.1 *Anticyclonic Eddy East of the Hainan Island (HAE)*

The anticyclonic eddy (AE) east of the Hainan Island (named as HAE in this chapter) is a recurrent feature. The analysis results derived from hydrological data from Chinese coastal waters and the temperature and salinity data from the US Oceanographic Data Center

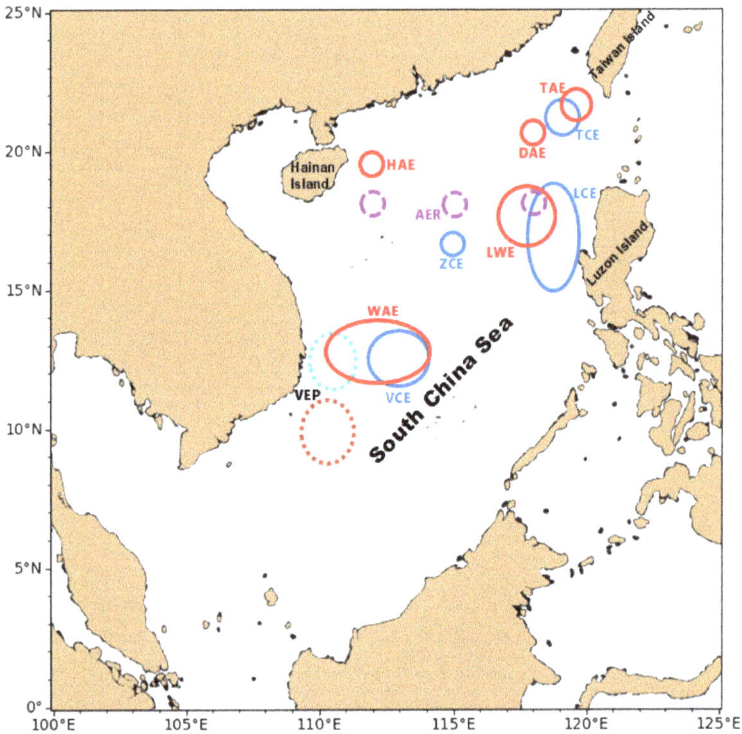

Fig. 9.1. Schematic diagram of 11 typical mesoscale eddies observed in the SCS.

indicate that HAE has a spatial scale of approximately 1° by 1.5°. Meanwhile, multi-cruise observations reveal significant interannual variability in its position and extent (Zhong, 1990). HAE forms due to the anticyclonic deflection of westward flows (e.g., the SCS branch of the Kuroshio Current) upon encountering the shelf-slope region in the northern SCS (NSCS). This deflection extends to form the SCS Warm Current, generating an AE on its right side. Alternatively, HAE can form when the SCS Warm Current meanders northeastward in an anticyclonic pattern (Guan, 1997).

HAE exhibits distinct seasonal variability, weakening in winter (from December to March of the following year) and strengthening in summer. During summer, the southwest monsoon induces upwelling along the eastern coast of Hainan Island and enhancing HAE. For example, observations in June 1959 showed the eddy centered near

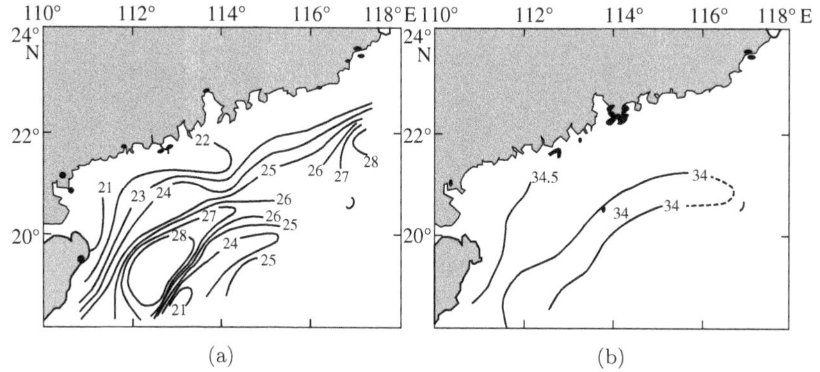

Fig. 9.2. (a) Temperature (°C) and (b) salinity distributions of the HAE at the depth of 30 m measured in June 1983.
Source: Redrawn from Chai *et al.* (2001).

(19°N, 113°E) with an elliptical shape and a major axis radius of ~110 km. The maximum surface current velocity reached 0.98 m s^{-1}, and its influence extended to the depth of 200 m. In June 1983, HAE was particularly strong, with its center at the depth of 30 m near (19.5°N, 112°E) (Figure 9.2). The core temperature was 28.93°C, about 7.8°C higher than that at the western periphery and about 4.5°C higher than that at the eastern periphery (Chai *et al.*, 2001).

9.2.2 *Anticyclonic Eddy East of the Dongsha Atoll (DAE)*

In winter, an anticyclonic warm eddy (named as DAE in this chapter) is observed in east of the Dongsha Atoll, centered near (20.75°N, 118°E) (Figure 9.3). The upper layer of DAE contains thick warm water, with its mixed layer and salinity maximum layer significantly deeper than surrounding waters (Liu *et al.*, 2013). Geostrophic flow fields and shipboard ADCP (Acoustic Doppler Current Profiler) measurements confirm its existence.

Sectional distributions of temperature and salinity (Figure 9.3(a, b)) reveal concave isotherms and isohalines within the upper 500 m. The mixed layer depth at DAE center reaches 250 m, far exceeding surrounding stations, while the salinity maximum occurs at the depth ~300 m. At 150 m, a low-salinity and high-temperature water tongue extends southwestward, with temperature >23°C and

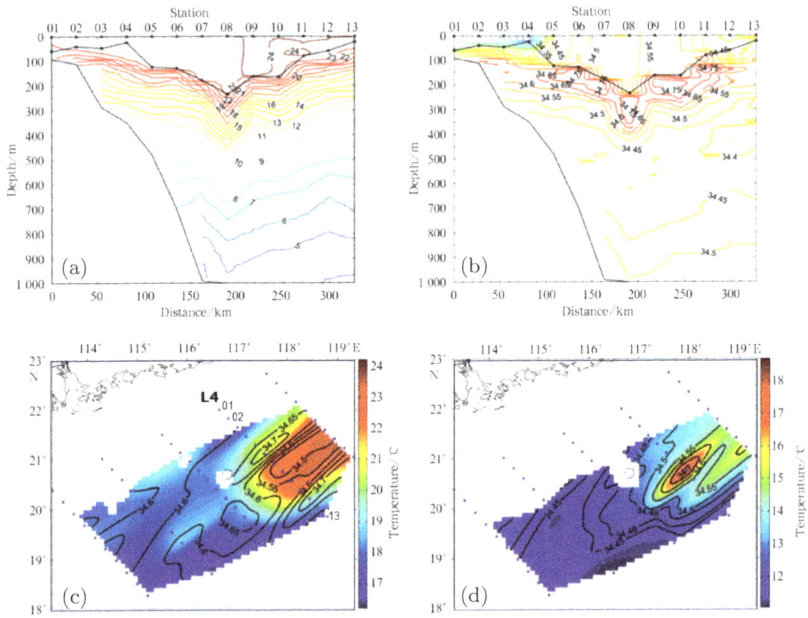

Fig. 9.3. (a) Temperature (°C) and (b) salinity distributions along Transect L4. Contour intervals: 1°C (temperature) and 0.05 (salinity). Black dotted lines in (a) and (b) denote mixed layer depth. Horizontal distributions of temperature (shading) and salinity (contours) at (c) 150 m and (d) 300 m, respectively. Blue dots mark Conductivity-Temperature-Depth (CTD) stations in January 2010.
Source: Redrawn from Liu *et al.* (2013).

salinity <34.5, resembling the Northwest Pacific water properties. This suggests that the central waters of DAE originate from east of the Kuroshio Current (Figure 9.3(c)).

9.2.3 *Anticyclonic Eddy Southwest of the Taiwan Island (TAE)*

The analysis results of 32 years of satellite-tracked drifter data indicate that mesoscale anticyclonic eddies (AEs) frequently occur in southwest of Taiwan Island (known as TAE) (Li *et al.*, 2011). For instance, an Argo float tracked the TAE near (22°N, 120°E) from May to July 2009. Observations reveal that TAE resulted from the Kuroshio intruding into the Luzon Strait and entering the SCS. Its salinity was higher than the typical SCS water but lower than that of the Kuroshio water, indicating strong mixing happened during its generation process (Liu *et al.*, 2015). Satellite altimetry confirms TAE diameter of 160–280 km and sea level anomaly (SLA) of 16–23 cm, persisting from May to June.

Hu *et al.* (2012) identified TAE using cruise conductivity-temperature-depth (CTD) profiler and ADCP data in winter. They found that a high-salinity prism (salinity >34.8) at depths of 170–200 m near (21°N, 118°E) (Figure 9.4). The T–S diagram analysis shows this prism resembles the Northwest Pacific water but with slightly lower salinity, likely due to mixing during propagation. Argo data trace its origin to the Luzon Strait's eastern side.

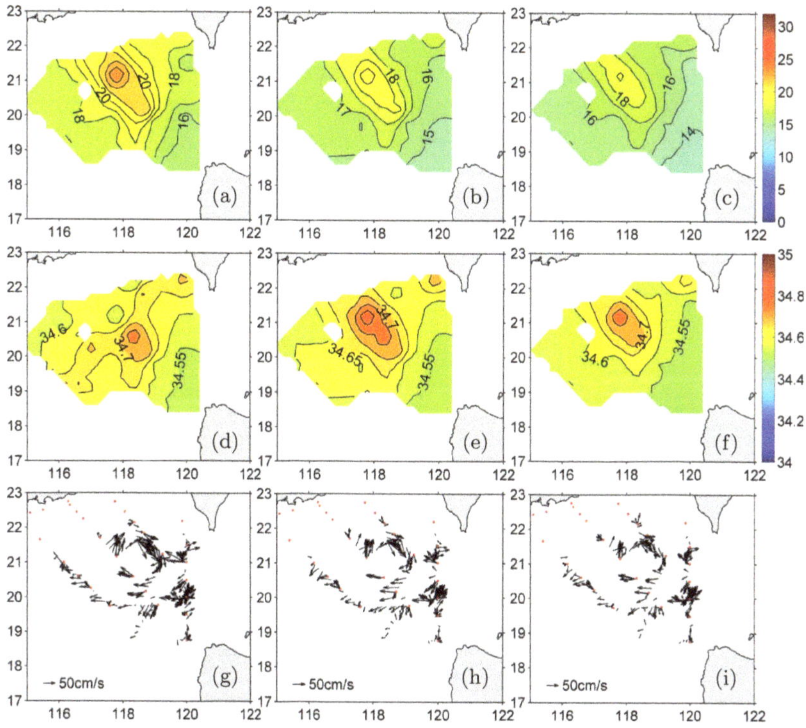

Fig. 9.4. Distributions of temperature (°C) and salinity southwest of the Taiwan Island at (a, d) 150 m, (b, e) 180 m, and (c, f) 200 m, respectively. ADCP measurements at (g) 148 m, (h) 180 m, and (i) 196 m during the cruise from 20 to 30 January 2010. The horizontal axis is in °E longitude and the vertical axis in °N latitude.

Source: Taken from Hu *et al.* (2012).

9.2.4 *Cyclonic Eddy Southwest of the Taiwan Island (TCE)*

Figure 9.5 shows a cyclonic cold eddy (known as TCE) southwest of Taiwan Island observed by the CTD and Argos data during 23–25 May 2017 (Huang *et al.*, 2019). One can see that isotherms and isohalines in the eddy core uplift by ~100 m within depths of 150–300 m, shoaling the thermocline. Geostrophic flows show that this TCE's evolution is linked to an AE east of the Luzon Strait.

9.2.5 *Cyclonic Eddy North of the Zhongsha Islands (ZCE)*

Figure 9.6 shows a cyclonic cold eddy (named as ZCE in this chapter) north of the Zhongsha Islands, which is characterized by low temperature and high salinity observed by Tian *et al.* (2016) in March 2014. One can see that at the depth of 10 m, eddy core temperature is 25°C (~1°C lower than surroundings), and salinity reaches 34.11 (~0.74 higher). ZCE spans ~1° × 1.5° horizontally and 30 m vertically, lasting for ~13 days (Tian *et al.*, 2016). In March

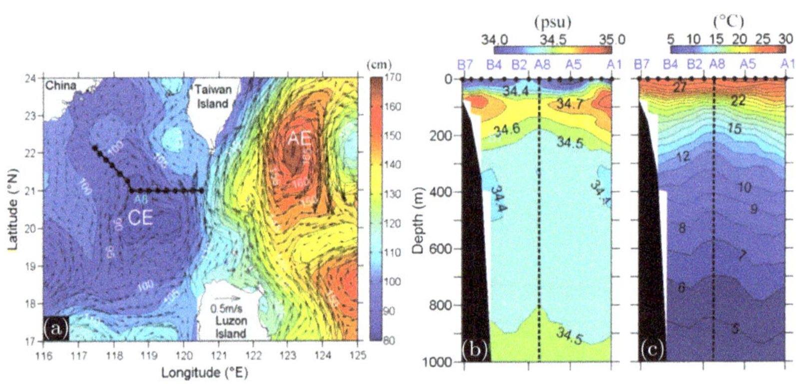

Fig. 9.5. (a) Absolute dynamic topography and geostrophic currents on both sides of the Luzon Strait during 23–25 May 2017. Black lines with dots mark cruise survey transects and stations. (b) Salinity and (c) temperature along Transects A and B (0–1000 m). Vertical dashed lines mark the location of Station A8. Black shading on the left side shows continental slope topography.

Source: Redrawn from Huang *et al.* (2019).

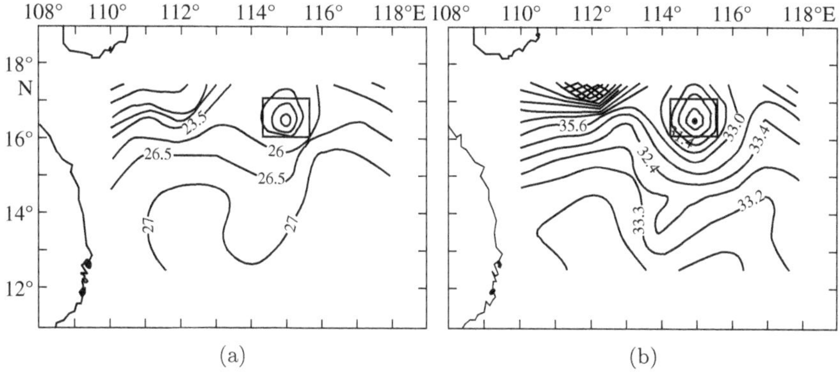

Fig. 9.6. Cruise observed (a) temperature (°C) and (b) salinity at the depth of 10 m in the central SCS in March 2014.
Source: Redrawn from Tian *et al.* (2016).

2014, a mesoscale cold eddy shed from the Luzon cold eddy system propagated westward, stalling at the continental shelf. Under topographic and monsoon forcing, ZCE separated from its southern edge near (16.5°N, 115°E) and propagated southward along the 3000-m isobath. It dissipated when northeasterly winds shifted to southerly, enhancing mixing with warm southern waters.

9.2.6 *Luzon Cold Eddy (LCE)*

The analysis results derived from 105 years of historical observation data confirm existence of Luzon cold eddy (LCE) northwest of the Luzon Island (Qu, 2000; Udarbe-Walker and Villanoy, 2001). LCE occurs in winter and spring, developing from October–November, peaking in December–January and decaying by May (Liu *et al.*, 2006). In January, a low-temperature water zone appears at the depth of 15 m near (17°N, 119.5°E) with a 0.5° latitudinal extent and 1°C temperature difference. At 100 m, the core temperature is <15°C with a 6–7°C temperature difference and 4–5°latitudinal range (Figure 9.7). It is believed that LCE forms due to interaction of wind stress and the Kuroshio, with wind dominance. Previous investigators also reported multi-core structures of LCE based on cruise observations (Yuan *et al.*, 2004; Sun and Liu, 2011).

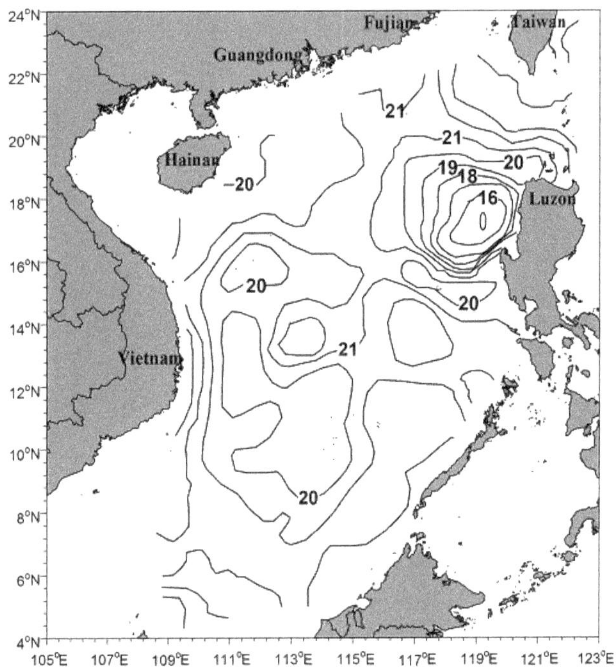

Fig. 9.7. Monthly mean temperature distribution (°C) at the depth of 100 m in the SCS in January.
Source: Redrawn from Liu *et al.* (2006).

9.2.7 *Luzon Warm Eddy (LWE)*

The Luzon Warm Eddy (LWE), located in west of the Luzon Island, extends vertically >500 m, with a maximum temperature anomaly of 5°C and minimum salinity anomaly of 0.5 near the thermocline. Peak currents (~0.6 m s^{-1}) occur in the upper 200 m. LWE typically initiates in July, intensifies in August–September, propagates westward along 18–19°N in October, weakens by December and dissipates by February of the next year (Chen *et al.*, 2009). In August 2012, LWE had a high-temperature (difference >2.12°C) and low-salinity (difference <0.25) core near 117.5°E, influencing the depth ~200 m (Figure 9.8) (Wang *et al.*, 2017). It is noted that the eddy is characterized by northwestward vertical tilt (>70 km) and flow field structure even affecting depths below 1000 m (Men *et al.*, 2020).

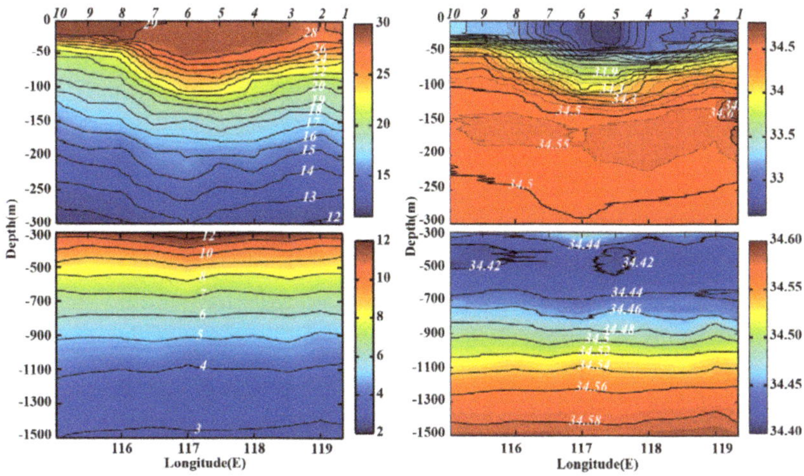

(a) (b)

Fig. 9.8. Sectional distributions of (a) temperature and (b) salinity near 18°N for the LWE during 12–14 August 2012. Color codes are in °C for temperature and psu for salinity, respectively.

Source: Taken from Wang *et al.* (2017).

9.2.8 *Anticyclonic Eddy Row (AER)*

An AE row (named as AER in this chapter) along 18°N in the NSCS comprises three long-lived eddies ACE1, ACE2 and ACE3 (Figure 9.9), lasting for 147, 168, and 210 days, respectively (Nan *et al.*, 2011). For the generation mechanisms, it has been confirmed that the long-lived eddy train comprises a portion of the Rossby normal modes of the SCS deep basin (see Chapter 5 for details).

9.2.9 *Vietnam Eddy Pair (VEP)*

The Vietnam Eddy Pair (known as VEP), east of Vietnamese coast, consists of a cold eddy and a warm eddy. It typically forms in late June–early July, intensifies in August–September, weakens in October and dissipates by late October. The AE dissipates earlier, possibly due to the cyclonic surface circulation in late October (Chu *et al.*, 2017).

Cruise observations in September 2007 show the CE core of the VEP near 12.2°N, 110.8°E with temperature of 2–3°C lower than that of surrounding water (Figure 9.10). One can see that strong

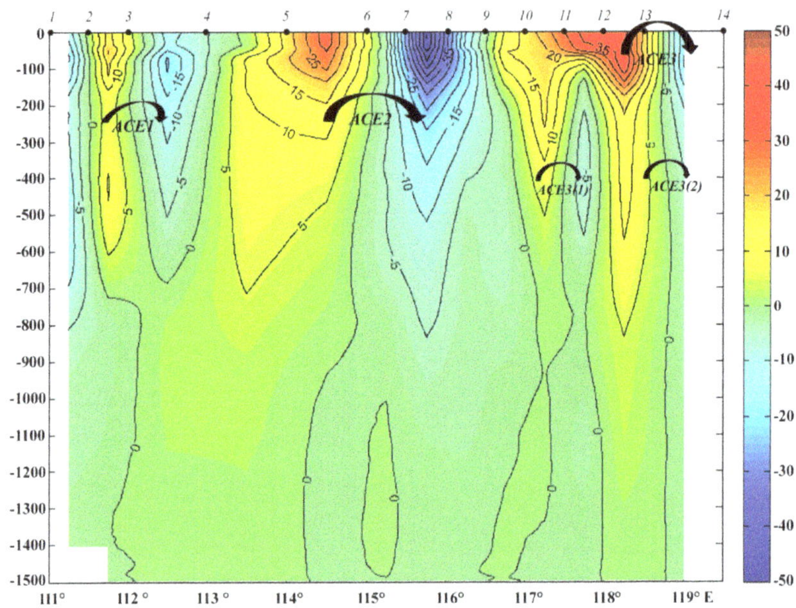

Fig. 9.9. Sectional distribution of geostrophic current velocity of AER measured by conductivity-temperature-depth (CTD) along 18°N between 111°E and 119°30′E during the period of 20–24 August 2007, Color cades are in cm s^{-1}. *Source*: Redrawn from Nan *et al.* (2011).

temperature fronts occur at 50–100 m, and weaken with the depth and disappear at 500 m. Flow fields show cyclonic rotation (max speed ~1.0 m s^{-1} on the southeast side). The thermohaline structure exhibits a dome-shaped uplift, tilting southwestward with the depth. The cold eddy together with the surrounding warm eddy forms the VEP and interacts with each other (Hu *et al.*, 2011). Cruise CTD data in September 2014 also show a cold eddy of the VEP with a horizontal scale of about 150 km and low-temperature and high-salinity core extending to the depth of 200 m (Xiang *et al.*, 2015).

9.2.10 *Western SCS Anticyclonic Eddy (WAE)*

Figure 9.11(a) shows the SLA image of the Western SCS Anticyclonic Eddy (WAE) during 20–25 May 2004. One can see that elliptical WAE was centered at about (13.5°N, 112.5°E) with the central SLA of 20 cm, the major axis of about 350 km and the minor axis

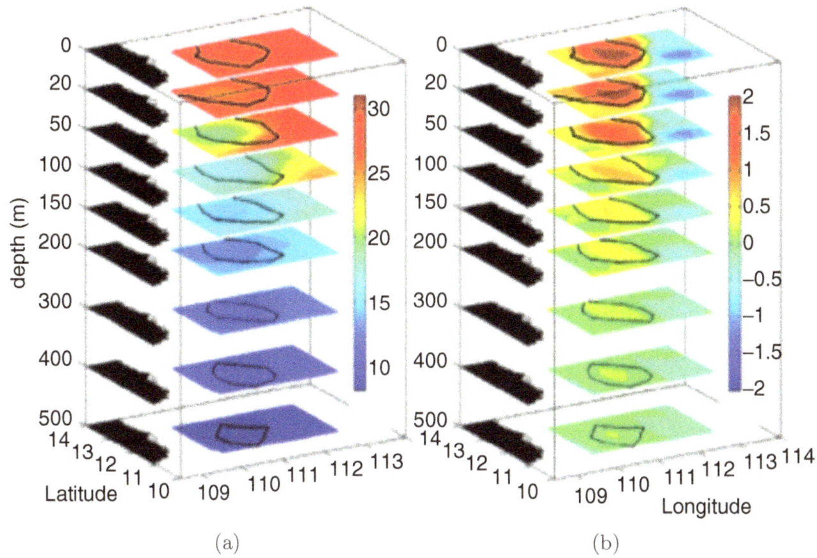

Fig. 9.10. 3D structures of (a) temperature (°C) and (b) vorticity (10^{-5} s^{-1}) of a cruise observed CE of the VEP near 12.2°N, 110.8°E in September 2007. Black contours mark eddy boundary at each depth.

Source: Redrawn from Hu *et al.* (2011).

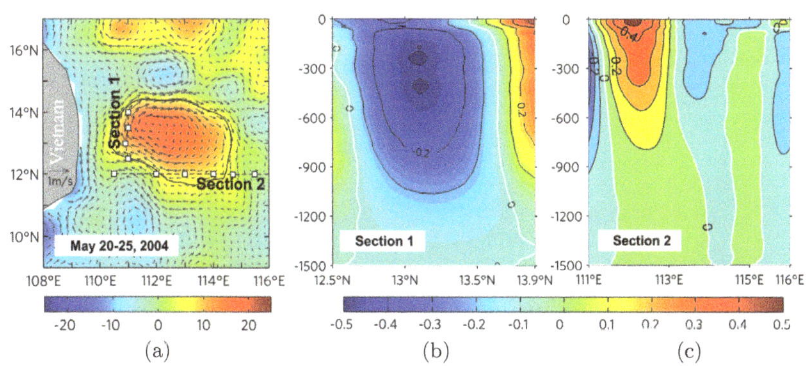

Fig. 9.11. (a) Mean SLA (cm; shading) image and geostrophic currents (vectors in m s^{-1}) of the western SCS during 20–25 May 2004. (b, c) Zonal and meridional geostrophic current components (m s^{-1}) derived from CTD measurements along Sections 1 and 2 marked by white squares in (a). Contour interval is 0.1 m s^{-1} with respect to zero velocity contour in white.

Source: Redrawn from Chu *et al.* (2020).

of about 250 km. Multiyear observations indicate that WAE is a seasonal feature, and forms in February–March induced by winter monsoon-driven cyclonic circulation in the SCS. It persists for about 5 months, dissipating in July. Cruise observations along the western and southern boundaries of WAE (Sections 1 and 2 in Figure 9.11(a)) in May 2004 show that the maximum zonal geostrophic flow core of -0.5 m s^{-1} was located at depths of 250–450 m (Figure 9.11(b)) and the maximum meridional geostrophic flow core of 0.5 m s^{-1} at the sea surface layer (Figure 9.11(c)). From Figures 9.11(b) and 9.11(c), one can see that the deepest depth of WAE reaches 1500 m, implying that it is a strong mesoscale eddy in the SCS (see Chapter 11 for details).

Figure 9.12(a–d) show another case of WAE observed during 6–13 June 2017. From the SLA image in Figure 9.12(a), one can see that

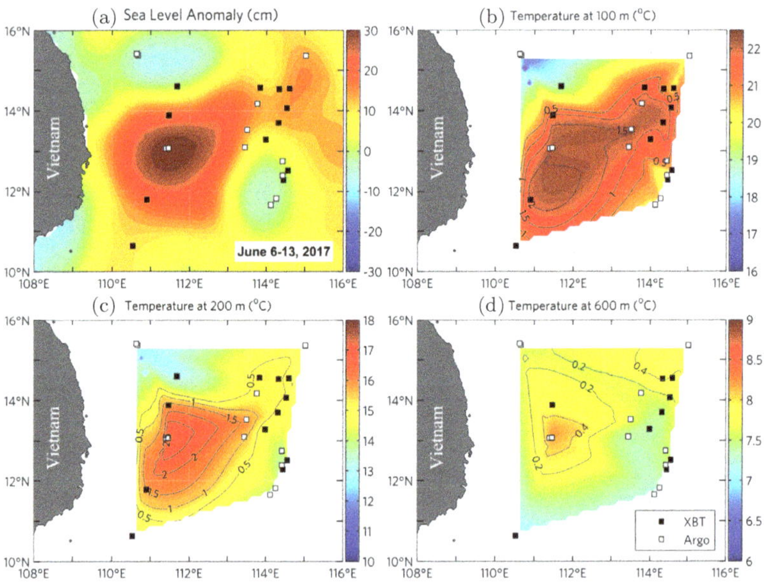

Fig. 9.12. (a) Mean SLA (cm; shading) image of the western SCS during 6–13 June 2017. (b–d) Temperature (°C; shading) and temperature anomaly (numerals on contours) at depths of 100 m, 200 m and 600 m from XBT and Argo measurements from stations marked by solid and open squares.

Source: Redrawn from Chu *et al.* (2020).

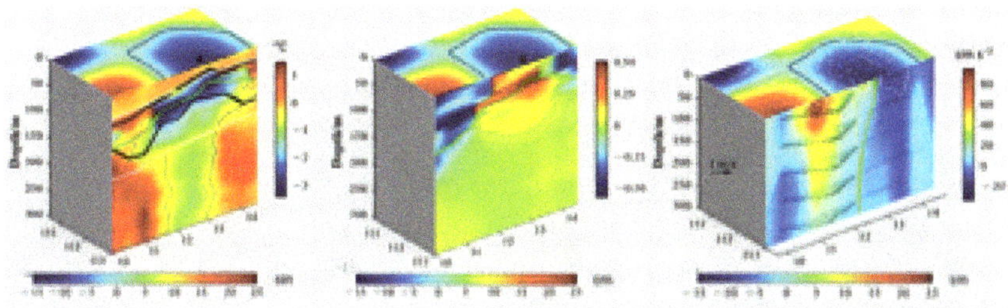

(a) (b) (c)

Fig. 9.13. 3D structures of (a) temperature anomalies, (b) salinity anomalies and (c) flow field of the VCE observed from 10 October to 21 November, 2012. Vertical color codes are in °C for temperature, psu for salinity and cm s^{-1} for current velocity, respectively. Horizontal color code is in cm for the SLA.
Source: Redrawn from Hu *et al.* (2017).

WAE appeared an irregular circle centered at about (13°N, 111.5°E) with the central SLA of 25 cm and the diameter of about 350 km. Figure 9.12(b–d) show 3D thermal structures of WAE derived from eXpendable Bathy Thermographs (XBT) and Argo float data at depths 100, 200 and 600 m (Chu *et al.*, 2020). One can see that the temperature at 100 m is 2.5°C higher than that in the surrounding water and the high temperature water is mainly concentrated in the upper 300 m.

9.2.11 *Vietnam Winter Cyclonic Eddy (VCE)*

Cruise observations during October–November 2012 reveal the Vietnam Winter Cyclonic Eddy (VCE) east of Vietnam, centered around (12.5°N, 113°E). It was generated on 10 October 2012, then moved northwestward and disappeared after 21 November, lasting for about 6 weeks. VCE is characterized by a cold core (max temperature anomaly of −4.7°C at 25–75 m) and a high-salinity core (max salinity anomaly: >0.60 at 75 m) as shown in Figure 9.13(a–b)). Its 3D flow field exhibits asymmetric cyclonic rotation, with higher velocities on the south side (Figure 9.13(c)) (Hu *et al.*, 2017).

9.3 Summary

This chapter summarizes the research results on mesoscale eddies in the SCS using cruise observations. The data, parameters and conclusions derived from such observations can be treated as the ground truth for determination of characteristic scales for dynamic analysis and validation of numerical modeling. The results of 11 cases of mesoscale eddies in the SCS observed from 1983 to 2017 are reviewed. Their major features are specified as follows.

- **HAE:** Strengthen in summer due to monsoon-driven upwelling; linked to anticyclonic deflection of westward flows in the northern SCS.
- **DAE:** Deep warm water with properties resembling the Northwest Pacific water.
- **TAE:** Originate from the Northwest Pacific, with the highest salinity at the subsurface layer.
- **TCE:** Originate from Kuroshio intrusions; affecting the northeastern SCS circulation.
- **ZCE:** Form from the shedding of the LCE system.
- **LCE:** Winter-spring feature driven by wind-Kuroshio interactions.
- **LWE:** Seasonal westward propagation along 18–19°N.
- **AER:** Long-lived eddies related to the SCS western boundary current instability and the Rossby normal modes of the SCS deep basin.
- **VEP:** Asymmetric cold–warm eddy pairs with distinct thermohaline signatures.
- **VCE:** Distinct negative temperature anomalies and positive salinity anomalies in winter.
- **WAE:** Monsoon-induced seasonal eddy with deep vertical influence.

Mesoscale eddies in the SCS exhibit complex spatiotemporal patterns and dynamic mechanisms, profoundly impacting regional oceanography. Future research should integrate multidisciplinary approaches to further elucidate their formation, evolution, and environmental effects.

References

Chai, F., Xue, H. J. and Shi, M. C. (2001). Hydrographic characteristics and seasonal variation of three anticyclonic eddies on the northern continental shelf of the South China Sea, in H. J. Xue *et al.* (ed.), *Oceanography in China*, Vol. 13, Ocean Press, Beijing, pp. 105–116.

Chen, G. X., Hou, Y. J., Chu, X. Q., Qi, P. and Hu, P. (2009). The variability of eddy kinetic energy in the South China Sea deduced from satellite altimeter data, *Chin. J. Oceanol. Limnol.*, 27(4), pp. 943–954, doi:10.1007/s00343-009-9297-6.

Chu, X. Q., Chen, G. X. and Qi, Y. Q. (2020). Periodic mesoscale eddies in the South China Sea, *J. Geophys. Res. Oceans*, 125, e2019JC015139, doi:10.1029/2019JC015139.

Chu, X. Q., Dong, C. M. and Qi, Y. Q. (2017). The influence of ENSO on an oceanic eddy pair in the South China Sea, *J. Geophys. Res. Oceans*, 122, pp. 1643–1652, doi:10.1002/2016JC012642.

Guan, B. X. (1997). Warm eddy in the open sea east of Hainan Island, *J. Oceanogr. Huanghai & Bohai Seas*, 15(4), pp. 1–7.

Hu, D., Chen, X., Li, Y., Mao, K. and Zhang, S. (2017). Characteristics of an observed cyclonic eddy off eastern Vietnam, *J. PLA Univ. Sci. Tech. (Nat. Sci. Ed.)*, 18(2), pp. 163–169.

Hu, J., Gan, J., Sun, Z., Zhu, J. and Dai, M. (2011). Observed three-dimensional structure of a cold eddy in the southwestern South China Sea, *J. Geophys. Res. Oceans*, 116, C05016, doi:10.1029/2010JC006810.

Hu, J., Zheng, Q., Sun, Z. and Tai, C.-K. (2012). Penetration of nonlinear Rossby eddies into South China Sea evidenced by cruise data, *J. Geophys. Res. Oceans*, 117, C03010, doi:10.1029/2011JC007525.

Huang, Z., Zhuang, W., Hu, J. and Huang, B. (2019). Observations of the Luzon Cold Eddy in the northeastern South China Sea in May 2017, *J. Oceanogr.*, 75, pp. 415–422, doi:10.1007/s10872-019-00510-z.

Li, J., Zhang, R. and Jin, B. (2011). Eddy characteristics in the northern South China Sea as inferred from Lagrangian drifter data, *Ocean Sci.*, 7, pp. 661–669, doi:10.5194/os-7-661-2011.

Liu, C. J., Du, Y., Zhuang, W., Xia, H. Y. and Xie, Q. (2013). Evolution and propagation of a mesoscale eddy in the northern South China Sea during winter, *Acta Oceanol. Sin.*, 32(7), pp. 1–7, doi:10.1007/s13131-013-0325-1.

Liu, J. F., Mao, K. X., Yan, M., Zhang, X. D. and Shi, Y. J. (2006). The general distribution characteristics of the Luzon Cold Eddy, *Mar. Forecasts*, 23(2), pp. 39–44.

Liu, Z., Hou, Y. J. and Xie, Q. (2015). Eddy formation and surface flow field in the Luzon Strait area during the summer of 2009, *Chin. J. Oceanol. Limnol.*, 33(5), pp. 1320–1333, doi:10.1007/s00343-015-4397-y.

Men, G. R., Ma, W. W. and Wan, X. Q. (2020). Observed 3D structure and simulation analysis of Luzon Warm Eddy, *J. Ocean Univ. China*, 50(1), pp. 11–20.

Nan, F., He, Z., Zhou, H. and Wang, D. (2011). Three long-lived anticyclonic eddies in the northern South China Sea, *J. Geophys. Res. Oceans*, 116, C05002, doi:10.1029/2010JC006790.

Qu, T. (2000). Upper-layer circulation in the South China Sea, *J. Phys. Oceanogr.*, 30, pp. 1450–1460, doi:10.1175/1520-0485(2000)030 <1450:ULCITS>2.0.CO;2.

Sun, C. X. and Liu, Q. Y. (2011). Double eddy structure of the winter Luzon cold eddy based on satellite altimeter data, *J. Trop. Oceanogr.*, 30(3), pp. 9–15.

Tian, Y. Q., Huang, H. H., Gong, X. Y. and Yu, S. M. (2016). The formation mechanism of the low temperature and high salinity water mass near the Zhongsha Islands in the South China Sea in March 2014, *J. Trop. Oceanogr.*, 35(2), pp. 1–9.

Udarbe-Walker, M. J. B. and Villanoy C. L. (2001). Structure of potential upwelling areas in the Philippines, *Deep Sea Res. Part I*, 48, pp. 1499–1518, doi:10.1016/S0967-0637(00)00100-X.

Wang, X., Fang, W., Chen, R. and Cai, S. (2017). The summer-fall anticyclonic eddy west of Luzon: Structure and evolution in 2012 and interannual variability, *J. Marine Systems*, 172, pp. 84–92, doi:10.1016/j.jmarsys.2017.03.004.

Xiang, R., Fang, W. D., Lu, Y. Z., Huang, X. R. and Zhou, S. Q. (2015). Observed three-dimensional structures of a cold eddy and an eastward jet in the western South China Sea during September 2014, *J. Trop. Oceanogr.*, 34(6), pp. 1–10.

Yuan, Y. C., Bu, X. W., Liao, G. H., Lou, R. Y., Su, L. and Wang, K. S. (2004). Diagnostic calculation of the upper-layer circulation in the South China Sea during the winter of 1998, *Acta Oceanol. Sin.*, 23(2), pp. 187–199.

Zhong, H. L. (1990). Structures of the density circulation, in Y. L. Ma *et al.* (ed.), *Report of Decadal Hydrographic Series Survey of the Shelf and Adjacent Waters of the Northern South China Sea, Ocean Press*, Beijing, pp. 215–241.

Chapter 10

Three-Dimensional Structures and Evolution of Mesoscale Eddies

10.1 Introduction

The ocean eddies are rotational water bodies relatively isolated from surrounding waters. During propagation, eddies carry the energy, the heat, their own water mass and ecologic environment matters (Wunsch, 1999; Qiu and Chen, 2005; Dong *et al.*, 2014; Zhang *et al.*, 2014; Cheng, *et al.*, 2020; Stanley *et al.*, 2020; Bian *et al.*, 2023). Meanwhile, they may grow, dissipate and modify due to interactions with surrounding waters and topography. To explore the above processes quantitatively, it is a must to know three-dimensional (3D) structures of eddies (Sandalyuk *et al.*, 2020). Zheng *et al.* (2011) used an approximate of cylinder eddy to calculate the total momentum of eddy approaching the Kuroshio. Zhang *et al.* (2013) suggested an empirical model of a universal structure to express 3D sea surface height (SSH) distribution function of ocean mesoscale eddies. Ren *et al.* (2023) constructed 3D structures of mesoscale eddies and explored their interaction with Kuroshio based on observations from a current pressure inverted echo sounder (CPIES) array deployed in the Kuroshio region.

The South China Sea (SCS) is a semi-enclosed marginal sea of the Pacific with a horizontal length scale of $O(2000)$ km. Thus, in its circulation system research and interdisciplinary research, mesoscale eddies with a horizontal length scale of $O(100–200)$ km occupy a prominent position, because of their multiplicity (Zheng *et al.*, 2014;

Xie and Zheng, 2017; Xie *et al.*, 2018). Thus, previous investigators have been dedicated to exploring 3D structures of mesoscale eddies. Hu *et al.* (2011) examined 3D structures and physical properties of a cold eddy in the southwestern SCS using cruise measurements and concurrent satellite altimeter sea surface height (SSH) images. Hu *et al.* (2012) observed a mesoscale anticyclonic eddy (AE) with a prism-shaped, high-salinity core in the subsurface layer west of the Luzon Strait and constructed its 3D structures of physical properties using the cruise conductivity-temperature-depth (CTD) profiler and acoustic Doppler current profiler (ADCP). Wang and Gan (2014) constructed 3D structures of a mesoscale cyclonic eddy (CE) west the Luzon Island by a high-resolution regional ocean model. Lin *et al.* (2015) constructed 3D structures of properties of mesoscale eddies in the SCS based on eddy-resolving model output. Zhang *et al.* (2016) obtained 3D current velocity structures of mesoscale eddies in the northeastern SCS using mooring array measurements. Li *et al.* (2019) constructed 3D structures of an AE in the SCS using multiple underwater gliders. Han *et al.* (2021) suggested two approximate expressions, circle, and ellipse, for description of two-dimensional (2D) horizontal section of mesoscale eddies in the SCS. This chapter summarizes major research results on 3D structures of mesoscale eddies in the SCS.

10.2 Eddy Structures on the Sea Surface

10.2.1 *2D horizontal structures of eddies on the sea surface*

The accurate description of 2D horizontal sectional shapes varying with the depth is of significance for 3D volume integral calculation of dynamic parameters of ocean eddies. As a simplified model, the symmetrical circle relative to the eddy center has been broadly used for describing 2D geometric shape of horizontal sections of mesoscale eddies (Zhang *et al.*, 2016; Zheng *et al.*, 2019). In fact, statistics of eddy eccentricity ($=\sqrt{1-b^2/a^2}$, here a is the length of the semi-major axis and b is the length of the semi-minor axis) derived from 26 years of Archiving, Validation and Interpretation of Satellite Oceanographic (AVISO) satellite SLA product

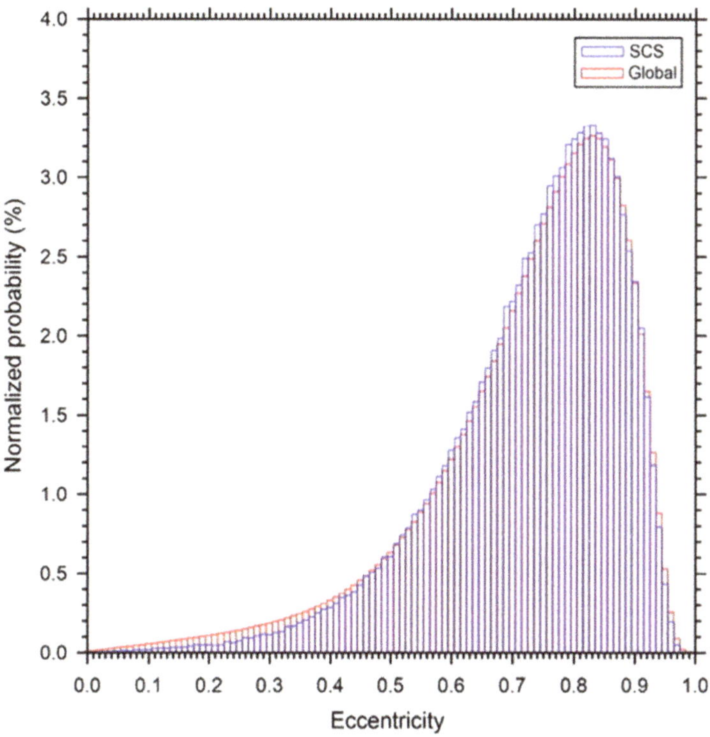

Fig. 10.1. Occurrence probability distribution histograms of the eddy eccentricity in the SCS (blue) and the global ocean (red).

Source: Taken from Han *et al.* (2021).

with a spatial resolution of 0.25° by 0.25° from January 1993 to August 2018 (https://www.aviso.altimetry.fr/en/home.html) reveal that very small proportion of eddies have a circular shape, i.e., eccentricity = 0, in the SCS and the global ocean as shown in Figure 10.1 (Han *et al.*, 2021). The peak values of both histograms are 0.83, and near 60% of the values are distributed in a range of 0.7 ~ 0.9, implying that most eddies in the SCS have an elliptic structure. The typical mean shape is a mathematic ellipse with a semi-major axis of 101.3 km and a semi-minor axis of 61.3 km.

Han *et al.* (2021) derived the evolutionary processes of mesoscale eddy eccentricities on the sea surface with dynamic parameters in the SCS from 26 years of AVISO satellite SLA product from January 1993 to August 2018. The results are shown in Figure 10.2. One can

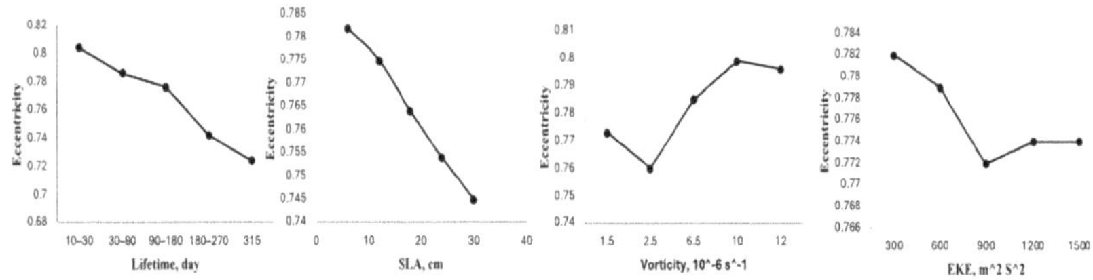

(a) (b) (c) (d)

Fig. 10.2. The evolutionary processes of mesoscale eddy eccentricities on the sea surface with dynamic parameters (a) lifetime (in day), (b) peak SLA (in cm), (c) vorticity (in 10^{-6} s^{-1}) and (d) EKE (in m^2 s^2) in the SCS derived from 26 years of AVISO satellite SLA product from January 1993 to August 2018.

Source: Data taken from Han *et al.* (2021).

see that eddy eccentricities become smaller with the increase of the lifetime and the SLA amplitude, implying that eddies become more circular with the time and the eddy intensity. Meanwhile, the eddy eccentricity appears as a general increase (decrease) trend with the vorticity (eddy kinematic energy, eddy kinetic energy (EKE)).

10.2.2 *3D SSH structures of mesoscale eddies*

The SSH of ocean dynamic processes is a key factor of sea surface pressure term P $(= g\eta$, where g is the gravitational acceleration and η is the SSH) in the equations of motion, though its value is generally not greater than $O(1)$ m. Updated satellite altimetry has been capable of providing the global ocean SSH data products with an SSH accuracy of 2 cm, a spatial resolution of 0.25° latitude by 0.25° longitude and a temporal resolution of one day. The data are suitable for detecting mesoscale eddies with the horizontal length scale (diameter) larger than 100 km. From the AVISO SSH data from October 1992 to January 2010 and about 600,000 Argo float temperature and salinity profiles in the upper 2000 m between 1996 and 2010, Zhang *et al.* (2013) derived an empirical model for 3D SSH distribution of mesoscale eddies, so called "the universal structure of mesoscale eddies". For an axially symmetric eddy in the cylindrical coordinates (r, z), where r is the radius vector and z the vertical axis upward positive with respect to a horizontal reference plane, the associated pressure anomaly is described by an empirical function of

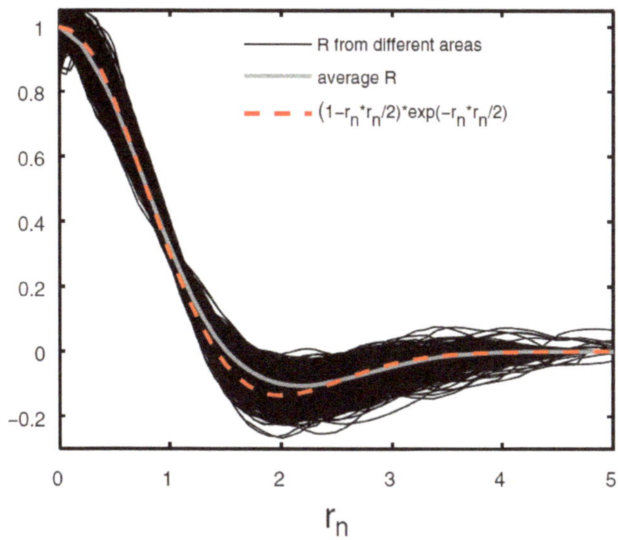

Fig. 10.3. Comparisons of the empirical model of the universal structure of mesoscale eddies (red dashed line) to eddy structures derived from the satellite SSH products and Argo float observations (black lines) as well as their mean (gray line).

Source: Taken from Zhang *et al.* (2013).

the normalized radius vector in normalized stretched coordinates,

$$R(r_n) = \left(1 - \frac{r_n^2}{2}\right) e^{-\frac{r_n^2}{2}}, \tag{10.1}$$

where $r_n = r/R_0$, in which R_0 is the horizontal length scale (radius) of eddies. Figure 10.3 shows comparisons of the empirical model of the universal structure of mesoscale eddies (10.1) to eddy structures derived from the satellite SSH products and Argo float observations as well as their mean. One can see that the empirical model fits the data quite well.

10.3 Eddy Classification by Vertical Structures

10.3.1 *Normal eddies*

Here normal eddies, i.e., general eddies, are relative to abnormal eddies defined in recent years. According to the vertical structures

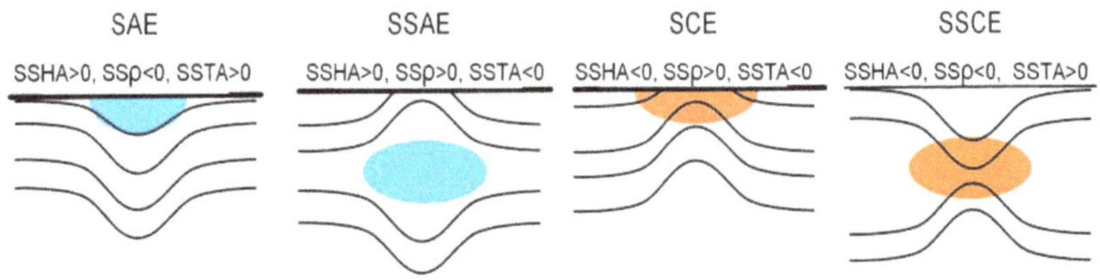

Fig. 10.4. Schematics of isopycnal displacements and IDACs (orange positive/cold/salty and light blue/warm/fresh) for SAE, SSAE, SCE and SSCE.
Source: Taken from Liu and Li (2023).

of the physical properties and rotation directions, normal mesoscale eddies, or just mesoscale eddies for simplicity, are classified into two categories: surface intensified eddies (SIEs) including surface intensified anticyclonic eddies (SAEs) and surface intensified cyclonic eddies (SCEs), and subsurface intensified eddies (SSIEs) including subsurface intensified anticyclonic eddies (SSAEs) and subsurface intensified cyclonic eddies (SSCEs) (Nan *et al.*, 2022; Liu and Li, 2023). Figure 10.4 shows schematics of four sub-categories of ocean eddies. One can see the following features:

(1) The vertical structure of SAE is featured by a negative intensified density anomaly core (IDAC) seating in the sea surface layer with sea surface height anomaly (SSHA) > 0, sea surface density anomaly (SSρ) < 0 and sea surface temperature anomaly (SSTA) > 0. Isopicnals present concave patterns. Here, the temperature and salinity anomalies are defined as the values at each point minus the average of each layer.
(2) The vertical structure of SSAE is featured by the negative IDAC seating in the subsurface layer with SSHA > 0, SS$\rho > 0$ and SSTA < 0. Isopicnals present convex (concave) patterns above (below) the IDAC.
(3) The vertical structure of SCE is featured by the positive IDAC seating in the sea surface layer with the SSHA < 0, SS$\rho > 0$ and SSTA < 0. Isopicnals present convex patterns.
(4) The vertical structure of SSCE is featured by the positive IDAC seating in the subsurface layer with SSHA < 0, SS$\rho < 0$ and SSTA > 0. Isopicnals present concave (convex) patterns above (below) the IDAC.

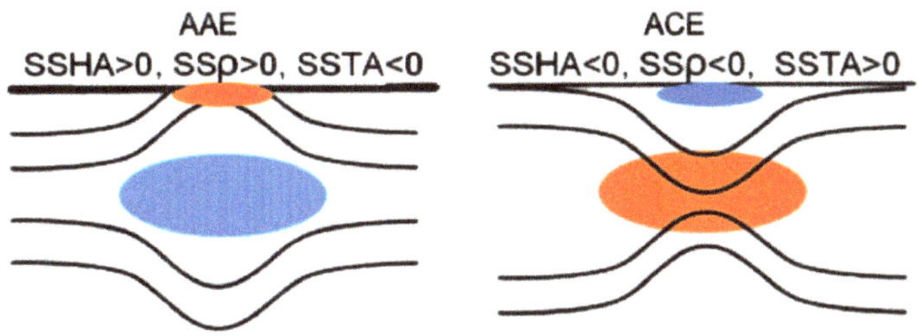

Fig. 10.5. Schematics of isopycnal displacements, IDACs (orange positive/cold/salty and light blue/warm/fresh) and SSTA (orange SS$\rho > 0$ and light blue SS$\rho < 0$) for AAE and ACE.

Source: Taken and re-drawn from Liu and Li (2023).

10.3.2 *Abnormal eddies*

General or normal cyclonic (anticyclonic) eddies have lower (higher) SST cores. In recent years, however, opposite cases, in which cyclonic (anticyclonic) eddies characterized by higher (lower) SST cores, so-called "abnormal" eddies, have been found in the global ocean (Liu *et al.*, 2021) and the SCS (Sun *et al.*, 2021). Figure 10.5 shows schematics of isopycnal displacements and IDACs (orange positive and blue negative) for abnormal anticyclonic eddy (AAE) and abnormal cyclonic eddy (ACE). Compared to Figure 10.4, one can see that in fact, abnormal eddies are special cases of SSAE and SSCE. AAE is the case, in which the convex isopicnals break and induce divergent surface flows, so that a low SST eddy core forms at the sea surface. ACE is the case, in which the concave isopicnals deepen and form convergent surface flows, so that a high SST eddy core forms at the sea surface.

10.3.3 *3D geometric models of eddies*

Based on the vertical distribution of eddy radius at 15 levels derived from eddy-resolving model output, Lin *et al.* (2015) constructed 3D geometric models of eddy vertical structures in the SCS: the bowl-shaped, lens-shaped, and cone-shaped eddies with larger eddy sizes on the surface, middle-depth and deep-water as shown in Figure 10.6. Statistical results show that over 80% of eddies are bowl-shaped ones. It is suggested that different vertical structures of eddies are attributed to different eddy generation mechanisms. Bowl-shaped

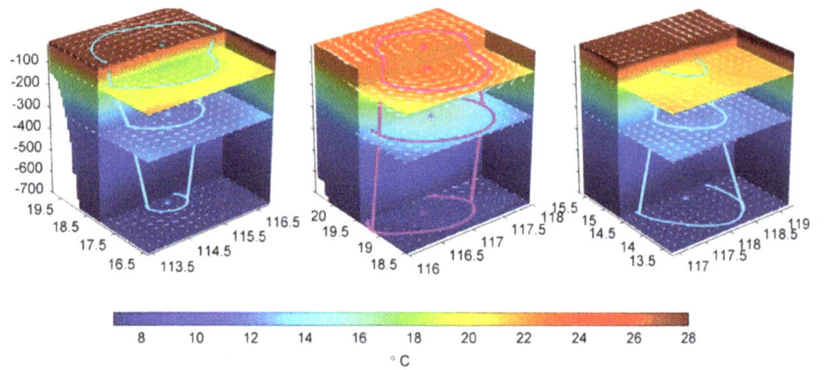

(a) (b) (c)

Fig. 10.6. 3D geometric models of vertical structures of abnormal and normal mesoscale eddies in the SCS: (a) bowl-shaped ACE, (b) lens-shaped AE and (c) cone-shaped ACE. White arrows represent the current velocity fields (in m s^{-1}). Cyan (magenta) outlines represent cyclonic (anticyclonic) eddy boundaries. Color codes represent the temperature in °C.

Source: Taken from Lin *et al.* (2015).

eddies result from the surface flow instability and wind stress curl input. Lens-shaped eddies caused by middle-depth intensification result from the current interact with the shelf slope. Cone-shaped eddies caused by deep-water intensification result from the flow instability in the bottom boundary layer. Note that eddies shown in Figures 10.6(a) and 10.6(c) are ACEs, while Figure 10.6(b) a normal AE.

10.4 3D Structures of Eddies Observed in the SCS

10.4.1 *Surface-intensified eddies*

Using the mooring array observations collected during the South China Sea Mesoscale Eddy Experiment (S-MEE) in the northern SCS (NSCS) in 2013 and 2014, Zhang *et al.* (2016) constructed full-depth 3D current field structures of an anticyclonic and cyclonic eddy pair. During the cruise observation, two bottom-anchored subsurface mooring arrays were deployed along the statistical pathway of mesoscale eddies in the test region as shown in Figure 10.7(a). The first array, cross-shaped and consisting of 10 moorings marked

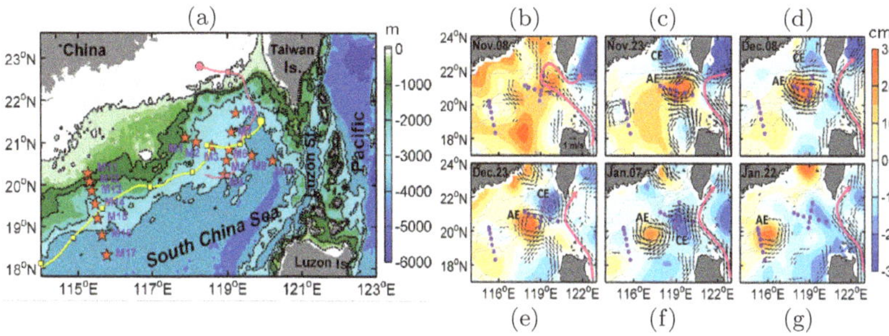

Fig. 10.7. (a) Locations of the S-MEE mooring arrays (red stars). Color-coded bathymetry with black dashed lines shows isobaths 100, 500, 1000, 2000 and 3000 m. Yellow and magenta lines denote trajectories of AE and CE, on which the dots and the squares represent the eddy generation site and locations every 15 days from 8 November 2013 to 22 January 2014. (b–g) Time series SLA images with imaging date labeled on the top right corner of each image. Color codes of SLA are in cm. Small arrows represent surface absolute geostrophic velocities >0.3 ms^{-1}. Purple dots represent the mooring locations. Magenta long arrows show outlines of the Kuroshio axis locations.

Source: Taken and re-labeled from Zhang *et al.* (2016).

as stars, was deployed in west of the Luzon Strait in the generation region of eddies in late October 2013, and recovered in early June 2014. The second array, consisting of 7 moorings along a line normal to the local isobaths, was deployed in the dissipation region of eddies in mid-January 2013, and recovered in mid-June 2014. All the moorings were equipped with ADCPs, recording current meters (RCMs), CTDs and temperature chains to monitor horizontal current velocity, pressure and temperature/salinity (T/S) in the whole water column. The moored velocity and T/S data are used to construct 3D structures of eddies. AVISO merged satellite altimeter SLA and absolute surface geostrophic velocity data are used to identify and track eddies.

Using the mooring-array measured water temperature data, Zhang *et al.* (2016) constructed 3D temperature structures of mesoscale eddies over the water column from 150 m to 800 m for AE0 (Figure 10.7(c)), CE (Figure 10.7(f)), and AE1 (Figure 10.7(d)) in the SCS as shown in Figure 10.8(a–c). One can see that all three cases belong to the SIEs with the maximum temperature anomaly of 6°C, −4°C and 4°C at the depth of 150 m, respectively. The temperature anomaly decreases sharply with the depth and reaches about 10%

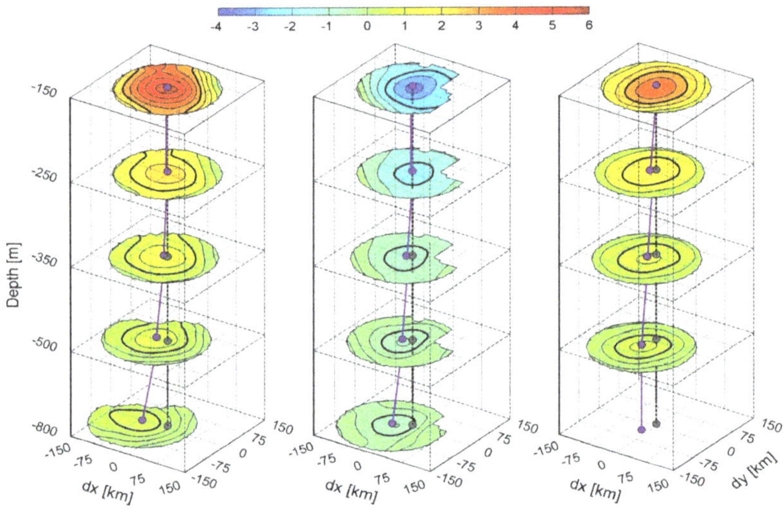

(a) (b) (c)

Fig. 10.8. 3D temperature anomaly (in °C) structures of ocean eddies (a) AE0, (b) CE and (c) AE1 in Figure 10.7(c), (f) and (d). Temperature anomaly at 500 and 800 m are amplified by 1.5 and 2. The contour intervals (black lines) at 150, 250, 350, 500 and 800 m are 0.6, 0.6, 0.3, 0.15 and 0.15°C, respectively. The black-thick lines at layers from top to bottom represent contours 3, 1.8, 0.9, 0.6 and 0.6°C for (a) and (c), and −1.8, −1.2, −0.6, −0.45 and −0.45°C for (b). Purple dots at each layer represent eddy centers defined as the points of maximum temperature anomaly. Purple line represents the eddy axis defined as the line connecting the eddy centers at each layer. Black dashed line represents the z-axis of the coordinates.

Source: Taken from Zhang *et al.* (2016).

at the lower layer. The eddy axes are tilting from the vertical axis of coordinates. The inclination angles with respect to the horizontal bottom plane are as small as 1.15°, 1.15° and 0.8° for AE0, CE and AE1, respectively.

Combining the mooring-array data and AVISO data, Zhang *et al.* (2016) constructed 3D structures of current velocity fields of mesoscale eddies over the full water column in the SCS. Figure 10.9(a–c) show 3D structures of horizontal velocity for AE0, CE and AE1, respectively. One can see the following features. Vertically, all three cases belong to the SIEs with the maximum velocity amplitude of 0.4 m s^{-1} in the surface layer. They penetrate the full water column from the surface through the near bottom (with depth

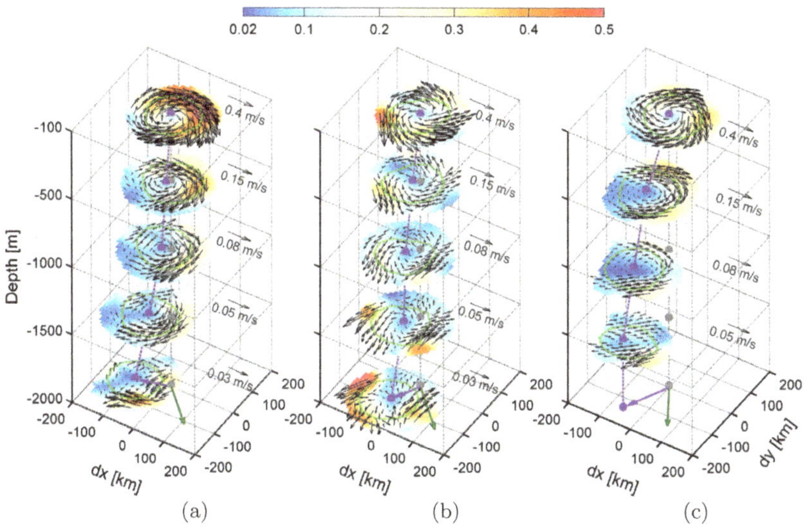

Fig. 10.9. Mooring array-derived 3D structures of the horizontal velocity for (a) AE0, (b) CE and (c) AE1. Color codes and black arrows denote the velocity magnitude (in m s^{-1}) and velocity vectors, respectively. Note the varying velocity amplitude scales with depths, i.e., the amplifying ratios of 1, 3, 5, 8 and 15 are applied for plotting the velocity fields at 100, 500, 1000, 1500 and 2000 m, respectively. Gray dotted line is the z-axis of the coordinates. Eddy centers at each layer, which are defined as the points of minimum velocity magnitudes, are marked by purple dots. Edges of eddy core at each layer are defined as the radial positions of maximum velocity magnitudes and marked by green dotted circles. Near vertical purple dotted lines denote the eddy axes defined as the lines connecting the eddy centers at each layer. Purple arrows show the drift of eddy center from the z-axis at 2000 m. The green arrow denotes the direction of the local topography gradient.

Source: Taken from Zhang *et al.* (2016).

over 2000 m) with sharply decreasing velocities. The maximum velocity magnitude at 2000 m exceeds 0.035 m s^{-1} (8.75% of that in the surface layer) and 0.05 m s^{-1} (12.5% of that in the surface layer) for the AE0 and CE, respectively, which are much stronger than the time-mean current of $O(0.01\text{–}0.02)$ m s^{-1} detected in the deep SCS circulation (Zhou *et al.*, 2020). Horizontally, the current field structures of eddy cores present swirl patterns, so that the velocity vector distributions are center-asymmetric, i.e., the velocities are enhanced on the northeastern side of AE0 and the southwestern side of CE. It implies that from the viewpoint of the current field, the eddy core does not have any enclosed boundary, the so-called circle and ellipse are only approximate models. The horizontal section area for CE change little from the sea surface to 2000 m, while that of AE0 and AE1 reduce about 10–20%.

From Figure 10.7(a), 10.7(c) and 10.7(d), one can see that the AE was well captured twice by the crossing mooring array with a time interval of 15 days from 23 November (AE0) to 8 December (AE1) 2013. Thus, it provides a good chance to examine the evolution of 3D structures of mesoscale AE during the early dissipation process. Within 15 days from AE0 to AE1, the peak SLA amplitude, EKE and available potential energy (APE) decreased by 40%, 47% and 35%, implying a quick dissipation process (Zhang *et al.*, 2016). During the dissipation process, 3D structures presented remarkable evolutions in comparison of Figure 10.9(c) to Figure 10.9(a). (1) 3D scales were shrunk: the horizontal section scale (major axis) at 100 m shrank for about 4% from about 135 km to about 130 km and for about 19% from about 135 km to about 115 km at 1500 m, the disturbed depth shrunk for at least 25% from 2000 m to 1500 m. (2) 3D vertical geometric shape evolved from a near cylinder to a bowl-shaped pattern. (3) The eddy axis was further tilting. The inclination angle with respect to the horizontal bottom plane was getting 30% lower from 1.15° to 0.8°.

10.4.2 *Subsurface-intensified eddies*

10.4.2.1 *3D structures of SSAE in the NSCS*

Li *et al.* (2019) constructed 3D structures of an AE in the NSCS using an underwater glider network consisting of 12 Petrel gliders to measure the high-resolution data in August 2017. The surveyed AE was centered at around 21°N, 118°E with the depths of 2000 $\sim$ 3000 m. Time series maps of absolute dynamic topography (MADT) as shown in Figure 10.10 and the glider-surveyed T-S diagram (Figure 8 in Li *et al.*, 2019) reveal that the surveyed AE was shed by the loop current (LOC) of the Kuroshio (see Chapter 7 for details).

Using the glider-surveyed data, 3D structures of potential temperature anomaly (PTA) and salinity anomaly (SA) of the AE from 50 m to 900 m are constructed as shown in Figure 10.11. From 3D PTA structures shown in Figures 10.11(a) and 10.11(b), one can see that the horizontal length scale (PTA $>$ 0) of eddy is of $O(200)$ km

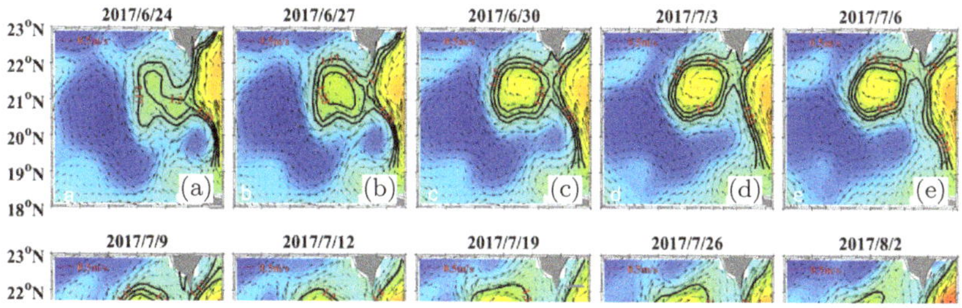

Fig. 10.10. Weekly MADT of the survey area in the NSCS west of the Luzon Strait from (a) 9 August to (e) 6 September 2017. Thick black contours represent the MADT isolines of 1.23 m. Color codes represent the MADT in m. Small arrows show the geostrophic current vector fields (in cm s^{-1}). Gray boxes denote the survey area of underwater gliders.
Source: Taken and re-labeled from Li *et al.* (2019).

at 100 m. Within the eddy horizontal section, the PTA distribution is center-asymmetrical, i.e., there is a high PTA core (>1°C) on the western side. The horizontal scale of high PTA core is of $O(100)$ km at the depth of 100 m and reduces to $O(50)$ km at 900 m. From 3D SA structures shown in Figure 10.8(c) and 10.8(d), one can see that the horizontal length scale of eddy high salinity core (SA > 0) is of $O(200)$ km at 50 m, reduces to $O(100)$ km at 150 m and disappears at 200 m. The glider-survey data give that the maximum PTA over 3°C appears at the thermocline of 100 m and the maximum SA of 0.8 psu exists at 50 m, implying that the surveyed AE is an SSAE.

10.4.2.2 *3D structures of eddies in the Kuroshio region near the SCS*

Ren *et al.* (2023) constructed 3D structures of mesoscale eddies based on observations from the CPIES array deployed in the Kuroshio region near the SCS from January 2018 and July 2019. As shown in Figure 10.12(a), one can see that the CPIES array consisting of 12 stations was located at the latitude band of 22°–23°N, where eddies from the Pacific frequently intrude and have significant impacts to eddy generation in the SCS (see Chapter 7 for details).

Figure 10.13 shows 2D (x–z) vertical profiles of temperature, salinity and meridional velocity derived from CPIES array measurements from the sea surface to 1000 m along the 22.3°N section across AE

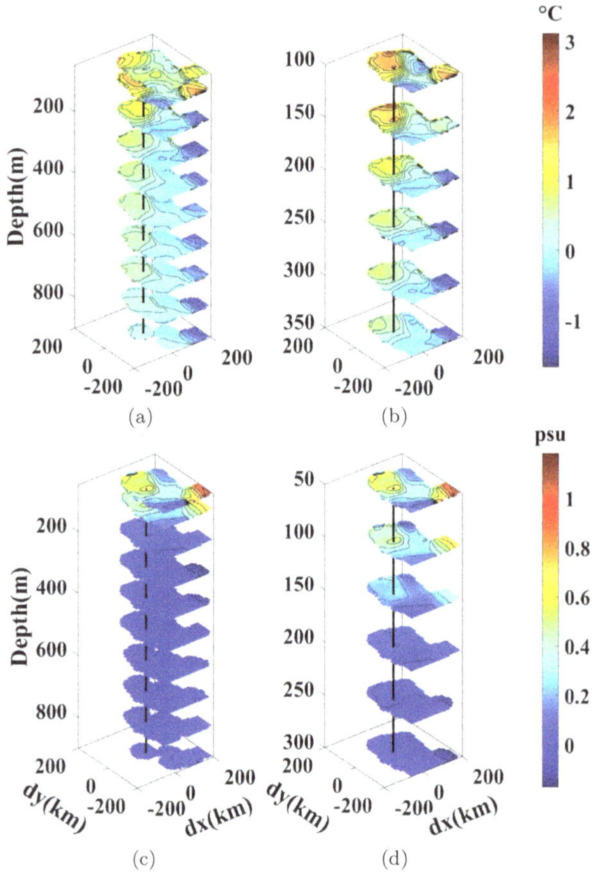

Fig. 10.11. Underwater glider-surveyed 3D structures of (a) eddy PTA (in °C) above 900 m and (b) above 350 m, (c) SA (in psu) above 900 m and (d) above 300 m.
Source: Taken from Li *et al.* (2019).

and CE centers shown in Figures 10.12(b) and 10.12(c) (Ren *et al.*, 2023). One can see that the AE centered at 22.35°N, 123.5°E has two positive temperature anomaly cores centered at about 200 and 500 m with the maximum values of about 0.6°C and 1.4°C at the eddy center and the thicknesses of 200 and 400 m as shown in Figure 10.13(a). The salinity profile (Figure 10.13(b)) shows one positive salinity anomaly core centered at 450 m with the maximum value of about 0.08 psu and the thickness of 300 m, and another negative salinity anomaly core centered at 850 m with the minimum value of

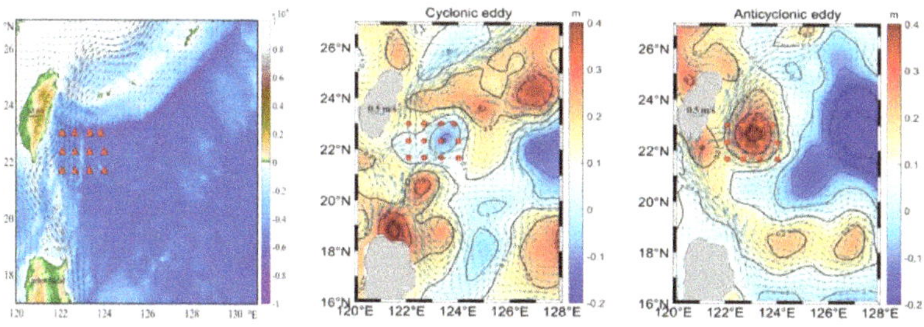

(a) (b) (c)

Fig. 10.12. (a) CPIES array (red triangles) deployed in the Kuroshio region near the SCS from January 2018 and July 2019. Concurrent average surface geotropic currents (black arrows in m s^{-1}) are derived from AVISO data. Bottom topography (color codes in 1×10^4 m) is derived from the ETOPO1 Global Relief Model (https://www.ngdc.noaa.gov/mgg/global/etopo1sources.html). (b) and (c) SLA images (color codes in m) with the surface geostrophic current (blue arrows in m s^{-1}) of AE and CE captured by the CPIES array.

Source: Taken and re-drawn from Ren *et al.* (2023).

about −0.06 psu and the thickness of 300 m. The meridional velocity profile (Figure 10.13(c)) shows that the eddy disturbance reaches 1000 m ($|v| = 0$) and the strong disturbance ($|v| > 0.2\,\mathrm{m\,s^{-1}}$) reaches the depth of 600 m. For the CE centered at 123.75°E, 22.35°N, the temperature profile (Figure 10.13(d)) shows two negative temperature anomaly cores centered at about 150 and 300 m with the minimum value of about −0.8°C at the eddy center and the thicknesses of 100 and 250 m. The salinity profile (Figure 10.13(e)) shows one negative salinity anomaly core centered at 350 m with the minimum value of about −0.06 psu and the thickness of 350 m, and another positive salinity anomaly core centered at 800 m with the maximum value of about 0.04 psu and the thickness of 300 m. The meridional velocity profile (Figure 10.13(f)) shows that the eddy disturbance reaches the depth of 1000 m ($|v| = 0$) and the strong disturbance ($|v| > 0.2\,\mathrm{m\,s^{-1}}$) reaches the depth of 300 m.

Figure 10.14 shows 3D structures of the current velocity of AE and CE (Figures 10.12(b) and 10.12(c)) derived from CPIES array measurements from the sea surface to 1000 m. The horizontal distributions of temperature and salinity anomalies as shown in Figures 10.12 are superposed to each layer. The eddy center at each layer is identified by the minimum velocity point (Ren *et al.*, 2023). One can see the following features of 3D structures of observed eddies.

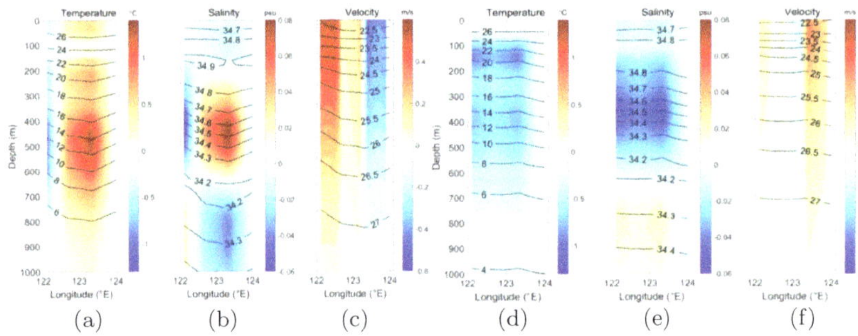

Fig. 10.13. 2D (x–z) vertical profiles of (a) temperature, (b) salinity and (c) meridional velocity from the sea surface to 1000 m along the 22.3°N section across the AE center as shown in Figure 10.12(b). Solid lines represent isothermal, isohaline and isopycnic lines. Color codes represent temperature anomaly (in °C), salinity anomaly (in psu) and meridional velocity (in m s^{-1}). (d–f) the same as (a–c) but for the CE in Figure 10.12(c).

Source: Taken and re-drawn from Ren *et al.* (2023).

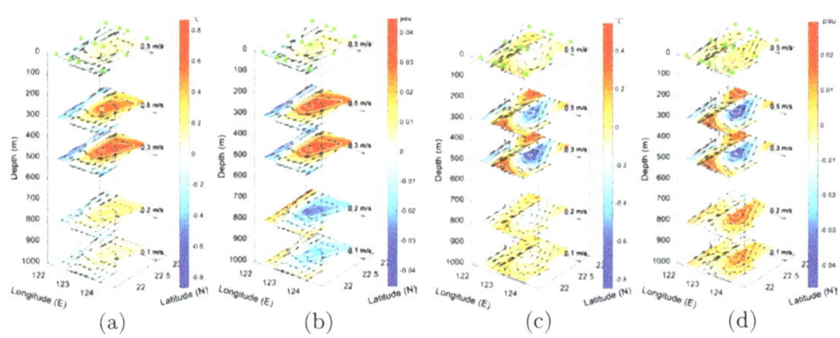

Fig. 10.14. CPIES array-derived 3D structures of the horizontal velocities of AE and CE in Figure 10.12(b) and 10.12(c). Black arrows denote the velocity vectors. Note the varying velocity amplitude scales (in m s^{-1}) with depths, i.e., the amplifying ratios of 1, 1.5, 2.5, 3.5 and 7.5 are applied for plotting the velocity fields at 0, 300, 500, 800 and 1000 m, respectively. Color codes represent the temperature anomaly (in °C) in (a) and (c) and the salinity anomaly (in psu) in (b) and (d). Green triangles shown on the surface layer represent the CPIES array.

Source: Taken and re-drawn from Ren *et al.* (2023).

Vertically, the two cases belong to the SSIEs characterized by intensified temperature-salinity abnormal cores distributed from 300 to 500 m. The maximum velocity amplitudes in the surface layer are 0.7 m s^{-1} for AE and 0.4 m s^{-1} for CE. The eddies penetrate the observed full water column from the surface through 1000 m. The maximum velocity magnitude at 1000 m linearly reduces to 0.083 m

s^{-1} (12% of that in the surface layer) for AE and 0.038 m s^{-1} (9.4% of that in the surface layer) for CE, respectively. The eddy axes are near parallel to the vertical axis of coordinates, implying little tilt.

Horizontally, the current field structures of eddy cores present swirl patterns, so that the velocity vector distributions are center-asymmetric, i.e., the velocities are enhanced on the western side of AE and the eastern side of CE, partially due to effects of the Kuroshio current field. In the two cases, the eddy radii of 70 km for AE and 60 km for CE change little from the sea surface to 1000 m.

10.5 Abnormal Eddies in the SCS

10.5.1 *Occurrence probability derived from modelling data*

Based on 9 years (2000–2008) of numerical modelling data, Sun *et al.* (2021) detected 9879 AAEs and 8620 ACEs in the SCS using the eddy snapshot counting method, accounting for 15.8% and 14.6% of the total eddy number, respectively. The average radius of abnormal eddies is around 50 km smaller than that of normal eddies. The average lifetime of AAEs (ACEs) accounts for 17.1% (16.3%) of the total average eddy lifetime, much shorter than that of normal eddies. Geographically, the highest concentration area of abnormal eddies is the northwest of Luzon Strait. The Kuroshio intrusion is considered as a key mechanism for the generation of abnormal eddies.

10.5.2 *A case of AAE detected by cruise survey*

Qi *et al.* (2022) observed an AAE in the NSCS by cruise survey and satellite SLA image on 7–9 September 2021. The survey area and CTD stations along survey sections S1 and S2 are shown in Figure 10.15. One can see that the AAE of the horizontal length scale of $O(300)$ km is centered at 17.97°N, 114.50°E in the SCS deep basin, where the depth is 3598 m.

Figure 10.16 shows 2D (x–z, y–z) structures of temperature, salinity, density and buoyancy frequency of the AE derived from cruise measurements. One can see that vertically, the AE is characterized by a lens-shaped, high-temperature, high-salinity and low-density core within 25–150 m between 26°C and 22°C isotherms. The SST at the

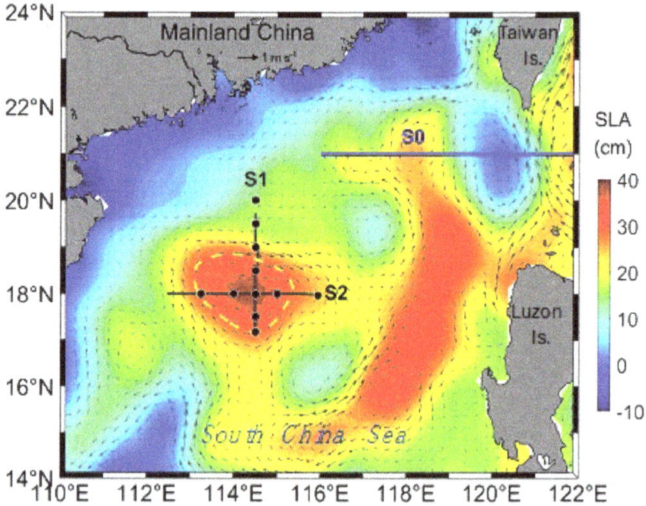

Fig. 10.15. AVISO SLA (in cm) image with the surface geostrophic velocity field (black arrows in m s^{-1}) in the NSCS on 8 September 2021. Black dots on the crossing cruise sections S1 and S2 represent CTD stations. Yellow dashed line shows the AAE centered at 17.97°N, 114.50°E.

Source: Taken and re-labeled from Qi *et al.* (2022).

AE core is 0.4°C lower than that in its peripheral region. The above features indicate that the observed AE has the AAE properties as shown in Figure 10.5.

10.6 Summaries

To analyze the eddy role in modulating of the basin circulation quantitatively, accurate knowledge of 3D structures of eddies is a must. Thus, the research in this field has gained important progresses in recent years. This chapter summaries the research results on 3D structures of eddies in the SCS. According to vertical structure features, eddies are categorized into SIEs and SSIEs with abnormal eddy as a special case of SSIEs. The major features of each category are summarized as follows.

On the sea surface, statistics of eddy eccentricity derived from 26 years of AVISO satellite SLA product from 1993 to 2018 reveal that 2D horizontal structures of most eddies in all categories appear

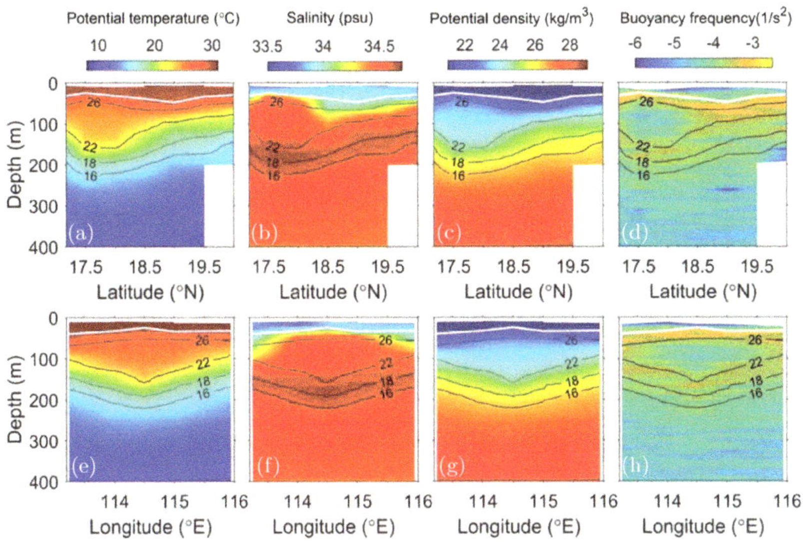

Fig. 10.16. 2D structures of thermohaline properties of AAE measured along meridional section S1 (y–z, top row) and zonal section S2 (x–z, bottom row), (a) and (e) potential temperature (in °C), (b) and (f) salinity (in psu), (c) and (g) potential density (in kg m^{-3}) and (d) and (h) squared buoyancy frequency (N^2) (in s^{-2}). White thick lines represent the mixed layer depth (MLD). Black thin lines represent isotherms of 26°C, 22°C, 18°C and 16°C.

Source: Taken from Qi *et al.* (2022).

as the elliptic structure with the eccentricity in a range from 0.7 to 0.9. The statistical mean shape is a mathematic ellipse with the semi-major axis of 101.3 km and a semi-minor axis of 61.3 km. The eddy eccentricity decreases with the lifetime and the SLA amplitude and appears as a general increase (decrease) trend with the vorticity (EKE). The 3D SSH distribution of mesoscale eddies in all categories can be described by the empirical model "universal structure of mesoscale eddies".

The vertical structures of SIEs have an IDAC seating in the sea surface layer. For SAEs, the sea surface density anomaly ($SS\rho$) < 0 and isopicnals present concave patterns. For SCEs, $SS\rho > 0$ and isopicnals present convex patterns. Mooring-array measurements of an SAE in the NSCS show that the eddy appears as a near cylinder pattern at the early growing stage and evolves to a shrunken bowl-shaped pattern at the dissipation stage. The eddy axis is tilting with the inclination angle of about $O(1^\circ)$ with respect to the horizontal bottom plane.

The vertical structures of SSIEs are featured by the IDACs seating in the subsurface layer. For SSAEs, SSHA > 0, SS$\rho > 0$, SSTA < 0 and isopicnals present convex (concave) patterns above (below) IDAC. For SSCEs, SSHA < 0, SS$\rho < 0$, SSTA > 0 and isopicnals present concave (convex) patterns above (below) IDAC. 3D structures of PTA and SA of an SSAE in the northeastern SCS are constructed by the glider-surveyed data, showing that the maximum PTA over 3°C appears at 100 m and the maximum SA of 0.8 psu exists at 50 m in the subsurface layer. In comparison, the cruise measurements of two SSIEs in the Kuroshio region show that intensified temperature-salinity abnormal cores are distributed at depths from 300 to 500 m.

The abnormal eddies are special cases of SSIEs. The AAE is the case, in which the convex isopicnals of an SSAE break at the sea surface and induce divergent surface flows and a low SST eddy core at the sea surface. The ACE is the case, in which the concave isopicnals of an SSCE deepen and form convergent surface flows and a high SST eddy core at the sea surface. The occurrence probability of abnormal eddies in the SCS is about 15% and the average lifetime accounts for 17% of the total average eddy lifetime. The highest concentration area is the northwest of Luzon Strait. An AAE in the NSCS deep basin was caught by the cruise survey and satellite SLA image. It is characterized by a lens-shaped, high-temperature, high-salinity and low-density core within 25–150 m. The SST at the AAE core is 0.4°C lower than that in its peripheral region.

References

Bian, C., Jing, Z., Wang, H., Wu, L., Chen, Z., Gan, B. and Yang, H. (2023). Oceanic mesoscale eddies as crucial drivers of global marine heatwaves, *Nat. Commun.*, 14, p. 2970, doi:10.1038/s41467-023-38811-z.

Cheng, Z., Meng, Z., Zhong, Y., Zhang, Z., Liu, H. and Zhou, L. (2020). Statistical characteristics of mesoscale eddies on the continental slope in the northern South China Sea, *Acta Oceanol. Sin.*, 39(3), pp. 36–44, doi:10.1007/s13131-019-1530-3.

Dong, C., McWilliams, J., Liu, Y. and Chen, D. (2014). Global heat and salt transports by eddy movement, *Nat. Commun.*, 5, pp. 3294, doi:10.1038/ncomms4294.

Han, G., Tian, F., Ma, C. and Chen, G. (2021). The geometry of mesoscale eddies in the South China Sea: characteristics and implications, *Intl. J. Digital Earth*, 14(4), pp. 464–479, doi:10.1080/17538947.2020.1842523.

Hu, J., Gan, J., Sun, Z., Zhu, J. and Dai, M. (2011). Observed three-dimensional structure of a cold eddy in the southwestern South China Sea, *J. Geophys. Res.*, 116, C05016, doi:10.1029/2010JC006810.

Hu, J., Zheng, Q., Sun, Z. and Tai, C.-K. (2012). Penetration of nonlinear Rossby eddies into South China Sea evidenced by cruise data, *J. Geophys. Res.*, 117, C03010, doi:10.1029/2011JC007525.

Li, S., Wang, S., Zhang, F. and Wang, Y. (2019). Constructing the three-dimensional structure of an anticyclonic eddy in the South China Sea using multiple underwater gliders, *J. Atmos. Ocean Tech.*, 36, pp. 2449–2469, doi:10.1175/JTECH-D-19-0006.1.

Lin, X., Dong, C., Chen, D., Liu, Y., Yang, J., Zou, B. and Guan, Y. (2015). Three-dimensional properties of mesoscale eddies in the South China Sea based on eddy-resolving model output, *Deep-Sea Res. I*, 99, pp. 46–64, doi:10.1016/j.dsr.2015.01.007.

Liu, Y. and Li, X. (2023). Impact of surface and subsurface-intensified eddies on sea surface temperature and chlorophyll a in the northern Indian Ocean utilizing deep learning, *Ocean Sci.*, 19, pp. 1579–1593, doi:10.5194/os-19-1579-2023.

Liu, Y., Zheng, Q. and Li, X. (2021). Characteristics of global ocean abnormal mesoscale eddies derived from the fusion of sea surface height and temperature data by deep learning, *Geophys. Res. Lett.*, 48, e2021GL094772, doi:10.1029/2021GL094772.

Nan, F., Yu, F., Xu, A. and Ding, Y. (2022). Progress and prospect of subsurface-intensified eddies in the northwestern Pacific Ocean, *Adv. Earth Sci.*, 37(11), pp. 1115–1126, doi:10.11867/j.issn.1001-8166.2022.061.

Qi, Y., Mao, H., Du, Y., Li, X., Yang, Z., Xu, K., Yang, Y., Zhong, W., Zhong, F., Yu, L. and Xing, H. (2022). A lens-shaped, cold core anticyclonic surface eddy in the northern South China Sea, *Front. Mar. Sci.*, 9, 976273, doi:10.3389/fmars.2022.976273.

Qiu, B. and Chen, S. (2005). Eddy-induced heat transport in the subtropical North Pacific from Argo, TMI and altimetry measurements, *J. Phys. Oceanogr.*, 35, pp. 458–473, doi:10.1175/JPO2696.1.

Ren, Q., Liu, Y, Wang, R., Nan, F., Wang, J., Diao, X., Chen, Z., Yu, F., Zhang, C., Zhao, R., Zheng, H. and Zhu, X.-H. (2023). Three-dimensional structure of mesoscale eddies and their interaction with Kuroshio based on observations from a CPIES array, *Front. Mar. Sci.*, 10, 1109894, doi:10.3389/fmars.2023.1109894.

Sandalyuk, N. V., Bosse, A. and Belonenko, T. V. (2020). The 3-D structure of mesoscale eddies in the Lofoten Basin of the Norwegian Sea: A composite analysis from altimetry and in situ data, *J. Geophys. Res. Oceans*, 125, 10, doi:10.1029/2020JC016331.

Stanley, Z., Bachman, S. D. and Grooms, I. (2020). Vertical structure of ocean mesoscale eddies with implications for parameterizations of tracer transport, *J. Adv. Model. Earth Sys.*, 12, e2020MS002151, doi:10.1029/2020MS002151.

Sun, W., Liu, Y., Chen, G., Tan, W., Lin, X., Guan, Y. and Dong, C. (2021). Three-dimensional properties of mesoscale cyclonic warm-core and anticyclonic cold-core eddies in the South China Sea, *Acta Oceanol. Sin.*, 40(10), pp. 17–29, doi:10.1007/s13131-021-1770-x.

Wang, L. and Gan, J. (2014). Delving into three-dimensional structure of the West Luzon Eddy in a regional ocean model, *Deep-Sea Res. I*, 90, pp. 48–61, doi:10.1016/j.dsr.2014.04.011.

Wunsch, C. (1999). Where do ocean eddy heat fluxes matter? *J. Geophys. Res.*, 104(C6), pp. 13235–13249, doi:10.1029/1999JC900062.

Xie, L. and Zheng, Q. (2017). New insight into the South China Sea: Rossby normal modes, *Acta Oceanol. Sin.*, 36(7), pp. 1–3, doi:10.1007/s13131-017-1077-0.

Xie, L., Zheng, Q., Zhang, S., Hu, J., Li, M., Li, J. and Xu, Y. (2018). Rossby wave normal modes in the South China Sea deep basin evidenced by satellite altimetry, *Intl. J. Rem. Sen.*, 39(2), pp. 399–417, doi:10.1080/01431161.2017.1384591.

Zhang, Z., Tian, J., Qiu, B., Zhao, W., Chang, P., Wu, D. and Wan, X. (2016). Observed 3D structure, generation, and dissipation of oceanic mesoscale eddies in the South China Sea, *Sci. Rep.*, 6, 24349, doi:10.1038/srep24349.

Zhang, Z., Wang, W. and Qiu, B. (2014). Oceanic mass transport by mesoscale eddies, *Science*, 345(6194), pp. 322–324.

Zhang, Z., Zhang, Y., Wang, W. and Huang, R. X. (2013). Universal structure of mesoscale eddies in the ocean, *Geophys. Res. Lett.*, 40, pp. 3677–3681, doi:10.1002/grl.50736.

Zheng, Q., Ho, C.-R., Xie, L. and Li, M. (2019). A case study of a Kuroshio main path cut-off event and impacts on the South China Sea in fall-winter 2013–2014, *Acta Oceanol. Sin.*, 38(4), pp. 1–8, doi:10.1007/s13131-019-1411-9.

Zheng, Q., Hu, J., Zhu, B., Feng, Y., Jo, Y.-H., Sun, Z., Zhu, J., Lin, H., Li, J. and Xu, Y. (2014). Standing wave modes observed in the South China Sea deep basin, *J. Geophys. Res.*, 119, pp. 4185–4199, doi:10.1002/2014JC009957.

Zheng, Q., Tai, C.-K., Hu, J., Lin, H., Zhang, R., Su, F.-C. and Yang, X. (2011). Satellite altimeter observations of nonlinear Rossby eddy-Kuroshio interaction at the Luzon Strait, *J. Oceanogr.*, 67, pp. 365–376, doi:10.1007/s10872-011-0035-2.

Zhou, M., Wang, G., Liu, W. and Chen, C. (2020). Variability of the observed deep western boundary current in the South China Sea, *J. Phys. Oceanogr.*, 50, pp. 2953–2963, doi:10.1175/JPO-D-20-0013.1.

Chapter 11

Eddy-Induced Transports in the South China Sea

11.1 Introduction

The South China Sea (SCS), the largest semi-enclosed marginal sea in the Asia-Pacific region, connects to the western Pacific Ocean through the Luzon Strait (LS) in the east and features a broad northern shelf and steep continental slope. The LS opens a pathway for Pacific waters to intrude the SCS seasonally. The upper-ocean circulation in the SCS is dominated by monsoon winds and intermittent Kuroshio intrusions, which drive strong seasonal variability (Su, 2005). In contrast, the northern deep-sea basin sustains a perennial cyclonic gyre (Hu *et al.*, 2000), accompanied by a southwestward slope current along its northern boundary (Shu *et al.*, 2018).

Mesoscale eddies are key dynamic features that substantially contribute the energy and material transport to the global ocean (Shu *et al.*, 2014; Zhang *et al.*, 2020). Satellite and *in situ* analyses reveal that the SCS exhibits particularly high eddy activity (Lin *et al.*, 2012; Zheng *et al.*, 2017), making the investigation of eddy-induced transport processes scientifically significant.

11.2 Characteristics of Eddy-Induced Transports in the SCS

Mesoscale eddy-induced transport processes in the SCS show pronounced seasonal variability and spatial heterogeneity, closely linked to asymmetric temperature and salinity distributions. Eddy heat transport (EHT) primarily occurs in the thermocline (upper 400 m), while salt transport is confined to the shallower halocline (upper 100 m) due to stronger vertical salinity gradients near the surface (Chen *et al.*, 2012). This contrast reflects the barrier-layer stratification, where strong vertical temperature gradients extend deeper than salinity gradients, resulting in spatial layering of transport pathways.

Spatially, EHT maxima (>180 MW m^{-1}) concentrate in the SCS western boundary current region, exhibiting zonal distribution patterns in both the northern SCS (NSCS) and southeastern Vietnamese waters (SEV). Seasonally, EHT peaks in fall in SEV but reaches maximum values during spring and winter in the NSCS, with consistently lowest summer values (Jiang *et al.*, 2016). Notably, regional variations exit: the SEV and western LS regions show strongest EHT within the upper 400 m, with seasonal maxima occurring in fall and winter respectively (Ding *et al.*, 2021). However, Gonaduwage *et al.* (2021) reported that EHT peaks occur in summer, particularly in the western SCS.

Eddy-induced turbulent heat flux anomaly displays distinct seasonal patterns: a dipole distribution in winter versus a monopole structure in summer (Zu *et al.*, 2019). These anomalies correlate linearly with sea surface temperature (SST) anomalies, with fitting slopes in winter significantly steeper than those in summer.

EHT in the SCS exhibits pronounced interannual variability. The SEV shows peak EHT about every 4 years (2003, 2007, 2011), while the NSCS follows an about 3-year cycle (2004, 2007, 2010) (Jiang *et al.*, 2016). The analysis of 1993–2015 data reveals an annual average of 4.6 cyclonic eddies (CEs) and 4.5 anticyclonic eddies (AEs), predominantly in the northern and western SCS, with mean lifetimes of 62 and 70 days, respectively. These eddies drive substantial heat (CE: 1.4×10^{12} W northward; AE: -3.1×10^{12} W southward) and salt transports (CE: -4.0×10^{4} kg s^{-1}; AE: 5.6×10^{4} kg s^{-1}), with >90% occurring in the upper 300 m (heat) and 100 m (salt) (He *et al.*, 2018).

Westward eddy propagation contributes an additional 1.4 Sv (1 Sv = 10^6 m^3 s^{-1}) water transport.

Contrastingly, meridional heat transports nearly cancel, but significant freshwater fluxes persist (CE: $373.6 \pm 9.7 \times 10^3$ m^3 s^{-1} eastward; AE: $384.2 \pm 10.8 \times 10^3$ m^3 s^{-1} westward) (Sun *et al.*, 2018). Extended analysis (1993–2018) identified 122 shed and 69 locally generated eddies, collectively transporting the volume of 0.77 Sv, the heat of 8.78×10^{-4} PW and the salt of 7.88×10^4 kg s^{-1} annually (Yang *et al.*, 2021).

EHT extremes correlate with the Summer Eastward Jet (SEJ) dynamics: strong EHT years (1994, 1999, 2002) featured enhanced SEJ-induced baroclinic instability, while weak EHT years (1995, 2000, 2001) showed reduced activity (Gonaduwage *et al.*, 2021). Recent work highlighted the influence of winter circulation patterns: "O-type" pattern weakens western boundary currents, enhancing Vietnamese coastal EHT, whereas "U-type" pattern shifts EHT maxima to the southwestern Taiwan Island and the SCS western boundary current regions, facilitating poleward heat transport (Liu *et al.*, 2023).

11.3 Cross-Slope Transport by Eddies in the NSCS

Eddies prevalent in slope regions drive along-slope currents, serving as critical conduits between the deep basin and the shelf. Their cross-slope transport facilitates shelf–open ocean exchange, significantly influencing regional physical and biological processes. In the NSCS, this cross-slope transport displays distinct spatial patterns (Ding *et al.*, 2020), with dynamics governed by topography, baroclinic effects, and Kuroshio intrusions. For instance, near the Dongsha Atoll, mid-layer flows traverse isobaths basin-ward under the combined influence of topographic and baroclinic forcings (Wang *et al.*, 2013). Model diagnostics confirm that topographic steering and baroclinic forcing are the primary drivers of the cross-slope transport. Kuroshio intrusions east of 118°E further intensify the baroclinic effects, enhancing cross-slope motions (Huang *et al.*, 2017). Additionally, the intraseasonal variability in the NSCS slope flows primarily correlates with mesoscale eddies west of the LS (Wang *et al.*, 2020).

Eddies modulate cross-slope flow through two primary mechanisms: (1) the asymmetric structure of eddies directly drives horizontal and vertical motions, generating significant cross-slope transport, and (2) the life cycle of eddies (from generation, development to dissipation) regulates the spatiotemporal patterns of the cross-slope flow. Long-term mooring observations reveal that eddies near the Dongsha slope generate net cross-slope transport through their horizontal/vertical asymmetry and evolution, though nonlinear effects suppress this transport (Wang *et al.*, 2018). Model analyses using the vorticity-deformation index reveal that as driven by the baroclinic instability, AEs transit from a circular shape to an oval one during deformation, meanwhile transport $\sim 10^{14}$ kg of water (Qiu *et al.*, 2019). Numerical simulations show that intruding eddies transfer the kinetic energy to the mean flow (Su *et al.*, 2020). The Regional Ocean Modeling System (ROMS) modeling further confirms that eddy asymmetry is the dominant control on cross-slope transport, quantifying its three-dimensional structure in the SCS (Lin *et al.*, 2023).

Cruise observations further elucidate the role of eddies in cross-slope flow dynamics. Using moored acoustic Doppler current profiler (ADCP) data from the NSCS, Li *et al.* (2014) observed that CEs induce onshore surface residual flow and offshore bottom flow, enhancing cross-slope volume transport by about 40% and generating upwelling ($\sim 5 \times 10^{-5}$ m s^{-1}). In contrast, AEs drive offshore surface flow and onshore bottom flow, promoting shelf-deep water mixing and causing salinity anomaly (± 0.3) and temperature anomaly ($\pm 1.5°$ C), with peak eddy-induced transport occurring in June–July. Tai *et al.* (2017) showed via moored observations that eddies northwest of the LS deflect slope flows northeastward (under AEs) or southwestward (under CEs).

Statistical analysis of slope-intruding eddies in the NSCS (1993–2018) revealed that 36 AEs and 22 CEs (15.7% of total eddies) had longer lifetimes (mean: 59 days) and higher energy density than ordinary eddies, with CEs persisting longer than AEs. These eddies predominantly form in winter, occur more frequently during La Niña than during El Niño events, and propagate westward or northwestward and intrude along the continental slope toward the Dongsha Atoll, east of the Hainan Island and north of the Xisha Islands.

Meanwhile, AEs continue to move onshore after reaching the slope, and CEs often dissipate near the Dongsha Atoll (Su *et al.*, 2020).

Using 11-year (2007–2017) high-resolution LASG/IAP Climate System Ocean Model (LICOM) simulations and Argo/mooring data, Qiu *et al.* (2022) investigated mesoscale eddy asymmetry and associated cross-slope heat/salt transport in the NSCS slope region. Their study revealed that slope-intruding eddies exhibit greater energy and asymmetry than non-intruding eddies, driving substantial cross-slope transports near the Xisha Islands and the Dongsha Atoll (heat of $\sim 10^{12}$ W and salt of $\sim 10^4$ kg s^{-1}; Figures 11.1 and 11.2). Potential vorticity budget analysis shows that eddy velocity asymmetry and potential vorticity asymmetry are primarily induced by eddy nonlinearity near the Dongsha Atoll and topographic β-effects near the Xisha Islands.

Wu (2025) identified three mesoscale eddies along the NSCS slope: two CEs (CE1 and CE2) and one AE (AE1) during 5 stations of mooring observations (see Figure 11.3 for mooring locations) from 8 April to 26 October 2014. For example, Figure 11.3 tracks AE1's evolution during 25 March–4 May 2014. Initially beyond M5, AE1's northern edge reached M3 by 3 April, then migrated southwestward (3–15 April; ≤ 11 cm s^{-1}), with M5 at its periphery. By 10 April, its center approached M2 while its edge touched M1. The Dongsha Atoll's topography deformed AE1, shifting its center westward in shallow waters (<1000 m). After detaching from M4 (15 April), AE1 accelerated southwestward (mean 12 cm s^{-1}). Correspondingly, core SLA and size declined after peaking near M2 in mid-April.

To examine cross-slope transport in the NSCS slope region, a unit-width cross-section perpendicular to a line between mooring stations M2–M3 was selected, and the cross-slope transport was calculated within the mooring observation depth ranges: M1 (30–225 m), M2 (50–460 m), M3 (55–445 m), M4 (25–400 m), and M5 (200–600 m). By correlating SLAs with cross-slope transport time series (Figure 11.4), Wu (2025) indicated that the passage of AE1 induced significant offshore transport, with intensity inversely proportional to the distance between AE1 center and mooring station. In contrast, during CE1 and CE2 as well as eddy-free periods, onshore transport dominates, suggesting that the background cross-slope flow is primarily onshore throughout the observation period.

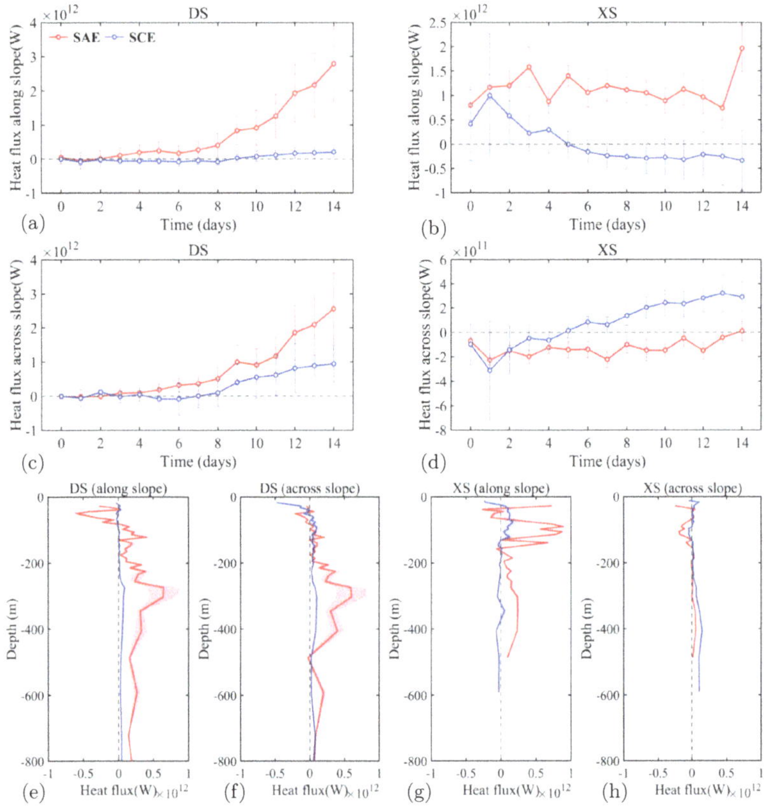

Fig. 11.1. (a, b) Along-slope and (c, d) cross-slope heat fluxes induced by slope-intruding AEs (SAEs; red lines) and slope-intruding CEs (SCEs; blue lines) near (a, c) the Dongsha Atoll and (b, d) the Xisha Islands. (e, g) Average along-slope and (f, h) cross-slope heat transport (in W) during the two weeks following eddy arrival at the slope near (e, f) the Dongsha Atoll and (g, h) the Xisha Islands. Shading denotes standard deviation.

Source: Taken from Qiu *et al.* (2022).

Table 11.1 compares the total cross-slope transport with the cross-slope transport occurring during eddy-influenced periods at each mooring station (Wu, 2025). During AE1, offshore transport was entirely dominated by eddy-induced flows at M4 and M5 (100% of the total), and accounted for 34%, 29%, and 26% of the total offshore transport at M1, M2 and M3, respectively. This underscores AE1's dominant capacity to drive offshore transport, particularly at station M2 and M4. In contrast, CEs exerted a comparatively modest

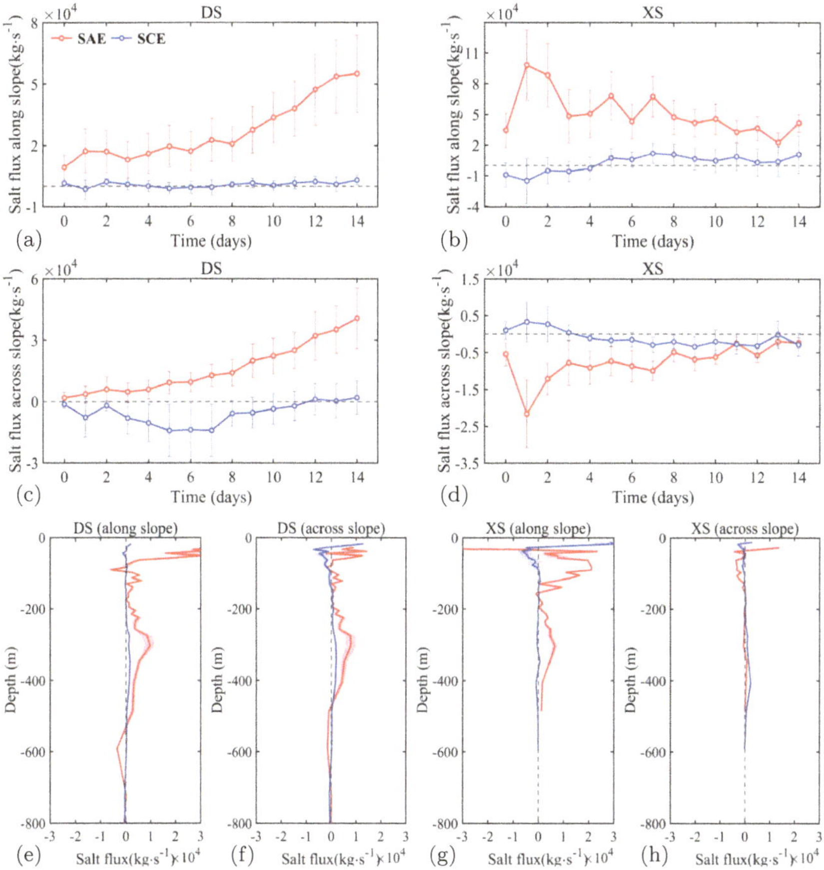

Fig. 11.2. (a, b) Along-slope and (c, d) cross-slope salt fluxes induced by SAEs (red lines) and SCEs (blue lines) near (a, c) the Dongsha Atoll and (b, d) the Xisha Islands. (e, g) Average along-slope and (f, h) cross-slope salt transport (in kg s^{-1}) during the two weeks following eddy arrival at the slope, near (e, f) the Dongsha Atoll and (g, h) the Xisha Islands. Shading denotes standard deviation. *Source*: Taken from Qiu *et al.* (2022).

effect on onshore transport: CE1 contributed 16%, 6%, 28%, 32%, and 46% of the total onshore transport at M1–M5, respectively, while CE2 affected only M1–M3 (31%, 18%, and 29%). The substantially greater offshore transport induced by AE1 compared to CEs' onshore transport suggests stronger energy in AE, enabling it to effectively reverse the background onshore-dominated flow.

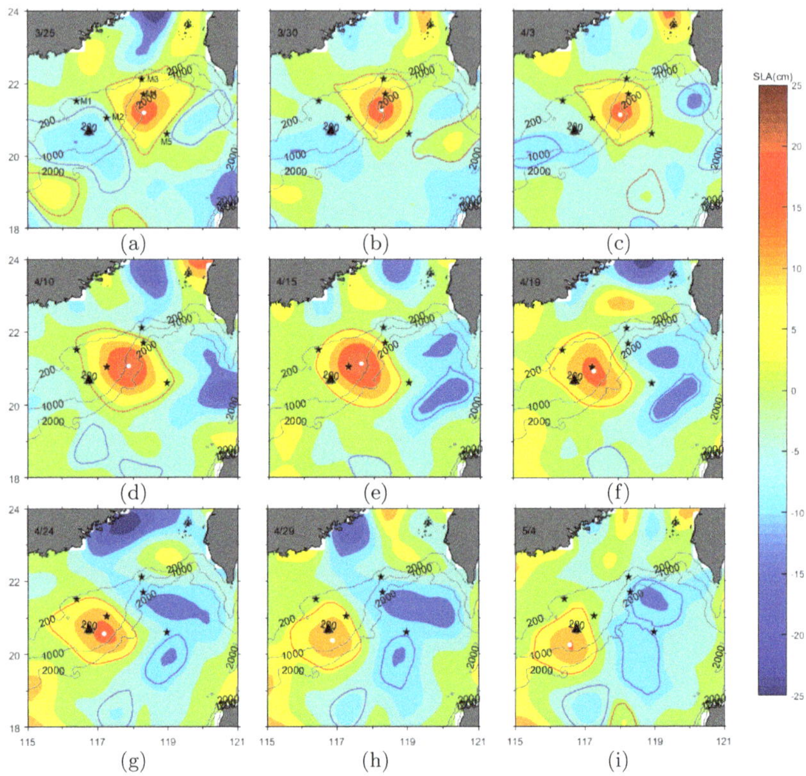

Fig. 11.3. Evolution of sea level anomaly (SLA, in cm) of AE1. Blue and red contours mark CE and AE edges, respectively. Color codes represent SLA magnitude, and white circles denote eddy centers. Black triangles mark the location of the Dongsha Atoll, with brown contours showing isobaths of 200 m, 1000 m, and 2000 m. Mooring stations M1–M5 are marked in (a).

Source: Taken from Wu (2025).

11.4 Transport by Eddies in the Northeastern SCS

The southwestern waters of Taiwan Island are characterized by intense mesoscale eddy activities. Observational studies have documented substantial eddy generation and propagation in this area. Zhang *et al.* (2013) observed the coexistence of an AE and a CE from November 2011 to January 2012, both propagating southwestwards. The AE exhibits significantly greater intensity and a longer

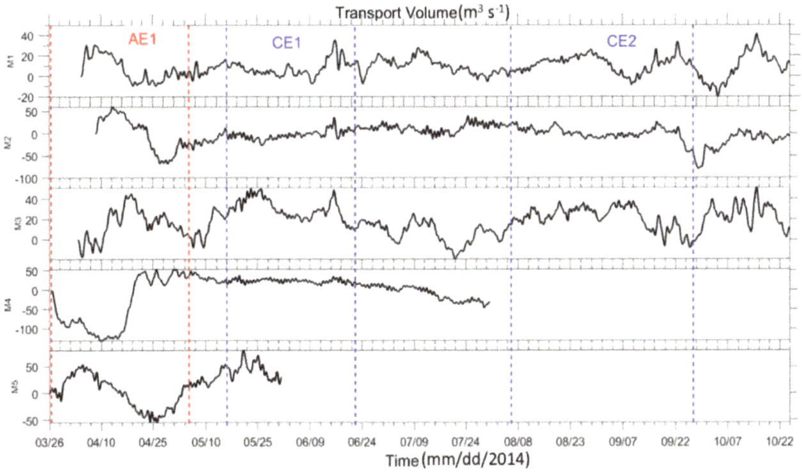

Fig. 11.4. Time series of daily cross-slope transport at mooring observation depths in the northeastern SCS (26 March–26 October 2014). Eddy events are numerically labeled. Red and blue dash lines denote periods influenced by AE and CEs, respectively. Positive (negative) values are defined as onshore (offshore) transport toward the shelf (basin).

Source: Taken from Wu (2025).

Table 11.1. Total cross-slope transport at all stations during the mooring observation period and during eddy events.

Unit: 10^3 m^3 s^{-1}	M1	M2	M3	M4	M5
AE1 (Offshore Transport only during eddy event)	−1.5	−5.8	−0.5	−50.4	−6.3
CE1 (Onshore Transport only during eddy event)	7.6	1.1	13.2	9.3	8.1
CE2 (Onshore Transport only during eddy event)	15.0	3.0	13.5	nan	nan
Total Onshore Transport during the mooring observation period	48.4	17.1	47.3	29.2	17.8
Total Offshore Transport during the mooring observation period	−4.4	−20.3	−1.9	−50.4	−6.3

Source: Taken from Wu (2025).

life cycle than the CE, with a maximum temperature anomaly of 7.5°C at 180 m depth, whereas the CE produces only weak negative temperature anomaly in the upper 200 m. Notably, a pronounced phase difference between AE temperature anomaly and velocity field results in a net westward heat transport of 228.5 ± 12.0 TW and an average annual volume transport of 0.86 ± 0.15 Sv.

Regarding eddy life cycles, Lin *et al.* (2018) documented an AE persisting from fall 2003 to spring 2004, lasting 121 days with an average radius of 50.4 km and traveling 1437.5 km (mean speed: 13.5 cm s^{-1}). The eddy propagates southwestward along the 1000–2000 m isobaths, exhibiting strong material conservation. However, during dissipation, its stability weakens due to topographic barriers and flow-field interactions, resulting in significant particle loss, so that only 18.6–35.6% of water parcels reach the waters south of the Hainan Island.

Stage-dependent transport has also been quantified. Wang *et al.* (2019) divided an AE in the northeastern SCS in November 2011 into three distinct life cycle stages: birth (22 days), growth (64 days), and decay (47 days). The eddy displayed pronounced vertical heterogeneity, featuring anticyclonic flow in the upper layer (<400 m) but cyclonic flow below 700 m, with a transitional zone between 400 and 700 m. Heat transport peaked during the growth stage (73.03% of the total), while deeper-layer (700–1500 m) heat transport increased notably during decay.

As for the material transport, Wang *et al.* (2014) indicated that an AE in January 2010 retained 93% of tracer particles over 55 days (7% escape rate) using particle tracking method. Using ROMS model with passive tracers, Zhao *et al.* (2022) quantified that an AE transports 51% of its initial water volume for 689 km during its 100-day lifetime (daily loss rate of 6‰). Their study also reveals vigorous horizontal water exchange during propagation, with 65% of initial mixed-layer water ultimately reaching the subsurface layer.

Sun *et al.* (2022) identified four subsurface mesoscale eddies (SMEs; two CEs and two AEs) in the northeastern SCS through mooring observations during the period of 2014–2016. These SMEs exhibit weak surface signals but reach the maximum velocity of 17 cm s^{-1} at the depth of 370 m, propagating westward at 4.3 cm s^{-1} on average. During 2-year observation, cyclonic SMEs (C-SMEs) transport 1.13 ± 0.27 Sv of water mass to the northeastern SCS from the northern West Pacific with 61.7% of the North Pacific Intermediate Water (NPIW), highlighting their importance in Pacific-SCS intermediate water exchange. Meanwhile, C-SMEs generate an eastward heat transport of 0.84 ± 0.25 TW by trapping, which is an order of magnitude larger than the 0.09 ± 0.03 TW induced by stirring.

11.5 Transport of Eddies Through the LS

Mesoscale eddies propagating from the Pacific across the LS represent a critical mechanism for modifying heat, salt and volume transports in the SCS. By assimilating multi-source data (satellite altimetry, SST, and hydrographic observations), Wang *et al.* (2012) reconstructed a 3D temperature-salinity dataset for the SCS (1993–2008) and quantified eddy contributions in the SCS heat and salt transports. CEs transport substantial negative heat anomaly (-6.77×10^{18} J) and freshwater (5.7×10^{-4} Sv), while AEs carry positive heat anomaly (4.67×10^{18} J) and freshwater (4.7×10^{-4} Sv), respectively. On the isopycnal of $25.0\sigma_\theta$, eddies mediate average transports of 3.7×10^{-5} PW (heat), 4.7×10^{3} kg s^{-1} (salt), and 0.3 Sv (volume). These transports show strong interannual variability, particularly linked to ENSO: El Niño summer favors positive heat transport, whereas La Niña summer exhibits negative or weakly positive transport.

Wang *et al.* (2021) investigated two high-salinity water ($\geq$34.7) intrusion events in the northeastern SCS during fall-winter of 2015/2016 using multi-platform observations (Argo floats, satellite altimetry, and reanalysis data). The study identified a dual transport mechanism involving mesoscale eddies and Kuroshio intrusion: both events were associated with Kuroshio-triggered AE–CE pairs, where the northern AE and southern CE formed a high-salinity pathway with their intervening westward jet. The first event was dominated by eddy capture (74%) supplemented by the jet (26%), while the second event showed reversed contributions (jet: 52%; AE: 43% and CE: 5%).

Using satellite altimetry data from 1993 to 2021, Zhao *et al.* (2025) identified 21 AEs and 14 CEs successfully crossing the LS into the SCS. Eddy polarity, intensity, and Kuroshio path variations emerge as key determinants of traversal capability. They

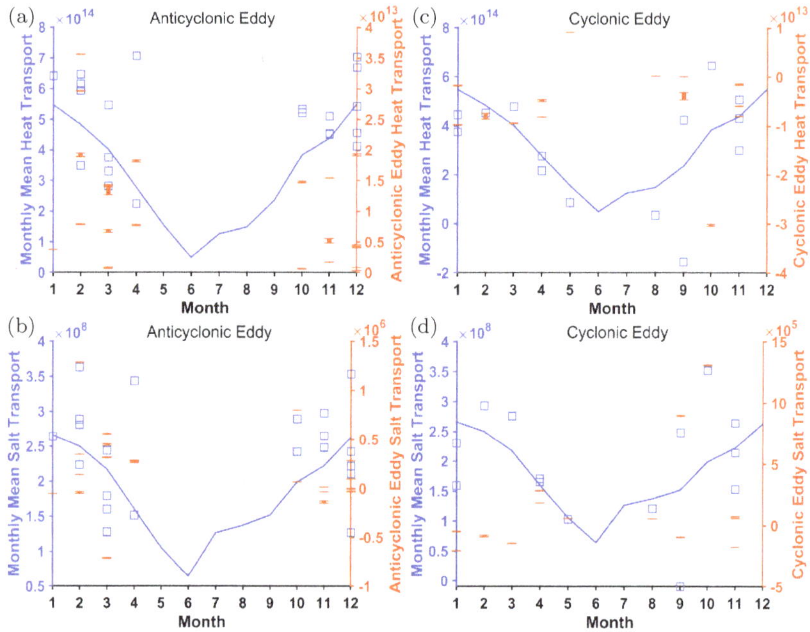

Fig. 11.5. Monthly heat and salt transports through the LS, comparing to eddy-induced transports. (a, c) AE and CE heat transport (in W); (b, d) corresponding salt transport (in kg s^{-1}). The signs "–" indicate eddy-induced transports, with orange error bars showing uncertainties from vertical resolution limitations of reanalysis data. Squares denote monthly mean transports when eddies cross the Kuroshio. Blue lines show the LS climatological monthly means.

Source: Taken from Zhao *et al.* (2025).

further quantified eddy-induced heat/salt transport using the Copernicus Marine Environment Monitoring Service (CMEMS) high-resolution reanalysis data, comparing these with total LS transports (Figure 11.5). Traversing eddies show pronounced seasonal occurrence from October to April, consistent with Hsin *et al.* (2013)'s finding that Kuroshio weakening facilitates eddy passage. During AE traversal months, heat and salt transports in the LS increase by 13% and 8% above climatological means, respectively. At 121° E, eddy-induced heat transport accounts for 1–10% of monthly mean values, while salt contributions are minimal (<1%). These findings advance understanding of eddy-mediated heat/salt balance in the SCS.

11.6 Summary

Recent studies have shown the critical role of eddy-induced transports in the SCS, particularly in energy transfer and material exchange. These findings highlight eddy's significant role in regional circulation dynamics and climate feedback, offering new insights into multi-scale processes in marginal seas. This chapter summarizes key findings regarding: (1) heat/salt transport, (2) cross-slope exchange, (3) regional transport difference, and (4) interannual variability.

11.6.1 *Key findings*

(1) Spatiotemporal heterogeneity of heat and salt transports: Mesoscale eddies in the SCS show distinct seasonal and spatial patterns in heat and salt transports. Eddy-induced heat transport is maximized in the thermocline (upper 400 m), while salt transport is confined to the halocline (upper 100 m). The western boundary current and southeastern Vietnam waters exhibit maximum heat transport, peaking in fall and winter, respectively. However, exact peak timings remain debated among studies.

(2) Dynamical mechanisms of cross-slope transport: Mesoscale eddies facilitate cross-slope transport through asymmetric horizontal structures and vertical motions, serving as critical shelf-basin connectors. AEs drive offshore transport, while CEs enhance onshore flow. AEs generally contribute more net cross-slope transport, generally contribute more net cross-slope transport. Near the Dongsha Atoll, eddy asymmetry arises primarily from topographic β-effects and nonlinearity, whereas eddy dynamics in the Xisha Islands are dominated by topographic influences. Slope-intruding eddies (e.g., SAEs and SCEs around the Dongsha Atoll) transport heat ($\sim 10^{12}$ W) and salt ($\sim$104 kg s^{-1}). Eddy propagation (typically southwestward) and lifetime (mean 59 days) further modulate transport spatiotemporal patterns.

(3) Regional differences in transport: Eddies in the SCS exhibit distinct regional transport characteristics: (1) Northeastern SCS eddies (e.g., AEs in southwest of Taiwan Island) sometimes show vertical heterogeneity (surface anticyclonic/subsurface cyclonic flow), with peak heat transport during growth stage. (2) Subsurface mesoscale eddies (SMEs) in the northeastern SCS mediate the transport of NPIW water into the SCS. (3) The LS eddy penetration correlates with Kuroshio's seasonal cycle (October–April peak). AE crossings enhance the LS heat/salt transport by 13% above climatological means. Kuroshio-triggered AE–CE pairs establish high-salinity pathways with dynamic partitioning of transport contributions.

(4) Interannual variability and climate linkages: Eddy transport exhibits ENSO-modulated interannual variability: El Niño summers favor positive heat transport, while La Niña summers yield negative r weak transport. Peak transport years (e.g., 1994, 2002) coincide with intensified surface Summer Eastward Jet (SEJ) and strengthened baroclinic instability. La Niña events additionally enhance slope-intruding eddy generation, underscoring climate forcing on eddy dynamics.

11.6.2 *Outstanding challenges and future directions*

Despite the significant achievements of existing research, the following challenges remain: (1) achieving fine-scale observation and numerical simulation of 3D eddy structures; (2) quantifying cross-scale interactions among eddies, boundary currents (e.g., Kuroshio), and monsoon forcings; and (3) elucidating how mesoscale eddies impact nutrient, carbon, and ecosystem dynamics (i.e., biogeochemical-transport coupling mechanisms).

Future studies should leverage high-resolution numerical modeling, integrated observational arrays (e.g., gliders, floats, radars), and machine-learning algorithms to advance our mechanistic understanding of eddy transport in marginal seas, thereby improving climate projections and informing marine ecosystem management.

References

Chen, G., Gan, J., Xie, Q., Chu, X., Wang, D. and Hou, Y. (2012). Eddy heat and salt transports in the South China Sea and their seasonal modulations, *J. Geophys. Res. Oceans*, 117, C05021, doi:10.1029/2011 JC007724.

Ding, R., Xuan, J., Zhang, T., Zhou, L., Zhou, F., Meng, Q. and Kang, I.-S. (2021). Eddy-induced heat transport in the South China Sea, *J. Phys. Oceanogr.*, 51, pp. 2329–2349, doi:10.1175/JPO-D-20-0206.1.

Ding, Y., Yao, Z., Zhou, L., Bao, M. and Zhang, Z. (2020). Numerical modeling of the seasonal circulation in the coastal ocean of the northern South China Sea, *Front. Earth Sci.*, 14, pp. 90–109, doi:10.1007/s11707-018-0741-9.

Gonaduwage, L.P., Chen, G., Priyadarshana, T., Wang, D. and Yao, J. (2021). Interannual variability of summertime eddy-induced heat transport in the western South China Sea and its formation mechanism, *Clim. Dyn.*, 57, pp. 451–468, doi:10.1007/s00382-021-05719-7.

He, Q., Zhan, H., Cai, S., He, Y., Huang, G. and Zhan, W. (2018). A new assessment of mesoscale eddies in the South China Sea: Surface features, three-dimensional structures, and thermohaline transports, *J. Geophys. Res. Oceans*, 123, pp. 4906–4929, doi:10.1029/2018JC014054.

Hsin, Y. C., Qiu, B., Chiang, T. L. and Wu, C. R. (2013). Seasonal to interannual variations in the intensity and central position of the surface Kuroshio east of Taiwan, *J. Geophys. Res. Oceans*, 118 (9), pp. 4305–4316, doi:10.1002/jgrc.20323.

Hu, J., Kawamura, H., Hong, H. and Qi, Y. (2020). A review on the currents in the South China Sea: Seasonal circulation, South China Sea warm current and Kuroshio intrusion, *J. Oceanogr.*, 56, pp. 607–624, doi:10.1023/A:1011117531252.

Huang, X. R., Wang, Q., Zhou, W. D. and Zhou, S. Q. (2017). Model diagnostic analysis of cross-shelf flow in the northern South China Sea, *J. Chin. Sci. Bull.*, 62, pp. 1059–1070, doi:10.1360/N972016-00570.

Jiang, W., Sun, Z. C., Xing, B. and Wang, J. (2016). Characteristic analysis of spatio-temporal variability of eddy heat transports in the South China Sea, *Hydrogr. Surv. Chart.*, 36(4), pp. 35–38.

Li, R., Chen, C., Xia, H., Beardsley, R., Shi, M., Lai, Z., Lin, H., Feng, Y., Liu, C. and Xu, Q. (2014). Observed wintertime tidal and subtidal currents over the continental shelf in the northern South China Sea, *J. Geophys. Res. Oceans*, 119(8), pp. 5289–5310, doi:10.1002/2014JC0 09931.

Lin, H., Hu, J. and Zheng, Q. (2012). Satellite altimeter data analysis of the South China Sea and the northwest Pacific Ocean: Statistical features of oceanic mesoscale eddies, *J. Oceanogr. Taiwan Str.*, 31(1), pp. 105–113.

Lin, X., Wang, G., Han, G., Liu, Y., Zhang, H., Liao, X. and Ji, Q. (2023). Cross-slope transport by a mesoscale anticyclone in the northern South China Sea, *J. Mar. Sci. Engin.*, 11, pp. 1–15, doi:10.3390/jmse11020305.

Lin, X., Dong, C. and Chen, D. (2018). Cross-basin particle transport by a warm eddy southwest of Taiwan Island, *J. Tropical Oceanogr.*, 37(3), pp. 9–18. (in Chinese with English abstract)

Liu, Q., Li, W., Shi, R, Che, J., Li, C. and Xie, Q. (2023). The characteristics of eddy in western boundary current of South China Sea and its relationship with winter circulation, *J. Trop. Oceanogr.*, 42(3), pp. 52–66. (in Chinese with English abstract)

Qiu, C., Mao, H., Liu, H., Xie, Q., Yu, J., Su, D., Ouyang, J. and Lian, S. (2019). Deformation of a warm eddy in the northern South China Sea, *J. Geophys. Res. Oceans*, 124(8), pp. 5551–5564, doi:10.1029/2019JC015288.

Qiu, C., Yi, Z., Su, D., Wu, Z., Liu, H., Lin, P., He, Y. and Wang, D. (2022). Cross-slope heat and salt transport induced by slope intrusion eddy's horizontal asymmetry in the northern South China Sea, *J. Geophys. Res. Oceans*, 127, e2022JC018406, doi:10.1029/2022JC018406.

Shu, Y., Chen, J., Yao, J., Pan, J., Wang, W., Mao, H. and Wang, D. (2014). Effects of the Pearl River plume on the vertical structure of coastal currents in the northern South China Sea during summer 2008, *Ocean Dyn.*, 64(12), pp. 1743–1752, doi:10.1007/s10236-014-0779-5.

Shu, Y., Wang, Q. and Zu, T. (2018). Progress on shelf and slope circulation in the northern South China Sea, *Sci. Chin. Earth Sci.*, 48(3), pp. 276–287, doi:10.1007/s11430-017-9152-y.

Su, D., Lin, P., Mao, H., Wu, J., Liu, H., Cui, Y. and Qiu, C. (2020). Features of slope intrusion mesoscale eddies in the northern South China Sea, *J. Geophys. Res. Oceans*, 125, e2019JC015349, doi:10.1029/2019JC015349.

Su, J. (2005). Overview of the South China Sea circulation and its dynamics, *Haiyang Xuebao*, 27(6), pp. 3–10.

Sun, W., Dong, C., Tan, W., Liu, Y., He, Y. and Wang, J. (2018). Vertical structure anomalies of oceanic eddies and eddy-induced transports in the South China Sea, *Rem. Sens.*, 10, 795, doi:10.3390/rs10050795.

Sun, Z., Zhang, Z., Qiu, B., Zhou, C., Zhao, W. and Tian, J. (2022). Subsurface mesoscale eddies observed in the northeastern South China Sea: Dynamic features and water mass transport, *J. Phys. Oceanogr.*, 52, pp. 841–855, doi:10.1175/JPO-D-21-0177.1.

Tai, J., Yang, K. and Gawarkiewicz, G. (2017). Subtidal current structure and variability of the continental shelf and slope of the northern South China Sea, *Terr. Atmos. Ocean. Sci.*, 28(3), pp. 411–423, doi:10.3319/TAO.2016.11.05.01.

Wang, D., Wang, Q., Zhou, W., Cai, S., Li, L. and Hong, B. (2013). An analysis of the current deflection around Dongsha Islands in the northern South China Sea, *J. Geophys. Res. Oceans*, 118(1), pp. 490–501, doi:10.1029/2012JC008429.

Wang, H., Xu, D., Chen, Z., Xu, M., Yang, L. and Chen, H. (2014). Exploration on the transportation capacity using an anticyclonic eddy of the South China Sea in January, 2010, *J. Mar. Sci.*, 32(4), pp. 1–10.

Wang, M., Zhang, Y., Liu, Z., Zhao, Y. and Li, J. (2019). Temporal and spatial evolution of a deep-reaching anticyclonic eddy in the South China Sea, *Sci. Chin. Earth Sci.*, 62, pp. 1002–1023, doi:10.1007/s11430-018-9318-6.

Wang, Q., Zeng, L., Li, J., Chen, J., He, Y., Yao, J., Wang, D. and Zhou, W. (2018). Observed cross-shelf flow induced by mesoscale eddies in the northern South China Sea, *J. Phys. Oceanogr.*, 48, pp. 1609–1627, doi:10.1175/JPO-D-17-0180.1.

Wang, Q., Zhou, W., Zeng, L., Chen, J. and Wang, D. (2020). Intraseasonal variability of cross-slope flow in the northern South China Sea, *J. Phys. Oceanogr.*, 50(7), pp. 2071–2084, doi:10.1175/JPO-D-19-0293.1.

Wang, X., Du, Y., Zhang, Y., Wang, A. and Wang, T. (2021). Influence of two eddy pairs on high-salinity water intrusion in the northern South China Sea during fall-winter 2015/2016, *J. Geophys. Res. Oceans*, 126, e2020JC016733, doi:10.1029/2020JC016733.

Wang, X., Li, W., Qi, Y. and Han, G. (2012). Heat, salt and volume transports by eddies in the vicinity of the Luzon Strait, *Deep-Sea Res. I*, 61, pp. 21–33, doi:10.1016/j.dsr.2011.11.006.

Wu, C. (2025). Characteristics of the eddy-induced cross-slope flow in the northeastern South China Sea, M.S. thesis, Xiamen University, 68pp.

Yang, Y., Zeng, L. and Wang, Q. (2021). How much heat and salt are transported into the South China Sea by mesoscale eddies? *Earth's Future*, 9, e2020EF001857, doi:10.1029/2020EF001857.

Zhang, M., Wu, Y., Wang, F, Xu, D. and Zhou, M. (2020). Hotspot of organic carbon export driven by mesoscale eddies in the slope region of the northern South China Sea, *Front. Mar. Sci.*, 7, pp. 1–15, doi:10.3389/fmars.2020.00444.

Zhang, Z., Zhao, W., Tian, J. and Liang, X. (2013). A mesoscale eddy pair southwest of Taiwan and its influence on deep circulation, *J. Geophys. Res. Oceans*, 118, pp. 6479–6494, doi:10.1002/2013JC008994.

Zhao, H., Bai, X. and Hu, J. (2025). Heat and salt transports in the Luzon Strait: Contributions of mesoscale eddies from the western Pacific to the South China Sea, *Progr. Oceanogr.*, in press.

Zhao, J., Wang, F., Gao, S., Hou, Y. and Liu, K. (2022). Mass transport of a mesoscale eddy in the South China Sea identified by a simulated passive tracer, *J. Oceanol. Limnol.*, 40, pp. 389–412, doi:10.1007/s00343-021-1069-y.

Zheng, Q., Xie, L., Zheng, Z. and Hu, J. (2017). Progress in research of mesoscale eddies in the South China Sea, *Adv. Mar. Sci.*, 35(2), pp. 131–158.

Zu, Y., Sun, S., Zhao, W., Li, P., Liu, B. and Fang, Y. (2019). Seasonal characteristics and mechanism of the sea surface heat fluxes associated with mesoscale eddies in the South China Sea, *Adv. Mar. Sci.*, 37(1), pp. 11–21.

Part 4

Statistical Features of Eddies in the South China Sea

Chapter 12

Statistics of Mesoscale Eddies in the SCS

12.1 Introduction

The statistical analysis is a key step to explore the properties of a natural phenomenon. For an ocean phenomenon, such as mesoscale eddies in the South China Sea (SCS), if the quantities of data samples are adequate, investigators can obtain spatiotemporal distribution, structure features, characteristic scales and values of dynamic parameters with the statistical analysis. These results not only provide important clues for exploring the genesis, evolution and internal linkage of eddies, but also serve as the basis for the dynamic analysis, empirical and numerical forecasts. This chapter aims to review progresses in statistical research of mesoscale eddies in the SCS and summarize the updated research results.

Since 1993, satellite altimeter sea surface height (SSH) image products have been available and created a possibility for statistical analysis of ocean mesoscale eddies. Wang *et al.* (2003) used the first 8 years (1993–2000) of satellite altimeter sea level anomaly (SLA) data for the statistical analysis of spatiotemporal distribution of mesoscale eddies in the SCS. The criteria they used for identifying mesoscale eddies are (1) closeness of SLA contours, (2) SLA at eddy center is 7.5 cm higher than that of surrounding area (>4.0 cm for continuous tracking), (3) eddy lifetime is longer than 30 days, and the depth of eddy center location is deeper than 1000 m. Total 86 mesoscale eddies including 28 cyclonic eddies (CEs) and

58 anticyclonic eddies (AEs) were found. The concentrative regions of eddies were the northeastern SCS deep basin west of the Luzon Strait and the western boundary area of deep basin.

Lin *et al.* (2007) used the 9 years (1993–2001) of TOPEX/Poseidon (US–France satellite) merged European remote sensing satellite (ERS) - 1/2 altimeter data for doing the statistical analysis of spatiotemporal distribution of mesoscale eddies in the SCS. Thei results show that there are 18 eddies emerged annually, related with El Niño/La Niña events. Eddies are concentrated along a northeast-southwest band from southwest of Taiwan Island to east of Vietnam. About 80% eddies move westward with zonal speeds ranging from -8 to 3 cm s^{-1}. Eddy radii range from 100 to 250 km with lifetimes from 30 to 60 days.

Xiu *et al.* (2010) developed a Pacific basin-wide three-dimensional physical-biogeochemical model and used the result of the SCS subdomain to quantify the eddy activities during a period of 15 years (1993–2007). On average, there are about 32.9 ± 2.4 eddies predicted by the model and 32.8 ± 3.4 eddies observed by satellite per year, and about 52% of them are CEs. The eddy radius ranges from about 46.5 to 223.5 km, with a mean value of 87.4 km. More than 70% of the eddies have a radius smaller than 100 km. The mean area covered by eddies is around 160,170 km^2, equivalent to 9.8% of the SCS area with water depths greater than 1000 m. Interannual variations in eddy numbers and the total eddy occupied area indicate that eddy activities in the SCS do not directly correspond to the El Niño–Southern Oscillation events. The wind stress curls are thought to be an important but not the only mechanism of eddy genesis in the SCS.

Chen *et al.* (2011) investigated mean properties and the spatiotemporal distribution of eddies in the SCS by analyzing more than 7000 eddies corresponding to 827 eddy tracks, identified by the winding angle method (Sadarjoen and Post, 2000) and 17 years (1992–2009) of satellite altimetry data. Their statistical results show that the mean radius and lifetime of eddies are 132 km and 8.8 weeks, respectively. In the northern SCS (NSCS), eddies mainly propagate southwestward along the continental slope with velocities of 5.0–9.0 cm s^{-1}, while in the central SCS, eddies tend to move westward with velocities of 2.0–6.4 cm s^{-1}. Eddy propagation in the western basin east of Vietnam is quite random, without uniform propagate direction.

Li *et al.* (2011) developed geometric eddy identification method to detect eddies from the large scale to the sub mesoscale based on the satellite-tracked Lagrangian drifters in the NSCS. Statistical analysis results show that there are in total 2208 eddies identified, 70% of which are AEs. The ratio of the mesoscale AE number (210) to the CE number (171) is 1.2:1. The spatial distribution of the eddies is concentrated in southwest of Taiwan Island. The temporal distribution of eddy number has a close relation with the Asian monsoon. The maximum and mean tangential velocities of AEs (CEs) are 40 (30) cm s^{-1} and 25 (15) cm s^{-1}, respectively.

Lin *et al.* (2012a, 2012b) used 17 years (1993–2010) of satellite altimeter SLA data for the statistical analysis of mesoscale eddies in the SCS. The criteria for identifying mesoscale eddies are the same as that by Wang *et al.* (2003). Their results show that the annual mean of total eddy number is 21 ± 4, of which CE and AE numbers are 10.3 ± 2.4 and 10.7 ± 2.4, respectively. The eddy radiuses are distributed in a range of 40–90 km.

Zheng *et al.* (2017) summarized advances in statistics of mesoscale eddies in the SCS. The eddies are mainly distributed in the deep basin with the depth deeper than 1000 m and concentrated in west of the Luzon Strait and the western boundary of the basin. The occurrence frequencies of AEs and CEs are near the same. The sea surface radii are mostly distributed within a range from 50 to 150 km, smaller than that in the Pacific. The eddy number has monthly variation, but not remarkable seasonal variation. The annual fluctuation of eddy number is about 20%, but not remarkable correspondence to the El Niño–Southern Oscillation (ENSO) events. The statistical results of eddy number per year are inconsistent. The minimum number is 11 per year, and the maximum is 49 per year. The rotating current vector reaches the maximum value of 40 cm s^{-1} at the surface layer of 100 m, decreases with the water depth to 3.5–5.0 cm s^{-1} at 2000 m. The eddy propagation velocities are within in a range from 2.0 to 9.0 cm s^{-1}. The eddy sea surface vorticity is within a range from 5.4 to 20×10^{-6} s^{-1}, higher than that in the Pacific.

He *et al.* (2018) refreshed statistics of surface eddy properties in the SCS using 22 years (1993–2015) of daily satellite eddy data. More than 7,000 of historical Argo profiles were collocated into eddy-centered coordinates to reveal the composite mean three-dimensional (3D) structures of eddies. The results indicate that eddies of both polarities have long conical shape, with a maximum (minimum) density anomaly of 0.55 kg m^{-3} (0.51 kg m^{-3}) at 60 m (90 m) in the composite cyclonic (anticyclonic) eddy. Temperature and salinity anomalies also peak at eddy cores, with values of 1.5°C and 0.15 psu in CE and 1.4°C and 0.16 psu in AE. The temperature and density anomalies extend vertically to depths of 400–500 m, while the salinity anomaly is apparent only in the upper 150 m. The temperature anomaly contributes about 90% of the density anomaly. Mixed layer depths in CEs are on average 15 m shallower than that in AEs. The rotation of the composite cyclonic (anticyclonic) eddy generates meridional heat transport of 1.4×10^{12} W (-3.1×10^{12} W) and salt transport of -4.0×10^{4} kg s^{-1} (5.6×10^{4} kg s^{-1}). More than 90% of the heat and salt transports are concentrated in the upper 300 and 100 m, respectively. Compared to the meridional transports, westward propagation of eddies results in zonal heat and salt transports on the same orders of magnitudes and a basin-scale westward water transport of 1.4 Sv, equivalent to about 30% of the annual-mean Luzon Strait transport.

In recent years, with the sustaining growth of satellite altimeter data products and observation data available, refreshed and novel statistical results of mesoscale eddies in the SCS have been published. Huang *et al.* (2020) analyzed the statistical characteristics of spatiotemporal distribution of mesoscale eddy propagation speeds in the SCS deep basin with depths >1000 m, using 25 years (1993–2017) of mesoscale eddy trajectory product derived from satellite altimeter data. The results show that climatologically, the zonal propagation speeds are westwards in the whole basin, while the meridional speeds are southwards in the northwestern basin, and northwards in the southeastern basin. The variation of meridional speeds with longitude is consistent with that of the background meridional currents with correlation coefficient R^2 of 0.96, while the variation of zonal speeds is related both to the background zonal currents and β effect. The propagation velocities are characterized by the significant seasonality with the minimum magnitude occurring in summer and the maximum in winter for zonal speed and total propagation speed. Interannually, larger values of speed components mostly occur in La Niña years and the negative phase of the Pacific Decadal Oscillation (PDO). Mesoscale eddies move fast at the beginning and the end of their lifetime, i.e., during their growth and dissipation periods, and slowly during their stable "midlife" period. This trend is negatively correlated with the rotating tangential velocity with R^2 of –0.93. Eddies with extreme propagation velocities defined as slower (faster) than 1.5 cm s^{-1} (15.4 cm s^{-1}) occupy 1.5% (1.9%) of the total eddy number. Extremely slow-moving (fast moving) eddies tend to appear in the middle (on the edge) of the basin and mostly occur in summer (winter).

Zhang *et al.* (2022) analyzed the statistical features of path types and temporal variability of eddies that entered the northern continental shelf of the SCS using 24 years (1993–2016) of AVISO (Archiving Validation and Interpretation of Satellite Data in Oceanography) mesoscale eddy trajectory atlas product (https://www.aviso.altimetry.fr/es/data/products/value-added-products/global-mesoscale-eddy-trajectory-product.html). A total of 184 mesoscale eddies are classified into four types according to the motion trajectories: along-the-isobath type (87), intrusion-of-continental-shelf type (38), local wandering type (23), and shelf-internal-generation type (36). respectively. The mean SLA amplitude, radius and lifetime are in ranges of 16–18 cm, 146–153 km and 74–93 days, respectively (see Chapter 12 for details).

Jin *et al.* (2024) refreshed statistics of mesoscale eddies in the SCS using 29 years (1993–2021) of satellite altimeter data and the vector geometry eddy detection method. On average, 230 eddies per year (lifetime > 7 days) are detected with AE and CE numbers accounting 47.8% and 52.2% of the total number, respectively.

12.2 Temporal Distribution

Up to date, the newest statistics of spatiotemporal distribution of mesoscale eddies in the SCS are derived by Jin *et al.* (2024) from the longest time series of 29 years (1993–2021) of satellite altimeter data products. They used the Euler vector geometry eddy detection method to identify and track eddies based on velocity vectors (Nencioli *et al.*, 2010). Compared with the Okubo–Weiss (O–W) function method and angle winding method, the Euler vector geometry eddy detection method can achieve a higher success of detection rate and a lower excess of detection rate (Beron-Vera *et al.*, 2013;

Lin *et al.*, 2007; Liu *et al.*, 2020; Sun *et al.*, 2019, 2020). Eddy detection and tracking procedures can be divided into three steps: (1) To determine the eddy center. The eddy center must satisfy four constraints (Sun *et al.*, 2022; Yang *et al.*, 2020). Two parameters A (=4) and B (=2) are determined to detect the eddy minimum size and adapt the algorithm to meshes of different resolutions. (2) To determine the eddy boundary. The eddy boundary is determined by defining the outermost closed curve around the eddy center (Xia and Shen, 2015). The eddy radius (size) is defined as the average distance from the boundary point to the eddy center (Nencioli *et al.*, 2010). (3) To determine the eddy trajectory. The eddy trajectory is determined by comparing the eddy centers at continuous time steps (Nencioli *et al.*, 2010). Based on this eddy detection method, a dataset including the center position, radius, lifetime, and eddy boundary of the SCS eddies is obtained.

The automatic eddy detection method was evaluated by comparing results with a mesoscale eddy expert identification library (where experts manually identify detected eddy data based on flow field data). The results show a success detection rate (SDR) of over 90% and an excess detection rate (EDR) of less than 10% in different regions of the global ocean. In the SCS, the SDR and EDR are 96.7% and 7.2%.

Figure 12.1(a) shows the statistics of interannual distribution of mesoscale eddies in the SCS derived from the longest time series of 29 years (1993–2021) of satellite altimeter data products (Jin *et al.*, 2024). On average, about 230 eddies (lifetime $\geq$ 7 days) per year are detected. The interannual variability of eddy numbers is significant, i.e., the highest and the lowest eddy numbers were 257 in 2015 and 185 in 2020, 12% more and 20% less than the long-term mean (230 per year), respectively. The average numbers per year of CEs and AEs are 120 (52.2%) and 110 (47.8%), close to the previous results (Xiu *et al.*, 2010). Linear regression of the eddy numbers with the time gives decreasing trends of -0.06 and -0.25 per year for CEs and AEs (R^2 = 0.085 and 0.071). The statistical trends and their significances are worth further confirming and pursuing.

Figure 12.1(b) shows statistics of monthly distribution of mesoscale eddies with the lifetime $\geq$ 7 days in the SCS from 1993 to 2021. On average, 10 CEs and 9 AEs were detected per month, respectively. One can see that the AE number is higher than that of

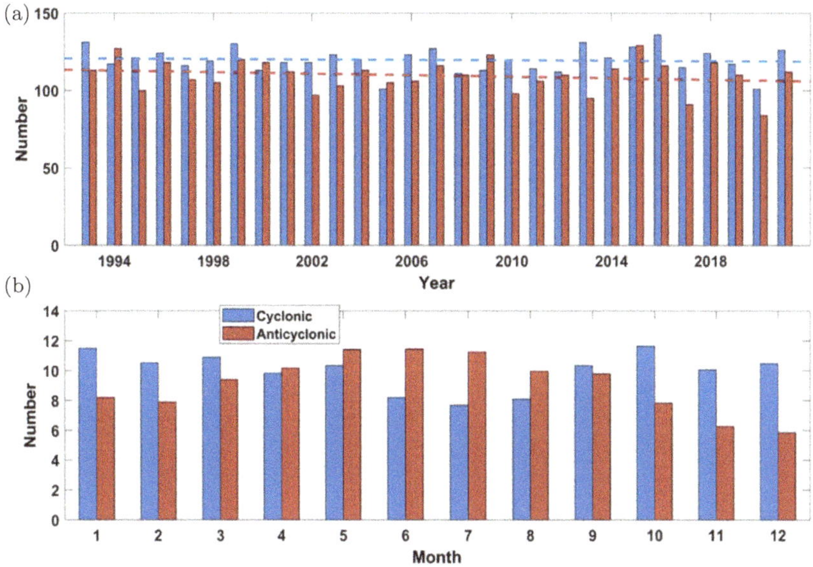

Fig. 12.1. (a) Histograms of interannual distribution of eddy numbers (lifetime $\geq$ 7 days) in the SCS from 1993 to 2021. Light blue and red dashed lines represent linear regression trends in variations of CE and AE numbers. (b) The same as (a), but for the monthly mean.
Source: Taken from Jin *et al.* (2024).

CEs in spring and summer from April to August, and the CE number is higher than that of AEs in fall and winter from September to March of the next year, implying that winter (summer) is a favorable season for the generation of CEs (AEs) (Lin *et al.*, 2007; Zeng *et al.*, 2019).

12.3 Spatial Distribution

12.3.1 *Annual eddy trajectories*

Figure 12.2 shows two-dimensional (2D) horizontal distributions of eddy trajectories (lifetime $\geq$ 7 days) in the SCS for every two years from 1993 to 2021 (Jin *et al.*, 2024). One can see that eddies are scattered in the whole SCS deep basin. Statistics indicate that 73.0% of all eddies propagate westward, and westward propagating CEs and

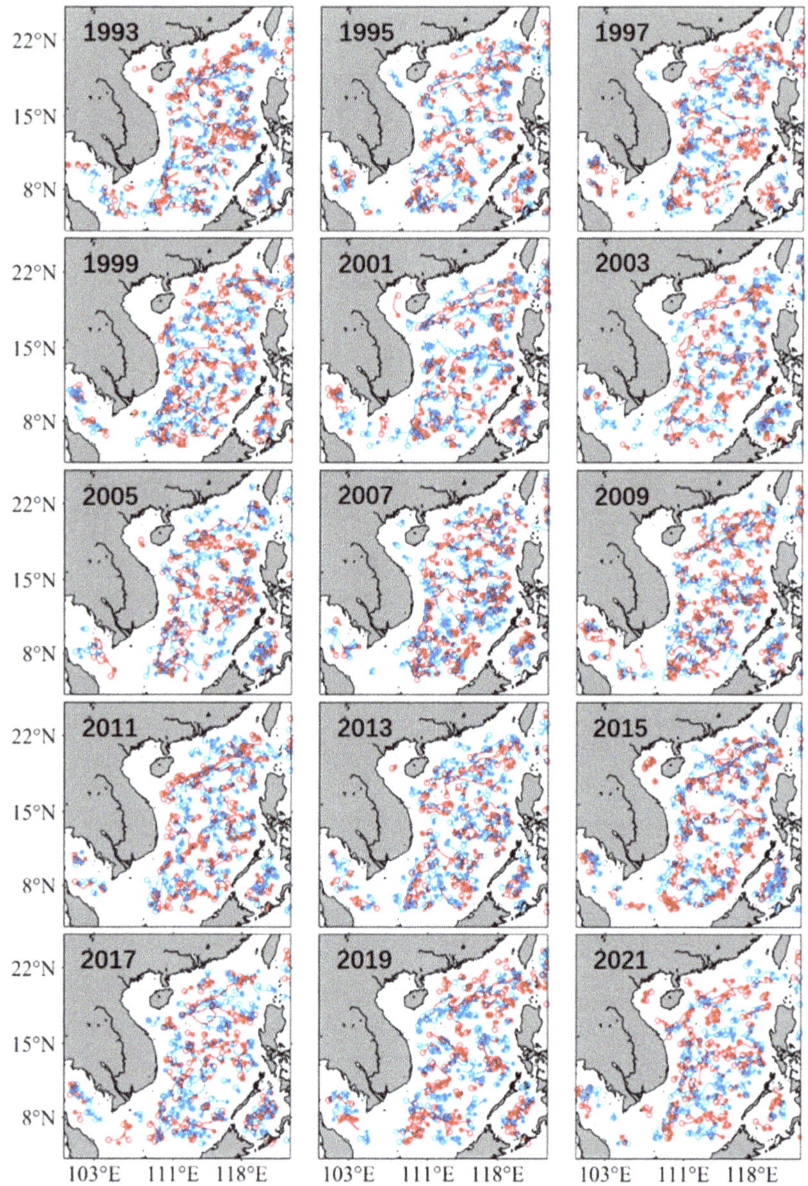

Fig. 12.2. 2D horizontal distributions of AE (red) and CE (light blue) trajectories (lifetime $\geq$ 7 days) in the SCS for every two years from 1993 to 2021. Symbols "*" and "o" mark positions of eddy sources and sinks, respectively.
Source: Taken from Jin *et al.* (2024).

AEs account for 73.1% and 72.8% of their total numbers, reflecting general features of the SCS circulation (Fang *et al.*, 2002; Wang *et al.*, 2004; Hu *et al.*, 2011). Meanwhile, the eddy trajectories can be classified according to their directions, sources, sinks, distribution features and temporal variability for further analyses (Zhang *et al.*, 2020; Chapter 12 of this volume).

12.3.2 *Occurrence, genesis and dissipation*

He *et al.* (2018) used 22.3 years (January 1993 and April 2015) of the Version 4 eddy data set developed by Chelton *et al.* (2011) to analyze statistical features of eddies in the SCS. A total number of 37,460 CE and 36,812 AE snapshots, corresponding to 625 and 534 eddies (total 52 eddies per year) with lifetimes longer than 28 days were detected. This means that there are on average 4.6 CEs and 4.5 AEs in the SCS at any given time. The ratio of the numbers of CEs and AEs is 53.9%:46.1% close to 52%:48% of Xiu *et al.* (2010) and 52.2%:47.8% of Jin *et al.* (2024) but reversed with 32.6%:67.4% by Wang *et al.* (2003), 44.9%:55.1% by Li *et al.* (2011) and 49.0%:51.0% by Lin *et al.* (2012a, 2012b). The 2D horizontal distributions of sea surface eddy occurrence, genesis and dissipation locations at each 0.5° by 0.5° latitude-longitude grid are shown in Figure 12.3.

Figure 12.3(a) shows that 40% of eddies are distributed in the NSCS north of 17°N, and concentrated in regions southwest of Taiwan Island, northwest of Luzon Island and the northwestern SCS deep basin. The western SCS is another region with high eddy occurrence. Approximately 26% of eddies are observed west of 113°E (7–17°N). The eastern SCS is also an eddy-active region with 11% of eddies east of 116°E (12–15°N). In contrast, the central and southeastern SCS are of sparse eddy activities. Figure 12.3(b) and 12.3(c) show that the spatial distribution patterns of CEs and AEs are almost the same each other and as that of all eddies in Figure 12.3(a).

Figure 12.3(d) shows the distribution patterns of eddy genesis locations, which are defined as where eddies are first detected. One can see that 34% of eddies originate from two major sources east of 118°E near the eastern boundary, the Luzon Strait and the Mindoro Strait. The western SCS is a relatively high eddy genesis region but much lower than the eastern SCS. Figure 12.2(e) and 12.2(f) show that the distributions of eddy genesis for CEs and AEs are generally

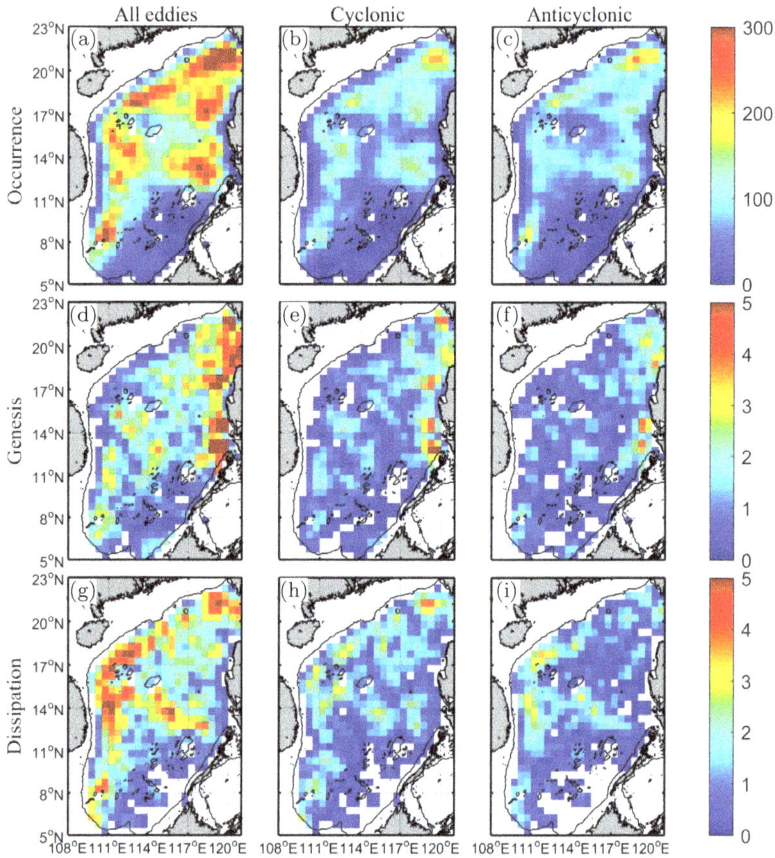

Fig. 12.3. 2D horizontal distributions of (a–c) eddy occurrence, (d–f) genesis, and (g–i) dissipation at each 0.5° by 0.5° pixel in the SCS from January 1993 to April 2015. Left, middle and right columns are for all eddies, CEs, and AEs, respectively. Black contours represent 200 m isobath. Color codes denote for eddy occurrence, the eddy numbers of all the detected eddy snapshots; for eddy genesis, that identified the first time; and for eddy dissipation, that detected the last time. *Source*: Taken from He *et al.* (2018).

similar, except in west of the Luzon Strait, where more AEs are observed (see Chapter 7 for details).

Figure 12.3(g) shows that eddies dissipate in the western SCS, especially along the northwestern and western boundaries, where 40% of eddies dissipate in west of 113°E. Figure 12.3(h) and 12.3(i) show that the dissipation regions of CEs depart somewhat from that

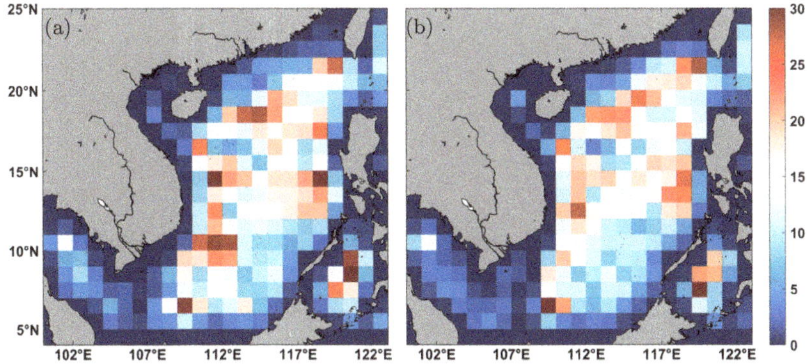

Fig. 12.4. 2D horizontal distributions of (a) CE and (b) AE numbers per year (lifetime $\geq$ 7 days) in the SCS averaged over 29 years from 1993 to 2021. Color code represents total mean of eddy numbers per year.
Source: Taken from Jin *et al.* (2024).

of AEs in the SCS. CEs dissipate all the way along the northwestern boundary, while most AEs dissipate in west of 113°E. The central SCS is also a region of high eddy dissipation for both CEs and AEs.

As a comparison, Figure 12.4(a) and 12.4(b) show 2D horizontal distributions of CEs and AEs with the lifetime $\geq$ 7 days in the SCS from 1993–2021 (Jin *et al.*, 2024). The color code represents the total mean number of yearly eddy centers appearing within the ith grid defined as

$$N_i(x, y) = \sum_{k=1993}^{2021} n_k(x, y)/\tau, \tag{12.1}$$

where (x, y) represent the central position of ith grid, n_k is the total eddy number appearing within the ith grid in the kth year, and τ (= 29 years) is the sampling interval (year).

From the figures, one can see the following features. (1) Eddies may occur in more than 90% of the SCS water area, including the continental shelf, slope and deep basin. (2) The high-density grids with eddy numbers more than 20 per year are mainly distributed along the eastern, northern and western boundary regions of deep basin. (3) The CE and AE distribution patterns are highly similar, indicating that there are internal linkages between the two types of eddies.

12.3.3 *Amplitude, radius and propagation speed*

Figure 12.5(a–i) show 2D horizontal distributions of eddy amplitude, radius and propagation speed in the SCS from 1993–2015 derived by He *et al.* (2018) from the same dataset. Figure 12.5(a) shows that large-amplitude eddies are concentrated in the northeastern and western SCS, where are the regions of energetic currents (Qu *et al.*, 2009; Su, 2004). Figure 12.5(b) and 12.5(c) show that absolute amplitudes of AEs are larger than that of CEs in most regions of the SCS.

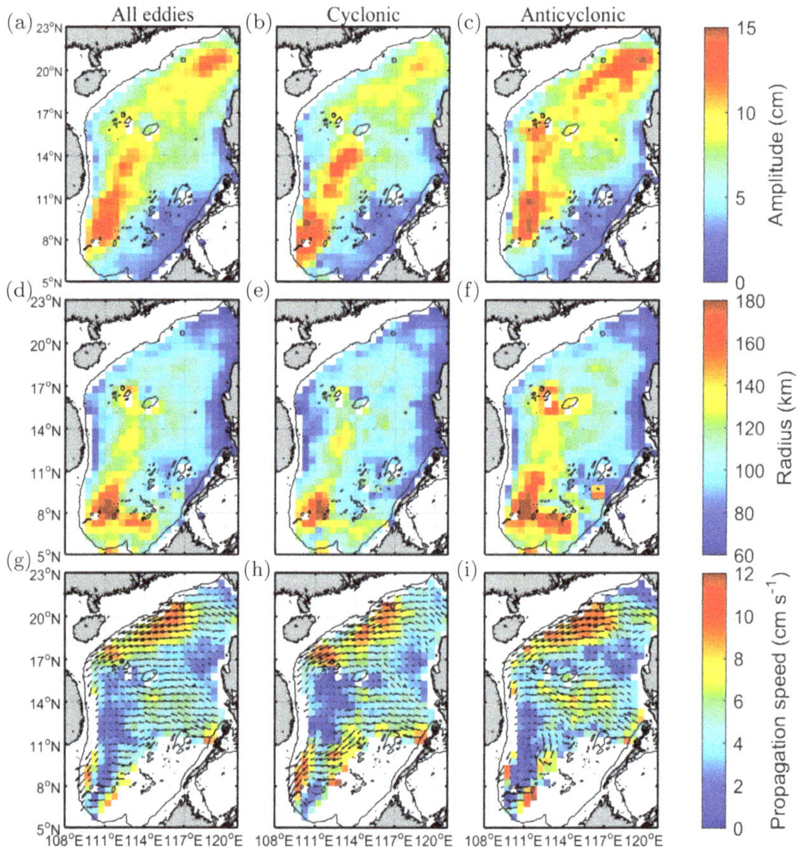

Fig. 12.5. 2D horizontal distributions of (a–c) eddy amplitude, (d–f) radius and (g–i) propagation speed in the SCS averaged over 23 years from 1993 to 2015. Left, middle, and right columns are for all eddies, CEs and AEs, respectively. Black contours mark 200 m isobath.
Source: Taken from He *et al.* (2018).

The difference is the most apparent in the NSCS, where the Kuroshio intrusion generates intense AEs (see Chapters 7 and 12 for details).

Figure 12.5(d) shows the spatial distribution of eddy radii, which is characterized by that eddies in the southern SCS (SSCS) generally have larger radii than that in the NSCS. This is true for both CES and AEs (Figure 12.5(e) and 12.5(f)). The decrease of eddy radius with the increase of latitude is consistent in trend with the decrease of Rossby radius of deformation (Chen *et al.*, 2011). Meanwhile, eddies near the eastern and western boundaries, where are locations of eddy genesis and dissipation (Figure 12.3(d) and 12.3(g)), have the radii smaller than that in the central SCS.

Figure 12.5(g) shows the spatial distribution of eddy propagation speeds, which is characterized by that eddies west of the Luzon Strait propagate westward or northwestward at an average speed of ~5 cm s^{-1}. The speed increases dramatically to ~10 cm s^{-1}, when eddies encounter with the shoaling topography. In addition, the propagation direction deflects to southwestward along the northwestern boundary of the SCS deep basin. Eddies in the eastern SCS propagate generally westward at an average speed of ~5 cm s^{-1}, while that in the western SCS generally have the speed slower than 3 cm s^{-1} because of the impediment of the western boundary. Eddies west of the Luzon Island also have propagation speeds slower than 3 cm s^{-1}. An important reason for this slow propagation is that in winter, Luzon Cold Eddy almost stations there (Qu, 2000; Yang and Liu, 2003; He *et al.*, 2016). Figure 12.5(h) and 12.5(i) show that the distribution patterns of propagation speeds of CEs and AEs are similar in most regions except the central SCS, where the westward propagation speeds of AEs are about 2 cm s^{-1} higher than that of CEs.

12.4 Distributions of Eddy Numbers vs Dynamic Parameters

Figure 12.6(a–e) show statistical distribution histograms of eddy numbers per year in the SCS vs dynamic parameters: the lifetime, radius (km), eddy kinetic energy (EKE) (m^2 s^{-2}), SLA deviation (relative eddy amplitude) (m), and maximum tangential velocity (m s^{-1}) from 1993 to 2021 (Jin *et al.*, 2024). One can see that in all the cases the histogram envelopes look like the Rayleigh distribution.

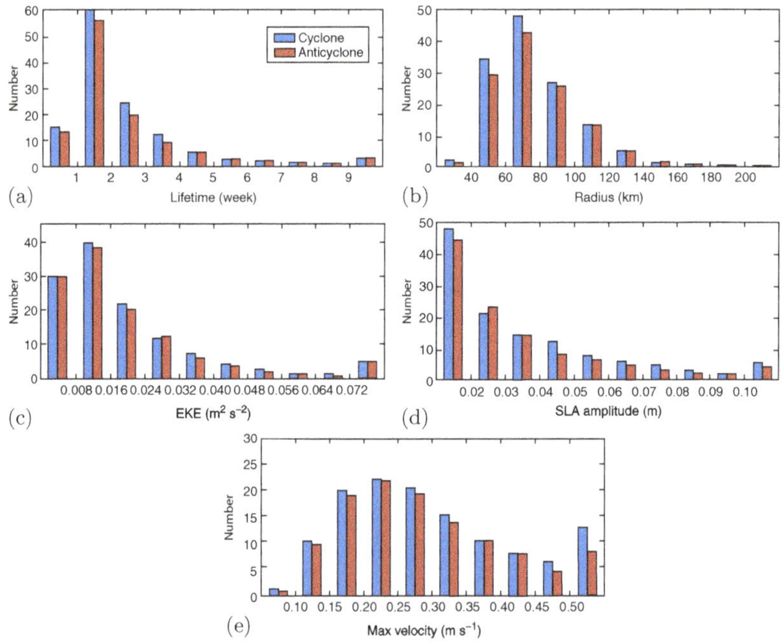

Fig. 12.6. Statistical histograms of eddy number per year vs dynamic parameters: (a) lifetime, (b) radius (km), (c) EKE (m^2 s^{-2}), (d) SLA amplitude (m) and (e) maximum tangential velocity (m s^{-1}) in the SCS from 1993 to 2021. *Source*: Taken from Jin *et al.* (2024).

Figure 12.6(a) shows that eddy numbers quickly peak at the lifetime of 1–2 weeks and decrease as the lifetime increases. Eddies with lifetimes in a range of 2–3 weeks account for 64.0% of the total number. On average, 40 eddies (51.0% CEs and 49.0% AEs) with a lifetime of at least 28 days are detected per year, accounting for about 17% of the total number.

Figure 12.6(b) shows that the eddy numbers quickly peak at the radius of 60–80 km (35.7%) and decrease as the radius increases. Eddies with radii within 60–100 km account for 81.6% of the total number, of which CEs and AEs account for 53.1% and 46.9%, respectively.

Figure 12.6(c) shows that the eddy numbers quickly peak at the EKE of 0.008–0.016 m^2 s^{-2} and decrease as the EKE increases. 78% of eddies are concentrated in the range of EKE lower than 0.24 m^2 s^{-2}.

Figure 12.6(d) shows that eddies are concentrated at the lowest level of SLA. Eddies with SLA amplitudes less than 0.02 m account for 37.5% of the total number. The eddy numbers decrease near exponentially with decreasing SLA amplitude. Figure 12.6(e) shows that the numbers of eddies gradually peak at the maximum tangential velocity of 0.20–0.25 m s^{-1} and gradually decrease as the velocity increases.

12.5 Statistical Relationships of Dynamic Parameters vs Eddy Lifetime

Figure 12.7(a–d) show statistical relationships of dynamic parameters: mean radius, EKE, SLA amplitude and maximum tangential velocity vs lifetime of eddy from 1993 to 2021 (Jin *et al.*, 2024). One can see that all dynamic parameters increase with the lifetime increase. These statistical relationships contain internal linkages among the parameters and are worth further pursuing. Note that different eddy detection methods would generate different data of mesoscale eddy boundaries and dynamic parameters (Zeng *et al.*, 2019; Yang *et al.*, 2020). Thus, the statistical relationships may look different.

12.6 Discussion

Of statistics of mesoscale eddies in the SCS, three important questions need to be answered. (1) How many eddies per year appear? (2) What is the ratio of CE numbers to AE numbers? (3) Do eddies have any response to climate change? This section tries to find convincing answers to these questions.

12.6.1 *Of eddy number per year*

The eddy number per year in the SCS is the most basic parameter, which has significant influences on the basin scale circulation, energy and mass balances, ecological environment and response to the climate change (see Chapter 2 for details). Since the early statistical result was published by Wang *et al.* (2003), however, the number has been keeping changing. The factors influencing eddy counting

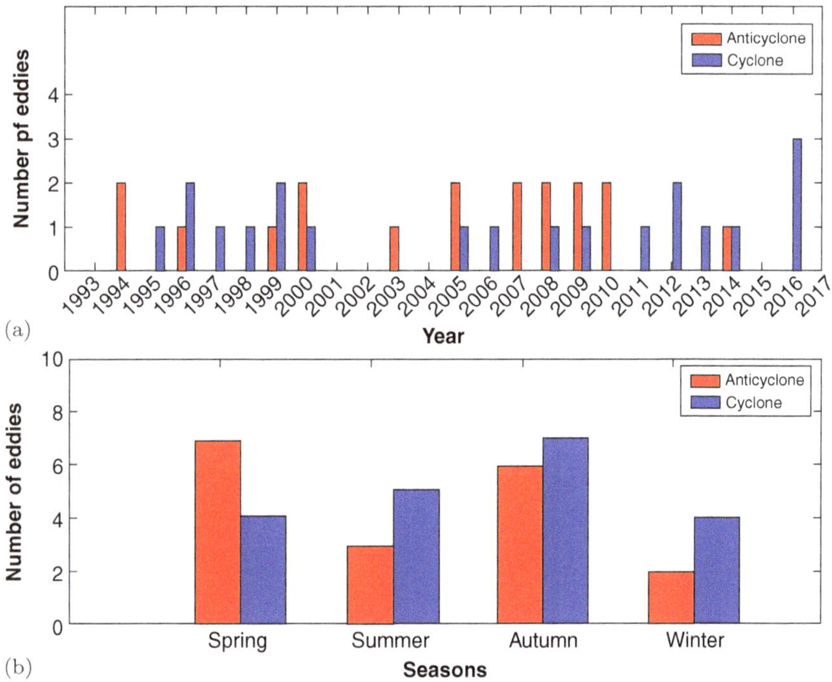

Fig. 12.7. Statistical relationships of mean (a) radius (km), (b) EKE (m^2 s^{-2}), (c) SLA amplitude (m), and (d) maximum tangential velocity (m s^{-1}) vs eddy lifetime in the SCS from 1993 to 2021.
Source: Taken from Jin *et al.* (2024).

include (1) eddy detection or identification methods, (2) eddy lifetime selection and (3) sampling time interval, the longer interval, the more accurate eddy number.

Among the three factors, the eddy lifetime selection for eddy counting is a common and comparable one for different detection methods. At present, there are no commonly accepted criteria for the eddy lifetime selection. Here an eddy lifetime index is defined as a criterion for eddy counting,

$$\alpha = T_l/T_e, \tag{12.2}$$

where T_l and T_e are the eddy lifetime selected for eddy counting and the eddy period, respectively. Define a threshold eddy lifetime (TEL) for $\alpha = 1$ to represent the counted eddy as a complete and mature

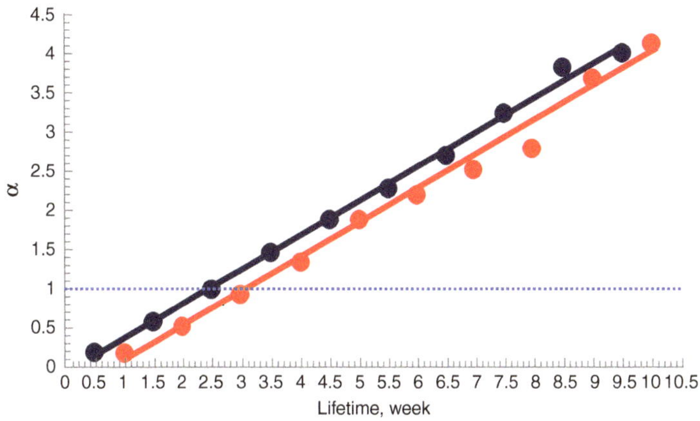

Fig. 12.8. The eddy lifetime index vs eddy lifetime. Data (blue for CEs and red for AEs) are derived from measurements shown in Figure 12.7(a) and 12.7(d).

one. Thus, eddies for $\alpha \geq 1$ are counted for the statistical analysis, while eddies for $\alpha < 1$ should be forgotten.

Figure 12.8 shows the statistical relationship of the eddy lifetime index vs the eddy lifetime selected for eddy counting derived from measurements shown in Figure 12.7(a) and 12.7(d). One can see that for $\alpha = 1$, the TELs for eddy counting are 2.5 and 3 weeks (17.5 and 21 days) for CEs and AEs, respectively. Fortunately, some previous statistical analyses of eddy characteristics in the SCS selected the eddy lifetime of at least 28 days for the eddy counting (Xiu *et al.*, 2010; Chen *et al.*, 2011; Li *et al.*, 2014).

Table 12.1 lists statistical results of the eddy number per year and the ratio of CE number to AE number in the SCS counted with the TEL > 28 days. These data have the comparability. Figure 12.9 shows the eddy number per year vs sampling interval derived from the data listed in Table 12.1. The data points separate automatically into two groups. The first group within the solid circle derived by the Okubo-Weiss method shows an increasing trend of eddy number per year with sampling interval increase of 9 years from 8 to 17 years. The maximum number reaches 32.8. The second group within the dash circle derived by three different methods: winding angle, snapshots and vector geometry methods shows a little changed eddy numbers per year (51.0 ± 1.4) with sampling interval increase of 12 years from

Table 12.1. Statistical results of the eddy number per year and the ratio of CE number to AE number in the SCS counted with the TEL > 28 days.

Author *et al.* (year)	Detection method	Sampling interval (year)	Lifetime > (day)	Eddy num. per year	CE%:AE%
Wang (2003)	Okubo–Weiss	8 (1993–2000)	30	10.8	32.6%:67.4%
Lin (2007)	Okubo–Weiss	9 (1993–2001)	30–60	18	
Xiu (2010)	Okubo–Weiss	15 (1993–2007)	30	32.8 ± 3.4	52%:48%
Chen (2011)	Winding Angle	17 (1992–2009)	28	49	46.9%:53.1%
Lin (2012a, 2012b)	Okubo–Weiss	17 (1993–2010)	30	21 ± 4	49%:51%
He (2018)	Snapshots	22 (1993–2015)	28	52	53.9%:46.1%
Jin (2024)	Vector Geometry	29 (1993–2021)	28	52	50%:50%

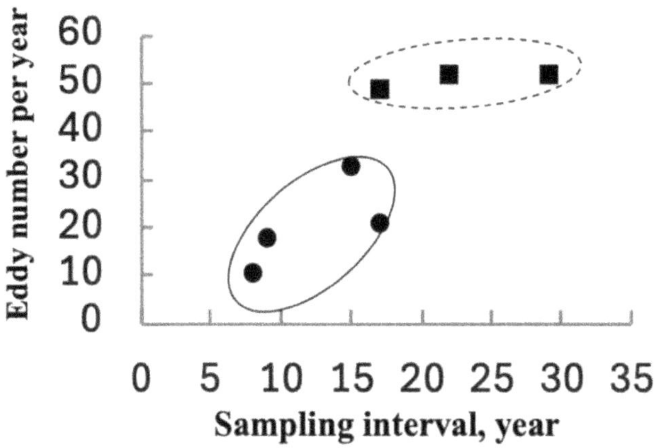

Fig. 12.9. Eddy number per year vs sampling interval. The data points within the solid circle are derived by the Okubo-Weiss method and within the dash circle by winding angle, snapshots and vector geometry methods, respectively.
Source: Data taken from Table 12.1.

17 to 29 years. Therefore, 51 eddies per year are a more reliable estimate for mesoscale eddies in the SCS. The eddy numbers derived with the Okubo-Weiss method seem to be seriously underestimated.

12.6.2 *Of ratio of CE to AE numbers*

The ratio of CE number per year (Nce) to AE number per year (Nae) is another important parameter, which has significant influences on the basin scale circulation, energy and mass balances, ecological environment and response to the climate change in the SCS. Figure 12.10 shows the ratio of CE number to AE number per year (Nce/Nae) vs sampling interval. The data points are taken from Table 12.1 for the sampling interval long enough (>15 years). One can see that the data point distribution appears a little scattering, but their linear regression trend is very close to the horizontal line, Nce/Nae =1. The long-term mean of Nce/Nae derived from 30 years of observations from 1992 to 2021 is 1.02, i.e., 50.5%:49.5%. Namely, the CE number per year is 1% more than that of AEs.

12.6.3 *Of responses of eddies to climate change*

In principle, as a marginal sea of the Pacific, dynamic processes including mesoscale eddies in the SCS should have responses to the

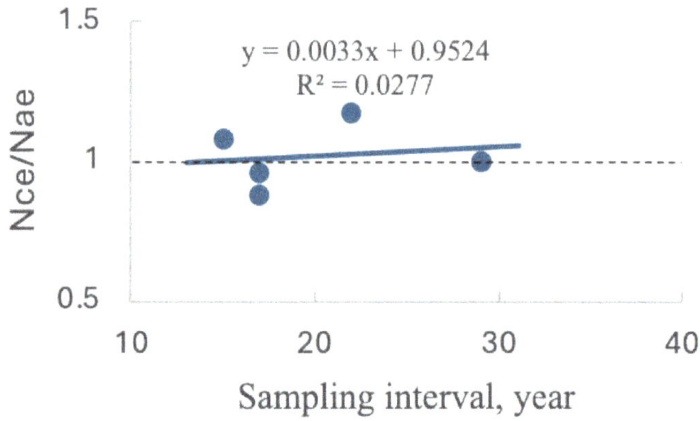

Fig. 12.10. Ratio of CE number per year to AE number per year (Nce/Nae) vs sampling interval.
Source: Data taken from Table 12.1.

interannual variability of the Pacific, such as the ENSO events and the PDO. However, early investigations gave the question whether eddy activities in the SCS have responses to the interannual variability of the Pacific two contrary answers: yes (e.g., Lin *et al.*, 2007) and no (e.g., Xiu *et al.*, 2010; Chen *et al.*, 2011; Lin *et al.*, 2012a, 2012b).

With extension of satellite data products and availability of new field observations, investigators find that the Kuroshio intrusion through the Luzon Strait into the SCS is a significant mechanism for eddy genesis in the SCS. Take eddy numbers in the NSCS (accounting for 40% of whole basin) as an example, statistics in Chapter 12 indicate that linear regression correlation coefficient (R^2) of total AE numbers from 1993 to 2016 with the Kuroshio forcing index as high as 0.89, implying close linkage between the two. Meanwhile, forced by big eddies originating from the WNPO, propagating westward and colliding with the western boundary, the Kuroshio flow path is frequently modified even cut off and forced to intrude into the SCS (see Chapters 5–7 for details). The behavior of big eddies in the Pacific is naturally modulated by the ENSO events and the PDO. So that, the eddy activities in the SCS must have responses to the interannual variability of the Pacific. For example, Huang *et al.* (2020) analyzed the statistical characteristics of mesoscale eddy propagation speeds in the SCS deep basin using 25 years (1993–2017) of satellite altimeter mesoscale eddy trajectory product. Their results show that larger values of propagation speeds mostly occur in La Niña years and the negative phase of the PDO.

12.7 Summary

The aims of this chapter are trifold: (1) to review progresses in statistical research of mesoscale eddies in the SCS, (2) summarize the updated research results and (3) look for the best answers to the questions and uncertainty of eddy statistics such as eddy number per year, ratio of CE number to AE number and eddy responses to climate change. Since Wang *et al.* (2003) published the statistical result of mesoscale eddies in the SCS in 2003, two decades have passed. During this period, the sampling interval of satellite altimeter data product has extended from 8 years to longer than 30 years. The data sources have extended to numerical model products and Argo drifter data for 3D eddy structure re-constructure (He *et al.*, 2018). The eddy detection and identification methods have extended to snapshot observation method (Chelton *et al.*, 2011), winding angle method (Sadarjoen and Post, 2000) and vector geometry method (Nencioli *et al.*, 2011). The technical advances promote continuous generation of refreshed and more reliable statistical results as summarized in Sections 12.2–12.5. For statistics of 3D eddy structures, see Chapter 10 for details.

The eddy number per year in the SCS is the most basic parameter, but its statistical results are keeping changing due to different eddy detection or identification method, eddy lifetime selection for eddy counting and sampling time interval. Among the three factors, the eddy lifetime selection for eddy counting is a comparable one for different detection methods. This chapter proposes a threshold eddy lifetime (TEL) as a criterion for the eddy counting, which are 2.5 and 3 weeks (17.5 and 21 days) for CEs and AEs derived from 29 years of satellite altimeter data product (Jin *et al.*, 2024). Based on the criterion, seven comparable statistical results of the eddy number per year in the SCS published from 2003 to 2024 are collected. The long-term variation trend analysis of these results gives 51 mesoscale eddies per year. The analysis results reveal that the Okubo-Weiss method seriously underestimates eddy numbers for more than 50% compared with the other three methods. Meanwhile, the long-term mean ratio of CE number to AE number is 1.02, i.e., 50.5%:49.5%, derived from 30 years of observations from 1992 to 2021. Of the question whether eddy activities in the SCS have responses to the ENSO events and the PDO has obtained a definite answer by recent research results that the Kuroshio modification and intrusion forced by the Pacific big eddies are significant mechanisms for eddy genesis in the SCS.

References

Beron-Vera, F. J., Wang, Y., Olascoaga, M. J., Goni, G. J. and Haller, G. (2013). Objective detection of oceanic eddies and the Agulhas leakage, *J. Phys. Oceanogr.*, 43, pp. 1426–1438.

Chelton, D. B., Schlax, M. G. and Samelson, R. M. (2011). Global observations of nonlinear mesoscale eddies, *Progr. Oceanogr.*, 91(2), pp. 167–216, doi:10.1016/j.pocean.2012.01.002.

Chen, G., Hou Y. and Chu X. (2011). Mesoscale eddies in the South China Sea: Mean properties, spatiotemporal variability, and impact on thermohaline structure, *J. Geophys. Res. Oceans*, 116(C6), pp. 102–108, doi:10.1029/2010JC006716.

Fang, W., Fang, G., Shi, P., Huang, Q. and Xie, Q. (2002). Seasonal structures of upper layer circulation in the southern South China Sea from in situ observations, *J. Geophys. Res. Oceans*, 107, pp. 23-1–23-12, doi:10.1029/2002JC001343.

He, Q., Zhan, H., Cai, S. and Li, Z. (2016). Eddy effects on surface chlorophyll in the northern South China Sea: Mechanism investigation and temporal variability analysis, *Deep Sea Res. Part I: Oceanogr. Res. Pap.*, 112, pp. 25–36, doi:10.1016/j.dsr.2016.03.004.

He, Q., Zhan, H., Cai, S., He, Y., Huang, G. and Zhan, W. (2018). A new assessment of mesoscale eddies in the South China Sea: Surface features, three-dimensional structures, and thermohaline transports, *J. Geophys. Res. Oceans*, 123, pp. 4906–4929, doi:10.1029/2018JC014054.

Hu, J., Gan, J., Sun, Z., Zhu, J. and Dai, M. (2011). Observed three-dimensional structure of a cold eddy in the southwestern South China Sea, *J. Geophys. Res. Oceans*, 116, C05016, doi:10.1029/2010 JC006810.

Huang, R., Xie, L., Zheng, Q., Li, M., Bai, P. and Tan, K. (2020). Statistical analysis of mesoscale eddy propagation velocity in the South China Sea deep basin, *Acta Oceanol. Sin.*, 39(11), pp. 91–102, doi:10.1007/s13131-020-1678-x.

Jin, Y., Jin, M., Wang, D. and Dong, C. (2024). Statistical analysis of multi-year South China Sea eddies and exploration of eddy classification, *Rem. Sens.*, 16, 1818, doi:10.3390/rs16101818.

Li, Y., Han, W., Wilkin, J. L., Zhang, W., Arango, H., Zavala-Garay, J., Levin, J. and Castruccio, F. S. (2014). Interannual variability of the surface summertime eastward jet in the South China Sea, *J. Geophys. Res. Oceans*, 119, pp. 7205–7228, doi:10.1002/2014JC010206.

Li, J., Zhang, R. and Jin, B. (2011). Eddy characteristics in the northern South China Sea as inferred from Lagrangian drifter data, *Ocean Sci. Discus.*, 7(5), pp. 661–669, doi:10.5194/os-7-661-2011.

Lin, H., Hu, J. and Zheng, Q. (2012a). Satellite altimeter data analysis of the South China Sea and the northwest Pacific Ocean: Statistical features of oceanic mesoscale eddies, *J. Oceanogr. Taiwan Str.*, 31(1), pp. 105–113.

Lin, H., Hu, J. and Zheng, Q. (2012b). Statistical analysis of the features of meso-scale eddies near the Luzon Strait, *Haiyang Xuebao*, 34(1), pp. 1–7.

Lin, P., Wang, F., Chen, Y. and Tang, X. (2007). Temporal and spatial variation characteristics on eddies in the South China Sea I. Statistical analyses, *J. Oceanogr.*, 29, pp. 14–22.

Liu, J., Tian, F. and Chen, G. (2020). Statistical study on the characteristics of sea surface temperature of mesoscale eddies in the South China Sea, *J. Ocean Univ. China Nat. Sci. Ed.*, 50, pp. 146–156.

Nencioli, F., Dong, C., Dickey, T., Washburn, L. and McWilliams, J. C. (2010). A vector geometry-based eddy detection algorithm and its application to a high-resolution numerical model product and high-frequency radar surface velocities in the Southern California Bight, *J. Atmos. Ocean. Technol.*, 27, pp. 564–579, doi:10.1175/2009JTECHO725.1.

Qu, T. (2000). Upper-layer circulation in the South China Sea, *J. Phys. Oceanogr.*, 30(6), pp. 1450–1460, doi:10.1175/1520-0485(2000)_030<1450:ULCITS>2.0.CO;2.

Qu, T., Song, Y. T. and Yamagata, T. (2009). An introduction to the South China Sea throughflow: Its dynamics, variability, and application for climate, *Dyn. Atmos. Oceans*, 47(1–3), pp. 3–14, doi:10.1016/j.dynatmoce.2008.05.001.

Sadarjoen, I. A. and Post, F. H. (2000). Detection, quantification and tracking of vortices using streamline geometry, *Comp. Graphics*, 24, pp. 333–341, doi:10.1016/S0097-8493(00)_00029-7.

Su, J. (2004). Overview of the South China Sea circulation and its influence on the coastal physical oceanography outside the Pearl River Estuary, *Cont. Shelf Res.*, 24(16), pp. 1745–1760, doi:10.1016/j.csr.2004.06.005.

Sun, W., An, M., Liu, J., Liu, J., Yang, J., Tan, W., Dong, C. and Liu, Y. (2022). Comparative analysis of four types of mesoscale eddies in the Kuroshio-Oyashio extension region, *Front. Mar. Sci.*, 9, 984244, doi:10.3389/fmars.2022.984244.

Sun, W., Dong, C., Tan, W. and He, Y. (2019). Statistical characteristics of cyclonic warm-core eddies and anticyclonic cold-core eddies in the North Pacific based on remote sensing data, *Rem. Sens.*, 11, 208, doi:10.3390/rs11020208.

Sun, W., Liu, Y., Chen, G., Tan, W., Lin, X., Guan, Y. and Dong, C. (2020). Three-dimensional properties of mesoscale cyclonic warm-core and anticyclonic cold-core eddies in the South China Sea, *Acta Oceanol. Sin.*, 40, pp. 17–29, doi:10.1007/s13131-021-1770-x.

Wang, G., Su, J. and Chu, P. (2003). Mesoscale eddies in the South China Sea observed with altimeter data, *Geophys. Res. Lett.*, 30(21), 2121, doi:10.1029/2003GL018532.

Wang, H., Yuan, Y., Guan, W., Lou, R. and Wang, K. (2004). Circulation in the South China Sea during summer 2000 as obtained from observations and a generalized topography-following ocean model, *J. Geophys. Res. Oceans*, 109, C07007, doi:10.1029/2003JC002134.

Xia, Q. and Shen, H. (2015). Automatic detection of oceanic mesoscale eddies in the South China Sea, *Chin. J. Oceanol. Limnol.*, 33, pp. 1334–1348, doi:10.1007/s00343-015-4354-9.

Xiu, P., Chai, F., Shi, L., Xue, H. and Chao, Y. (2010). A census of eddy activities in the South China Sea during 1993-2007, *J. Geophys. Res. Oceans*, 115, C03012, doi:10.1029/2009JC005657.

Yang, H. and Liu, Q. (2003). Forced Rossby wave in the northern South China Sea, *Deep Sea Res. Part I: Oceanogr. Res. Pap.*, 50(7), pp. 917–926, doi:10.1016/s0967-0637(03)_00074-8.

Yang, X., Xu, G., Liu, Y., Sun, W., Xia, C. and Dong, C. (2020). Multi-source data analysis of mesoscale eddies and their effects on surface chlorophyll in the Bay of Bengal, *Rem. Sens.*, 12, 3485, doi:10.3390/rs12213485.

Zeng, W., Zhang, S., Ma, Y. and Wang, T. (2019). Characterization of mesoscale eddies in the South China Sea 1993–2017, *J. Guangdong Ocean Univ.*, 39, pp. 96–106.

Zhang, T., Li, J., Xie, L. and Zheng, Q. (2022). Statistical analysis of mesoscale eddies entering the continental shelf of the northern South China Sea, *J. Mar. Sci. Eng.*, 10, 206, doi:10.3390/jmse10020206.

Zheng, Q., Xie, L., Zheng, Z. and Hu, J. (2017). Progress in research of mesoscale eddies in the South China Sea, *Adv. Mar. Sci.*, 35, pp. 131–158.

Chapter 13

Statistics of Eddies on the Northern SCS Continental Shelf

13.1 Introduction

The Northern Continental Shelf of South China Sea (NCSS) refers to the northernmost shallow water of the SCS, which is surrounded by the Hainan Island, the South China and the Taiwan Island on the west, north and east and adjacent to the SCS deep basin on the south. It has total area of about 2×10^5 km^2 with a length of about 1000 km and a width of 200 km as shown in Figure 13.1(a). The circulation system of NCSS consists of the southwestward slope current, the SCS warm current between the slope and the shelf, the coastal currents with opposite directions in winter and summer (Shu *et al.*, 2018). Meanwhile, the NCSS is a dynamically active shallow water area with energetic mesoscale processes resulting from the exchanges with connected Taiwan Strait, Luzon Strait, SCS deep basin and atmospheric forcings of typhoon and seasonally reversed monsoon winds. Among them, mesoscale eddies entering NCSS play a significant role in regional mass, heat and momentum exchanges, because they carry their own water mass, and their energy is one order of magnitude higher than that of the background flow field (Chelton *et al.*, 2007, 2011; Le Traon and Morrow, 2001; Xiu *et al.*, 2010; Zhang *et al.*, 2013; Zheng *et al.*, 2014; Xie *et al.*, 2018; Wang *et al.*, 2019).

Statistics indicate that one-third of mesoscale eddies in the SCS reach the NCSS bounded by the depth of 1000 m as shown in Figure 13.1(b). This chapter focuses on statistical features of

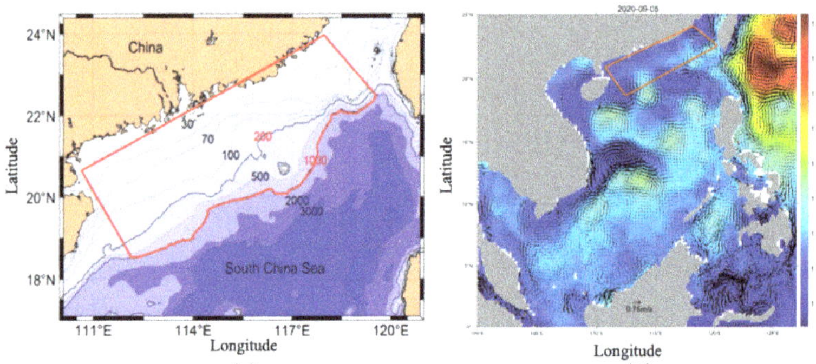

(a) (b)

Fig. 13.1. (a) Topographic map of the NCSS. The numerals on isobaths are in meter. (b) An example of Maps of Absolute Dynamic Topography (MADT) product of the SCS on 5 September 2020. Color codes are in cm. The red boxes mark outlines of the NCSS.

Sources: (a) Taken from Zhang *et al.* (2022). (b) Drawn with AVISO MADT dataset.

mesoscale eddies entering the NCSS (Zhang *et al.*, 2022). The daily mesoscale eddy trajectory atlas product from 1993 to 2016 is used for the statistical analysis. The mesoscale eddy trajectory atlas product is derived from the two-satellite daily delayed time gridded sea level anomaly (SLA) product with 0.25° of spatial resolution, distributed by AVISO (https://www.aviso.altimetry.fr/es/data/products/value-added-products/global-mesoscale-eddy-trajectory-product.html). The dataset was processed using algorithms by Chelton *et al.* (2011). The eddies are identified as features with diameters of 100–300 km, and a filter is applied to remove short tracks (less than 28 days). The dataset includes information such as mesoscale eddy amplitude (defined as the maximum SLA), rotational speed, radius, position and observation time.

13.2 Statistics of the Temporal Distribution of Eddy Numbers

13.2.1 *Interannual variability of the mesoscale eddy numbers*

From the daily mesoscale eddy trajectory atlas product from 1993 to 2016, total 184 mesoscale eddies entering the NCSS are detected.

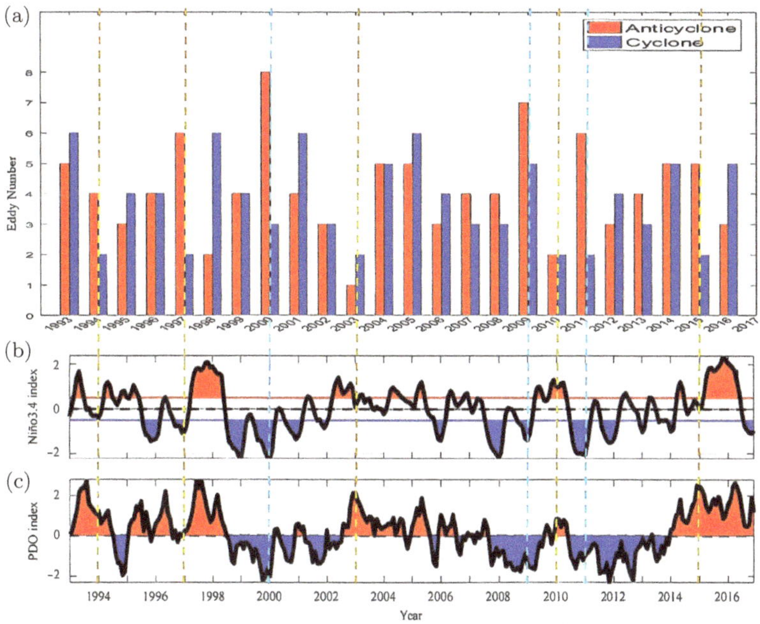

Fig. 13.2. (a) Histogram of numbers of mesoscale eddies in the NCSS from 1993 to 2016. Red and blue bars represent the numbers of AEs and CEs. (b) and (c) Niño 3.4 index and PDO index time series. Vertical orange and light blue dash lines mark El Niño and La Niña years.

Sources: (a) Taken from Zhang *et al.* (2022). (b) and (c) Taken from Huang *et al.* (2020).

Figure 13.2(a) shows the temporal distribution histogram of the yearly numbers of eddies entering the NCSS. Among them, there are 96 anticyclonic eddies (AEs) and 88 cyclonic eddies (CEs) accounted for 52.2% and 47.8% of the total number, respectively. The mean yearly numbers of total, AEs and CEs are 8.0 ± 2.0, 4.2 ± 1.6 and 3.7 ± 1.4, respectively. The histogram shows multiple time scale variability of the yearly numbers of eddies. Compared with corresponding interannual variability indexes, the El Niño (Niño 3.4) index in Figure 13.2(b) and the Pacific Decadal Oscillation (PDO) index in Figure 13.2(c), one can see that during the El Niño events (positively peaked Niño 3.4 and PDO indexes) occurred in 1994, 1997, 2003, 2010 and 2015, the mean yearly eddy number is 5.6 ± 1.9, 30% less than the total mean of 8.0 ± 2.0, and the mean yearly AE number is 3.6 ± 1.9, 24% less than the total mean of 4.2 ± 1.6. During the La Niña events (negatively peaked Niño 3.4 and PDO indexes) occurred in 2000, 2009 and 2011, the mean yearly eddy number is 10.3 ± 1.7, 29% more than the total mean of 8.0 ± 2.0, and the mean yearly AE number is 7.0 ± 0.8, 67% more than the total mean of 4.2 ± 1.6. It implies that the NCSS has strong responses to the interannual and long-term variabilities of the Pacific, i.e., in the El Niño years, the mean eddy number is 30% less than the total mean, while in the La Niña years, that is 29% more, particularly the AE number is 67% more, than the mean.

13.2.2 *Seasonal variability of the mesoscale eddy numbers*

The histogram of seasonal distribution of mesoscale eddy numbers entering NCSS from 1993 to 2016 is shown in Figure 13.3(a). One can see the following statistical features. (1) Among the total 184 eddies, the AE number is 4.4% more than that of CEs. In contrast, for the entire SCS, AE number occupies 47.8% and CE number 52.2% of the total number, i.e., the AE number is 4.4% less than that of CEs (Jin *et al.*, 2024). (2) The numbers of AEs and CEs are remarkably unbalaced. In spring, the AE number is 2.2 times of that of CEs. In fall, it is upsidedown that the CE number is 1.9 times of that of AEs. In winter, it is upsidedown again that the AE number is 1.8 times of that of CEs. (3) The seasonal variability trends of AEs and CEs are out of sync. The CE number is peaked in fall and the AE number in winter. It indicates that the numbers of AEs and CEs are probably controlled separately by their own single-polarity generators or modulators. All the statistical features reflect the uniqueness and complexity of continental shelf sea circulation in comparison to the deep ocean.

In the NCSS, major external dynamic processes that are capable of inflencing the seasonal variability of numbers of basin-scale distributed mesoscale eddies include typhoon, Kuroshio intrusion and monsoon. As a cyclone in the atmosphere, typhoon is a single-polarity, cyclonic driver for the ocean dynamic processes in the northern hemisphere, so that it is favorable for generation of CEs (Lin *et al.*, 2025). Meanwhile, the NCSS is a region where the typhoon trajectories are concentrated with a yearly mean of 10.4 trajectories (Zheng *et al.*, 2015; Shi *et al.*, 2019; Wang, 2020; Xie and Li, 2020;

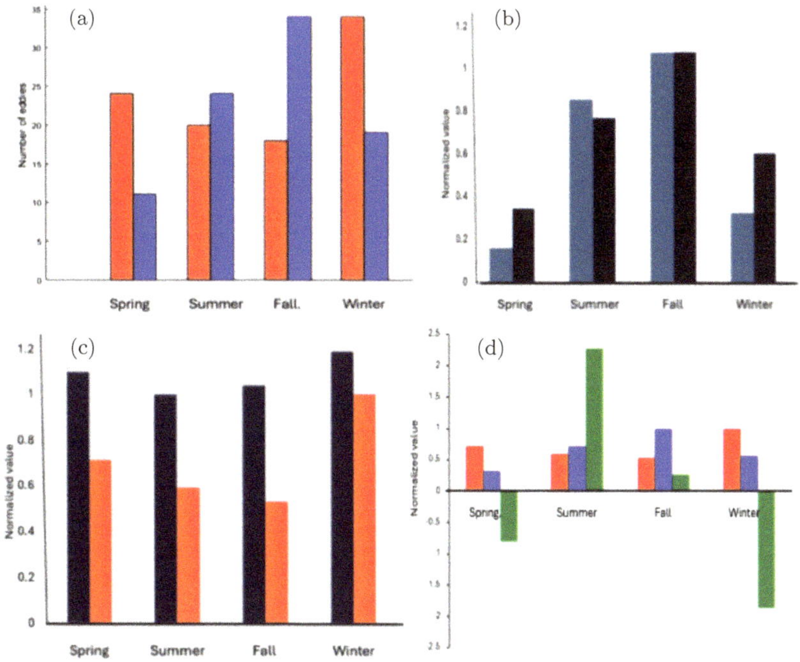

Fig. 13.3. (a) Histogram of seasonal distribution of the number of mesoscale eddies in the NCSS. Red and blue bars represent AE and CE. (b) Comparison of seasonal variability of CE numbers (black bars) with that of TFI_s (gray bars) derived from the typhoons passing over the SCS from September 2002 to August 2018. (c) Same as (b), but for AE numbers (orange bars) with simultaneous KFIs (black bars). (d) Same as (b), but for normalized AE (red bars) and CE (blue bars) numbers with MIIs (green bars).

Sources: (a) Taken from Zhang *et al.* (2022), typhoon data in (b) taken from (Wang, 2020), Kuroshio intrusion data in (c) taken from Wu *et al.* (2017) and monsoon data in (d) taken from Liu *et al.* (2023).

Liu *et al.*, 2023). To explore the statistical correlationship of eddy numbers with typhoons passing over the SCS, a parameter, typhoon forcing index (TFI), is defined as

$$TFI_{m,s} = \sum_{i=1}^{5} iN_i / TFI_p, \qquad (13.1)$$

where m and s represent month or season, N is the typhoon occurrence location number during the statistical time interval (month or season), i is the typhoon intensity category from 1 to 5, and TFI_p is the peak value in a year. Figure 13.3(b) shows comparison of seasonal variability of CE numbers in the NCSS with that of TFI_s derived from the typhoons in the SCS that occurred between September 2002 and August 2018 (Wang, 2020). The linear regression correlation coefficient (R^2) of CE numbers and TFI_s reaches as high as 0.91, implying a close causality between the two.

In contrast to typhoon, Kuroshio intrusion is a single-polarity, AE generator (Zheng *et al.*, 2011, 2019; Hu *et al.*, 2012; Wu *et al.*, 2017; Zhang *et al.*, 2017). Like the case of typhoon, define a nondimensional parameter, Kuroshio intrusion forcing index (KFI) as

$$KFI = \bar{\eta}_p/\bar{\eta}, \tag{13.2}$$

where $\bar{\eta}$ is the mean sea level in the characteristic area, Kuroshio Intrusion South (KIS), (115–120°E, 15–20°N) in the SCS deep basin (Wu, 2013; Wu *et al.*, 2017), and $\bar{\eta}_p$ is a peak value of $\bar{\eta}$ during the statistical time interval. Figure 13.3(c) shows comparison of seasonal variability of AE numbers in the NCSS with that of KFI derived from the simultaneous sea level data in KIS (Wu *et al.*, 2017). The linear regression R^2 of AE numbers and seasonal mean KFIs reaches as high as 0.89, implying a close causality between the two.

Monsoon is a seasonally reversed, basin scale, external forcing process for the SCS, so that it is a dual-polarity modulator for ocean eddies. Previous investigators have defined an SCS monsoon summer intensity index as (Wu and Liang, 2002; Liu *et al.*, 2023)

$$MII \equiv I_{Vsw} + I_{OLR}, \tag{13.3}$$

where I_{Vsw} is the standardized southwest wind index and I_{OLR} is the standardized outgoing longwave radiation (OLR) index. Figure 13.3(d) shows comparison of seasonal variability of AE and CE numbers normalized by their peak values with that of MII derived from monthly I_{Vsw} and I_{OLR} data covering the period from 1979 to 2020 (Liu *et al.*, 2023).

To explore statistical correlationships of seasonal variabilities of eddy numbers with external forcings, linear regression equations and correlation coefficients (R^2) are calculated. The results are shown as a matrix of Table 13.1. One can see that the linear regression R^2 of CE numbers and seasonal mean MIIs reaches 0.61, implying a good causality between the two. However, R^2 of AE numbers and MIIs

Table 13.1. Linear regression equations and correlation coefficient (R^2) matrix of total eddy numbers with typhoon, Kuroshio intrusion and monsoon forcings.

	TFI	KFI	MII
Total AE	$y = -0.3779x + 0.9191$ $R^2 = 0.526$	$y = 2.3797x - 1.8686$ $R^2 = 0.886$	$y = 0.0708x + 0.6498$ $R^2 = 0.193$
Total CE	$y = 0.6771x + 0.2683$ $R^2 = 0.9095$	$y = -1.7228x + 2.5125$ $R^2 = 0.25$	$y = -0.0923x + 0.7045$ $R^2 = 0.609$

reaches 0.19 only, implying a fair causality. The Kuroshio intrusion is the primary forcing for total AEs and typhoon forcing is the secondary one (negative correlation). For total CEs, the typhoon forcing is the primary forcing and the monsoon forcing is the secondary one (negative correlation).

13.3 Eddy Motion Trajectory Classification

Ocean eddy motion trajectory (EMT) results from complex interaction of external forcings and its background dynamic environments. Thus, it is possible to extract the important data and information on eddy generation and evolution through the statistical analysis of EMTs. According to EMT features, total 184 mesoscale eddies in the NCSS from 1993 to 2016 are classified into four types: along isobath (AI), intruding shelf (IS), local wandering (LW) and continental shelf generation (CG) types. The statistical features of the four types of eddies are examined as follows.

13.3.1 *Type of along isobath (AI)*

The along isobath (AI) type of eddies accounts for 87, 47.3% of the total mesoscale eddies during the 24-year period. The trajectories of AI type of eddies are shown in Figure 13.4. One can see that the origins of mesoscale eddies are concentrated in the northeastern SCS, i.e., source region AI_{source} as labeled in the figure. After entering the NCSS, eddies move along the isobaths of 1000 m or so and finally dissipate in the northwest of SCS deep basin, i.e., sink region AI_{sink} as labeled in the figure.

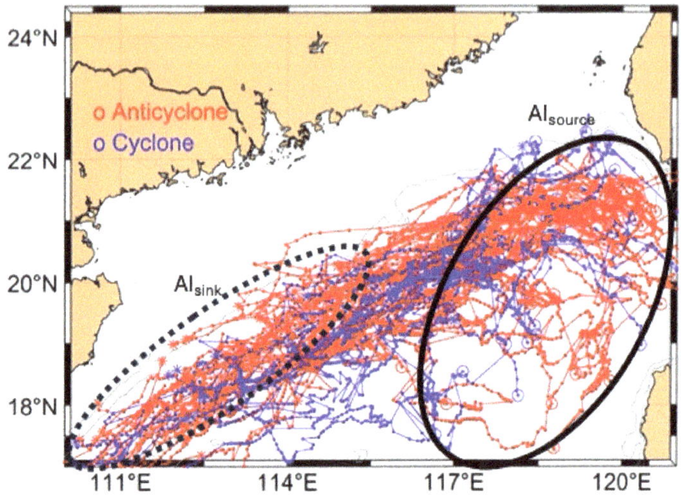

Fig. 13.4. Trajectories of AI-type of AEs (in red) and CEs (in blue) from 1993 to 2016. Circles and asterisks on both end of trajectories represent eddy generation and dissipation locations. Solid and dash ellipses represent the source region and sink region of eddies.

Source: Taken and re-drawn from Zhang *et al.* (2022).

Figure 13.5(a) shows statistical histogram of yearly numbers of AI type of eddies from 1993 to 2016. One can see that among the total 87 eddies, there are 51 AEs and 36 CEs occupying 58.6% and 41.4% of the total number, respectively. The AE number is 17.2% more than that of CEs. The mean yearly numbers of total, AEs and CEs are 3.6 ± 1.8, 2.1 ± 1.2 and 1.5 ± 0.9, respectively. During the El Niño events in 1994, 1997, 2003, 2010 and 2015, the mean yearly eddy number is 2.4 ± 1.2, 33% less than the total mean of 3.6 ± 1.8, the mean yearly AE number is 1.4 ± 1.2, 33% less than the total mean of 2.1 ± 1.2 and the mean yearly CE number is 1.0 ± 0.0, 52% less than the total mean of 2.1 ± 1.2. During the La Niña events in 2000, 2009 and 2011, the mean yearly eddy number is 6.0 ± 0.8, 67% more than the total mean of 3.6 ± 1.8, the mean yearly AE number is 4.3 ± 0.5, 105% more than the total mean of 2.1 ± 1.2 and the mean yearly CE number is 1.7 ± 0.5, 13% more than the total mean of 1.5 ± 0.9. It implies that AI type of eddies has strong responses to the interannual and long-term variability of the Pacific, i.e., in the El Niño years, the mean yearly eddy number is 33% less than the total

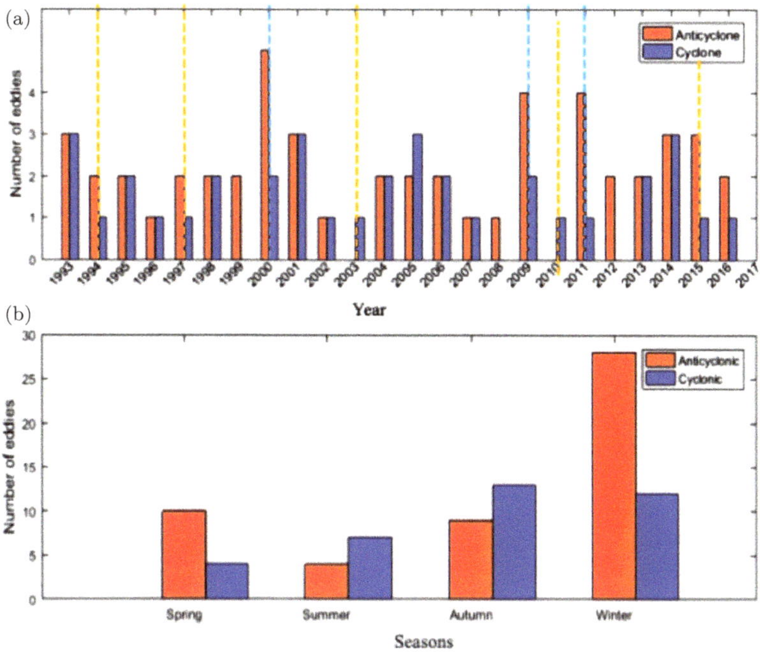

Fig. 13.5. Statistical histograms of the numbers of AI type of AEs (in red) and CEs (in blue) in the NCSS from 1993 to 2016. (a) Interannual and (b) seasonal variability.

Source: Taken and re-drawn from Zhang *et al.* (2022).

mean, while in the La Niña years, that is 67% more, particularly the AE number is 105% more than the total mean.

Figure 13.5(b) shows the statistical histogram of seasonal distribution of AI type of mesoscale eddy numbers from 1993 to 2016. One can see that the numbers of AEs and CEs are remarkably unbalanced. In spring, the AE number is 2.5 times of that of CEs. In summer and fall, the ratios decrease to 0.57 and 0.69. In winter, it increases to 2.3. Meanwhile, the seasonal variability trends of AEs and CEs are out of sync. The CE and AE numbers are peaked in fall and winter, respectively. Linear regression equations and correlation coefficients (R^2) of seasonal variabilities of eddy numbers with external forcings are shown as a matrix of Table 13.2. One can see that the Kuroshio intrusion is the primary forcing for AI–AE and monsoon forcing is the secondary one (negative correlation). For AI–CE, the typhoon

Table 13.2. Linear regression equations and correlation coefficient (R^2) matrix of AI type of eddies with typhoon, Kuroshio intrusion and monsoon forcings.

	TFI	KFI	MII
AI–AE	$y = -0.4695x + 0.7129$ $R^2 = 0.2486$	$y = 4.3126x - 4.2184$ $R^2 = 0.8906$	$y = -0.1794x + 0.4442$ $R^2 = 0.7047$
AI–CE	$y = 0.4104x + 0.4627$ $R^2 = 0.2571$	$y = 0.8437x - 0.2208$ $R^2 = 0.0461$	$y = -0.0409x + 0.6912$ $R^2 = 0.0495$

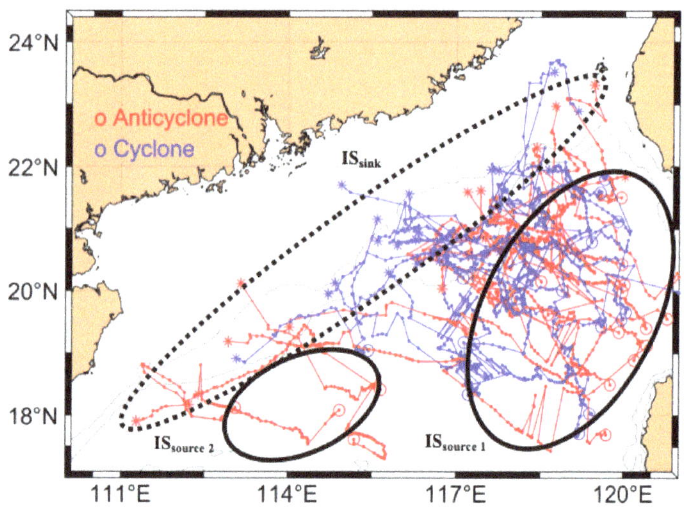

Fig. 13.6. Trajectories of IS-type of AEs (in red) and CEs (in blue) from 1993 to 2016. Circles and asterisks on both end of trajectories represent eddy generation and dissipation locations. Solid and dash ellipses represent the source region and sink region of eddies.

Source: Taken and re-drawn from Zhang *et al.* (2022).

forcing is the weakly primary forcing and Kuroshio intrusion and monsoon forcings seem to be ignorable.

13.3.2 *Type of intruding shelf (IS)*

The number of intruding shelf (IS) type of mesoscale eddies accounts for 38, 20.7% of the total eddy number from 1993 to 2016. The trajectories are shown in Figure 13.6. One can see that the eddy source region consists of two sub-regions: $IS_{source1}$ east of 117°E contains

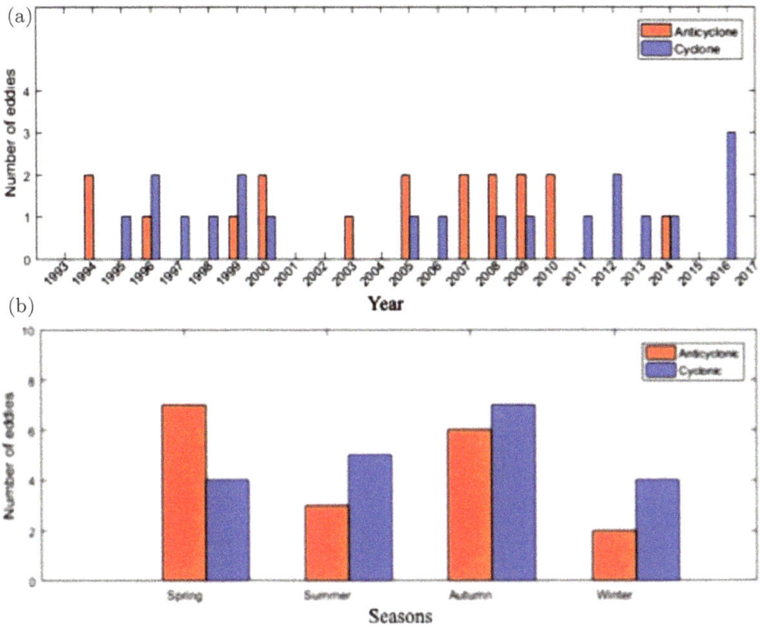

Fig. 13.7. Statistical histograms of the numbers of IS type of AEs (in red) and CEs (in blue) in the NCSS from 1993 to 2016. (a) Interannual and (b) seasonal variability.

Source: Taken from Zhang *et al.* (2022).

87% of eddies, and $IS_{source2}$ west of 115.5°E contains rest 13%. All eddies move across isobaths and dissipate in the IS_{sink} region on the slope and shelf. Majority (87%) dissipate within the continental slope with depths between 200 m and 1000 m. The probability for intruding onto the shelf with depths shallower than 200 m reaches 13% of IS type of eddies, i.e., 2.7% of total eddy number (184). Although the percentages seem to be low, this type of eddies has special significance for the research on exchanges between the deep basin and continental shelf.

The interannual variability of the IS type of eddy numbers is shown as statistical histogram in Figure 13.7(a). One can see that the yearly mean eddy number intruding into the NCSS is 1.6 ± 1.1. AEs and CEs occupy 47% and 53%, respectively. However, confident interannual variability trend cannot be obtained because of the small dataset.

Table 13.3. Linear regression equations and correlation coefficient (R^2) matrix of IS type of eddies with typhoon, Kuroshio intrusion and monsoon forcings.

	TFI	KFI	MII
IS–AE	$y = -0.0168x + 0.6544$ $R^2 = 0.0004$	$y = -1.1404x + 1.8795$ $R^2 = 0.0773$	$y = -0.0054x + 0.6448$ $R^2 = 0.0008$
IS–CE	$y = 0.4591x + 0.4554$ $R^2 = 0.8246$	$y = -1.4567x + 2.2893$ $R^2 = 0.3524$	$y = 0.0483x + 0.7141$ $R^2 = 0.1774$

Figure 13.7(b) shows the statistical histogram of seasonal distribution of IS type of mesoscale eddy numbers from 1993 to 2016. One can see that like the case of AI type, the numbers of AEs and CEs are remarkably unbalanced. In spring, the AE number is 1.75 times of that of CEs. In summer, fall and winter, however, the ratios decrease to 0.60, 0.86 and 0.50. Meanwhile, the seasonal variability trends of AEs and CEs are out of sync. The AE and CE numbers are peaked in spring and fall, respectively. Linear regression equations and correlation coefficients (R^2) of seasonal variabilities of eddy numbers with external forcings are shown as a matrix of Table 13.3. One can see that for IS-AE, all three external forcings are not correlated, implying their local generation property. For IS-CE, the typhoon forcing is the primary forcing, Kuroshio intrusion is secondary one (negative correlation), and monsoon forcing is weakly correlated.

13.3.3 *Type of local wandering (LW)*

The number of local wandering (LW) type of mesoscale eddies accounts for 23, 13.5% of the total eddies from 1993 to 2016. The trajectories are shown in Figure 13.8. One can see that there are eddy wandering regions LW1 and LW2 separated by the slope topography protruding into the deep basin at around 117°E, which contain 61% and 29% of eddies. All eddies generate and dissipate within their wandering regions.

The interannual variability of the LW type of eddy numbers is shown as statistical histogram in Figure 13.9(a). One can see that the yearly mean eddy number is 0.93 ± 0.84, and the numbers of AEs and CEs are 6 and 17 accounting for 26% and 74%, respectively. However, confident interannual variability trend cannot be

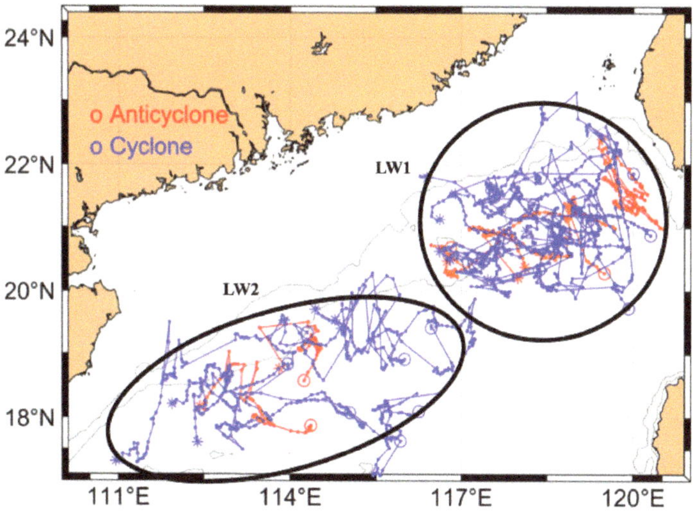

Fig. 13.8. Trajectories of LW type of AEs (in red) and CEs (in blue) from 1993 to 2016. Circles and asterisks on both end of trajectories represent eddy generation and dissipation locations. Solid ellipses represent two wandering regions of eddies. *Source*: Taken and re-drawn from Zhang *et al.* (2022).

determined by the small dataset. Figure 13.9(b) shows the statistical histogram of seasonal distribution of LW type of mesoscale eddy numbers. One can see that the total numbers of AEs and CEs are remarkably unbalanced, i.e., the CE number is 2.83 times of that of AEs. Linear regression equations and correlation coefficients (R^2) of seasonal variability of LW-CE (excluding LW-AE due to small dataset) numbers with external forcings are shown as a matrix of Table 13.4. One can see that the monsoon forcing is the primary forcing, Kuroshio intrusion and typhoon forcing are the secondary (negative correlation) and third one, respectively.

13.3.4 *Type of continental shelf generation (CG)*

The number of continental shelf generation (CG) type of mesoscale eddies accounts for 36, 19.5% of the total eddies from 1993 to 2016. The trajectories are shown in Figure 13.10. One can see that eddies concentrate in three regions: CG_{AE} in the Taiwan Strait is an AE single-polarity concentrated region, in contrast, CG_{CE} east of the Hainan Island is a CE single-polarity concentration region,

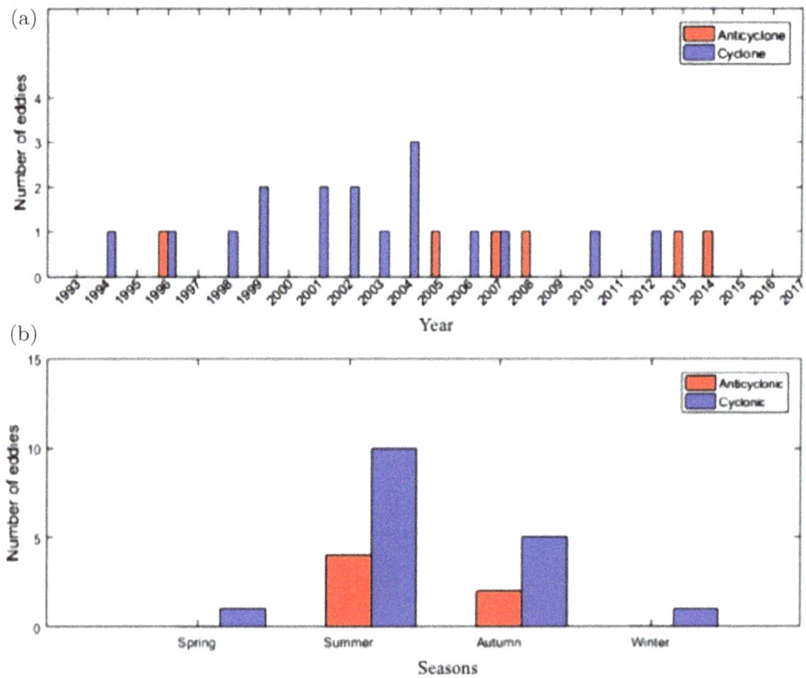

Fig. 13.9. Statistical histograms of the numbers of LW type of AEs (in red) and CEs (in blue) in the NCSS from 1993 to 2016. (a) Interannual and (b) seasonal variability.

Source: Taken from Zhang *et al.* (2022).

Table 13.4. Linear regression equations and correlation coefficient (R^2) matrix of LW type of CEs with typhoon, Kuroshio intrusion and monsoon forcings.

	TFI	KFI	MII
LW–CE	$y = 0.7943x - 0.0198$ $R^2 = 0.5556$	$y = -4.4567x + 5.2493$ $R^2 = 0.7428$	$y = 0.2344x + 0.4326$ $R^2 = 0.9388$

and between the two, CG_{AE+CE} is an AE and CE dual-polarity concentration region. Three regions are basically independent of each other with a few eddies exchanged.

The interannual variability of the CG type of eddy numbers is shown as statistical histogram in Figure 13.11(a). One can see that the yearly mean eddy number is 1.46 ± 0.96,

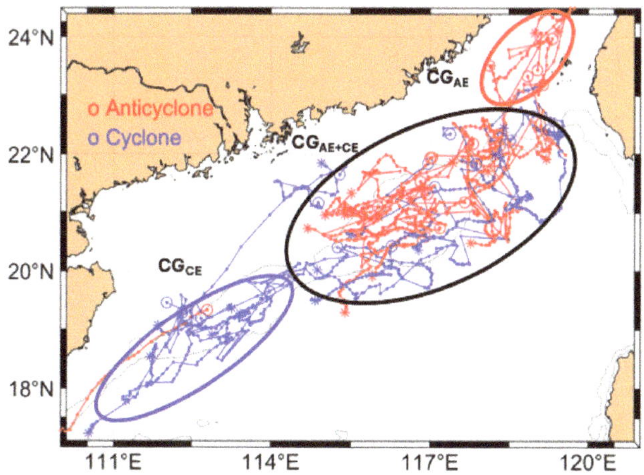

Fig. 13.10. Trajectories of CG-type of AEs (in red) and CEs (in blue) from 1993 to 2016. Circles and asterisks on both end of trajectories represent eddy generation and dissipation locations. Solid ellipses represent three concentration regions of eddies.

Source: Taken and re-drawn from Zhang *et al.* (2022).

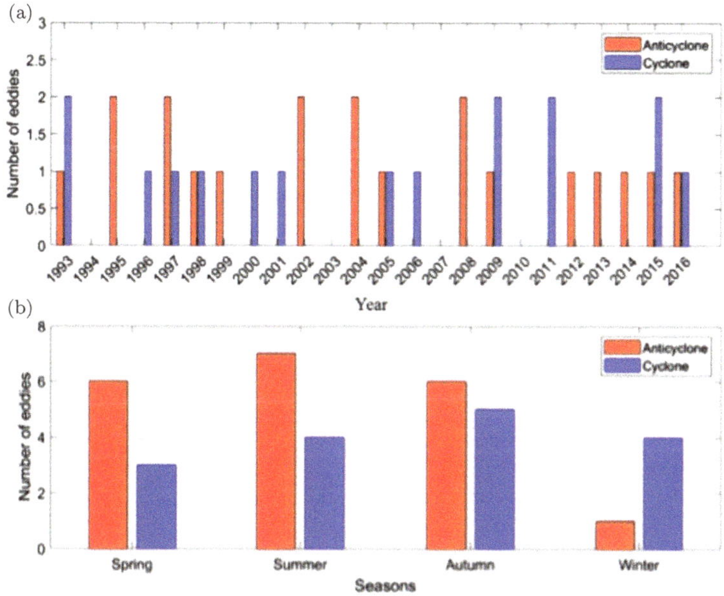

Fig. 13.11. Statistical histograms of the numbers of CG type of AEs (in red) and CEs (in blue) in the NCSS from 1993 to 2016. (a) Interannual and (b) seasonal variability.

Source: Taken from Zhang *et al.* (2022).

Table 13.5. Linear regression equations and correlation coefficient (R^2) matrix of CG type of eddies with typhoon, Kuroshio intrusion and monsoon forcings.

	TFI	KFI	MII
CG–AE	$y = 0.455x + 0.4602$	$y = -4.3443x + 5.4177$	$y = 0.1758x + 0.7207$
	$R^2 = 0.2199$	$R^2 = 0.8513$	$R^2 = 0.6373$
CG–CE	$y = 0.3526x + 0.6026$	$y = -0.5861x + 1.4344$	$y = 0.0229x + 0.8007$
	$R^2 = 0.7492$	$R^2 = 0.0879$	$R^2 = 0.0612$

Table 13.6. Statistical means of characteristic parameters of four types of eddies.

Trajectory type	Occurrent probability	Primary forcing for AE/CE	Lifetime (day)	Amplitude (cm)	Radius (km)
AI	47.3%	Kuroshio/Typhoon	93 ± 55	18.0 ± 8.0	154 ± 43
IS	20.7%	NA/Typhoon	82 ± 53	17.0 ± 7.4	149 ± 41
LW	13.5%	NA/Monsoon	85 ± 53	16.9 ± 7.8	149 ± 41
CG	19.5%	Kuroshio/Typhoon	74 ± 54	16.0 ± 7.9	146 ± 42
Total mean			83.5	17.0	149.5

and the numbers of AEs and CEs account for 56% and 44%, respectively. However, confident interannual variability trend cannot be determined by the small dataset. Figure 13.11(b) shows the statistical histogram of seasonal distribution of CG type of mesoscale eddy numbers. One can see that the total numbers of AEs and CEs are remarkably unbalanced, i.e., the AE number is 1.25 times of that of CEs. Linear regression equations and correlation coefficients (R^2) of seasonal variability of eddy numbers with external forcings are shown as a matrix of Table 13.5. One can see that for CG–AE, the Kuroshio intrusion is the primary forcing (negative correlation) and the monsoon forcing is the secondary one. For CG–CE, the typhoon forcing is the primary forcing, and Kuroshio intrusion and monsoon forcings are ignorable.

13.3.5 *Statistics of characteristic parameters of eddies*

From the AVISO data, the statistical mean lifetime, amplitude, and radius for four types of eddies are calculated. The results are listed in

Table 13.6. One can see that among four types of eddies, AI type is the strongest one with the mean lifetime, mean amplitude and mean radius 11%, 5.9% and 3.0% greater than the total means. In contrast, CG type is the weakest one with that 11%, 5.9% and 2.3% smaller than the total means. IS and LW types are between the two.

13.4 Summaries

This chapter contains statistical analysis results for 184 mesoscale eddies that entered the northern continental shelf of the SCS (NCSS) from 1993 to 2016. The results indicate that statistical features of eddies in the shallow continental shelf of marginal sea are distinct from that of the deep ocean. Among the total 184 eddies, there are 96 AEs and 88 CEs accounting for 52.2% and 47.8% of the total number. The AE number is 4.4% more than that of CEs,while it is 4.4% less for the entire SCS. The numbers of AEs and CEs are remarkably unbalanced with ratios of 1:0.46 in spring, 1:1.2 in summer, 1:1.9 in fall and 1:0.56 in winter, implying that the numbers of AEs and CEs are controlled separately by their own single-polarity generators or modulators. Meanwhile, eddy numbers in the NCSS have strong responses to the interannual and long-term variability of the Pacific. In the El Niño years, the mean yearly eddy number is 30% less than the total mean, particularly the mean yearly AE number is 24% less than the total mean, while in the La Niña years, that is 29% and 67% more than the total mean.

According to the spatial distribution and morphologic features of trajectories, eddies are classified into four types: along isobath (AI), intruding shelf (IS), local wandering (LW) and continental shelf generation (CG) types with occurrence probabilities of 47.3%, 20.7%, 13.5% and 19.5%. Using the typhoon forcing index (TFI), Kuroshio intrusion forcing index (KFI) and monsoon intensity index (MII) as parameters of three major external forcings for eddies and linear regression correlation coefficients of eddy numbers and parameters as a criterion, primary external forcings for the four types of eddies are determined. For AI and CG types, Kuroshio intrusion and typhoon are primary forcings for AEs and CEs, respectively. For IS and LW types, typhoon and monsoon are primary forcings for CEs, while none is determined for AEs.

The statistical analysis is an important step for understanding general features and potential generation and modulation mechanisms for a natural process. The above statistics of mesoscale eddies in the NCSS reveal their spatiotemporal distribution features and primary external forcing mechanisms. These results provide a baseline for empirical forecasting of mesoscale eddies in the NCSS and raise a series of problems on mechanisms to be solved by theoretical studies.

References

Chelton, D. B., Schlax, M. G., Samelson, R. M. and de Szoeke, R. A. (2007). Global observations of large oceanic eddies, *Geophys. Res. Lett.*, 34, L15606, doi:10.1029/2007GL030812.

Chelton, D. B., Schlax, M. G. and Samelson, R. M. (2011). Global observations of nonlinear mesoscale eddies, *Progr. Oceanogr.*, 91, pp. 167–216, doi:10.1016/j.pocean.2011.01.002.

Hu, J., Zheng, Q., Sun, Z. and Tai, C. K. (2012). Penetration of nonlinear Rossby eddies into South China Sea evidenced by cruise data, *J. Geophys. Res. Oceans*, 117, C03010, doi:10.1029/2011JC007525.

Huang, R., Xie, L., Zheng, Q., Li, M., Bai, P. and Tan, K. (2020). Statistical analysis of mesoscale eddy propagation velocity in the South China Sea deep basin, *Acta Oceanol. Sin.*, 39, pp. 91–102, doi:10.1007/s13131-020-1678-x.

Jin, Y., Jin, M., Wang, D. and Dong, C. (2024). Statistical analysis of multi-year South China Sea eddies and exploration of eddy classification, *Rem. Sens.*, 16, 1818, doi:10.3390/rs16101818.

Le Traon, P. Y. and Morrow, R. (2001). Ocean currents and eddies, *Int. Geophys.*, 69, pp. 171–215, doi:10.1016/S0074-6142(01)80148-0.

Lin, J. Y., Ho, H., Gopalakrishnan, G., Zheng, Z.-W., Tseng, R.-S., Pan, J., Ho, C.-R. and Zheng, Q. (2025). Typhoon induced mesoscale cyclonic eddy a long neglected linkage between atmosphere ocean and climate, *npj Clim. Atmos. Sci.*, 8, 64, doi:10.1038/s41612-025-00946-9.

Liu, S., Xu, J., Tu, S., Zheng, M. and Chen, Z. (2023). Relationship between South China Sea summer monsoon and Western North Pacific tropical cyclones linkages with the interaction of Indo-Pacific Pattern, *Atmosphere*, 14, 645, doi:10.3390/atmos14040645.

Shi, Y., Xie, L., Zheng, Q., Zhang, S., Li, M. and Li, J. (2019). Unusual coastal ocean cooling in the northern South China Sea by a katabatic cold jet associated with Typhoon Mujigea (2015), *Acta Oceanol. Sin.*, 38, pp. 62–75, doi:10.1007/s13131-019-1440-4.

Shu, Y., Wang, Q. and Yan, T. (2018). Research progress on the continental shelf flow system in the northern South China Sea, *Sci. China Earth Sci.,* 48, pp. 276–287.

Wang, M., Zhang, Y., Liu, Z. and Wu, J. (2019). Temporal and Spatial characteristics of mesoscale eddies in the northern South China Sea: Statistics analysis based on altimeter data, *Adv. Earth Sci.*, 34, pp. 1069–1080.

Wang, Y. (2020). Composite of typhoon induced sea surface temperature and chlorophyll-a responses in the South China Sea, *J. Geophys. Res. Oceans*, 125, e2020JC016243, doi:10.1029/2020JC016243.

Wu, C. R. (2013). Interannual modulation of the Pacific Decadal Oscillation (PDO) on the low-latitude western North Pacific, *Progr. Oceanogr.,* 110, pp. 49–58, doi:10.1016/j.pocean.2012.12.001.

Wu, C.-R., Wang, Y.-L., Lin, Y.-F. and Chao, S.-Y. (2017). Intrusion of the Kuroshio into the South and East China Seas, *Sci. Rep.,* 7, 7895, doi:10.1038/s41598-017-08206-4.

Wu, S. and Liang, J. (2002). Intensity index of South China Sea monsoon and its variation characteristics, *J. Trop. Meteorol.*, 8, pp. 1–9.

Xie, L. and Li, M. (2020). Response of coastal upwelling east of Hainan Island in the South China Sea to sudden impact and long-term variability of atmospheric forcing, in J. Pan and A. Devlin (eds.), *Estuaries and Coastal Zones - Dynamics and Response to Environmental Changes,* IntechOpen, doi:10.5772/intechopen.88828.

Xie, L., Zheng, Q., Zhang, S., Hu, J., Li, M., Li, J. and Xu, Y. (2018). The Rossby normal modes in the South China Sea deep basin evidenced by satellite altimetry, *Intl. J. Rem. Sens.*, 39, pp. 399–417, doi:10.1080/01431161.2017.1384591.

Xiu, P., Chai, F., Shi, L., Xue, H. and Chao, J. (2010). A census of eddy activities in the South China Sea during 1993–2007, *J. Geophys. Res. Oceans,* 115, C03013, doi:10.1029/2009JC005657.

Zhang, T., Li, J., Xie, L. and Zheng, Q. (2022). Statistical analysis of mesoscale eddies entering the continental shelf of the northern South China Sea, *J. Mar. Sci. Eng.*, 10, 206, doi:10.3390/jmse10020206.

Zhang, Z., Zhao, W., Qiu, B. and Tian, J. (2017). Anticyclonic eddy shedding from Kuroshio loop and the accompanying cyclonic eddy in the northeastern South China Sea, *J. Phys. Oceanogr.,* 47, pp. 1243–1259, doi:10.1175/JPO-D-16-0185.1.

Zhang, Z., Zhao, W., Tian, J. and Liang, X. (2013). A mesoscale eddy pair southwest of Taiwan and its influence on deep circulation, *J. Geophys. Res. Oceans,* 118, pp. 6479–6494, doi:10.1002/2013JC008994.

Zheng, Q., Ho, C.-R., Xie, L. and Li, M. (2019). A case study of a Kuroshio main path cut-off event and impacts on the South China Sea in fall-winter 2013–2014, *Acta Oceanol. Sin.*, 38, pp. 1–8, doi:10.1007/s13131-019-1411-9.

Zheng, Q., Hu, J., Zhu, B., Feng, Y., Jo, Y.-H., Sun, Z., Zhu, J., Lin, H., Li, J. and Xu, Y. (2014). Standing wave modes observed in the South China Sea deep basin, *J. Geophys. Res. Oceans*, 119, pp. 4185–4199, doi:10.1002/2014JC009957.

Zheng, Q., Tai, C.-K., Hu, J., Lin, H., Zhang, R., Su, F.-C. and Yang, X. (2011). Satellite altimeter observations of nonlinear Rossby eddy-Kuroshio interaction at the Luzon Strait, *J. Oceanogr.*, 67, pp. 365–376, doi:10.1007/s10872-011-0035-2.

Zheng, Q., Zhu, B., Li, J., Sun, Z., Xu, Y. and Hu, J. (2015). Growth and dissipation of typhoon-forced solitary continental shelf waves in the northern South China Sea, *Clim. Dyn.*, 45, pp. 853–865, doi:10.1007/s00382-014-2318-y.

Part 5

Inter-scale Interactions

Chapter 14

Mesoscale Eddy Modulation to Sea Surface Waves

14.1 Introduction

Sea surface waves are the first important indicator of the sea state and play crucial role in ocean navigation and ocean engineering. Wind driving is a dominant force for the generation and propagation of sea surface waves, while background currents may modulate wave fields (Phillips, 1984; Romero *et al.*, 2017; Zhang *et al.*, 2021; Tan *et al.*, 2023). As the sea surface waves encounter a background current, exchanges of momentum and energy may occur between the waves and current, resulting in significant variations in wave features (Holthuijsen and Tolman, 1991). The wave-current interaction has become a hot topic in ocean dynamics (Cheng *et al.*, 2022; Echevarria *et al.*, 2021; Tamura *et al.*, 2008; Wang *et al.*, 2022; Wolf and Prandle, 1999; Xie *et al.*, 2001).

Using in situ observations, remote sensing and numerical models, previous studies found modulation of wave features by tidal currents in the coastal ocean (Ardhuin *et al.*, 2012; Bolaños *et al.*, 2014; Pearman *et al.*, 2014; Sorourian *et al.*, 2020) and strong background currents in the open ocean (Kudryavtsev *et al.*, 1995; Romero *et al.*, 2017; Wang *et al.*, 1994; Wu *et al.*, 2025). They found that the wave energy increases, and the wave direction shifts as the waves propagate against the current. The deformation and strain of the nonuniform current and the nonlinear interaction are found to be mechanisms of wave variation (Wang *et al.*, 2020). In the case of hurricanes, reported a 5% increase in the significant wave height (SWH) on the left-hand side and an 11% decrease on the right-hand side of the hurricane ground track due to current-induced wind energy input based on model results. Echevarria *et al.* (2021) revisited the influence of ocean surface currents on the global wind-wave field with the WaveWatchIII (WW3) model. They found that the simulated SWHs decrease in most ocean basins with surface current forcing due to a decreased relative wind given by co-flowing winds and currents, and wave directions deflect due to current-induced refraction in the western boundary and tropical regions across the eddy.

Mesoscale eddies are ubiquitous in the global ocean and have energetic rotating currents. As surface waves propagate through an eddy, the wave features should be significantly modulated by the eddy current and the front at the eddy edge (Bühler and McIntyre, 2005; Melville *et al.*, 2005). Liu *et al.* (1994) observed wave refraction in eddy area in SAR images. By reconstructing the surface currents derived from satellite altimetry, Gallet and Young (2014) found that mesoscale vorticity in the Southern Ocean significantly deflects the swells away from their original paths when crossing the Pacific. Using the Regional Ocean Modeling System (ROMS) and WW3 models, Ardhuin *et al.* (2017) found that mesoscale currents contribute 75% of the spatial variability of wave heights at the scale of 100 km in the Gulf Stream, where the flow intensity and structure can affect the wave height gradients. The WW3 model results by Villas Bôas *et al.* (2020) indicate that the spatial variability of SWH is highly sensitive to the underlying current, especially at scales from 15 to 200 km. Using the same model, Marechal and de Marez (2021) described variations in wave bulk parameters such as wave height, mean wave period, and wave direction between the inside and outside of an isolated eddy. They found that the unstable sub-mesoscale processes around mesoscale eddies contribute most to the wave variation, and variations are larger as incident waves are shorter. Furthermore, mesoscale eddies may modulate the surface wind field (Chelton, 2013; Frenger *et al.*, 2013).

As a semi-enclosed marginal sea of the western North Pacific (WNP) (Figure 14.1), the SCS is characterized by active mesoscale eddies and surface waves due to seasonally reversed monsoons, complex bottom topography, and disturbances from the Pacific (Chen *et al.*, 2011; Fang *et al.*, 2006; Xie *et al.*, 2018, 2020; Zheng

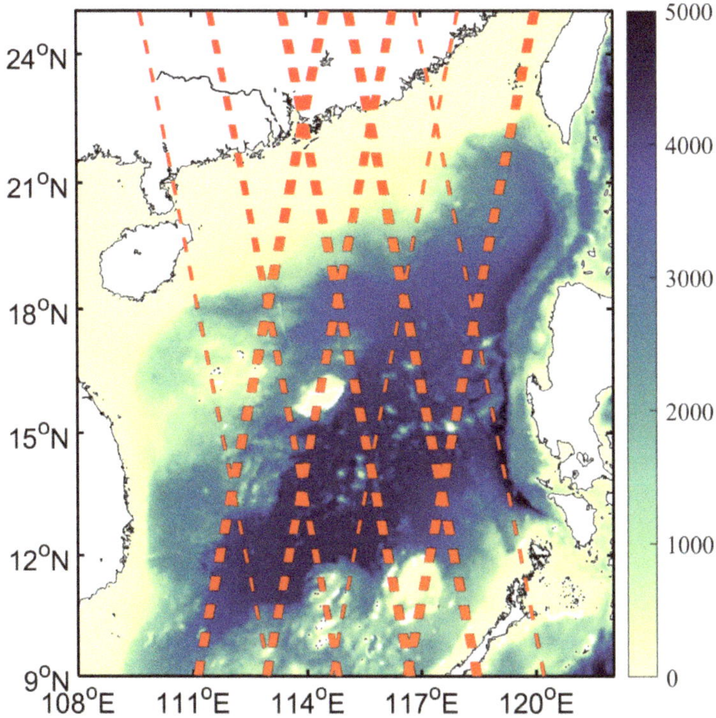

Fig. 14.1. Red dash lines mark CFOSAT orbit ground tracks (some tracks repeat with a little shift).

Source: Taken from Tan *et al.* (2023).

et al., 2017, 2019; Zhu *et al.*, 2015). The concurrent energetic eddies and waves in the SCS provide a good opportunity for investigating modulation effects of eddies on surface wave fields. Using a numerical model, Wang *et al.* (2022) investigated effects of larger warm eddies on surface waves in the western SCS near Vietnam coast using Chinese–French Oceanic SATellite (CFOSAT) data and found the wave height varied by 20–30%. Tan *et al.* (2023) examined modulation effects of mesoscale eddies on sea surface wave fields in the SCS using data measured by the Surface Wave Investigation and Monitoring (SWIM) instrument onboard the CFOSAT (Xu *et al.*, 2019). They found that variations of wave height, wavelength and direction across strong warm eddies reach 18%, 24%, and 28.4°, respectively.

This chapter summarizes the research results on sea surface wave variations across mesoscale eddies in the SCS derived from the updated satellite observations by SWIM onboard CFOSAT. Meanwhile, the wave-eddy interaction theories and the mechanism for the wave variation across mesoscale eddies are briefly introduced.

14.2 Sea Surface Wave Variations Across Mesoscale Eddies

14.2.1 *Wave parameters variations*

Using the Simulating Waves Nearshore (SWAN) wave model, which is validated by the CFOSAT observations during the period of October–November 2019, Wang *et al.* (2022) investigated mesoscale eddy effects on surface waves in the western SCS near the Vietnam coast. They found that mesoscale eddies can affect wave heights, period and steepness by 20%–30% and modify the wave direction by 30–40°.

Using CFOSAT observations in the whole SCS from July 2019 to August 2021, Tan *et al.* (2023) statistically analyzed wave parameter variations across mesoscale eddies. They found that in the total 33 cases of mesoscale eddies, the mean SWH is 1.81 m inside the eddy and decreases by 0.03 m and 0.05 m at the edge and outside of the eddy, respectively (Figure 14.2(a)). The wavelength has larger values of approximately 142 m and 140 m inside and outside of the eddy, respectively, while is evidently shorter than 137 m at the eddy edge (Figure 14.2(b)). The wave direction decreases from 216° inside to 212° at the eddy edge and increases to 214° outside of the eddy (Figure 14.2(c)).

For different eddy polarities, the SWH in anticyclonic eddies (AEs) is generally lower than that in cyclonic eddies (CEs), while the wavelength and wave direction are higher than those in CEs. The mean SWH decreases from the inside to the outside in cases of AEs, while the mean SWH decreases from the inside to the edge and increases outside the eddy in cases of CEss (Figure 14.2(a)). The minimum and maximum values of the mean wavelength (Figure 14.2(b)) occur at the eddy edges and in the insides of warm eddies. The maximum-minimum differences are above 15 m in CEs, while only

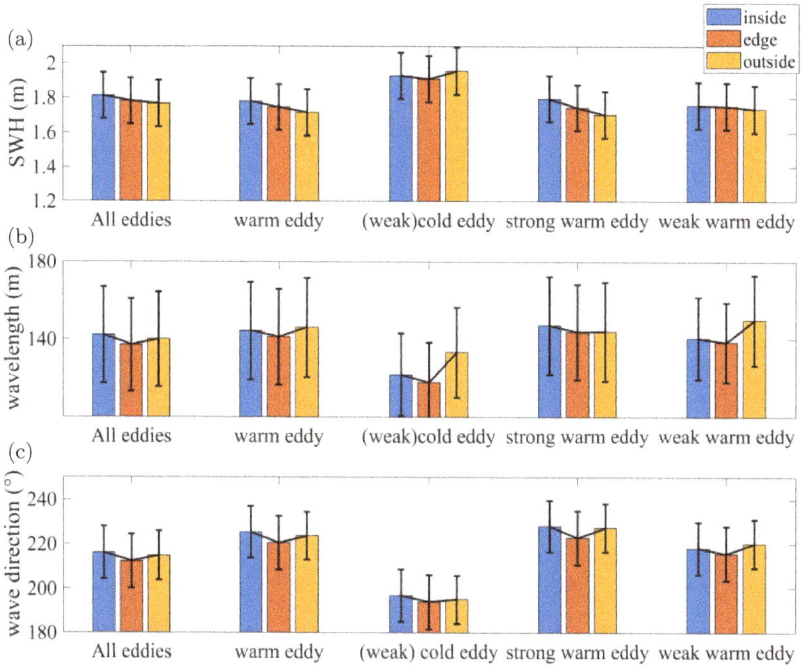

Fig. 14.2. Comparison of (a) mean SWH (in m), (b) mean wavelength (in m), and (c) mean wave direction (in degree) of the full spectrum at different eddy locations (blue: inside, red: edge, and orange: outside) for all eddies, warm eddies, cold eddies, strong warm eddies and weak warm eddies.

Source: Taken from Tan *et al.* (2023).

3–5 m in the insides of AEs. Comparing weak and strong AEs, the variation in SWH is larger in strong eddy cases, and the direction and wavelength variations are more significant in weak eddies.

The variations at different eddy locations in each case were also analyzed by Tan *et al.* (2023). As shown in Table 14.1, the maximum variation in SWH of all eddy cases is 0.18 m occurs between the outside and the inside, which represents 16% of the averaged SWH across the eddy. The largest variations in wavelength and direction occur between the inside and the outside, with values of 24.1 m (21% of the average wavelength) and 23.3°, respectively. The above feature is more evident in cases of strong eddies, where the difference in SWH between the outside and the inside is 0.22 m (18%), and the differences in wavelength and wave direction between the inside and

Table 14.1. Modulations of wave parameters by Eddiesat.

		Difference of SWH $\|\Delta h\|$ (m)	Percentage variation of SWH ($\|\Delta h\|/h$)	Difference of wavelength $\|\Delta\lambda\|$ (m)	Percentage variation of wavelength ($\|\Delta\lambda\|/\lambda$)	Deflection of wave direction $\|\Delta\theta\|$ (°)
All eddies	Inside–outside	*0.18 ± 0.05	16% ± 3%	24.1 ± 1.2	28% ± 9% 9% 21% ± 3%	23.3 ± 2.4
	Inside–edge	0.1 ± 0.03	10% ± 2%	21.5 ± 1.1	16% ± 3%	14.7 ± 2.6
	Edge–outside	0.12 ± 0.03	14% ± 2%	21.1 ± 1.6	16% ± 2%	17.6 ± 2.4
AE	Inside–outside	0.19 ± 0.05	16% ± 2%	24.7 ± 1.6	16% ± 2%	26.4 ± 2.6
	Inside–edge	0.11 ± 0.03	10% ± 2%	23.8 ± 1.8	16% ± 3%	15.8 ± 2.5
	Edge–outside	0.12 ± 0.03	15% ± 1%	23.2 ± 0.8	15% ± 3%	18.4 ± 2.4
Weak CE	Inside–outside	0.15 ± 0.04	12% ± 2%	21.9 ± 4.9	16% ± 1%	20.1 ± 2.8
	Inside–edge	0.08 ± 0.05	8% ± 2%	12.8 ± 3.8	19% ± 4%	12.8 ± 2.1
	Edge–outside	0.14 ± 0.03	11% ± 3%	12.2 ± 4.3	10% ± 3%	15.3 ± 2.8
Strong AE	Inside–outside	*0.22 ± 0.08	18% ± 1%	28.8 ± 6.7	24% ± 5%	28.4 ± 3.1
	Inside–edge	0.14 ± 0.03	13% ± 1%	27.3 ± 2.2	18% ± 2%	17.6 ± 2.4
	Edge–outside	0.12 ± 0.03	8% ± 1%	24.3 ± 3.6	16% ± 2%	20.4 ± 2.6
Weak AE	Inside–outside	0.14 ± 0.03	13% ± 3%	21.5 ± 6.4	14% ± 4%	19.6 ± 2.5
	Inside–edge	0.08 ± 0.06	7% ± 2%	15.7 ± 1.1	10% ± 2%	10.9 ± 2.7
	Edge–outside	0.11 ± 0.03	12% ± 2%	20.2 ± 3.6	14% ± 3%	11.0 ± 2.5

Note: *Maximum values.

the outside are 28.8 m (24%) and 28.4°, respectively. In cases of weak warm eddies, the differences between the inside and outside are 0.14 m (13%), 21.5 m (14%), and 19.6°, respectively. From the view of eddy polarities, the differences are 0.15 m (12%), 21.9 m (19%), and 20.1° in cases of weak cold eddies, which are larger than those in cases of weak warm eddies. In general, the differences between the eddy edge and the inside of the eddy are the smallest.

Tan *et al.* (2023) analyzed the dependence of the above statistical results on the incoming wave conditions. Based on the mean SWH in each case, they classified the sea states into three categories: low sea state (LSS) for SWH $\leq$ 2.5 m; middle sea state (MSS) for 2.5 m $<$ SWH $<$ 4 m, and high sea state (HSS) for SWH $\geq$ 4 m. They found that in the case of LSS, eddies have the greatest modulation on wavelength and wave direction, that is, 30.0 m and 25.8° occurring between the inside and outside of the eddy, respectively. As the sea state increases, the variations in wavelength and wave direction decrease, while the effects on the SWH increase. The maximum variation in SWH between the inside and outside of the eddy reaches 0.23 m (8%). From MSS to LSS, the variation in SWH decreases from 0.23 to 0.16 m, but the percentage increases from 8% to 16%, indicating that the eddy modulation on wave parameters is stronger at low sea states. Meanwhile, the average wavelengths at different eddy locations indicate that regardless of the sea state, the wavelength at the edge of the eddy is always the shortest.

14.2.2 *Correlation analysis of wave parameter variations with eddy parameters*

14.2.2.1 *Wave parameter variations vs. eddy strength*

Tan *et al.* (2023) analyzed the correlation coefficients between the wave parameters variations and the eddy features. As shown in Figure 14.3(a) and 14.3(e), variations in SWH ($|\Delta \mathrm{h}|$) and wave direction ($|\Delta\theta|$) increase with ΔSLA (SLA difference between the peak value within the closed contour and that at the outermost closed contour) for inside-outside differences in AEs (red dots and red plus). The normalized linear fits show that the regression rates are 0.54 and 0.51 for $|\Delta \mathrm{h}|$ and $|\Delta\theta|$, with correlated coefficients of 0.41 and 0.36, respectively, which are much larger than the value of 0.09 for the variation in wavelength $|\Delta\lambda|$ with a correlated coefficient of 0.05. For the inside-edge differences, however, the regression rates are 0.11 and 0.20 for $|\Delta h|$ and $|\Delta\theta|$, and 0.22 for $|\Delta\lambda|$. The correlated coefficient of $|\Delta\lambda|$ and $|\Delta SLA|$ is 0.25, larger than 0.09 and 0.18 for $|\Delta h|$ and $|\Delta\theta|$. The results indicate that (1) the eddy intensity generally has stronger effects on the wave height than on the wave direction and wavelength and (2) the inside-outside wave parameter change with the eddy intensity is more significant than that of the inside-edge, except for the wavelength. In cases of weak cold eddies (blue plus), $|\Delta\theta|$ between inside and outside seems to increase with negative ΔSLA with a regression rate of 0.14 (Figure 14.3(e)).

14.2.2.2 *Wave parameter variations vs. eddy current*

Previous studies have shown that the current-wave interaction could modulate the SWH, wavelength and wave direction (Yuan and Feng, 1991; Pleskachevsky *et al.*, 2009; Warner *et al.*, 2010; Ardhuin *et al.*, 2017). Hisaki (2023) found the SWH is notably underestimated/overestimated up to 20% in the case of the opposing/following current to the mean wave direction. The wave periods can be changed duo to the Doppler shift in opposing/following currents. Owing to the rotating velocity of mesoscale eddies, surface wave parameters are modulated by eddy currents in different ways based on wave direction and eddy quadrant. Wang *et al.* (2022) found that along the direction of wave propagation, both wave height and wave steepness increase on the right side of AE in the Northern Hemisphere, while decrease on the left side.

Tan *et al.* (2023) compared the along-track variation in the averaged SWH, wavelength and wave direction in eddy cases. As shown in Figure 14.4(a), the SWH increases from the southern to the northern edge of the eddy both in the AE (red line) and CE (blue line) cases, as well as in all cases (black line), which is close to the AE cases due to more cases of AEs. The along-track increase of the SWH from the southern to the northern edge is 0.12 m in Aecases and 0.15 m in CE cases. The increase in Aes suggests the effects of eddy polarity, while the SWH increases even in Ces, implying that eddy parameters other than polarity dominate the SWH variation. Similarly, the wave direction rotates clockwise from south to north of the eddy (Figure 14.4(c)), with a rotation angle of 13° in Ces

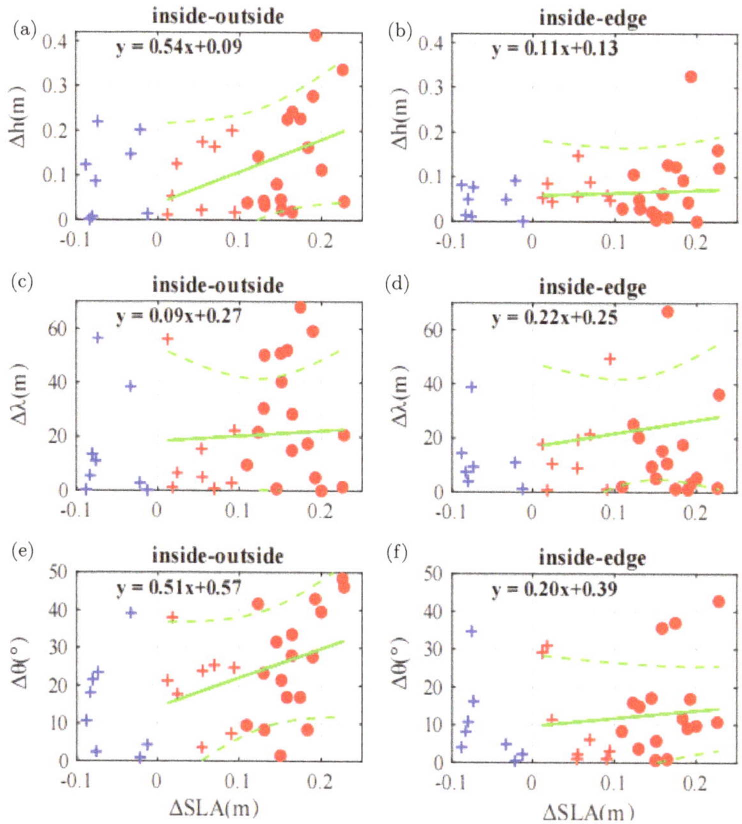

Fig. 14.3. Variations of sea surface wave parameters vs. SLA difference for 33 studied eddies. Red and blue colors represent the cases of AEs and CEs, respectively. Plus (+) and dots (.) represent the cases of weak and strong eddies, respectively. The left column of panels (a), (c) and (e) shows the difference between the inside and outside of the eddy, and the right column of panels (b), (d) and (f) shows the difference between the inside of the eddy and the eddy edge. Green solid lines are the linear fits of the wave parameter variations vs. SLA differences in cases of AEs. Green dashed lines represent the 95% confidence interval. Function y is the normalized linear regression equation.

Source: Taken from Tan *et al.* (2023).

and 11.9° in Aes. The wavelength has minimum values at the two edges of eddies (Figure 14.4(b)), which are 12 m and 8 m shorter than that at the southern inside in AEs and CEs, respectively. The along-track variations here are consistent with the mean values shown in Figure 14.2.

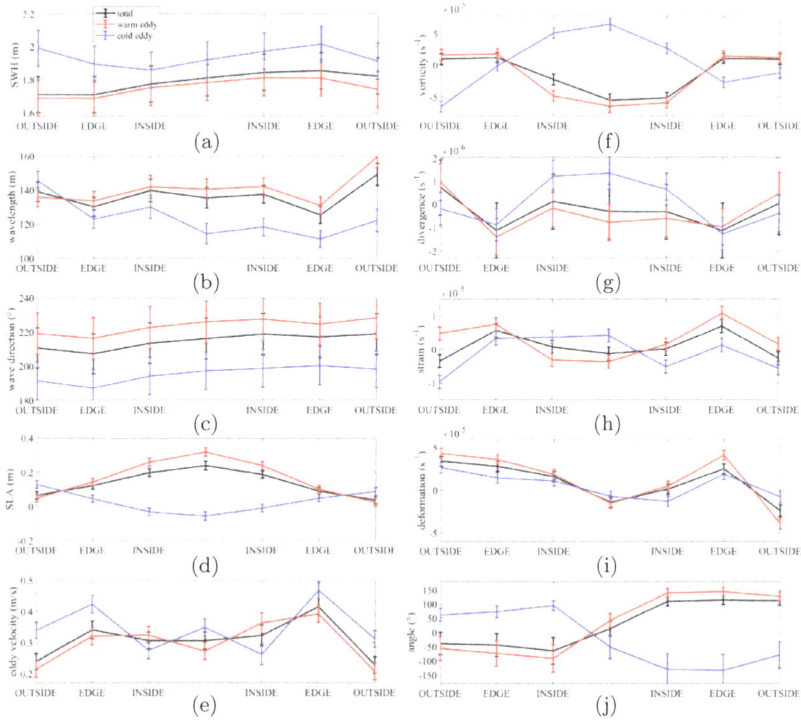

Fig. 14.4. Variation in averaged (a–c) wave parameters and (d–j) eddy parameters along the south-northward CFOSAT ground tracks across mesoscale eddies in all cases (black line), AE cases (red line), and CE cases (blue line). (a) SWH (in m), (b) wavelengths (in m), (c) wave direction (in degrees) and (d) sea level anomaly (in m), (e) eddy velocity magnitude (in m s^{-1}), (f) eddy current vorticity (in s^{-1}), (g) divergence (in s^{-1}), (h) strain (in s^{-1}), (i) deformation (in s^{-1}) and (j) angle (in degrees) of wave propagation direction relative to eddy velocity direction.

Source: Taken from Tan *et al.* (2023).

There are also similar trends for specific wave and eddy parameters. For example, the variation of the SWH (Figure 14.4(a)) is highly related to that of the wave-eddy angle (Figure 14.4(j)) with a correlated coefficient of 0.72. Additionally, the SWH seems to be related to the eddy velocity (Figure 14.4(e)), eddy velocity strain (Figure 14.4(h)) and deformation (Figure 14.4(i)), of which large values occur at the northern edge of the eddy. For the wavelength (Figure 14.4(b)), its variation trend is almost the same as the divergence (Figure 14.4(g)), i.e., shorter wavelengths are concurrent with convergence (negative divergence) at eddy edges. The correlation coefficient of the wavelength and the divergence reaches 0.74 in all eddy cases. The wave direction variation (Figure 14.4(c)) seems to be related to the wave-eddy angle shown in Figure 14.4(j), but not the vorticity (Figure 14.4(f)) as Gallet and Young (2014) proposed. The correlation coefficient of the wave direction and the vorticity is 0.27 in all eddy cases, while the value is 0.84 for the wave direction and the wave-eddy angle. As shown in Figure 14.4(j), the wave-eddy angles of AEs and CEs both change signs with values increasing from less than 90° to larger than 90° from the southern side to the northern side due to the eddy rotating velocity. On the northern side, the wave and current directions are nearly opposite with angles of −130° and 145° for CEs and AEs, respectively. The eddy currents tend to block waves and increase the SWH.

14.3 Mechanisms of Eddy Current Modulation Effects on Sea Surface Wave Fields

14.3.1 *Dynamics of eddy-current modulation on SWH*

The ratio of the current effects on waves (CEW) to total energy can be derived from the wave-current interaction source function (Tan *et al.*, 2023)

$$S = -\left[\frac{c_g}{c}(1+\cos^2\theta) - \frac{1}{2}\right]\frac{\partial u}{\partial x} - \frac{c_g}{c}\sin\theta\cos\theta\left(\frac{\partial u}{\partial y} + \frac{\partial v}{\partial x}\right) - \left[\frac{c_g}{c}(1+\sin^2\theta) - \frac{1}{2}\right]\frac{\partial v}{\partial y}, \tag{14.1}$$

where S is the ratio of the CEW to the total wave energy. c_g and c represent the group velocity and wave phase speed, respectively. θ is the wave direction (angle measured from the positive x-axis anticlockwise), and (u, v) are the components of the surface eddy current velocity. The ratio S can be separated into three terms, which characterize the current spatial variations:

$$S_1 = -\frac{c_g}{c}\sin\theta\cos\theta\left(\frac{\partial u}{\partial y} + \frac{\partial v}{\partial x}\right), \tag{14.2}$$

$$S_2 = -\left(\frac{3c_g}{2c} - \frac{1}{2}\right)\left(\frac{\partial u}{\partial x} + \frac{\partial v}{\partial y}\right), \tag{14.3}$$

$$S_3 = -\frac{c_g}{2c}\left(\cos^2\theta - \sin^2\theta\right)\left(\frac{\partial u}{\partial x} - \frac{\partial v}{\partial y}\right). \tag{14.4}$$

All three terms are the products of two parts related to the wave and current. The wave parts $\frac{c_g}{c}\sin\theta\cos\theta$, $(\frac{3c_g}{2c} - \frac{1}{2})$ and $\frac{c_g}{2c}(\cos^2\theta - \sin^2\theta)$ are dependent on the wave direction and wave phase speed. The other parts $(\frac{\partial u}{\partial y} + \frac{\partial v}{\partial x})$, $(\frac{\partial u}{\partial x} + \frac{\partial v}{\partial y})$ and $(\frac{\partial u}{\partial x} - \frac{\partial v}{\partial y})$ represent strain, divergence and deformation rates, respectively, which reflect the kinematic parameters of an oceanic current. One can see that the current strain, divergence and deformation, but not vorticity, affect the surface wave energy and the SWH.

Tan *et al.* (2023) calculated S_1, S_2, and S_3 with the daily wave parameters from CMEMS (Copernicus Marine Environment Monitoring Service) data in the SCS. As shown in Figure 14.5, the divergence-related term S_2 is generally two orders smaller than S_1 and S_3 in the whole SCS, indicating that the divergence contributes little to the wave energy variation. The terms related to the eddy current strain and deformation are relatively important in the CEW in the SCS. Focusing on the warm eddy near the Luzon Strait, one can see alternating positive and negative S around the eddy with maximum values over 2.1×10^{-5}, of which S_3 is dominant compared to S_1 since the eddy deformation rate is much stronger than the strain rate. Figure 14.6 further shows the along-track distributions of S, S_1, S_2, and S_3 across eddies. In all eddy cases, S has a significantly high value of 2.42×10^{-5} s^{-1} at the eddy edge and small values of 1.0×10^{-5} s^{-1} and 0.37×10^{-5}s^{-1} inside and outside the eddy, respectively. For the three parts, the deformation and strain terms S_3 and S_1 are at least one order larger than the divergence term S_2, consistent with the low values of the divergence rate shown in Figure 14.4. S_1, S_2, and S_3 are positive at all locations except S_1, which is negative outside of the eddy. S_1, S_2, and S_3 all have the largest values at the eddy edge. It is clear that S_3 associated with the current deformation is the dominant component, especially at the eddy edge. The average contribution of S_3 to S reaches 70% and 77% at the eddy edge and outside, respectively, while S_1 contributes 53% to S inside the eddy. The larger values of S and S_3 around the eddy in Figure 14.5 are also shown by the higher bars at the edge of

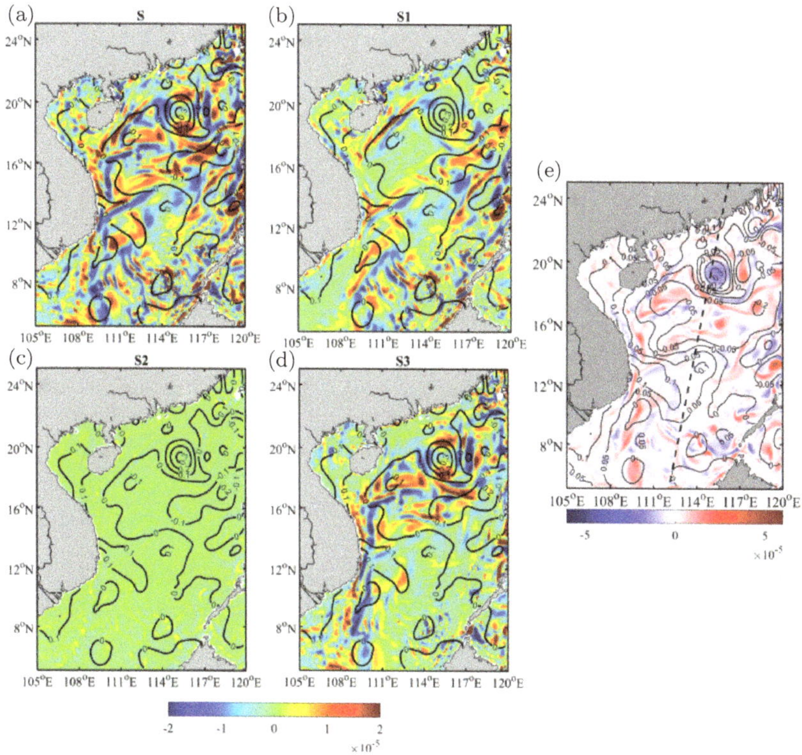

Fig. 14.5. (a) Current effects on waves (CEW) energy ratio S and its three parts (b) S_1, (c) S_2 and (d) S_3 calculated from eddy geostrophic current velocity and the CMEMS wave data on 4 March 2020. In panel (e), the AE is framed in a magenta box. Shading colors show the total current vorticity from the CMEMS current velocity. Contours represent the SLA (in m) Tan *et al.* (2023) is comparable to 8–20% in strong currents in global ocean (Echevarria *et al.*, 2021) and smaller than 30% induced by small-scale local energetic frontal jets (Romero *et al.*, 2017; Wang *et al.*, 2022).

Source: Taken from Tan *et al.* (2023).

the eddy. Comparing the CEW in AEs and CEs (Figure 14.5(b) and 14.5(c)), their variation patterns are both like that in total cases, of which S_1 and S_3 are the dominant terms and S has the largest values at the eddy edge. However, the highest value of S in AE cases (2.47×10^{-5} s^{-1}) is almost twice that of CEs (1.37×10^{-5} s^{-1}). Meanwhile, the divergence term S_2 has positive values in AE cases but negative values in CE cases.

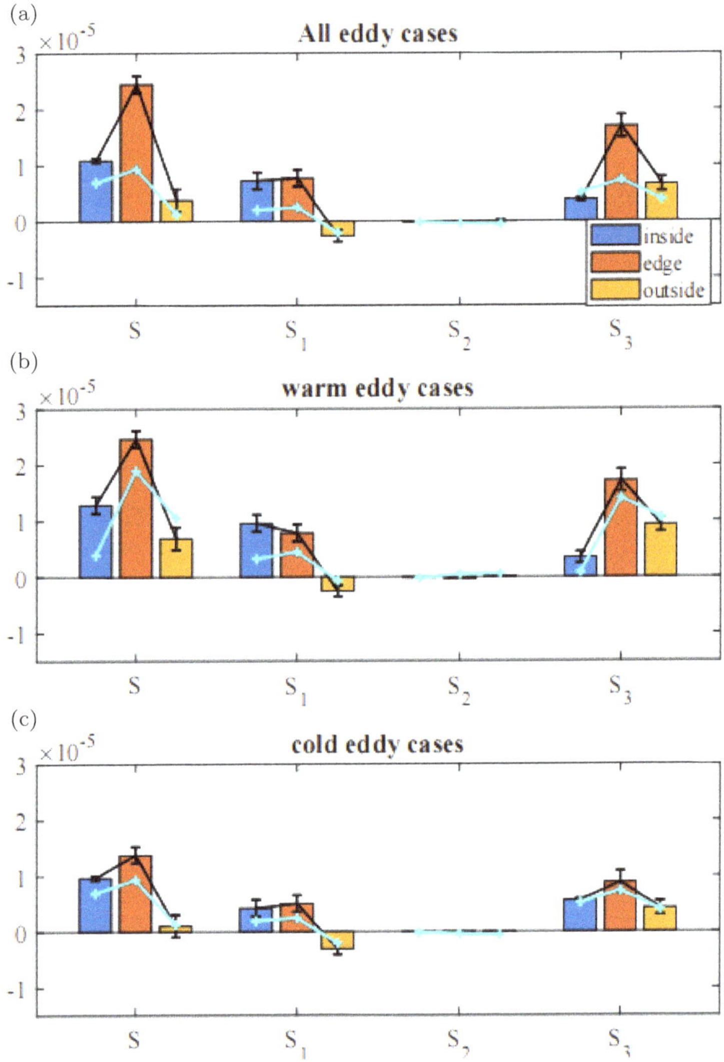

Fig. 14.6. Averaged S, S_1, S_2, and S_3 at different eddy locations (inside, edge, and outside) derived from eddy geostrophic current velocity and the SWIM wave data (a) in all studied eddy cases, (b) in AE cases and (c) in CE cases. Cyan lines represent the same variation but are calculated from eddy geostrophic current velocity and the CEMES wave data.

Source: Taken from Tan *et al.* (2023).

14.3.2 *Eddy current effects on wavelength*

As shown in Figure 14.4(b), the minimum wavelengths appear at the northern and southern edges of mesoscale eddies, where the turning points of minimum negative divergence occur (Figure 14.4(g)). On the other hand, the minimum wavelengths also appear coherent with the maximum geostrophic velocity anomaly (Figure 14.4(e)) and are negatively correlated with the largest current strain rate (Figure 14.4(h)). These relationships indicate that the variation in wavelength is associated with the eddy current shear. Ardhuin *et al.* (2017) pointed out that the variation in the wavenumber (wavelength) is associated with the water depth and surface current velocity. For deep-water waves, there is (Tan *et al.*, 2023)

$$\frac{dk}{dt} = -\left(k_x\frac{\partial u}{\partial x} + k_y\frac{\partial v}{\partial y}\right), \tag{14.5}$$

where $k = \frac{2\pi}{\lambda}$ is the magnitude of the wavenumber, $k_x = k \cdot \sin\theta$ and $k_y = k \cdot \cos\theta$ are two components of the wavenumber vector, and u and v are two components of the surface current velocity.

Tan *et al.* (2023) calculated the right-hand side term in Eq. (14.5) along the ground tracks across mesoscale eddies. They found that $\mathrm{d}k/\mathrm{d}t$ has significant positive values at the two edges of the eddy (Figure 14.7), indicating that wavenumber k increases and thus wavelength λ decreases there. The along-track distribution of $\mathrm{d}k/\mathrm{d}t$ is highly consistent with that of the wavelength shown in Figure 14.4(c). As $\mathrm{d}k/\mathrm{d}t$ increases, the wavelength decreases (Figure 14.7). Their correlation coefficient reaches −0.68.

14.3.3 *Eddy effects on wave direction*

Gallet and Young (2014) pointed out that the wave direction could be modulated by current vorticity. The deflection angle $\Delta\theta$ as the wave passes through an eddy can be calculated as follows:

$$\Delta\theta \sim \frac{4\pi\zeta_0 l}{\mathrm{g}} f, \tag{14.6}$$

where f is the wave frequency, g is the gravitational acceleration, ζ_0 is the vertical vorticity of the current and l is the eddy diameter.

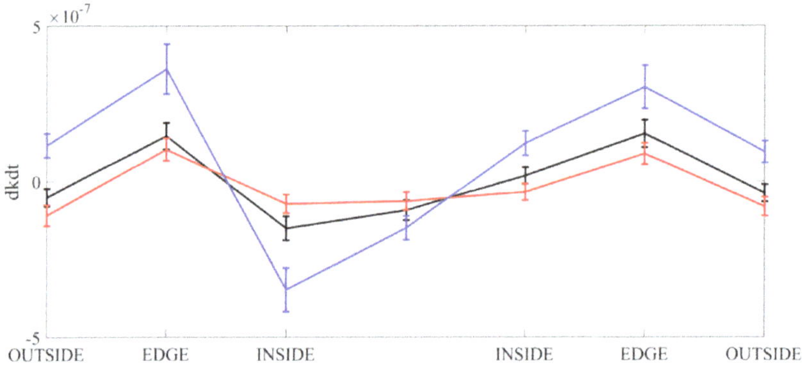

Fig. 14.7. Variations in composite dk/dt (in $\mathrm{m^{-1}s^{-1}}$) along the south-northward CFOSAT ground tracks across mesoscale eddies in all cases (black line), warm eddy cases (red line), and cold eddy cases (blue line).

Source: Taken from Tan *et al.* (2023).

Using the daily eddy current velocity from the AVISO data, Tan *et al.* (2023) calculate the deflection angles according to Eq. (14.6) in the mesoscale eddy cases studied. They found that the average theoretical deflection angle is 3.22° in all eddy cases, increasing to 3.61° in strong warm eddies and decreasing to 2.20° in weak cold eddies. Compared to the absolute variations shown in Table 14.1, the theoretical values are only 11–14% of the absolute variations. Furthermore, the variation trend of vorticity (Figure 14.4(f)) differs greatly from that of the wave direction (Figure 14.4(c)). Thus, the theoretical results based on eddy vorticity itself cannot explain all the variations in wave direction across eddies in the study cases.

According to Ardhuin *et al.* (2017), the dynamic terms controlling the variation in wave direction in a background current can be calculated as follows

$$\frac{d\theta}{dt} = \sin^2\theta\frac{\partial v}{\partial x} - \cos^2\theta\frac{\partial u}{\partial y} + \sin\theta\cos\theta\left(\frac{\partial u}{\partial x} - \frac{\partial v}{\partial y}\right) \qquad (14.7)$$

where u and v are the components of surface velocity and θ is the wave direction (from eastward). The equation has been simplified considering the local area and deep-water waves and ignoring the water depth variation and the earth sphericity. The variation in the wave propagation direction dθ/dt is controlled by three terms on the right-hand side of Eq. (14.7), the first two terms together are

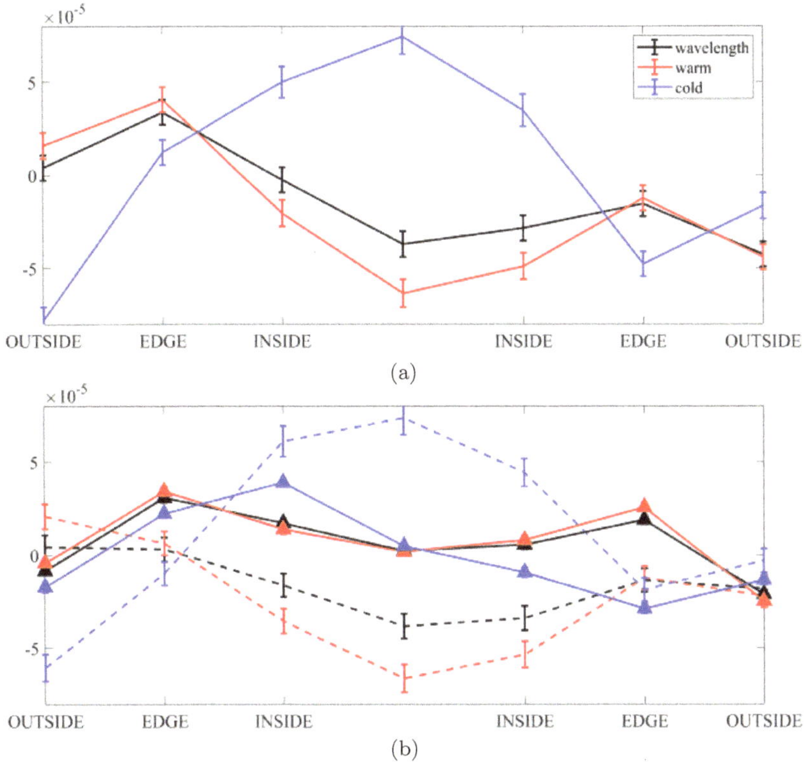

Fig. 14.8. Variations of (a) composite $d\theta/dt$ (in s^{-1}) and (b) its two parts (vorticity term, dash line; deformation term, solid line) along the south-northward CFOSAT ground tracks across mesoscale eddies in all cases (black line), AE cases (red line) and CE cases (blue line).

Source: Taken from Tan *et al.* (2023).

related to the vorticity, and the third term is associated with the deformation rate.

Figure 14.8(a) shows the variation in wave direction $d\theta/dt$ derived from the geostrophic velocity anomaly and the wave data from SWIM observations by Tan *et al.* (2023). They found that $d\theta/dt$ is dominated by the vorticity-related terms inside eddy but not by the deformation term at the edge and outside of eddy (Figure 14.8(b)). The vorticity term seems to be negatively related to the along-track variation of the wave direction (Figure 14.4(c)). The correlation coefficient reaches 0.37 for all cases and 0.41 in AE cases, much larger than the

Table 14.2. Equatorial area, North Brazil and Gulf Stream regions.

Study area	Wave parameter	Variation	Mechanism	Author (yyyy)
Equatorial area North Brazil and Gulf Stream regions	SWH	$8 \sim 15\%$	Exchanges of energy between waves and currents	Echevarria *et al.* (2021)
	Wave direction	$\sim 10°$	Current induced refraction	
Bodega Bay and the Gulf of Mexico	SWH	30%	Wave-current interaction	Romero *et al.* (2017)
	Wave direction	$\leq 45°$	Frontal jet induced refraction	
East China Sea	SWH	$10 \sim 20\%$	Wave-current interaction	Wang *et al.* (2020)
Antarctic Circumpolar Current	Wave direction	$\leq 9°$	Deflection by eddy vorticity	Gallet and Young (2014)
South China Sea	SWH	16–18%	Eddy current-wave interaction,	Tan *et al.* (2023)
	Wave direction	23.3–29.8°	Eddy current vorticity and eddy-induced wind direction,	
	Wavelength	21–24%	Eddy current divergence	

value of 0.27 between wave direction and eddy vorticity. The correlation coefficient between the wave direction and deformation term is 0.49. This indicates that the product of eddy vorticity and deformation with incident wave direction, instead of eddy vorticity itself, contributes to the variation of wave direction.

14.4 Comparisons with Other Oceans

Comparisons of the eddy effects on surface waves in the SCS to that in the other oceans are listed in Table 14.2. One can see that the mean SWH variation of 18 ± 1% across strong AEs observed in the SCS by Tan *et al.* (2023) is comparable to 8–20% in strong currents in global ocean (Echevarria *et al.*, 2021) and smaller than 30% induced by a small-scale local energetic frontal jet (Romero *et al.*, 2017; Wang *et al.*, 2022). The variation mechanisms are similarly attributed to the current-wave energy transfer due to current deformation and strain. For the variation in wave direction, the results of 23.3–28.9° (Tan *et al.*, 2023) and 30–40° (Wang *et al.*, 2022) in the SCS is larger than 9° deflection caused by a typical eddy in the Antarctic Circumpolar Current region (Gallet and Young, 2014) and 10° in the Equatorial Current region (Echevarria *et al.*, 2021), but smaller than 30–45° at the strong frontal jet (Romero *et al.*, 2017). Compared to the variations in SWH and wave direction induced by strong AEs near the Vietnam coast (Wang *et al.*, 2022), the statistical results of wave modulation derived from 33 eddies distributed in the SCS by Tan *et al.* (2023) are one-half weaker. The current vorticity and wave-eddy angle are found to be factors affecting the wave direction, while eddy-induced wind direction variation also plays an important role in the eddy area. For the wavelength variation, the CFOSAT observations show result of 21–24%. It is found that the eddy current divergence and the wave direction determine the wavelength variation as well.

14.5 Summary

As a critical process in air-sea interactions, variations of surface waves induced by mesoscale eddies directly affect the exchange between the ocean and the atmosphere. Meanwhile, the wavelength reduction at the eddy edge favors wave breaking and intensifies the energy and mass transported into the ocean. If the direction of strong waves is deflected by eddies, navigation and activities on the sea should carefully adjust. Thus, the role of mesoscale eddies played in sea surface wave forecasting and simulations of air-sea interaction is worth pursuing.

This chapter summarizes the updated research results on modulation of mesoscale eddies to the sea surface waves in the SCS. The sea surface wave and wind field data measured by the SWIM and the SCAT onboard CFOSAT provide the best opportunity for this purpose, i.e., to simultaneously observe wavelength and wave direction variations as well as SWH variation in the cases of co-existence of mesoscale eddies in the SCS with the macroscale coverage and high spatial resolution. The statistical results indicate that the SWH decreases (increases) from inside to outside of AEs (CEs). while the wavelengths and wave directions have minimum values at the eddy edge. The eddy modulation to wavelength and wave direction is stronger at low sea states, while the modulation to the SWH is stronger at higher sea states. The wave parameter variations across eddies mainly result from the eddy current shear and the direction difference between the eddy current and the waves. The wave-eddy angle and eddy-modulated wind variation also contribute to the SWH and wave direction variations, rather than the eddy vorticity.

References

Ardhuin, F., Gille, S. T., Menemenlis, D., Rocha, C. B., Rascle, N., Chapron, B., Gula, J. and Molemaker, J. (2017). Small-scale open ocean currents have large effects on wind wave heights, *J. Geophys. Res. Oceans*, 122(6), pp. 4500–4517, doi:10.1002/2016JC012413.

Ardhuin, F., Roland, A., Dumas, F., Bennis, A.-C., Sentchev, A., Forget, P., Wolf, J., Girard, F., Osuna, P. and Benoit, M. (2012). Numerical wave modeling in conditions with strong currents: Dissipation, refraction, and relative wind, *J. Phys. Oceanogr.*, 42(12), pp. 2101–2120, doi:10.1175/JPO-D-11-0220.1.

Bolaños, R., Brown, J. M. and Souza, A. J. (2014). Wave-current interactions in a tide dominated estuary, *Cont. Shelf Res.*, 87, pp. 109–123, doi:10.1016/j.csr.2014.05.009.

Bühler, O. and McIntyre, M. E. (2005). Wave capture and wave-vortex duality, *J. Fluid Mech.*, 534, pp. 67–99, doi:10.1017/S0022112005004374.

Chelton, D. (2013). Mesoscale eddy effects, *Nat. Geosci.*, 6(8), pp. 594–595.

Chen, G., Hou, Y. and Chu, X. (2011). Mesoscale eddies in the South China Sea: Mean properties, spatiotemporal variability, and impact on thermohaline structure, *J. Geophys. Res. Oceans*, 116(C6), C06018, doi:10.1029/2010JC006716.

Cheng, T., Chen, Z., Li, J., Ma, X., Wen, Q. and Wu, L. (2022). Surface wave height regulated by ocean currents: An observational perspective, *Deep Sea Res. Part I: Oceanogr. Res. Pap.*, 179, 103666.

Collard, F., Ardhuin, F. and Chapron, B. (2005). Extraction of coastal ocean wave fields from SAR images, *IEEE J. Ocean. Eng.*, 30(3), pp. 526–533, doi:10.1109/JOE.2005.857503.

Echevarria, E. R., Hemer, M. A. and Holbrook, N. J. (2021). Global implications of surface current modulation of the wind-wave field, *Ocean Modelling*, 161, 101792, doi:10.1016/j.ocemod.2021.101792.

Fang, G., Chen, H., Wei, Z., Wang, Y., Wang, X. and Li, C. (2006). Trends and interannual variability of the South China Sea surface winds, surface height, and surface temperature in the recent decade, *J. Geophys. Res. Oceans*, 111(C11), C11S16, doi:10.1029/2005JC003276.

Frenger, I., Gruber, N., Knutti, R. and Münnich, M. (2013). Imprint of Southern Ocean eddies on winds, clouds and rainfall, *Nat. Geosci.*, 6(8), pp. 608–612, doi:10.1038/ngeo1863.

Gallet, B. and Young, W. R. (2014). Refraction of swell by surface currents, *J. Mar. Res.*, 72(2), pp. 105–126, doi:10.1357/002224014813758959.

Hisaki, Y. (2023). Use of drifting buoys for wave observation: Effect of current on wave data, *Deep Sea Res. Part I: Oceanogr. Res. Pap.*, 195, 103986, doi:10.1016/j.dsr.2023.103986.

Holthuijsen, L. H. and Tolman, H. L. (1991). Effects of the Gulf Stream on ocean waves, *J. Geophys. Res. Oceans*, 96(C7), pp. 12755–12771, doi:10.1029/91JC00901.

Kudryavtsev, V. N., Grodsky, S. A., Dulov, V. A. and Bol'Shakov, A. N. (1995). Observations of wind waves in the Gulf Stream frontal zone, *J. Geophys. Res. Oceans*, 100(C10), pp. 20715–20727, doi:10.1029/95JC00425.

Liu, A. K., Peng, C. Y. and Schumacher, J. D. (1994). Wave-current interaction study in the Gulf of Alaska for detection of eddies by synthetic aperture radar, *J. Geophys. Res. Oceans*, 99(C5), pp. 10075–10085, doi:10.1029/94JC00422.

Marechal, G. and de Marez, C. (2021). Variability of wind wave field by realistic mesoscale and submesoscale eddy field, *Ocean Sci. Disc.*, pp. 1–18, doi:10.5194/os-2021-53.

Melville, W. K., Romero, L., Kleiss, J. M. and Swift, R. N. (2005). Extreme wave events in the Gulf of Tehuantepec, In *Rogue Waves: Proc. 14th 'Aha Huliko'a Hawaiian Winter Workshop*, pp. 23–28.

Pearman, D. W., Herbers, T. H. C., Janssen, T. T., van Ettinger, H. D., McIntyre, S. A. and Jessen, P. F. (2014). Drifter observations of the effects of shoals and tidal-currents on wave evolution in San Francisco Bight, *Cont. Shelf Res.*, 91, pp. 109–119, doi:10.1016/j.csr.2014.08.011.

Phillips, O. M. (1984). On the response of short ocean wave components at a fixed wavenumber to ocean current variations, *J. Phys. Oceanogr.*, 14(9), pp. 1425–1433, doi:10.1175/1520-0485(1984)014,1425:OTROSO.2.0.CO;2.

Pleskachevsky, A., Eppel, D. P. and Kapitza, H. (2009). Interaction of waves, currents and tides, and wave-energy impact on the beach area of Sylt Island, *Ocean Dyn.*, 59(3), pp. 451–461, doi:10.1007/s10236-008-0174-1.

Romero, L., Lenain, L. and Melville, W. K. (2017). Observations of surface wave-current interaction, *J. Phys. Oceanogr.*, 47(3), pp. 615–632, doi:10.1175/jpo-d-16-0108.1.

Sorourian, S., Huang, H., Li, C., Justic, D. and Payandeh, A. R. (2020). Wave dynamics near Barataria Bay tidal inlets during spring–summer time, *Ocean Modelling*, 147, 101553, doi:10.1016/j.ocemod.2019.101553.

Tamura, H., Waseda, T., Miyazawa, Y. and Komatsu, K. (2008). Current-induced modulation of the ocean wave spectrum and the role of non-linear energy transfer, *J. Phys. Oceanogr.*, 38(12), pp. 2662–2684, doi:10.1175/2008JPO4000.1.

Tan, K., Xie, L., Bai, P., Zheng, Q., Li, J., Xu, Y. and Li, M. (2023). Modulation effects of mesoscale eddies on sea surface wave fields in the South China Sea derived from a wave spectrometer onboard the China-France Ocean Satellite, *J. Geophys. Res. Oceans*, 128(1), e2021JC018088, doi:10.1029/2021JC018088.

Villas Bôas, A. B., Cornuelle, B. D., Mazloff, M. R., Gille, S. T. and Ardhuin, F. (2020). Wave–current interactions at meso-and submesoscales: Insights from idealized numerical simulations, *J. Phys. Oceanogr.*, 50(12), pp. 3483–3500, doi:10.1175/JPO-D-20-0151.1.

Wang, D. W., Liu, A. K., Peng, C. Y. and Meindl, E. A. (1994). Wave-current interaction near the Gulf Stream during the surface wave dynamics experiment, *J. Geophys. Res. Oceans*, 99(C3), pp. 5065–5079, doi:110.1029/93jc02714.

Wang, G., Su, J. and Chu, P. (2003). Mesoscale eddies in the South China Sea observed with altimeter data, *Geophys. Res. Lett.*, 30(21), doi:10.1029/2003GL018532.

Wang, J., Dong, C. and Yu, K. (2020). The influences of the Kuroshio on wave characteristics and wave energy distribution in the East China Sea, *Deep Sea Res. Part I: Oceanogr. Res. Pap.*, 158, 103228, doi:10.1016/j.dsr.2020.103228.

Wang, J., Bethel, B. J., Dong, C., Li, C. and Cao, Y. (2022). Numerical simulation and observational data analysis of mesoscale eddy effects on surface waves in the South China Sea, *Rem. Sens.*, 14(6), 1463, doi:10.3390/rs14061463.

Warner, J. C., Armstrong, B., He, R. and Zambon, J. B. (2010). Development of a coupled ocean–atmosphere–wave–sediment transport (COAWST) modeling system, *Ocean Modelling*, 35(3), pp. 230–244, doi:10.1016/j.ocemod.2010.07.010.

Wolf, J. and Prandle, D. (1999). Some observations of wave-current interaction, *Coastal Eng.*, 37(3–4), pp. 471–485, doi:10.1016/S0378-3839(99)00039-3.

Wu, H., Li, J., Chen, Z. and Ma, X. (2025). Observed surface wave variations in the background current field of the Kuroshio Extension, *Ocean Dyn.*, 75(2), p. 14.

Xie, L., Wu, K., Pietrafesa, L. and Zhang, C. (2001). A numerical study of wave-current interaction through surface and bottom stresses: Wind-driven circulation in the South Atlantic Bight under uniform winds, *J. Geophys. Res. Oceans*, 106(C8), pp. 16841–16855, doi:10.1029/2000JC000292.

Xie, L., Zheng, Q., Li, J., Li, M. and Ho, C.-R. (2020). Responses of the South China Sea to mesoscale disturbances from the Pacific, In *Regional Oceanography of The South China Sea*, 10, pp. 243–287, doi:10.1142/9789811206917_0010.

Xie, L., Zheng, Q., Zhang, S., Hu, J., Li, M., Li, J. and Xu, Y. (2018). The Rossby normal modes in the South China Sea deep basin evidenced by satellite altimetry, *Intl. J. Rem. Sens.*, 39(2), pp. 399–417, doi:10.1080/01431161.2017.1384591.

Xu, Y., Liu, J., Xie, L., Sun, C. and Xian, D. (2019). China-France Oceanography Satellite (CFOSAT) simultaneously observes the typhoon-induced wind and wave fields, *Acta Oceanol. Sin.*, 38(11), pp. 158–161, doi:10.1007/s13131-019-1506-3.

Yuan, Y. and Feng, H. (1991). LAGFD-WAM numerical wave model-?. basic physical model, *Acta Oceanol. Sin.*, (4), pp. 483–488, doi:CNKI:SUN:SEAE.0.1991-04-000.

Zhang, X., Simons, R., Zheng, J. and Zhang, C. (2021). A review of the state of research on wave-current interaction in nearshore areas, *Ocean Eng.*, 243(2022), 110202.

Zheng, Q., Ho, C.-R., Xie, L. and Li, M. (2019). A case study of a Kuroshio main path cut-off event and impacts on the South China Sea in fall-winter 2013–2014, *Acta Oceanol. Sin.*, 38(4), pp. 12–19, doi:10.1007/s13131-019-1411-9.

Zheng, Q., Xie, L., Xiong, X., Hu, X. and Chen, L. (2020). Progress in research of submesoscale processes in the South China Sea, *Acta Oceanol. Sin.*, 39(1), pp. 1–13, doi:10.1007/s13131-019-1521-4.

Zheng, Q., Xie, L., Zheng, Z. W. and Hu, J. (2017). Progress in research of mesoscale eddies in the South China Sea, *Adv. Mar. Sci.*, 35(2), pp. 131–158.

Zhu, G., Lin, W., Zhao, S. and Cao, Y. (2015). Spatial and temporal variation characteristics of ocean waves in the South China Sea during the boreal winter, *Acta Oceanol. Sin.*, 34(1), p. 23, doi:10.1007/s13131-015-0592-0.

Chapter 15

Mesoscale Eddy Modulation to Near Inertial Waves

15.1 Introduction

Parts 2–4 of this book have shown that that the SCS circulation system is characterized by broadly distributed energetic mesoscale eddies. Meanwhile, the SCS is also characterized by strong, typhoon-forced near inertial waves (NIWs) (Liang *et al.*, 2005; Zhu and Li, 2007; Alford, 2008; Zheng *et al.*, 2020). It is the atmospheric forcing, such as wind stress by the seasonally reverse monsoon and frequent passages of typhoons, that provides favorable conditions for generation and evolution of the NIWs in the SCS (Sun *et al.*, 2011a, 2011b; Chen *et al.*, 2013; Xu *et al.*, 2013; Guan *et al.*, 2014; Zhang *et al.*, 2016; Xiao *et al.*, 2016; Shu *et al.*, 2016; Cao *et al.*, 2018; Liu *et al.*, 2018; Li *et al.*, 2021). Meanwhile, the geographic location of the SCS provides the possibility for the special NIWs induced by the parametric subharmonic instability (PSI) occurring at critical latitudes of diurnal tides, 13.44°N for O_1 tide and 15.52°N for K_1 tide (Alford, 2008; Xie *et al.*, 2016). This chapter focuses on modulation of mesoscale eddies to the NIWs observed by moorings combined with concurrent satellite sea level data.

15.2 Theoretical Aspects

15.2.1 *Dispersion relation of NIWs in geostrophic shear flow*

Under assumption of WKB (Wentzel–Kramers–Brillouin) approximation that wave scales are much smaller than mean flow scales, the NIW variables are written in the form of a plane wave (Kunze, 1985)

$$\psi = \psi_0 \exp[i(k_x x + k_y y + k_z z - \omega t)], \tag{15.1}$$

where ω is the fixed frequency in a Eulerian frame and $\boldsymbol{k} = (k_x, k_y, k_z)$ is the wavevector in a Cartesian coordinate system with x-axis eastward positive, y-axis northward positive and z-axis upward positive. Further assume that the NIWs propagate in weak geostrophic shear (i.e., with horizontal shears smaller than the Coriolis frequency f (small Rossby number), and vertical shear much smaller than the buoyancy frequency N (large geostrophic Richardson number)), Kunze (1985) derived an approximate dispersion relation of the NIWs in the form of

$$\begin{aligned}\omega_0 = \omega - (\boldsymbol{k}\cdot\boldsymbol{V}) \approx & \left[f_{eff} + \frac{N^2 k_H^2}{2f k_z^2} + \frac{1}{k_z}\left(\frac{\partial U}{\partial z}k_y - \frac{\partial V}{\partial z}k_x\right)\right] \\ & + i\left\{\frac{\zeta}{2fk_z}\left(\frac{\partial U}{\partial z}k_x + \frac{\partial V}{\partial z}k_y\right) + \frac{N^2}{2f^2k_z^2}\left[\frac{\partial U}{\partial x}k_y^2\right.\right. \\ & \left.\left. - \left(\frac{\partial U}{\partial y} + \frac{\partial V}{\partial x}\right)k_x k_y + \frac{\partial V}{\partial y}k_x^2\right]\right\},\end{aligned} \tag{15.2}$$

where $V = (U, V)$ is the horizontal velocity vector of mean flow, ω_0 is the intrinsic frequency, $(\boldsymbol{k}\cdot\boldsymbol{V})$ is the Doppler shift, $\boldsymbol{k_H} = (k_x, k_y)$ is the horizontal wavenumber and

$$f_{eff} \approx f + \frac{\zeta}{2} = f + \frac{1}{2}\left(\frac{\partial V}{\partial x} - \frac{\partial U}{\partial y}\right) \tag{15.3}$$

is defined as the effective Coriolis frequency, in which the term in brackets is the relative vorticity of the shear mean flow. The real part of (15.2) describes frequency modulation of NIWs by the shear mean flow, and the imaginary part describes the energy exchange between the NIWs and the mean flow.

15.2.2 *Mesoscale eddy as a mean flow component*

This chapter is interested in a special case of the mean flow, which consists of two components, i.e.,

$$\boldsymbol{V} = \boldsymbol{V_b} + \boldsymbol{V_e}, \tag{15.4}$$

where $\boldsymbol{V_b}$ and $\boldsymbol{V_e}$ are a background flow and a mesoscale eddy flow, respectively. It is reasonable to suppose that the spatial scale of $\boldsymbol{V_b}$ is much larger than that of $\boldsymbol{V_e}$, thus, we have

$$\left|\frac{\partial V_b}{\partial x} - \frac{\partial U_b}{\partial y}\right| \ll \left|\frac{\partial V_e}{\partial x} - \frac{\partial U_e}{\partial y}\right|. \tag{15.5}$$

Therefore, we have an approximate expression of (3)

$$f_{eff} \approx f + \frac{\zeta}{2} = f + \frac{1}{2}\left(\frac{\partial V_e}{\partial x} - \frac{\partial U_e}{\partial y}\right). \tag{15.6}$$

A cylindrical coordinate system is more convenient to describe dynamic features of eddies as shown in Figure 15.1(a). The variable transform relations with that in the Cartesian coordinate system are

$$\begin{aligned} x &= r\cos\theta, \\ y &= r\sin\theta, \\ z &= z, \end{aligned}$$

where r is the radius vector originating from center o outward positive, θ is the radial direction angle anticlockwise positive with respect to the eastward direction, and z is the vertical axis upward positive. In the cylindrical coordinate system, the relative vorticity of the eddy flow $\boldsymbol{V_e}$ is expressed as

$$\zeta = \frac{v_\theta}{r} + \frac{\partial v_\theta}{\partial r}, \tag{15.7}$$

where v_θ is the tangential velocity anticlockwise positive.

Zhang *et al.* (2013) gave an empirical expression of the sea surface dynamic height of mesoscale eddy as

$$\eta = \eta_0 \left(1 - \frac{r_n^2}{2}\right) e^{-\frac{r_n^2}{2}}, \tag{15.8}$$

where $r_n = \frac{r}{R}$, in which R is the eddy radius. From the quasi-geostrophic momentum equation (Pedlosky, 1987), we have

$$v_\theta = -\frac{g}{f}\frac{\partial \eta}{\partial r}, \tag{15.9}$$

where g is the gravitational acceleration, f is the Coriolis parameter and η is the sea surface dynamic height. Substituting (15.8) into (15.9) yields

$$v_\theta = -\frac{g\eta_0}{fR} r_n \left(\frac{r_n^2}{2} - 2\right) e^{-\frac{r_n^2}{2}}. \tag{15.10}$$

Figure 15.1(b) shows the normalized tangential velocity vs. the normalized radius vector in the case of cyclonic eddy. One can see that the radial structure of v_θ is characterized by the minimum value 0 at the eddy center ($r_n = 0$), the peak value near 1 at $r_n = 0.8$, and almost vanishes beyond $r_n = 2$.

15.2.3 *Frequency shift*

From the real part of (15.2), we have

$$\omega \approx \left[f_{eff} + \frac{N^2 k_H^2}{2fk_z^2} + \frac{1}{k_z}\left(\frac{\partial U}{\partial z}k_y - \frac{\partial V}{\partial z}k_x\right)\right] + (\boldsymbol{k} \cdot \boldsymbol{V}). \tag{15.11}$$

According to estimates by Kunze (1985), f_{eff} is a dominant component among terms in the square brackets, if the relative frequency shift is smaller than 50%, i.e., the case observed in the SCS. Thus, we have an approximate form of (15.11)

$$\omega \approx f + \frac{1}{2}\left(\frac{\partial V}{\partial x} - \frac{\partial U}{\partial y}\right) + (\boldsymbol{k} \cdot \boldsymbol{V}). \tag{15.12}$$

One can see that the frequency shift consists of two components, the relative vorticity and the Doppler shift that includes contributions

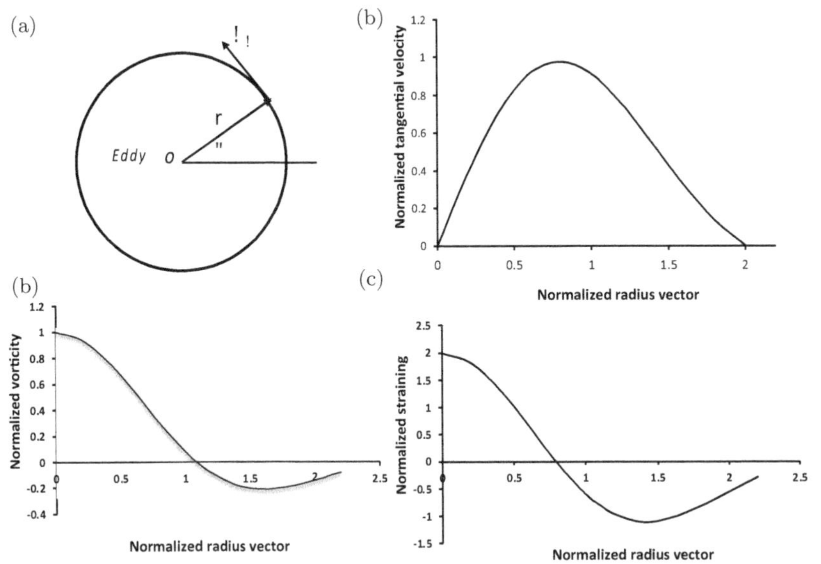

Fig. 15.1. A cylindrical cyclonic eddy in the cylindrical coordinate system. (a) Vertical view. (b) Normalized eddy tangential velocity vs. normalized radius vector. (c) Normalized relative vorticity vs. normalized radius vector. (d) Normalized growth rate vs. normalized radius vector.

of the background flow and the mesoscale eddy flow as shown in (15.4). Namely, mesoscale eddy has twofold modulations to the NIW frequency by its current velocity and vorticity, respectively.

Substituting (15.10) into (15.7) yields

$$\zeta = \frac{g\eta_0}{fR^2}\left(\frac{r_n^4}{2} - 4r_n^2 + 4\right)e^{-\frac{r_n^2}{2}}. \tag{15.13}$$

Figure 15.1(c) shows the normalized vorticity vs. normalized radius vector in the case of cyclonic eddy. One can see the important features of radial structure of $\zeta(r_n)$: it takes the maximum value of 4 at the eddy center ($r_n = 0$) and the minimum value of -0.18 at $r_n = 1.6$, changes the sign at around $r_n = 1.1$ and almost vanishes beyond $r_n = 2$. The absolute value of positive peak is 22 times larger than that of negative peak, implying that strong eddy flow vorticity is concentrated within a range of $r_n < 1$. Thus, the NIW frequency shift is not only dependent on the eddy polarities, but also on the observed location with respect to the eddy center.

15.2.4 *Amplitude growth rate*

On the other hand, the imaginary part of dispersion relation (15.2) is the exponential growth rate of the NIW amplitude, i.e.,

$$\psi = \psi_0 \exp\left\{\frac{\zeta}{2fk_z}\left(\frac{\partial U}{\partial z}k_x + \frac{\partial V}{\partial z}k_y\right) + \frac{N^2}{2f^2k_z^2}\left[\frac{\partial U}{\partial x}k_y^2 - \left(\frac{\partial U}{\partial y} + \frac{\partial V}{\partial x}\right)k_xk_y + \frac{\partial V}{\partial y}k_x^2\right]\right\}t. \tag{15.14}$$

According to estimates by Kunze (1985), the first term of exponent is negligible at all frequencies, because $|\partial U/\partial z, \partial V/\partial z| \ll N$. Thus, we have an approximate expression of (15.14)

$$\psi = \psi_0 \exp[\beta t], \tag{15.15}$$

where

$$\beta = \left(\frac{N^2}{2f^2}\right)\frac{1}{k_z^2}\left[\frac{\partial U}{\partial x}k_y^2 - \left(\frac{\partial U}{\partial y} + \frac{\partial V}{\partial x}\right)k_xk_y + \frac{\partial V}{\partial y}k_x^2\right]. \tag{15.16}$$

In the cylindrical coordinate system, (15.16) has the form of

$$\beta(r,k,\theta) = \left(\frac{N^2}{2f^2}\right)\frac{1}{k_z^2}\left(\tan\theta k_y^2 - 2k_xk_y + \cot\theta k_x^2\right)\frac{\partial v_\theta}{\partial r}. \tag{15.17}$$

Substituting (15.10) into (15.17) yields

$$\beta(r_n,k,\theta) = \left(\frac{N^2}{2f^2}\right)\left(\frac{g\eta_0}{fR^2}\right)\beta_1(r_n)\,\beta_2(k,\theta), \tag{15.18}$$

where

$$\beta_1(r_n) = \left(\frac{r_n^4}{2} - \frac{7r_n^2}{2} + 2\right)e^{-\frac{r_n^2}{2}}, \tag{15.19}$$

and

$$\beta_2(k,\theta) = \frac{1}{k_z^2}\left(\tan\theta k_y^2 - 2k_xk_y + \cot\theta k_x^2\right). \tag{15.20}$$

Figure 15.1(d) shows the radial structure of normalized $\beta_1(r_n)$. One can see that $\beta_1(r_n)$ is characterized by the maximum value of 2 at the eddy center ($r_n = 0$) and the minimum value of -1.12 at $r_n = 1.4$, changes the sign at around $r_n = 0.8$, and almost vanishes beyond $r_n = 2$. For an anticyclonic eddy, i.e., $\eta_0 > 0$, strong amplitude growth rate for the NIWs happens within the range of $r_n < 0.8$, implying that anticyclonic eddy may serve as an amplifier or a source for the NIWs. On the other hand, for a cyclonic eddy, i.e., $\eta_0 < 0$, strong amplitude decay rate happens, implying that cyclonic eddy may serve as a damper or sink for the NIWs.

Furthermore, from (15.20) one can see that $\beta_2(k, \theta)$ takes positive and negative infinities across lines of $\theta = \frac{n\pi}{2}$, $n = 0, 1, \ldots 4$. It means that there exist potential barriers along the central latitude and the central longitude of eddy, i.e., the NIWs may not penetrate the two lines. In the other words, the NIWs may propagate only within one quadrant of eddy. This inference needs to be further verified.

15.3 Mooring-observed Modulation of Mesoscale Eddy to the NIW Frequency

15.3.1 *A case near the SCS western boundary*

15.3.1.1 *Mooring observation*

In 2004, the South China Sea Institute of Oceanography, Chinese Academy of Sciences, deployed an upward-looking acoustic Doppler current profiler (ADCP) mooring near the SCS western boundary for the hydrographic survey. The ADCP of the broadband frequency of 75 kHz is a product of Teledyne RD Instruments, Inc., USA. The mooring station was located at 13.99°N, 110.52°E with the depth of 1800 m as shown in Figure 15.2. The ADCP was moored at the depth of 500 m. The effective measurements were from 48 m to 480 m with a vertical resolution of 8 m. The sampling interval for the velocity observation was 30 min. Typhoon Chanchu center passed over the mooring station on 12 June 2004 and generated the strong NIWs. Thus, a sub-dataset of ADCP data from 27 May to 30 July 2004 (about 30 inertial periods) was selected as a baseline for analyzing the NIW frequency shifts induced by mesoscale eddy modulation.

Figure 15.3 shows mooring-measured depth- time series of ocean current components u (zonal) and v (meridional). A low-pass-filterer has been applied to minimize the high-frequency fluctuation of raw data.

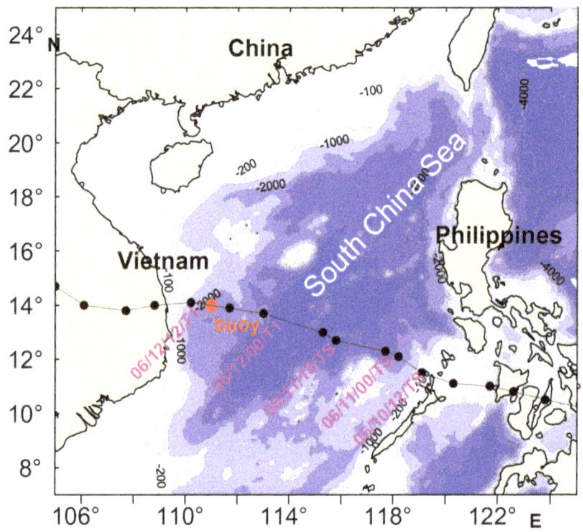

Fig. 15.2. The SCS deep basin chart with isobaths in m. ADCP mooring station is marked by a red dot at 13.99°N, 110.52°E. Typhoon Chanchu track is labeled by data strings of date/time (mm/dd/hh 2004 UTC)/category of tropical cyclone (Saffir-Simpson scales: TD for tropical depression, TS for tropical storm, and T1 for typhoon of category 1).

Source: Taken from Sun *et al.* (2011a, 2011b).

15.3.1.2 *Evolutionary process of NIWs during typhoon passage*

To analyze evolution of the wave-like signals shown in Figure 15.3(a) and 15.3(b) with the typhoon passage process, the observation period was divided into three stages: before-typhoon stage from 27 May to 12 June, typhoon-wake stage from 13 June to 10 July and after-typhoon stage from 11 July to 30 July 2004. Meanwhile, depths of 80, 208 and 400 m are chosen to represent the upper, sub-upper and lower layers of the ocean The kinetic-energy power spectra (KEPS) are used to analyze the constituents of the wave-like signals and their evolution at three stages. The results are shown in Figure 15.4(a–c). The frequencies of diurnal tides K_1 and O_1, semidiurnal tide M_2, subharmonics of diurnal tides 0.5 K_1 and 0.5 O_1 within the NIW band are marked by vertical lines in the figures.

Figure 15.4(a) shows the KEPS at before-typhoon stage. One can see that the NIW band in the upper layer has formed a spectral peak with peak power density (PPD) of $2.0 \times 10^2\,\text{cm}^2\,\text{s}^{-2}\,\text{cph}^{-1}$ although

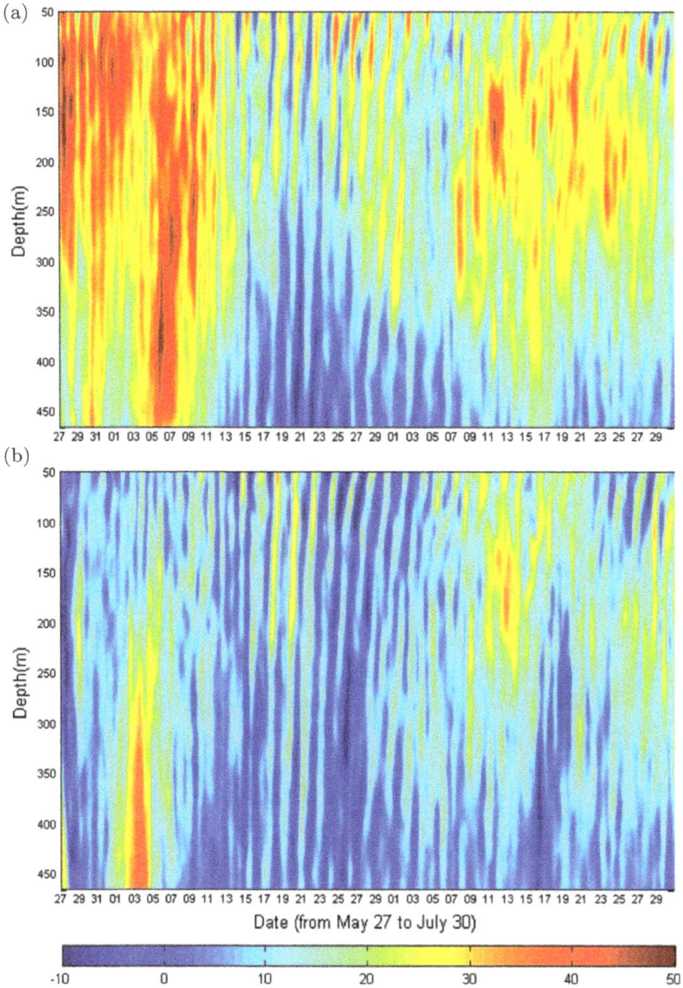

Fig. 15.3. Mooring-measured depth-time series of ocean current components (a) u and (b) v from 27 May to 30 July 2004. Color codes are in $\mathrm{cm\,s^{-1}}$. *Source*: Redrawn from Sun *et al.* (2011a, 2011b).

a little below 90% statistical significance level. The central frequency (CF) of 0.022 cph is near 0.02015 cph of the local inertial frequency and 0.02089 cph of the subharmonic frequency of $0.5K_1$. The mooring station is located at 13.99°N latitude, which is near 15.52°N of the critical latitude of $0.5K_1$ and on the equator-ward side, so that the parametric subharmonic instability (PSI) mechanism of diurnal

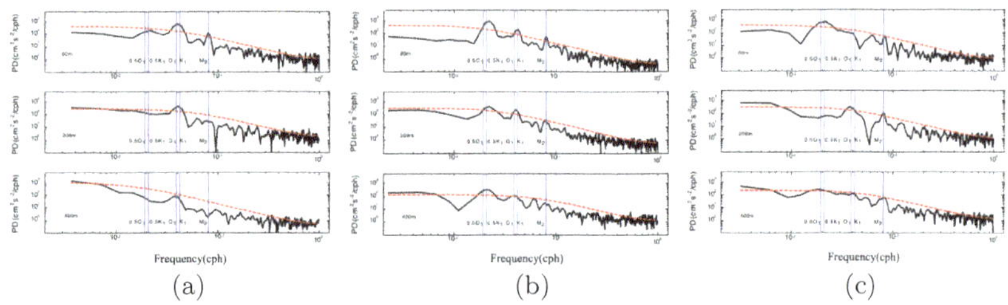

Fig. 15.4. Kinetic-energy power spectra at depths of 80, 208 and 400 m. (a) Before-typhoon stage from 27 May to 12 June 2004. (b) Typhoon-wake stage from 13 June to 10 July 2004. (c) After-typhoon stage from 11 July to 30 July 2004. Dashed red lines represent a 90% statistical significance level.

Source: Redrawn from Sun *et al.* (2011a, 2011b).

internal tides should play a key role in generating the NIWs. Meanwhile, the frequency shift (FS) of $0.09f$ implies occurrence of frequency modulation of the mean flow to the NIWs. In the sub-upper and lower layers, however, the spectral peaks of the NIW band are absent. These situations just represent a local normal case in summer, i.e., weak seasonal NIWs exist in the upper layer only.

Figure 15.4(b) shows the KEPS at typhoon-wake stage. One can see that in the upper layer, the PPD of the NIW band increases to $9.0 \times 10^2\,\mathrm{cm}^2\,\mathrm{s}^{-2}\,\mathrm{cph}^{-1}$, 4.5 times of that at before-typhoon stage. The CF keeps 0.022 cph with the blueshift of 0.09 f. In the sub-upper layer, the CF and the FS are the same as that in the upper layer, while the PPD is $4.0 \times 10^2\,\mathrm{cm}^2\,\mathrm{s}^{-2}\,\mathrm{cph}^{-1}$, implying that typhoon passage significantly strengthens the NIWs and deepens the upper ocean mixed layer. In the lower layer, the PPD still reaches $3.0 \times 10^2\,\mathrm{cm}^2\,\mathrm{s}^{-2}\,\mathrm{cph}^{-1}$ and the CF decreases to 0.021 cph with the blueshift of 0.04 f, implying that modulation of the mean flow becomes weaker with the depth.

Figure 15.4(c) shows the KEPS at after-typhoon stage. One can see that in the upper layer, the PPD decreases to $7.0 \times 10^2\,\mathrm{cm}^2\,\mathrm{s}^{-2}\,\mathrm{cph}^{-1}$, and the CF decreases to 0.021 cph with the blueshift of 0.04 f. In the sub-upper layer, the KEPS peak is absent. In the lower layer, the PPD appears as $2.0 \times 10^2\,\mathrm{cm}^2\,\mathrm{s}^{-2}\,\mathrm{cph}^{-1}$, and the CF decreases to 0.019 cph with the redshift of 0.06 f. It seems that the NIWs here are independent of that in the upper layer. The above results are summarized in Table 15.1.

15.3.1.3 *Decomposition of frequency shifts of NIWs*

From Table 15.1, one can see that in all observed cases, the NIW band has the FS from −6% to 9%, implying that modulation of the mean flow to the NIW frequencies always occur. Section 15.2.2 has derived theoretical solutions of the FS from the dispersion relation of the NIWs in the case of the mean flow consisting of the background flow and the mesoscale eddy flow. Thus, the FS can be decomposed into two components: vorticity-induced, ξ_v, and Doppler effect-induced, ξ_d, i.e.,

$$FS = \xi_v + \xi_d. \tag{15.21}$$

Suppose that the horizontal scale of background flow is much larger than that of the mesoscale eddy flow, then

$$\xi_v = \frac{g\eta_0}{2fR^2}\left(\frac{r^4}{2} - 4r^2 + 4\right)e^{-\frac{r^2}{2}}, \tag{15.22}$$

and

$$\xi_d = \boldsymbol{k} \cdot \boldsymbol{V} \tag{15.23}$$

where V is a vector sum of the background flow velocity and the tangential velocity of eddy at mooring station. Note that (15.22) and (15.23) indicate that mesoscale eddy is a dominant contributor

Table 15.1. PPD, CF and FS of the NIW band derived from the KEPS in Figures 15.4(a–c).

Depth (m)	Parameter	Before-typhoon stage (27 May–12 June)	Typhoon-wake stage (13 June–10 July)	After-typhoon stage (11–30 July 2004)
80	PPD	2.0	9.0	7.0
	CF (cph/f)	0.022 (1.09 ± 0.05)	0.022 (1.09 ± 0.05)	0.021 (1.04 ± 0.05)
	FS (f)	0.09	0.09	0.04
208	PPD		4.0	
	CF (cph/f)	N/A	0.022 (1.09 ± 0.05)	N/A
	FS (f)		0.09	
400	PPD		3.0	2.0
	CF (cph/f)	N/A	0.021 (1.04 ± 0.05)	0.019 (0.94 ± 0.05)
	FS (f)		0.04	−0.06

Note: PPD is in $10^2\,cm^2\,s^{-2}\,cph^{-1}$ with a relative error of 10% (est.)

to ξ_v, while both the background flow and the mesoscale eddy flow are contributors to ξ_d

Satellite altimeter data products, Maps of Absolute Dynamic Topography (MADT), with a spatial resolution of 1°/4 by 1°/4 downloaded from the website of Archiving Validation and Interpretation of Satellite Data in Oceanography (AVISO), Centre National d'Etudes Spatiales of France, are used for calculating ξ_v in the upper layer using (15.22). The MADT of the SCS on 10 June 2004, 13 June 2004 and 14 July 2004 are used to represent before-typhoon, typhoon-wake and after-typhoon stages as shown in Figure 15.5(a–c). One can see that the mooring station located at 13.99°N, 110.52°E is sandwiched by the Vietnam Offshore Current (VOC), an eastward extension of the Vietnam coastal current (Li *et al.*, 2014) and anticyclonic eddy AE. The parameters extracted from the MADT and calculated results of the FS components are listed in Table 15.2. The measurements by previous investigators indicate that mesoscale eddies and the VOC in that area might exist in the upper 250 m (Hu *et al.*, 2011; Li *et al.*, 2014). Thus, calculated results are suitable to interpret the measured results only in the upper layer.

From the results listed in Table 15.2, one can see that at the mooring station, the Doppler effect is a major contributor for the NIW frequency shift occupying 95.5%, 96.7% and 72.5% of the total observed FS, and the eddy vorticity is a minor one occupying 4.5%,

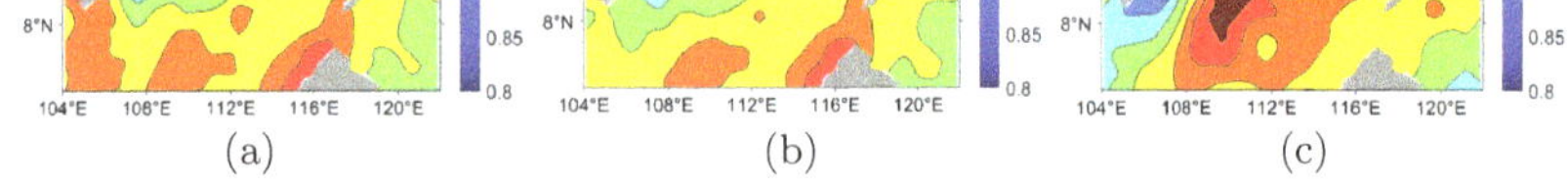

Fig. 15.5. MADT images of the SCS on (a) 10 June 2004 at before-typhoon stage, (b) 13 June 2004 at typhoon-wake stage and (c) 14 July 2004 at after-typhoon stage. AE and VOC represent anticyclonic eddy and the Vietnam Offshore Current near the ADCP mooring station (red star). Sea level color codes are in m.

Source: Downloaded from website of AVISO, Centre National d'Etudes Spatiales of France.

Table 15.2. Parameters extracted from MADT and calculated results of the FS components.

Parameter	10 June 2004 Before-typhoon	13 June 2004 Typhoon-wake	14 July 2004 After-typhoon
Measured FS in upper layer (f)	0.09	0.09	0.04
SSH η_0(cm)	15	15	5
Eddy radius R (km)	210	230	120
Coriolis parameter f(s^{-1})	3.5148×10^{-5}	3.5148×10^{-5}	3.5148×10^{-5}
Relative mooring position r (R)	1.18	1.17	1.54
Normalized vorticity $\varsigma(r)$	0.32	0.30	0.80
FS by vorticity ξ_v (f)	0.0043	0.0034	0.011
FS by Doppler effect ξ_d (f)	0.086	0.087	0.029

3.3% and 27.5%. Note that the results are sensitive to the mooring location, i.e., the radius vector with respect to the eddy center. Namely, for different mooring locations, the observed NIW FS may be completely different.

15.3.2 *A case on the northern SCS deep basin*

15.3.2.1 *Mooring observation program*

Yang *et al.* (2021) observed the NIWs in the northern SCS deep basin using three ADCP moorings in a period of September–October 2020. The 300-kHz ADCPs are up-looking and have a temporal resolution of 10 min and a vertical resolution of 8 m. The three moorings are coded as M1, M2 and M3 according to the deployment time as marked in Figure 15.6. The locations and detailed information of moorings are listed in Table 15.3. During the observation period, the tropical storm Noul passed over the SCS with a maximum wind speed of 23 m s^{-1} and a 15 m s^{-1} wind radius of 330 km. The closet distance between the center of Noul and the moorings were 300, 430 and 270 km for M1, M2 and M3 on 17 September 2020, respectively.

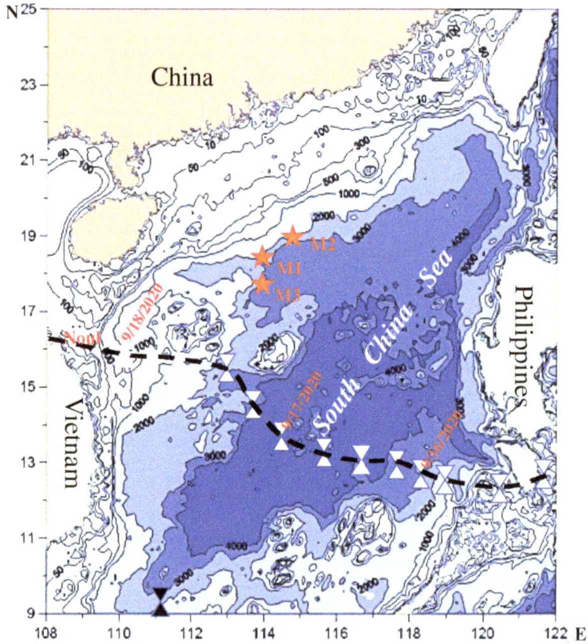

Fig. 15.6. The SCS deep basin chart. ADCP mooring stations M1, M2 and M3 for the NIW observation are marked by red stars. Tropical storm Noul's track downloaded from the Japanese Meteorology Agency is labeled by data strings of mm/dd/yyyy. Bathymetry is in m.

Table 15.3. Detailed designs of mooring observation.

Mooring station	M1	M2	M3
Position	18.41°N, 113.90°E	18.95°N, 115.80°E	17.88°N, 113.80°E
Inertial frequency (cpd)	0.6316	0.6495	0.6139
Water depth (m)	2205	2450	2485
Observed depths of ADCP (m)	24–96	24–120	24–104
Observed period (dd/mm/2020)	01/09–03/10	02/09–01/10	06/09–14/10

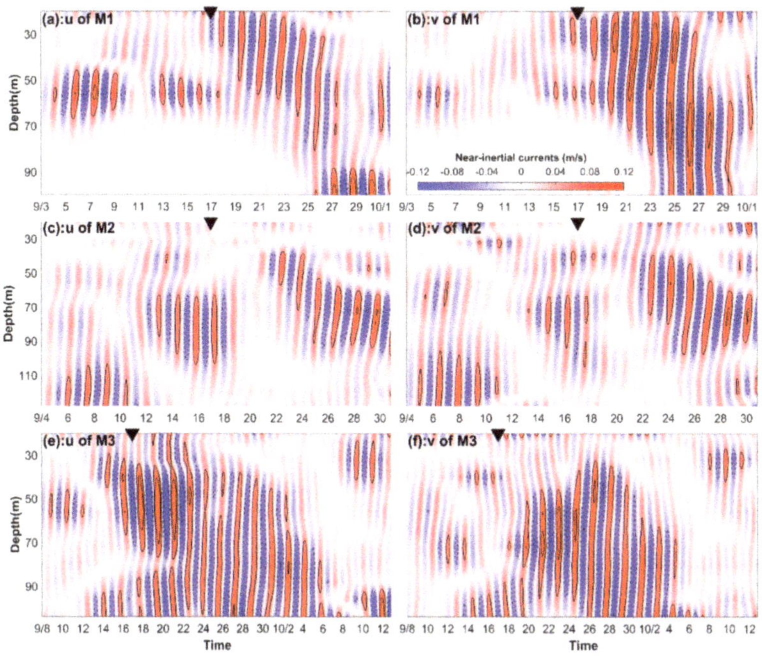

Fig. 15.7. Mooring-measured depth-time series of band-passed filtered zonal and meridional near-inertial currents at stations M1 (a and b), M2 (c and d) and M3 (e and f) in September and October 2020. Dash contours represent −0.05 and $-0.10\,\mathrm{m\,s^{-1}}$, and solid contours represent 0.05 and $0.10\,\mathrm{m\,s^{-1}}$. Inverse triangles denote the time for the nearest passage of tropical storm Noul over the moorings. Color codes of the inset in (b) are in $\mathrm{m\,s^{-1}}$.

Source: Taken from Yang *et al.* (2021).

15.3.2.2 *Mooring-measured depth- time series of ocean current velocity components*

Figure 15.7 shows mooring-measured depth-time series of ocean current components u and v in upper layer of 100 m at M1, M2 and M3 from 3 September to 12 October 2020. The observed raw data were de-tided by removing the barotropic tidal currents and filtered with a frequency passband of $(1.00 \pm 0.15)\ f$. One can see that during the observation period, the NIWs were broadly but discontinuously distributed in the spatiotemporal domain. In the most cases, they occurred in the form of wave packages with variable amplitudes, implying they were modulated by multiple processes. The near-inertial kinetic energy (NIKE $= \rho(u^2 + v^2)/2$, where ρ is the density of sea water) time series derived from three mooring observations show that their peak values dramatically increased 3–7 times from $O(1)\,\mathrm{kg\,m^{-1}\,s^{-2}}$ before 17 September 2020 and lasted for about two weeks, implying a tropical-storm-forced NIW event (Yang *et al.*, 2021).

15.3.2.3 *Modulation of mesoscale eddies to the NIW frequency*

Figure 15.8 shows time series images of the vorticity and geostrophic velocity vectors in the surrounding area of mooring stations derived from the AVISO dataset during the observation period from 1 September to 13 October 2020 with a temporal resolution of four days. One can see that the area was occupied by energetic mesoscale eddies with diameters of about 100–200 km all the time, providing

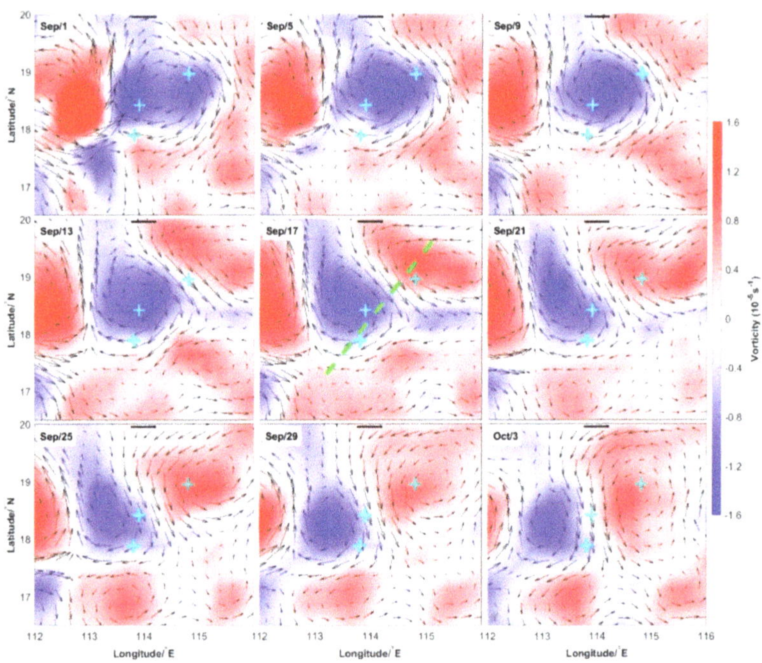

Fig. 15.8. Time series images of vorticity (color codes) and geostrophic velocity vectors (arrows) in the surrounding area of mooring stations (cyan crosses) derived from the AVISO dataset from 1 September to 3 October 2020 with a temporal resolution of four days. Color codes are in $10^{-5}\,\mathrm{s^{-1}}$. Velocity scale of $0.5\,\mathrm{m\,s^{-1}}$ is marked on the top of image.

Source: Taken from Yang *et al.* (2021).

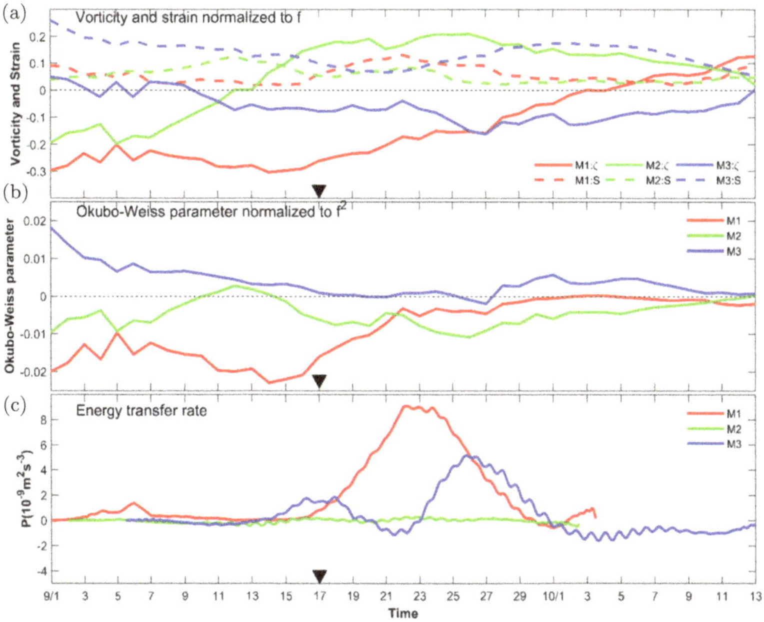

Fig. 15.9. Mooring-observed time series of (a) normalized vorticity and normalized strain in f, (b) normalized OW parameter in f^2, and (c) depth-mean energy transfer rate from mesoscale eddies to NIWs from 1 September to 13 October 2020. Black triangles denote approach time of typhoon Noul.

Source: Taken from Yang *et al.* (2021).

the best opportunity to examine modulation of mesoscale eddies to the NIWs.

Figure 15.9 shows to the behavior of eddy current vorticity at the mooring stations. One can see that the observation periods can be divided into two time-segments. For M1, Segment 11 from 1 September to 17 September 2020 was characterized by a steady vorticity value of about $-0.30\ f$, and Segment 12 from 17 September to 13 October by monotonically increasing values from $-0.30\ f$ to $0.12\ f$. For M2, Segment 21 from 1 September to 12 September was characterized by the negative vorticity, and Segment 22 from 12 September to 13 October by changing to the positive vorticity. For M3, Segment 31 from 1 September to 9 September was characterized by the positive vorticity and Segment 32 from 9 September to 13 October by changing to the negative vorticity. Using theories derived in Section 15.2 and Section 15.3.1, the observed NIW FSs are decomposed into two components as listed in Table 15.4.

Table 15.4. NIW FS components in the northern SCS deep basin in September–October 2020.

Mooring station	M1		M2		M3	
Time-segment	S11	S12	S21	S22	S31	S32
(dd/mm/2020)/13/10	01/09–17/09	17/09–13/10	01/09–12/09	12/09–13/10	01/09–09/09	09/09–13/10
MOF*(f)	0.93	0.95	1.07	1.03	1.08	0.98
FS(f)	−0.07	−0.05	0.07	0.03	0.08	−0.02
SOV**(f)	−0.28	−0.10	−0.10	0.10	0.00	−0.15
$\xi_v^\dagger(f)$	−0.14	−0.05	−0.05	0.05	0.00	−0.075
$\xi_d^\ddagger(f)$	0.07	0.00	0.12	−0.02	0.08	0.055

Notes: *Mooring-observed frequency, **Satellite-observed vorticity, †FS by vorticity, ‡FS by Doppler Effect.

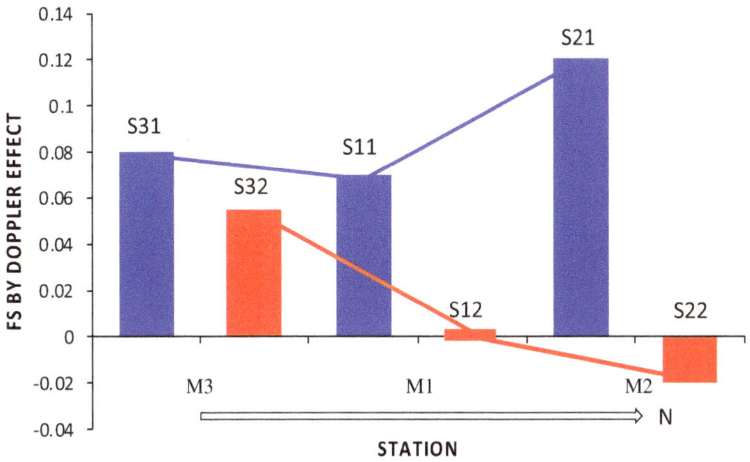

Fig. 15.10. Comparison of Doppler-effect-induced NIW FS component before and after passage of tropical storm Noul over three mooring stations in September–October 2020.

From Table 15.4, one can see following spatiotemporal distribution features. (1) The mean absolute values of the NIW FSs by the vorticity and by the Doppler effect are the same as $(0.06 \pm 0.04)\ f$, implying that contributions of two components are almost equal. It is different from the case of the western boundary (Section 15.3.1), in which the FS by the Doppler effect is always a dominant component. (2) For time-segment 1 without tropical storm influence, the NIW FSs by the Doppler effect $\xi_d (= \boldsymbol{k} \cdot \boldsymbol{V}$, in which $\boldsymbol{V}$ is the mean flow velocity) all take positive values and increase northward as shown in Figure 15.10. Suppose that the change in NIW wavenumber $\boldsymbol{k}$ is ignorable within the mooring station range of $O(\pm 50)$ km, the distribution of ξ_d agrees with the velocity field distribution features of the northern SCS circulation (Li *et al.*, 2014). (3) For time-segment 2 after tropical storm passage, the mean flow structure reflected by ξ_d is completely disturbed: weakened at station M3, disappeared at M1, and reversed at M2.

15.4 Mooring-Observed Modulation of Mesoscale Eddy to the NIW Amplitude

15.4.1 *Mooring observation and concurrent satellite data*

Shu *et al.* (2016) observed the NIWs driven by the summer monsoon onset using ADCP mooring data deployed in the SCS deep basin during the SCS Monsoon Experiment (SCSMEX) from January 1998 to October 1999. Mooring stations M1-3 are located at 115.6°E, 18.0°N, 115.0°E, 15.3°N and 115.4°E, 13.0°N, respectively, where the water depths are deeper than 2000 m as shown in Figure 15.11(a). Figure 15.11(b) and 15.11(c) show the averaged absolute dynamic topography (ADT) and geostrophic current images within the red dashed box in Figure 15.11(a) from 21 May to 1 June 1998 and from 1 June to 10 June 1999. Figure 15.11(d) and 15.11(e) show the wind speed anomaly (WSA) distributions on 25 May 1998 and 5 June 1999 relative to the averaged wind fields from 23 May to 27 May 1998 and from 3 June to 7 June 1999, respectively. One can see that in the case of 1998, the WSA region was distributed in the western SCS deep basin, while in 1999, it moved to the northern central SCS deep

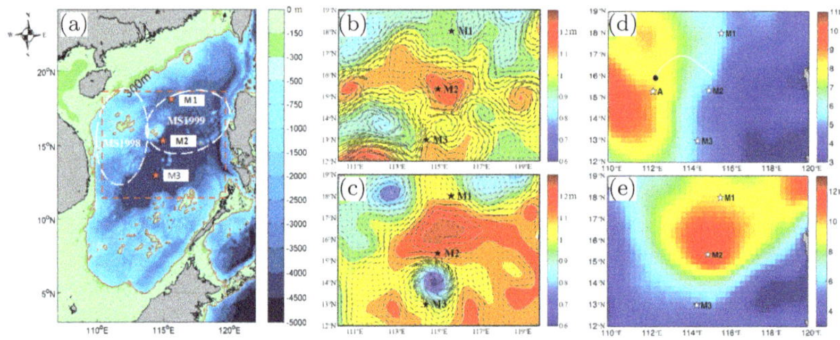

Fig. 15.11. (a) Chart of the study area with bathymetry in m. M1, M2 and M3 show mooring station locations. (b) Averaged ADT (m) and geostrophic current ($m\,s^{-1}$) images within the red dashed box in (a) from 21 May to 1 June 1998 and (c) from 1 June to 10 June 1999. (d) WSA ($m\,s^{-1}$) distribution on 25 May 1998 relative to the averaged wind field from 23 May to 27 May 1998. White line represents the ray path of the NIWs. (e) Same as (d), but on 5 June 1999 relative to that from 3 to 7 June 1999.

Source: Figures (b–e) are redrawn from Shu *et al.* (2016).

basin. In both cases, the maximum value of WSA reaches 10 m s^{-1}, showing a spatial distribution feature of summer monsoon onset.

15.4.2 *Mooring observed NIWs*

At mooring stations M1–M3, the 75 kHz ADCPs are moored upward looking with a bin size of 10 m and a sampling interval of 1 h. ADCP velocity records from 20 to 170 m at station M2 are used for extracting the NIW signals within a frequency band from 0.85 to 1.15f. The depth-time series of band-passed filtered zonal and meridional near-inertial current components as well as the near inertial kinetic energy (NIKE) from 20 May to 19 June 1998 and from 27 May to 25 June 1999 are shown in Figures 15.12 and 15.13, respectively. One can see that the two cases show similar episode events of intensified NIWs, which are characterized by a duration of about 3 weeks and the NIKE concentrated in the upper ocean above 150 m. The rotary power spectra of the currents indicate that the peak frequencies of NIWs are 0.89f and 1.06f in cases of 1998 and 1999, respectively. The red shift of 0.11f (11%) in 1998 and the blue shift of 0.06f (6%) in 1999 imply occurrence of the strong frequency modulation of NIWs by the mean flow. As shown in Figure 15.11(b) and 15.11(c), mooring station M2 was located near the center of anticyclonic eddy in the case of 1998, while outside an anticyclonic eddy in the case of 1999. Thus, it is reasonable to speculate that eddies play a key role in the mean flow modulation to the FS and the amplitude growth of NIWs.

15.4.3 *Growth rate of NIW amplitude*

From the mooring-measured current velocity data as shown in Figures 15.12 and 15.13, the growth processes of NIW amplitude ($\psi = \sqrt{u^2 + v^2}$) at initial stages of the events are obtained as shown in Figure 15.14. One can see that in all the cases, the NIW amplitude growth process obeys an exponential law, i.e.,

$$\psi = \psi_0 e^{\beta_e t}, \tag{15.24}$$

where β_e is an empirical growth rate, t is the time counted from the wave initial date. Fitting parameters are listed in Table 15.5.

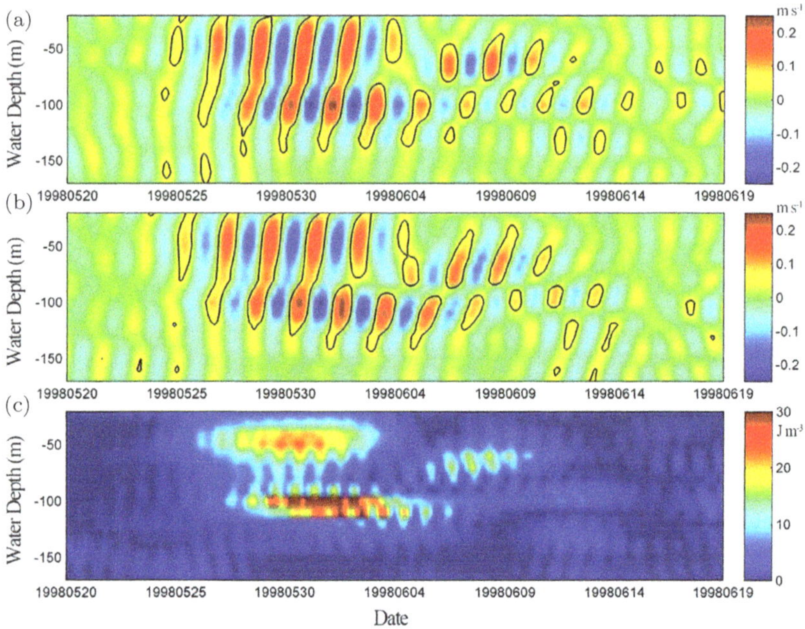

Fig. 15.12. Mooring-measured depth-time series of band-passed filtered (a) zonal and (b) meridional near-inertial currents (in $\mathrm{m\,s^{-1}}$) and (c) the NIKE (in $\mathrm{J\,m^{-3}}$) at station M2 from 20 May to 19 June 1998. Black contours in (a) and (b) delineate the speed of $0.05\,\mathrm{m\,s^{-1}}$.

Source: Taken from Shu *et al.* (2016).

On the other hand, Section 15.2 has obtained the approximate solution of exponential growth rate of the NIW amplitude if the waves propagate within a mesoscale eddy, i.e.,

$$\psi = \psi_0 \exp[\beta t], \tag{15.25}$$

where

$$\beta = \left(\frac{N^2}{2f^2}\right)\left(\frac{g\eta_0}{fR^2}\right)\beta_1(r_n)\,\beta_2(k,\theta), \tag{15.26}$$

where

$$\beta_1(r_n) = \left(\frac{r_n^4}{2} - \frac{7r_n^2}{2} + 2\right)e^{-\frac{r_n^2}{2}}, \tag{15.27}$$

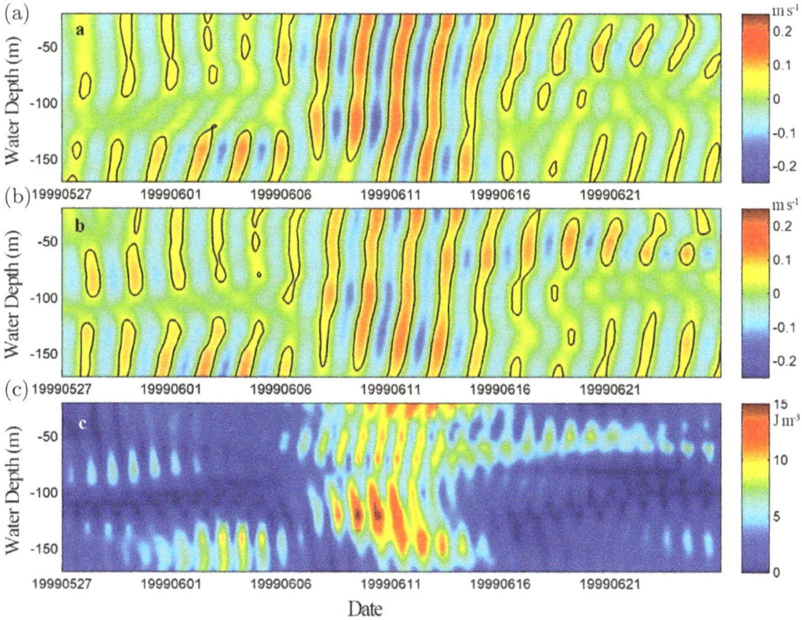

Fig. 15.13. Same as Fig. 15.12 but for the period from 27 May to 25 June 1999.

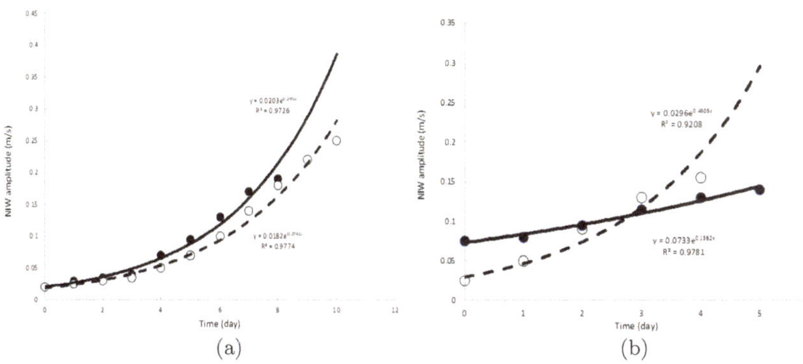

Fig. 15.14. Growth processes of NIW amplitudes during the initial stages of episode NIW events at M2. (a) The case of 1998. Solid and open circles represent the data measured at depths of 50 and 110 m, respectively. (b) The case of 1999. The same as (a), but at 30 and 120 m.

Source: Data taken from Figure 8 in Shu *et al.* (2016).

Table 15.5. Fitting parameters of the empirical exponential function to mooring-measured NIW amplitude data at initial stages of episode NIW events.

Time	Depth, m	$\psi_0, \mathrm{m\,s^{-1}}$	$\beta_e, \mathrm{d^{-1}}$	R^2
21–31 May 1998	50	0.020	0.30	0.97
	110	0.018	0.27	0.98
6–11 June 1999	30	0.073	0.14	0.98
	120	0.030	0.46	0.92

and

$$\beta_2(k,\theta) = \frac{1}{k_z^2}\left(\tan\theta k_y^2 - 2k_x k_y + \cot\theta k_x^2\right). \tag{15.28}$$

Thus, it is possible to estimate $\beta_1(r_n)$ or $\beta_2(k,\theta)$ using the empirical exponential growth rate derived from mooring-measured current velocities and satellite sea level data. Taking the case measured at the depth of 50 m on 21–31 May 1998 shown in Figure 15.14(a) as an example, the dynamic parameters are as follows: N^2: $2.5 \times 10^{-4}\,\mathrm{s^{-2}}$ (Chen *et al.*, 2013), f: $3.76 \times 10^{-5}\,\mathrm{s^{-1}}$ at 15°N; η_0: 1.2 m, R: 1×10^5 m, and r_n: 0 determined from the averaged ADT image in Figure 15.10(b), and $\beta_1(r_n = 0)$: 2 from (19). Let $\beta = \beta_e (= 0.30\,d^{-1})$, we obtain $\beta_2(k,\theta) = 0.61 \times 10^{-6}$. Consider that the horizontal and vertical length scales of NIWs are $O(100)$ km and $O(100)$ m, respectively, therefore, the value of $\beta_2(k,\theta)$ is within a reasonable range.

Furthermore, adopting the above dynamic parameters (except η_0 of 1.3 m) and the value of $\beta_2(k,\theta)$ to the case measured at the depth of 30 m on 6–11 June 1999 shown in Figure 15.13(b) yields $\beta_1(r_n) = 0.54$ for the empirical exponential growth rate $\beta_e = 0.15$. From (15.27), we obtain $r_n = 0.7$, which is a very good estimate of M2 location relative to the anticyclonic eddy center if measured from the averaged ADT image in Figure 15.11(c).

The above analysis results of two mooring measured NIW events in the SCS deep basin indicate that theoretical solutions of exponential growth rate of NIWs if propagating within mesoscale eddies explain mooring and satellite observations very well.

15.5 Energy Exchange between Mesoscale Eddies and NIWs

The energy exchange between low-frequency flows and internal waves is a classical problem of fluid dynamics. Previous investigators have developed theoretical frames and derived general solutions for the problem (Jones, 1967, 1969; Frankignoul, 1974, 1976; Brown and Owens, 1981; Bühler and McIntyre, 2005; Polzin, 2004, 2008, 2010). The energy exchange between mesoscale eddies and NIWs is a specific case of the problem in a shear, rotating geostrophic fluid (Kunze, 1985; Mied *et al.*, 1987, 1990; Whitt and Thomas, 2015; Jing *et al.*, 2017, 2018; Zhou and Hong, 2022). In the case, the most important result is the energy transfer rate from the low frequency mean flow consisting of the background flow and the eddy flow (U, V) to the NIWs (u, v) through the horizontal wave stress. From the internal wave energy equation (Müller, 1976) and the limit that $\omega \to f$ (Ruddick and Joyce, 1979), the energy transfer rate from the low frequency mean flow to the NIWs is derived in the form of

$$P = -(\langle uu \rangle - \langle vv \rangle)\frac{S_n}{2} - \langle uv \rangle S_s, \tag{15.29}$$

where $S_n = U_x - V_y$ and $S_s = U_y - -V_x$ are the normal and shear components of the strain of low frequency flow (subscripts denote the spatial derivatives), respectively. The $\langle \cdot \rangle$ represents the running mean over time (Polzin, 2010; Jing *et al.*, 2018). One can see that the energy transfer occurs if the value of P is nonzero, i.e., if the mean flow is sheared ($S_n \neq 0$ or $S_s \neq 0$), and the NIW field is anisotropic ($(\langle uu \rangle - \langle uv \rangle \neq 0)$). Meanwhile, the value of P can be both positive and negative. The sign of Okubo–Weiss (OW) parameter defined as $W = S_n^2 + S_s^2 - \zeta^2$ is used as a criterion for occurrence of the energy transfer (Okubo, 1970; Provenzale, 1999). A permanent energy transfer from the mean flow to the NIWs occurs only under the positive OW parameter condition. If the OW parameter is negative, the mean P value is near zero. Thus, a permanent energy transfer only occurs in periods if the total strain dominates the relative vorticity in the mean flows (Jing *et al.*, 2018).

In the case of the SCS, Yang *et al.* (2021) used definition in (15.29) to calculate the time series of P using the ADCP mooring array measurements of current velocities as shown in Figure 15.9. The results show that the peak value of $|P|$ reaches $8.5 \times 10^{-9}\,\mathrm{m^2\,s^{-3}}$, which is almost one order higher than $1.1 \times 10^{-9}\,\mathrm{m^2\,s^{-3}}$ in the Gulf of Mexico derived by Jing *et al.* (2018), indicating that the SCS is an ocean area with stronger interaction of mesoscale eddies with the NIWs compared to the Gulf of Mexico, a comparable case of marginal seas.

15.6 Summaries

In the SCS, there are high possibilities for concurrence of the typhoon-forced NIWs. Parts 2–4 of this book specify details of hydrography, statistics and genesis of mesoscale eddies in the SCS. This chapter focuses on advances in the research of modulation of mesoscale eddies to the NIWs in the SCS. The key points are summarized as follows.

From the approximate dispersion relation of the NIWs derived by Kunze (1985) and the empirical expression of the sea surface dynamic height of mesoscale eddy derived by Zhang *et al.* (2013), the approximate dispersion relation of the NIWs and the exponential growth rate in the cylindric coordinate system for the NIWs propagating within mesoscale eddies are derived. The theoretical solutions are applied to reanalyze the real cases observed by mooring systems and satellite altimetry in the SCS. The results indicate that (1) the NIW frequency shift (FS) always occurs as propagating into mesoscale velocity fields. Within the main body of cyclonic (anticyclonic) eddy, positive or blue (negative or red) FS occurs, and the FS value is dependent upon the radius vector of observed location. Beyond the eddy main body, the FSs change the signs. (2) The exponential growth rate of the NIW amplitude is dependent upon the eddy polarity and the radius vector of observed location. Within the main body of anticyclonic (cyclonic) eddy, the growth rate is positive (negative), implying that anticyclonic eddy may serve as an amplifier or a source of the NIW, while cyclonic eddy as a dumper or a sink. Beyond the eddy main bodies, the growth rates change the signs.

A brief introduction to theories of the energy exchange between low-frequency flows and internal waves is included in this chapter. The theories have been adopted in the research of energy exchange of mesoscale eddies with the NIWs in the SCS. Yang *et al.* (2021) calculated the time series of the energy transfer rate from the low frequency mean flow to the NIWs using the ADCP mooring array measurements of current velocities in the SCS deep basin. The results indicate that the interaction of mesoscale eddies with the NIWs in the SCS is much stronger than that in the comparable Gulf of Mexico.

References

Alford, M. H. (2008). Observations of parametric subharmonic instability of the diurnal internal tide in the South China Sea, *Geophys. Res. Lett.*, 35, L15602, doi:10.1029/2008GL034720.

Brown, E. D. and Owens, W. B. (1981). Observations of the horizontal interactions between the internal wave field and the mesoscale flow, *J. Phys. Oceanogr.*, 11, pp. 1474–1480, doi:10.1175/1520-0485(1981)011 <1474:OOTHIB>2.0.CO;2.

Bühler, O. and McIntyre, M. E. (2005). Wave capture and wave–vortex duality, *J. Fluid Mech.*, 534, pp. 67–95, doi:10.1017/S002211200500 4374.

Cao, A., Guo, Z., Song, J., Lv, X., He, H. and Fan, W. (2018). Near-inertial waves and their underlying mechanisms based on the South China Sea internal wave experiment (2010–2011), *J. Geophys. Res. Oceans*, 123, pp. 5026–5040, doi:10.1029/2018JC013753.

Chen, G., Xue, H., Wang, D. and Xie, Q. (2013). Observed near-inertial kinetic energy in the northwestern South China Sea, *J. Geophys. Res. Oceans*, 118, pp. 4965–4977.

Frankignoul, C. (1974). Observed anisotropy of spectral characteristics of internal waves induced by low-frequency currents, *J. Phys. Oceanogr.*, 4, pp. 625–634, doi:10.1175/1520-0485(1974)004<0625: OAOSCO>2.0.CO;2.

Frankignoul, C. (1976). Observed interaction between oceanic internal waves and mesoscale eddies, *Deep-Sea Res.*, 23, pp. 805–820, doi:10.1016/0011-7471(76)90848-2.

Guan, S., Zhao, W., Huthnance, J., Tian, J. and Wang, J. (2014). Observed upper ocean response to typhoon Megi (2010) in the Northern South China Sea, *J. Geophys. Res. Oceans*, 119, pp. 3134–3157, doi:10.1002/2013JC009661.

Hu, J., Gan, J., Sun, Z., Zhu, J. and Dai, M. (2011). Observed three-dimensional structure of a cold eddy in the southwestern South China Sea, *J. Geophys. Res.*, 116, C05016, doi:10.1029/2010JC006810.

Jing, Z., Wu, L. and Ma, X. (2017). Energy exchange between the mesoscale oceanic eddies and wind-forced near-inertial oscillations, *J. Phys. Oceanogr.*, 47, pp. 721–733, doi:10.1175/JPO-D-16-0214.1.

Jing, Z., Chang, P., Dimarco, S. F. and Wu, L. (2018). Observed energy exchange between low-frequency flows and internal waves in the Gulf of Mexico, *J. Phys. Oceanogr.*, 48, pp. 995–1008, doi:10.1175/JPO-D-17-0263.1.

Jones, W. L. (1967). Propagation of internal gravity waves in fluids with shear flow and rotation, *J. Fluid Mech.*, 30, p. 4391017/S, doi:10.1017/S0022112067001521.

Jones, W. L. (1969). Ray tracing for internal gravity waves, *J. Geophys. Res.*, 74, pp. 2028–2033, doi:10.1029/JB074i008p02028.

Kunze, E. (1985). Near-inertial wave propagation in geostrophic shear, *J. Phys. Oceanogr.*, 15(5), pp. 544–565, doi:10.1175/1520-0485(1985)015<0544:NIWPIG>2.0.CO;2.

Li, R., Chen, C., Dong, W., Beardsley, R. C., Wu, Z., Gong, W., Liu, Y., Liu, T. and Xu, D. (2021). Slope-intensified storm-induced near-inertial oscillations in the South China Sea, *J. Geophys. Res. Oceans*, 126, e2020JC016713, doi:10.1029/2020JC016713.

Li, Y., Han, W., Wilkin, J. L., Zhang, W., Arango, H., Zavala-Garay, J., Levin, J. and Castruccio, F. S. (2014). Interannual variability of the surface summertime eastward jet in the South China Sea, *J. Geophys. Res. Oceans*, 119, pp. 7205–7228, doi:10.1002/2014JC010206.

Liang, X., Zhang, X. and Tian, J. (2005). Observation of internal tides and near-inertial motions in the upper 450 m layer of the northern South China Sea, *Chin. Sci. Bull.*, 50(24), pp. 2890–2895, doi:10.1360/982005-210.

Liu, J., He, Y., Li, J., Cai, S., Wang, D. and Huang, Y. (2018). Case study of nonlinear interaction between near-inertial waves induced by typhoon and diurnal tides near Xisha Islands, *J. Geophys. Res. Oceans*, 123, pp. 2768–2784, doi:10.1029/2017JC013555.

Mied, R. P., Lindemann, G. J. and Trump, C. L. (1987). Inertial wave dynamics in the North Atlantic Subtropical Zone, *J. Geophys. Res.*, 92, pp. 13063–13074, doi:10.1029/JC092iC12p13063.

Mied, R. P., Shen, C. Y. and Kidd, M. J. (1990). Current shear-inertial wave interaction in the Sargasso Sea, *J. Phys. Oceanogr.*, 20, pp. 81–96, doi:10.1175/1520-0485(1990)020<0081:CSWIIT>2.0.CO;2.

Mülle, P. (1976). On the diffusion of momentum and mass by internal gravity waves, *J. Fluid Mech.*, 77, pp. 7891017/S doi:10.1017/S0022112076002899.

Okubo, A. (1970). Horizontal dispersion of floatable particles in the vicinity of velocity singularities such as convergences, *Deep Res. Oceanogr. Abstr.*, 17, pp. 4451016/0 doi:10.1016/0011-7471(70)90059-8.

Pedlosky, J. (1987). *Geophysical Fluid Dynamics*, Springer, New York, pp. 86–216, doi:10.1007/978-1-4612-4650-3.

Polzin, K. L. (2004). A heuristic description of internal wave dynamics, *J. Phys. Oceanogr.*, 34, pp. 214–230, doi:10.1175/1520-0485(2004)034<0214:AHDOIW>2.0.CO;2.

Polzin, K. L. (2008). Mesoscale eddy–internal wave coupling, Part I: Symmetry, wave capture and results from the Mid-Ocean Dynamics Experiment, *J. Phys. Oceanogr.*, 38, pp. 2556–2574, doi:10.1175/2008JPO3666.1.

Polzin, K. L. (2010). Mesoscale eddy–internal wave coupling, Part II: Energetics and results from PolyMode, *J. Phys. Oceanogr.*, 40, pp. 789–801, doi:10.1175/2009JPO4039.1.

Provenzale, A. (1999). Transport by coherent barotropic vortices, *Annu. Rev. Fluid Mech.*, 31, pp. 55–93, doi:10.1146/annurev.fluid.31.1.55.

Ruddick, B. R. and Joyce, T. M. (1979). Observations of interaction between the internal wavefield and low-frequency flows in the North Atlantic, *J. Phys. Oceanogr.*, 9, pp. 498–517, doi:10.1175/1520-0485(1979)009.

Shu, Y., Pan, J., Wang, D., Chen, G., Sun, L. and Yao, J. (2016). Generation of near-inertial oscillations by summer monsoon onset over the South China Sea in 1998 and 1999, *Deep Sea Res. Part I: Oceanogr. Res. Pap.*, 118, pp. 10–19, doi:10.1016/j.dsr.2016.10.008.

Sun, L., Zheng, Q., Wang, D., Hu, J., Tai, C.-K. and Sun, Z. (2011a). A case study of near-inertial oscillation in the South China Sea using mooring observations and satellite altimeter data, *J. Oceanogr.*, 67, pp. 677–687, doi:10.1007/s10872-011-0081-9.

Sun, Z., Hu, J., Zheng, Q. and Li, C. (2011b). Strong near-inertial oscillations in geostrophic shear in the northern South China Sea, *J. Oceanogr.*, 67, pp. 377–384, doi:10.1007/s10872-011-0038-z.

Whitt, D. B. and Thomas, L. N. (2015). Resonant generation and energetics of wind-forced near-inertial motions in a geostrophic flow, *J. Phys. Oceanogr.*, 45, pp. 181–208, doi:10.1175/JPO-D-14-0168.1.

Xiao, J., Xie, Q., Wang, D., Yang, L., Shu, Y., Liu, C., Chen, J., Yao, J. and Chen, G. (2016). On the near-inertial variations of meridional overturning circulation in the South China Sea, *Ocean Sci.*, 12(1), pp. 335–344, doi:10.5194/os-12-335-2016.

Xie, X., Liu, Q., Shang, X., Chen, G. and Wang, D. (2016). Poleward propagation of parametric subharmonic instability-induced inertial waves, *J. Geophys. Res. Oceans*, 121(3), pp. 1881–1895, doi:10.1002/2015JC011194.

Xu, Z., Yin, B., Hou, Y. and Xu, Y. (2013). Variability of internal tides and near-inertial waves on the continental slope of the northwestern South China Sea, *J. Geophys. Res.*, 118, pp. 197–211.

Yang, B., Hu, P. and Hou, Y. (2021). Observed near-inertial waves in the Northern South China Sea, *Rem. Sens.*, 13, 3223, doi:10.3390/rs13163223.

Zhang, H., Chen, D., Zhou, L., Liu, X., Ding, T. and Zhou, B. (2016). Upper ocean response to Typhoon Kalmaegi (2014), *J. Geophys. Res. Oceans*, 121(8), pp. 6520–6535, doi:10.1002/2016JC012064.

Zhang, Z., Zhang, Y., Wang, W. and Huang, R. (2013). Universal structure of mesoscale eddies in the ocean, *Geophys. Res. Lett.*, 40, pp. 3677–3681, doi:10.1002/grl.50736.

Zheng, Q., Xie, L., Xiong, X., Hu, X. and Chen, L. (2020). Progress in research of submesoscale processes in the South China Sea, *Acta Oceanol. Sin.*, 39(1), pp. 1–13, doi:10.1007/s13131-019-1521-4.

Zhou, L. and Hong, W. (2022). The propagation and dissipation of wind-generated near-inertial waves: a review, *Adv. Mar. Sci.*, 40(4), pp. 763–776.

Zhu, D. and Li, L. (2007). Near inertial oscillations in shelf-break of northern South China Sea after passage of typhoon Wayne, *J. Trop. Oceanogr.*, 26(4), pp. 1–7.

Chapter 16

Mesoscale Eddy Modulation to Internal Solitary Waves

16.1 Introduction

The South China Sea (SCS) is featured by coexistence of energetical mesoscale eddies (see Parts 2–4 of this book for details) and intensive internal solitary waves (ISWs) (Cai *et al.*, 2012; Zheng, 2017) due to its closeness of water body and complexity of bottom topography. For the regional circulation research, it is of significance to clarify interaction mechanisms and implications of the two processes with different scales and physical properties.

Zheng *et al.* (2007) derived a soliton amplitude growth ratio (SAGR) as the ISW propagates along a shoaling thermocline from the PKdV equation. The solution was used to explain the seasonal variability of ISW occurrence probability, and the spatial variation of mooring observed ISW amplitudes as passing over mesoscale eddies in the SCS (Huang *et al.*, 2017; Xu *et al.*, 2020).

Liao *et al.* (2012) investigated how a mesoscale eddy induces deformation of the ISWs. They used one-dimensional nonlinear variable-coefficient extended KdV (eKdV) equation to simulate an ISW passing through the mesoscale eddy field. The results show that the mode vertical structures of the ISWs are modified by the mesoscale eddy field. Meanwhile, anticyclonic eddy (AE) and cyclonic eddy (CE) have different influences on ISW propagation.

Xie *et al.* (2015) presented the visual evidence for deformed crestlines of ISWs after passing over an AE on concurrent synthetic aperture radar (SAR) image and satellite altimeter sea level anomaly (SLA) image in the northeastern SCS on 9 May 2001. To explore the dynamic mechanisms behind the phenomena shown on satellite images, they further carried out numerical experiments using the fully nonhydrostatic and nonlinear MITgcm model to examine the evolution process of a single ISW induced by perturbations of isolated AE and CE.

Li *et al.* (2016) investigated the effects of the Kuroshio and mesoscale eddies on the ISWs, using hybrid coordinate ocean model (HYCOM) reanalysis simulation. The results show that the Kuroshio enhances the zonal tilt of the thermocline and induces intruding flow in the Luzon Strait. The eddies change local stratification and induce additional background currents, thus affecting the nonlinear evolution of internal tides. In addition, the mesoscale eddies dramatically affect the observed amplitude changes in the ISWs.

Xie *et al.* (2016) reported a case of deformed crestlines of ISWs in the northeastern SCS shown on SAR image and used the nonlinear refraction model (NRM) to simulate effects of realistic bottom topography, current and stratification on the ISW propagation. Simulated results show that deformation of ISW crestlines is mainly caused by the strong current shear at the periphery of AE.

Huang *et al.* (2017, 2018) examined impacts of mesoscale eddies on the ISWs that evolved from the steepening internal tide generated in the Luzon Strait and propagated in the northern SCS (NSCS) deep basin, using the cross-mooring array data from November 2013 to January 2014. During the observation period, an energetic mesoscale eddy pair consisting of one AE and one CE propagated across the mooring array. Observation results reveal that the amplitude, propagation direction and speed of the ISWs are significantly modulated by the eddy pair.

Xu *et al.* (2020) observed that the deepened thermocline of an AE damped the ISW amplitudes using the measurements of a single mooring deployed in the NSCS in August and September 2014. They found that the ISW amplitude reduced to nearly zero as passing over the eddy center, while increased to the maximum at the eddy edge.

Yang *et al.* (2022) investigated the three-dimensional (3D) structures of ISWs and their temporal variability using measurements by a linear mooring array of five subsurface moorings deployed along the direction of ISW crestlines in the NSCS from October 2013 to June 2014. They found that the ISW crestlines could deform into convex, concave and S-shaped patterns caused by mesoscale eddies and the Kuroshio intrusion. Meanwhile, the intensities of 3D ISWs could temporally vary out of phase between the northern and southern portions of crests due to modulations by intensive eddies.

Sun *et al.* (2022) statistically analyzed modulation of mesoscale eddies to the propagation directions of ISWs in the SCS using satellite observations from 2010 to 2016. Their results show that spatially the co-occurrence of eddies and ISWs is concentrated along the northern continental slope and the western boundary of the SCS off the eastern Indo-China Peninsula, and seasonally from March to October. The purpose of this chapter is to summarize progress in observation and research on mesoscale eddy modulation to the ISWs in the SCS.

16.2 Evidence of Modulation of Eddy to ISWs in the SCS

16.2.1 *Satellite observations*

16.2.1.1 *RADARSAT ScanSAR observation on 9 May 2001*

Figure 16.1(a) and 16.1(b) show a RADARSAT ScanSAR image and concurrent satellite sea level anomaly (SLA) image of the northeastern SCS on 9 May 2001. The thick bright lines on the SAR image are interpreted as the imagery of ISW packets (Zheng *et al.*, 2001; Liu *et al.*, 2004; Xie *et al.*, 2015). The leading soliton crestline is as long as 230 km. Based on the separation distance between two adjacent ISW packets, the orientation of the wave crests and the assumption that the generation mechanism is the semidiurnal internal tide, the wave speed and propagation direction are estimated from the image as 1.7 m s^{-1} toward 298° true north (Liu *et al.*, 2004). From the SLA image, one can see that the ISW packets are located at the shortest distance of about 100 km from the AE center at 20°30″N, 119°10″E and within the velocity shear zone of the AE and the shoaling bottom topography (also shoaling thermocline) from 3000 m to 1000 m. Thus, it is reasonable to attribute deformation of ISW crestlines to the modulation of AE and the nonlinear effects of shoaling bottom topography.

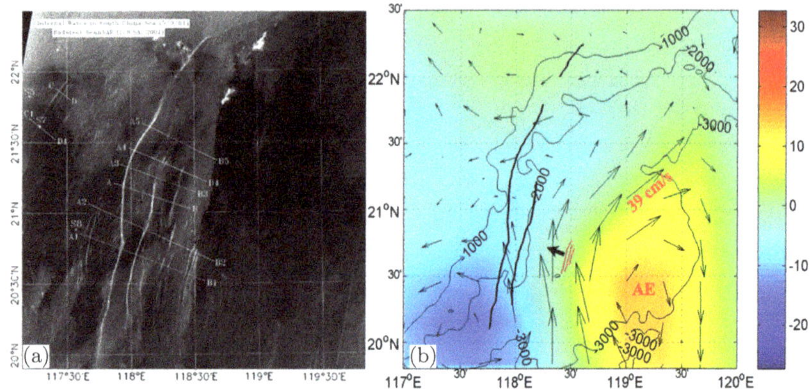

Fig. 16.1. Satellite images of ISW packets and a concurrent AE in the northeastern SCS. (a) The RADARSAT ScanSAR image centered at around 21°10′N, 118°30′E acquired at 21:57:34 UTC on 9 May 2001. Bright lines are the imagery of crestlines of westward-propagating ISW packets (Liu *et al.*, 2004). (b) The SLA image (color codes in cm) with the geostrophic current field (black arrows in cm s^{-1}). The AE with a peak SLA of about 15 cm centered at 20°30′N, 119°10′E co-exists with the ISW packets (thick black lines and thin short red lines). Thin black lines show isobaths with numbers in m.

Source: Taken from (a) Liu *et al.* (2004) and (b) Xie *et al.* (2015).

16.2.1.2 *Envisat ASAR observation on 18 June 2008*

A case of multiple eddy modulation to a single ISW is shown in Figure 16.2(a). One can see an Envisat advanced synthetic aperture radar (ASAR) image acquired at 02:13:30 UTC on 18 June 2008 superimposed on a concurrent satellite SLA image of the northeastern SCS. The ASAR image is centered at around 21°00′N, 116°45′E near Dongsha Atoll on the NSCS continental shelf. The image shows an S-shaped, thick bright line, which is interpreted as the imagery of a single ISW (Xie *et al.*, 2016). The horizontal length of soliton crestline is as long as 200 km. On the other hand, the SLA image shows that the single soliton is surrounded by three mesoscale eddies, AE1 and CE on the east side and AE2 on the west side, thus, its S-shaped pattern results from combined modulation effects of three eddies. Reconstructed current fields along sections 20.23°N and 21.43°N (Figure 16.2(b) and 16.2(c)) using the method by Wang *et al.* (2013) show that the water depths are from about 400 m to 1000 m, and thermocline depths are about 300 m at the single soliton

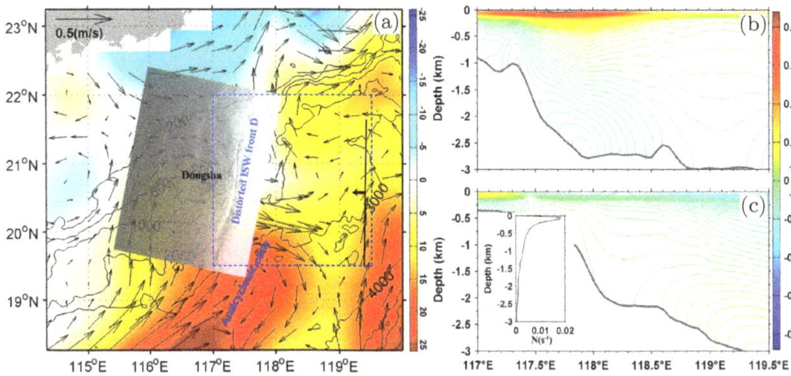

Fig. 16.2. (a) An Envisat ASAR image acquired at 02:13:30 UTC on 18 June 2008 superimposed on concurrent satellite SLA image in the northeastern SCS. Color codes of SLA are in cm. Black arrows represent the sea surface geostrophic current vectors in $\mathrm{m\,s^{-1}}$. Thin black lines with numbers represent isobaths in m. (b) Reconstructed current fields along section 20.23° N. Dash line on the top left shows the single soliton position. (c) Same as (b), but for section 21.43° N.
Source: Taken and re-drawn from Xie *et al.* (2016).

location. The protrusion features of current fields at around 117.4°E further confirm the existence of CE.

16.2.1.3 *RADARSAT-2 SAR observation on 25 June 2013*

Figure 16.3 shows an image of a RADARSAT-2 satellite SAR image acquired at 10:16:43 UTC on 25 June 2013 superimposed on concurrent satellite SLA image in the northeastern SCS. The SAR image is centered at around 21°00′N, 118°30′E west of the Luzon Strait. The image shows two arc-shaped, thick bright lines, which are interpreted as the imagery of leading soliton crestlines in two ISW packets near the Luzon Strait (Sun *et al.*, 2022). The horizontal length of soliton crestlines is as long as 200 km. Based on the separation distance of 58.2 km between two ISW packets, the orientation of the wave crests and the assumption that the generation mechanism is the semidiurnal internal tide, the wave speed and direction are estimated from the image as $1.3\,\mathrm{m\,s^{-1}}$ and toward 285° true north (WNW). On the other hand, the SLA image shows that the two ISW packets are within the current shear zone of an intensive AE with a peak SLA of 0.5 m centered at around 20°00′N, 119°00′E. The arc shaped ISW crestlines coincide with the sea surface geostrophic rotating current

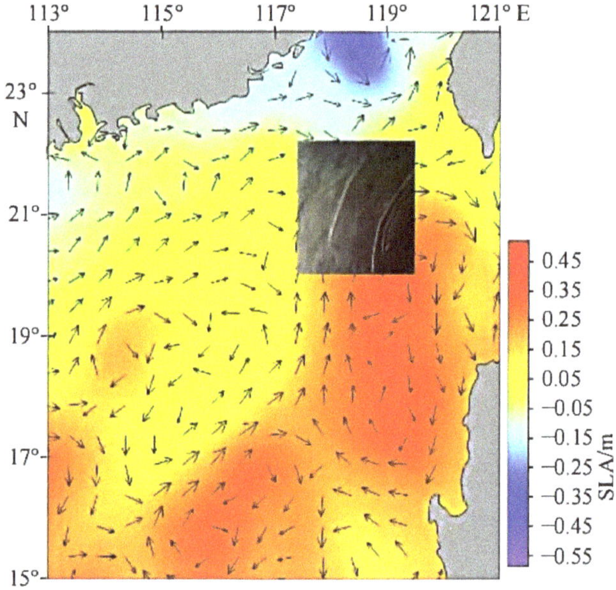

Fig. 16.3. A RADARSAT-2 satellite SAR image acquired at 10:16:43 UTC on 25 June 2013 superimposed on concurrent satellite SLA image in the northeastern SCS. The SAR image is centered at around 21°00′N, 118°30′E west of the Luzon Strait. Color codes of SLA are in m. Black arrows represent the sea surface geostrophic current vectors in $\mathrm{m\,s^{-1}}$. AE labels an anticyclonic eddy concurrent with the ISW packets on the SAR image.

Source: Taken and re-labeled from Sun *et al.* (2022).

field of the AE, implying that the ISW crestlines are modulated by the AE.

16.2.2 *Mooring observations*

16.2.2.1 *Cross-mooring array survey*

Previous investigations have confirmed that the intensive ISWs observed in the northeastern SCS are generated by the nonlinearized wave fronts of internal tides radiating from the Luzon Strait (Zhao *et al.*, 2004; Zhao, 2014). Therefore, the crestline patterns and occurrence positions of ISWs must coincide with that of wave fronts of internal tides. Huang *et al.* (2018) examined the role of mesoscale eddies in modulating the semidiurnal internal tide (SIT), so that in

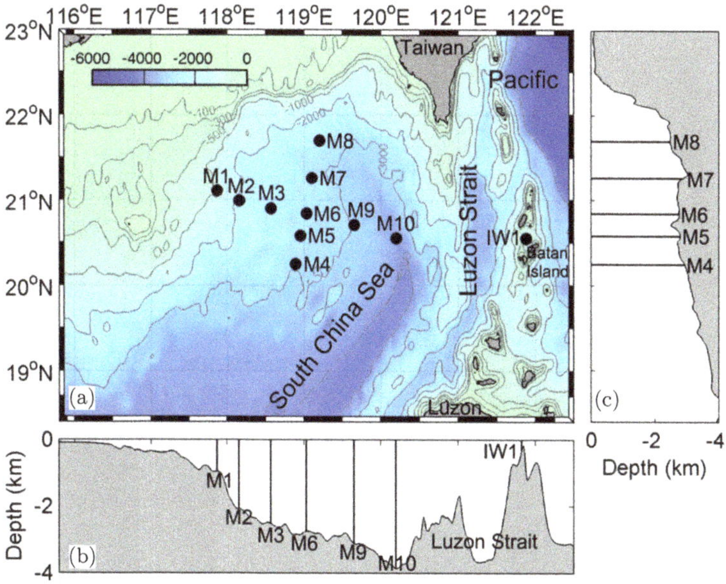

Fig. 16.4. Cross-mooring array survey area for the study of mesoscale eddy modulating the SIT in the northeastern SCS and the Luzon Strait from November 2013 to January 2014. (a) Bottom topography. Color codes of isobaths are in m. Solid circles denote locations of moorings M1 to M10 and IW1. (b) Bathymetric profile along the line of moorings M1-M3, M6, M9-M10 and IW1 deployed along the SIT beam direction. (c) The same as (b) but across the SIT beam direction (M4 to M8).

Source: Taken from Huang *et al.* (2018).

modulating the crestline patterns and occurrence positions of ISWs, using the data surveyed by a cross-mooring array consisting of 11 moorings M1 to M10 and IW1 in the NSCS from November 2013 to January 2014 as shown in Figure 16.4. The survey area is centered at 21°00′N, 119°00′E between the Luzon Strait and the NSCS continental shelf. Statistics have indicated that the area is an initial source region for internal tides (Zhao, 2014), mesoscale eddies and intensive ISWs (Zheng *et al.*, 2008, 2020; Huang *et al.*, 2016). Thus, it is an ideal place for exploring the interaction mechanisms among the above processes with different length scales.

During the observation period, an AE-CE pair crossed the survey area and met the westward SIT beam originating from the Luzon Strait. A time series of the evolution process of SIT wave crest patterns are shown in Figure 16.5. In the figure, the information of mesoscale eddies is shown by SLA images with the sea surface geostrophic current vectors. The mode-1 SIT wave component obtained from the harmonic analysis of the tidal data measured by the ADCP at mooring IW1 deployed near the Batan Island at the middle underwater ridge across the Luzon Strait, which is believed as the primary radiation source of ITs in the NSCS. Owing to the current and stratification modulations by the eddy pair, the propagation speed of the mode-1 SIT speed up (slow down) by up to $0.7\,\mathrm{m\ s^{-1}}$ ($0.4\,\mathrm{m\,s^{-1}}$) within the southern portion of AE (CE). As a result of the spatially varying phase speeds, the mode-1 SIT wave crests clockwise (counterclockwise) rotate within the AE (CE) core and exhibit convex and concave (concave and convex) patterns on the southern and northern peripheries of AE (CE), respectively. The SIT energy and flux are computed by the same procedure as described in Nash *et al.* (2005) and Zhao *et al.* (2010, 2012). As above mentioned, the intensive ISWs in the northeastern SCS are initiated by the nonlinearized wave fronts of internal tides radiating from the Luzon Strait, i.e., the crestline patterns and occurrence positions of wave fronts of internal tides coincide with that of ISWs. Thus, the observation results are significant for development of the local ISW prediction models.

16.2.2.2 *Linear mooring array survey*

Yang *et al.* (2022) examined the 3D structures of ISWs, and their temporal variability modulated by mesoscale eddies. The data they used were collected from a linear mooring array consisting of five moorings M4-M8 from October 2013 to June 2014. Figure 16.6 shows 3D bottom topography of their survey area and configuration of mooring system. One can see that the five moorings were deployed along the longitudinal line of 119°E centered at 21°N across the middle continental slope of the northeastern SCS deep basin, where the depths are about 2000 m.

Figure 16.7(a–c) and 16.7(e) show vertical profiles of daily average zonal currents (ISW signals filtered out) derived from measurements at moorings M8–M6 and M4 and temperature at mooring M5 during the survey period from 1 November 2013 to 1 June 2014 (sea surface data obtained from the E.U. Copernicus Marine Service). One can

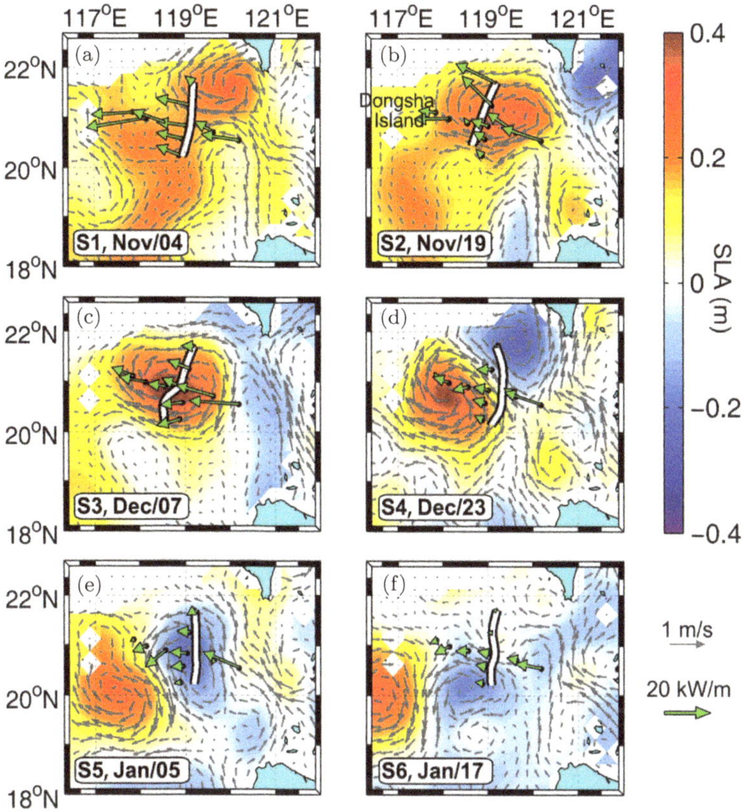

Fig. 16.5. A time series of the evolution process of SIT wave crest patterns modulated by an AE-CE pair in the northeastern SCS from 4 November 2013 to 17 January 2014. White curves denote the mode-1 SIT wave crest patterns. SLA color codes are in m. The gray arrows denote the absolute geostrophic currents (in m s^{-1}) derived from the Archiving, Validation and Interpretation of Satellite Oceanographic (AVISO) data. Green arrows denote depth-integrated energy flux vectors of the mode-1 SIT.

Source: Taken from Huang *et al.* (2018).

see that the background dynamic processes with multiple time scales are extremely active. Figure 16.7 S1–S4 show ISW crestlines derived from mooring observations superimposed on concurrent satellite SLA images derived from the AVISO data during four spring tide periods S1 on 21 November 2013, S2 on 22 December 2013, S3 on 3 February 2014 and S4 on 19 March 2014, respectively. One can see that

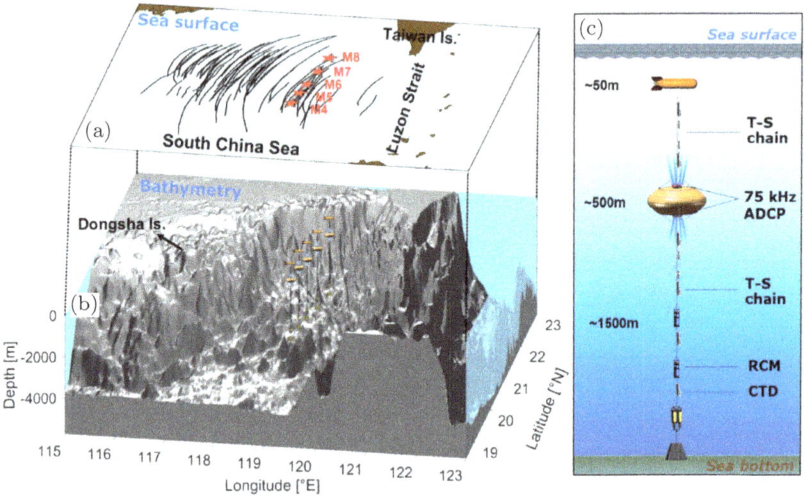

Fig. 16.6. (a) Sea surface distribution of the linear mooring array. Five moorings M4-M8 (red stars) were deployed along the longitudinal line of 119°E. Black lines represent the leading ISW crestlines extracted from the SAR images from August 2010 to October 2013. (b) 3D bottom topography in the northeastern SCS and the Luzon Strait. Mooring depths are around 2000 m. (c) Schematic configurations of the mooring system.

Source: Taken from Yang *et al.* (2022).

during the observation period, single AE or CE or AE–CE pair propagated through the linear mooring array from northeast to southwest (Huang *et al.*, 2017). The observations indicate that the AE and CE deepen and uplift the thermocline by more than 70 m at the moorings and intensify background current velocities up to $\pm 0.8\,\mathrm{m\,s^{-1}}$ as shown in Figure 16.7(a) and 16.7(e). Modulated by spatially varying eddy shear flows, the wave fronts of internal tides, i.e., ISW crestlines patterns, are deformed as convex (Figure 16.7(S1)), concave (S2 and S3) and S-shaped (S4).

Figure 16.8 shows 3D structures of convex, S-shaped, and concave ISWs reconstructed using the data measured by the linear mooring array at depths of 200, 500, 1000, 1500 and 2000 m on 13 May, 2 November and 21 December 2014, respectively. One can see that the ISW crestlines sustain the similar patterns vertically, but the amplitudes change along the crest directions. In the case of convex ISW, the maximum (minimum) amplitude is 123 (44) m at M5 (M8). The average amplitude at the southern moorings M4 and M5 is 47%

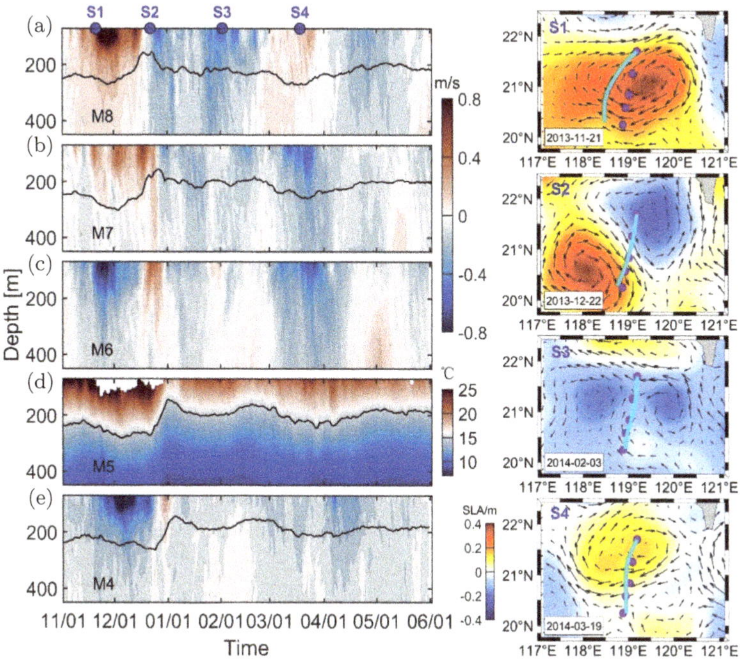

Fig. 16.7. Linear mooring array observed time series of (a)–(c) and (e) vertical profiles of daily average zonal currents (in $\mathrm{m\,s^{-1}}$) at moorings M8–M6 and M4 and (d) temperature (in °C) at mooring M5. Black curves in (a), (b), (d) and (e) represent the 15°C isotherm. Blue solid circles on top of (a) mark four spring tide periods S1–S4. The dates along horizontal axis started from 1 November 2013 and ended on 1 June 2014. (S1)–(S4) ISW crestlines (thick cyan lines) derived from mooring observations superimposed on concurrent satellite SLA images during four spring tide periods S1–S4. Color codes of SLA are in m. Black arrows represent the sea surface geostrophic current vectors in $\mathrm{m\,s^{-1}}$. Purple dots mark locations of moorings M4–M8 ordering northward.

Source: Taken from Yang *et al.* (2022).

larger than that at the northern M7 and M8. In the cases of convex and S-shaped, the ISW widths appear as a narrowing trend from south to north, while opposite trend in the case of concave ISW.

16.2.2.3 *Single mooring survey*

Xu *et al.* (2020) observed mesoscale eddy modulation to the ISWs using the data surveyed by a single mooring in August and September 2014. The mooring station was at 20°44.2′N, 117°44.7′E

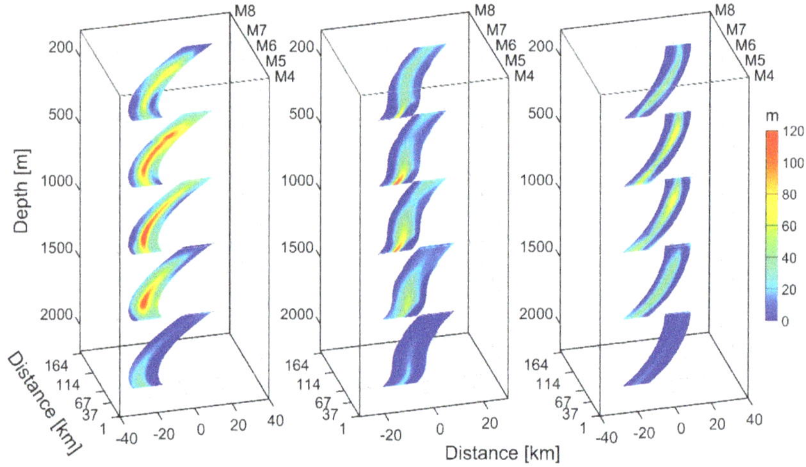

(a) (b) (c)

Fig. 16.8. Linear mooring array observed 3D structures of (a) convex, (b) S-shaped and (c) concave ISWs at depths of 200, 500, 1000, 1500 and 2000 m on 13 May, 2 November and 21 December 2014, respectively. Color codes mark the vertical amplitudes of ISWs in m.

Source: Taken from Yang *et al.* (2022).

on the continental slope of the northeastern SCS deep basin about 130 km east of Dongsha Atoll. The water depth is 1200 m as shown in Figure 16.9(a). From Figure 16.9(a) and 16.9(b), one can see that an AE with the horizontal length scale of $O(150)$ km traveled southwestward and crossed over the mooring with its northern portion from 23 September to 28 September 2014. Meanwhile, four ISW packets traveled westward and arrived at shear flow zone of the AE from 25 September to 28 September 2014. Thus, the mooring caught the best opportunity to detect the detailed modulation processes of the AE to the ISWs.

On 23 September 2014, the mooring was close to the eddy center. Observations show that the depth of 20°C isotherm was 183 m, the background current velocity was 1 m s^{-1}, and no ISW signals were detected. In comparison to the case of no eddy on 26 August 2014, the amplitude of the leading ISW reached 120 m and the background current velocity was 1.98 m s^{-1}. On 25 September 2014, the mooring was about 40 km from the eddy center, the depth of 20°C isotherm

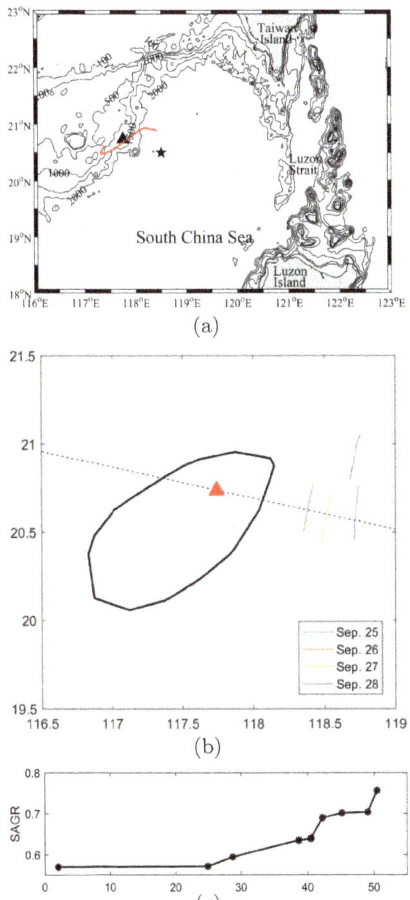

Fig. 16.9. (a) Mooring location and bottom topography in the northeastern SCS. Numbers on isobaths are in m. Solid triangle marks the mooring location 20°44.2′N, 117°44.7′E. Red line shows the track of the AE center from 16 September to 28 September 2014. (b) Color lines show upstream leading ISW crestlines taken from Aqua/Terra satellite moderate resolution imaging spectroradiometer (MODIS) images from 25 September to 28 September 2014. Red triangle marks the mooring station. Black contour line is the outermost closed SLA of the AE on 25 September 2014. Dashed line denotes the propagation direction of ISWs. (c) Soliton amplitude growth rate (SAGR) vs the distance between the eddy center and ISWs.

Source: Taken and re-drawn from Xu *et al.* (2020).

rose to 100 m. The single ISW amplitude was 80 m and the background current velocity reached 1.5 m s^{-1}. In comparison to the case of no eddy on 28 August 2014, the two parameters were 110 m and 2.08 m s^{-1}. On 27 September 2014, the mooring was located near the eddy edge about 50 km from the eddy center. The depth of 20°C isotherm was 75 m, the amplitude of the leading ISW reached 170 m and the background current velocity reached 1.6 m s^{-1}. In comparison to the case of no eddy on 30 August 2014, the two parameters were 120 m and 1.5 m s^{-1}. In summary, when the ISWs pass from the eddy center to the edge, the amplitude of the leading ISW rises from nearly zero to 170 m, and the wave-induced background current velocities increase with the amplitude increase.

Figure 16.9(c) shows the most important result on the mesoscale eddy modulation to the ISWs, i.e., the increase trend of ISW amplitude with eddy thermocline shoaling. The dynamic analysis for this phenomenon will be given in the next section based on Zheng *et al.* (2007).

16.3 Theoretical Solutions and Modeling

16.3.1 *PKdV equation solutions*

From the perturbed Korteweg-de Vries (PKdV) equation (Newell, 1985), Zheng *et al.* (2007) derived the theoretical solution for the problem how does a shoaling thermocline force the ISW amplitude growing? The physical model is shown in Figure 16.11, in which the x-axis is westward positive, and the y-axis is upward positive. The model ocean is vertically stratified into two layers by a thermocline. The deeper layer has an infinite depth. The upper layer depth is a constant h_0 for $x < x_0$, and variable as $h_0 - H(x)$ for $x \geq x_0$. An initial solitary disturbance with an amplitude N and a width l is formed at $x = x_0$ and propagated along the thermocline. Based on the observations from the northeastern SCS, the characteristic scales are analyzed as follows. The depth of 15°C isotherm is defined as the depth of thermocline. The scale of upper layer depth is determined as $h_0 = O(200\,\text{m})$. From SAR images, the characteristic half width (horizontal length scale) of the leading soliton is determined as $l = O(2\,\text{km})$ (Zheng *et al.*, 2001; Yuan *et al.*, 2006). The amplitude scale of soliton is estimated as $a = O(30)\,\text{m}$ (Zheng *et al.*, 2001). Thus, these characteristic scales satisfy the scale balance relations $h_0^2/l^2 = \varepsilon = 0.01 \ll 1$, and $a/h_0 = \mu = 0.15 \ll 1$. In this case, the soliton amplitude satisfies the PKdV equation (Newell, 1985),

$$q_\tau + 6qq_\theta + q_{\theta\theta\theta} = -\frac{9}{4}\frac{D_\tau}{D}q, \tag{16.1}$$

where τ is a rescaled distance coordinate,

$$\tau = \frac{1}{6}\int^X D^{1/2}dX, \tag{16.2}$$

θ is the retarded time,

$$\theta = -t + \frac{1}{\epsilon}\int^X \frac{dX}{D^{1/2}}, \tag{16.3}$$

where t is the time, D is the nondimensional depth of the upper layer, $D = 1 - h$, in which $h = H(x)/h_0$, $X = \varepsilon x$, and the subscript represents partial differential with respect to that variable.

Solutions of (16.1) can be expressed as

$$q\,(\theta,\tau) = q^{(0)} + \sigma q^{(1)} + \cdots, \tag{16.4}$$

where

$$q^{(0)} = 2\eta^2\text{sech}^2\eta\left(\theta - 4\eta^2\tau - \theta_0\right), \tag{16.5}$$

in which η is the amplitude parameter, a slowly varying function of the distance, and θ_0 is the initial retarded time, and

$$\begin{aligned} q^{(1)} &= \frac{\Gamma}{6\eta}\{-1 + \tanh\eta\xi + \text{sech}^2\eta\xi(3 - 3\eta\xi\tanh\eta\xi + 2\eta\xi \\ &\quad - 2\eta^2\xi^2\tanh\eta\xi)\} + \frac{\Gamma}{4}(1 - \tanh\eta\xi)\text{sech}^2\eta\xi, \end{aligned} \tag{16.6}$$

where

$$\Gamma = \frac{9}{4}\frac{D_\tau}{D}, \tag{16.7}$$

and $\xi = \theta - \bar{\theta}$ is a time variable in a coordinate system moving with the solitary wave, and σ is of the order of Γ [$= O(0.3)$ in the studied case] satisfying $0 < \varepsilon \ll \sigma \ll 1$. Solutions (16.5) and (16.6) represent a large soliton with a small oscillatory tail. As seen later, the initial solitary disturbance will grow and gradually evolve into a solitary wave packet as propagating along the shoaling thermocline.

The solvability of equation (16.1) demands

$$\frac{\partial \eta}{\partial \tau} = -\frac{2}{3}\Gamma \eta, \tag{16.8}$$

Transforming (16.8) into the x-y coordinates yields

$$\frac{d\eta}{dx} = \left[-\frac{3}{2D(x)}\frac{dD(x)}{dx}\right]\eta. \tag{16.9}$$

The solution of (16.9) is

$$\eta = \eta_0 D(x)^{-3/2}, \tag{16.10}$$

where η_0 is the initial solitary disturbance amplitude at x_0. Defining the soliton amplitude growth ratio (SAGR) as η/η_0 yields

$$\text{SAGR} = D(x)^{-3/2}. \tag{16.11}$$

Comparing to parameters defined in the physical model shown in Figure 16.10, the SAGR is calculated by dimensional parameters,

$$\text{SAGR} = \left(\frac{h_0}{d}\right)^{3/2}, \tag{16.12}$$

where h_0 and $d[= h_0 - H(x)]$ are the upper layer depths on the deep side and shallow side of the shoaling thermocline.

Based on solution (16.11), the behavior of the ISW amplitudes modulated by the thermoclines of normal AE and CE models can be predicted as schematically shown in Figure 16.11. One can see that the ISW amplitudes experience a dropping — rising process during passing over a normal AE, while a reverse process during passing over a normal CE.

Yang *et al.* (2022) measured the spatial variation of ISW amplitudes along the crestlines using the linear mooring array as shown in Figure 16.7. One can see that cases S1 on 21 November 2013 and S4 on 19 March 2014 show the leading ISW crestlines across AEs centered at 20°00′N, 119°30′E and 21°30′N, 119°00′E, respectively. The

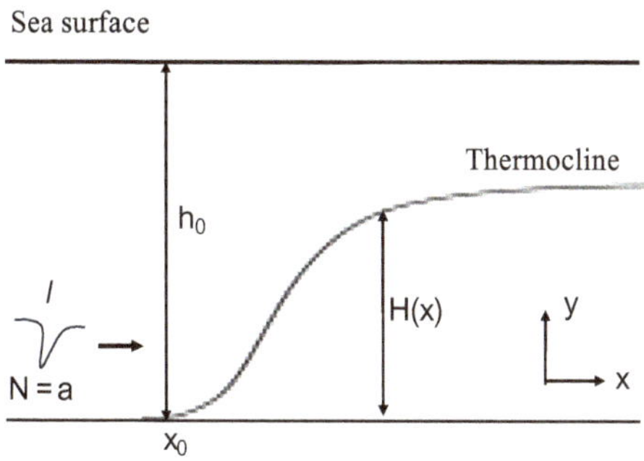

Fig. 16.10. A schematic of the physical model for the ISW to propagate along a shoaling thermocline. The x-axis is westward positive, and the y-axis is upward positive.

Source: Taken and re-drawn from Zheng *et al.* (2007).

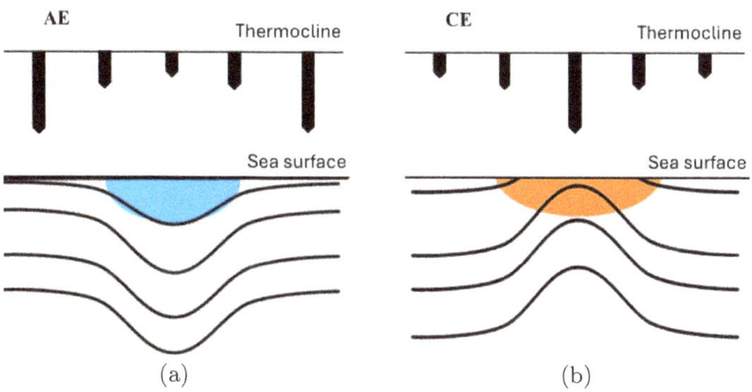

Fig. 16.11. Schematics of behavior of the ISW amplitudes modulated by the thermoclines of (a) normal AE and (b) CE models.

ISW amplitudes measured by the linear mooring array vs the radial distances from the eddy centers measured from the SLA images are shown in Figure 16.13(a). One can see the increase trends of ISW amplitudes from the eddy centers to the eddy edges, which qualitatively validate the prediction by solution (16.12) derived from the PKdV equation as schematically shown in Figure 16.12(a).

Meanwhile, the speed of the single soliton has the form of

$$v = c + \frac{\alpha \eta_0}{3}, \tag{16.13}$$

where c is the linear phase speed, α is the nonlinear coefficient and η_0 is the soliton amplitude (Zheng *et al.*, 2001). From (16.13), the

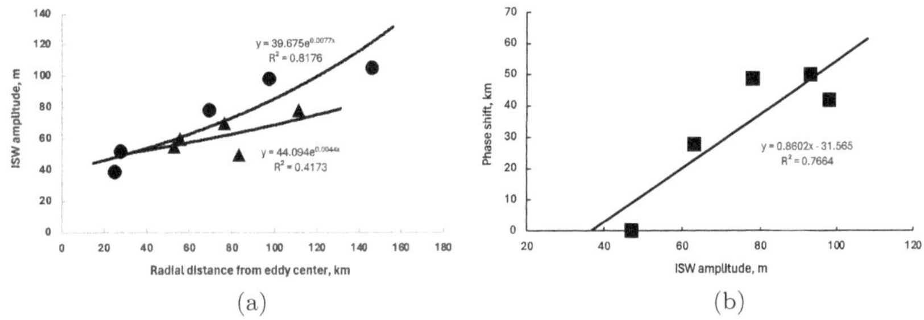

Fig. 16.12. (a) ISW amplitudes measured by the linear mooring array vs the radial distances from the eddy centers measured from SLA images. Solid triangles (dots) and fitted curve represent case S1 (S4) on 21 November 2013 (19 March 2014). (b) Phase shifts measured from the SLA images vs the ISW amplitudes measured by the linear mooring array in case S1.
Source: Data taken from Yang *et al.* (2022).

traveling distance of the soliton within time interval T is derived as

$$s = \int_0^T v dt = cT + \frac{\alpha T}{3}\eta_0, \tag{16.14}$$

in which c, T and α can be treated as constants, and η_0 is a 2D spatial distribution function. Thus, s is a 2D spatial distribution function of the phase shift linearly related to η_0, i.e., the crestline patterns. Namely, solution (16.11) can be used to construct the ISW crestline patterns during passing over mesoscale eddies. Figure 16.12(b) shows leading ISW phase shifts measured from the SLA images vs the ISW amplitudes measured by the linear mooring array in case S1, which validates the linear relation (16.14).

16.3.2 *Nonlinear refraction models*

Small (2001) proposed the NRM to describe shoaling and refraction of the ISWs in the ocean. The model is based on the KdV equation in the form of

$$\frac{\partial \eta}{\partial t} + c(x)\frac{\partial \eta}{\partial x} + \alpha(x)\eta\frac{\partial \eta}{\partial x} + \beta(x)\frac{\partial^3 \eta}{\partial x^3} + \nu(x)\eta + S(x)\eta = 0, \tag{16.15}$$

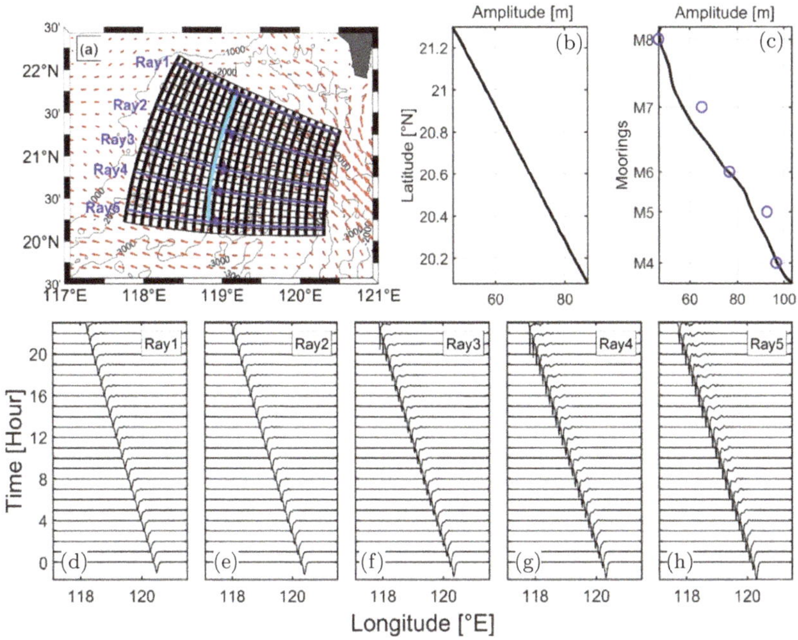

Fig. 16.13. (a) The NRM calculated results of ray paths and reconstructed 2D ISW crestline for an average current field in the northeastern SCS from October 2013 to June 2014. Thick and thin black lines represent ISW crestlines and radial rays, respectively. Five blue lines Rays 1–5 pass through five along-crest moorings marked by purple pentagrams. Thick cyan curve shows the observed wave crestline averaged over the whole observation period. Red vectors are the surface currents obtained from the HYCOM dataset averaged over the whole mooring observation period. (b) Initial wave amplitudes along the crestline. (c) Calculated along-crest wave amplitudes vs observed wave amplitudes averaged over the whole observation period (blue circles) at the moorings. (d)–(h) Evolution processes of the ISWs along Rays 1–5.

Source: Taken from Yang *et al.* (2022).

where η is the ISW amplitude, x is the spatial range along a ray path, c is the linear phase speed determined by solving the eigenvalue problem,

$$\frac{d}{dz}\left[(c-U)^2\frac{d\phi}{dz}\right]+N^2(z)\phi=0, \quad \phi(-H)=\phi(0)=0, \qquad (16.16)$$

where ϕ is the eigenfunction, U and N are the background current velocity and buoyancy frequency, z is the vertical coordinate upward positive, H is the water depth, α is the nonlinear coefficient calculated via (Grimshaw *et al.*, 2004)

$$\alpha = \frac{3}{2} \frac{\int_{-H}^{0} (c-U)^2 \left(\frac{d\phi_0}{dz}\right)^3 dz}{\int_{-H}^{0} (c-U) \left(\frac{d\phi}{dz}\right)^2 dz}, \tag{16.17}$$

β is the dispersion parameter calculated via

$$\beta = \frac{1}{2} \frac{\int_{-H}^{0} (c-U)^2 \phi^2 dz}{\int_{-H}^{0} (c-U) \left(\frac{d\phi}{dz}\right)^2 dz} \tag{16.18}$$

$\nu(x)$ summarizes effects of spatially varying environmental parameters and is defined as

$$\nu(x) = \frac{c}{Q} \frac{dQ}{dx}, \tag{16.19}$$

where Q is defined as the amplification factor calculated via

$$Q = \frac{c_0^2 \int_{-H}^{0} (c_0 - U_0) \left(\frac{d\phi_0}{dz}\right)^2 dz}{c^2 \int_{-H}^{0} (c-U) \left(\frac{d\phi}{dz}\right)^2 dz}, \tag{16.20}$$

where subscript 0 represents a fixed position x_0 on the initial ISW crestline, and $S(\text{x})$ is the spreading term defined as

$$S(x) = \frac{c}{2\Delta} \frac{d\Delta}{dx}, \tag{16.21}$$

where Δ is the separation distance between rays along the ISW crestline.

The idea to use the NRM for predicting the evolution of the ISWs as propagating in a spatially varying background current field, such as mesoscale eddies, is to reduce the horizontal 2D wave motion problem to the 1D problem, i.e., to slice the initial ISW crestline into a group of fine pieces that serve as the radiation ray sources and then use the NRM to predict the behavior of wave radiation along the rays. The new 2D ISW patterns and parameters at any time can be obtained by the data taken from the radiation rays.

The NRM has been successfully used to examine the detailed modulation processes of mesoscale eddies to the ISWs in the northeastern SCS (Xie *et al.*, 2016; Huang *et al.*, 2018; Yang *et al.*, 2022). As an example, Figure 16.13 shows the NRM calculated results of ray paths and reconstructed 2D ISW crestline for an average current field in the northeastern SCS derived from the HYCOM dataset from October 2013 to June 2014 (Yang *et al.*, 2022). One can see that the NRM calculated ISW amplitudes fit the observed data quite well (Figure 16.13(c)). Reconstructed horizontal 2D ISW crestline appears as a convex pattern. In comparison to the initial ISW, the reconstructed ISW crestline length at the moorings spread 1.4 times and amplitudes amplified about 10%, implying that the ISW was at the growth stage. These calculated results coincide with the general features of ISWs in the source region west of the Luzon Strait (Zhao and Alford, 2006; Zheng *et al.*, 2007; Cai *et al.*, 2012).

16.3.3 *MITgcm model*

Adopting the 3D MITgcm model (Marshall *et al.*, 1997), which is suitable to solve the fully nonhydrostatic and nonlinear equations, Xie *et al.* (2015) examined how the ISW characteristics are modulated by ocean eddies in the case of the ISW crestline length much larger than the horizontal scale of eddies, and the characteristic speed of ISWs much higher than the travel speed of eddies.

Configurations of eddy and ISW in MITgcm are as follows. The stream function ψ is expressed in the form of

$$\psi = \psi^h \phi(z), \tag{16.22}$$

where ψ^h and ϕ are the horizontal and vertical structures of a single isolated eddy or an ISW (Carton, 2001; Dunphy and Lamb, 2014). The vertical structures of eddy ϕ_E and ISW ϕ_I are derived from solutions of the eigenvalue problems,

$$\frac{d}{dz}\frac{f^2}{N^2(z)}\frac{d\phi_E}{dz} + \frac{1}{R_E^2}\phi_E(z) = 0, \quad \frac{d\phi_E}{dz} = 0, \text{ at } z = 0, -H, \tag{16.23}$$

and

$$\frac{d^2\phi_I}{dz^2} + \frac{N^2(z)}{c_I}\phi_I(z) = 0, \quad \phi_I = 0, \text{ at } z = 0, -H, \tag{16.24}$$

where $N(z)$ is the buoyancy frequency, f is the Coriolis parameter, H is the water depth, R_E is the baroclinic Rossby deformation radius of eddy, and c_I. is the phase speed of linear internal waves. The horizontal structure of eddy is a Gaussian function,

$$\psi_E^h = M \exp\left\{-\left[(x^2 + y^2)/L_E^2\right]\right\}, \quad (16.25)$$

where M is the eddy intensity factor, and L_E is the scale of eddy radius. The horizontal structure of ISW is a single soliton solution of the KdV equation,

$$\psi_I^h = c_I \eta^0 sech^2\left(\frac{x - Vt}{\Delta}\right), \quad (16.26)$$

where η^0, V and Δ are the amplitude, propagation speed and characteristic width of the ISW.

The numerical simulations are run in a planetary coordinate system (x, y, z), thus, the zonal and meridional velocities u and v are used for initializing the eddy in the MITgcm, and $u = -\partial\psi/\partial y$ and $v = -\partial\psi/\partial x$, respectively. Meanwhile, the zonal and vertical velocities u and w are used for initializing the ISW, and $u = \partial\psi/\partial z$ and $w = -\partial\psi/\partial x$, respectively. In addition, the sea surface flow and the isopycnal displacement are initialized into the MITgcm via the sea surface elevation field and the temperature and salinity fields, respectively (see details in Xie *et al.*, 2015).

Figure 16.14 shows MITgcm simulation results of dynamic processes of a single soliton modulated by a typical CE in the northern SCS (Xie *et al.*, 2015). Parameters in the modeling test are as follows. A flat topography with a constant depth $H = 1000\,\text{m}$ is assumed. The Coriolis parameter $f = 5.23 \times 10^{-5}\,\text{s}^{-1}$ at 21°N. The buoyancy frequency N(z) is calculated from the World Ocean Atlas 2009 (WOA09) data. The pycnocline depth h of the stratification is 100 m. The linear internal wave speed c_I is $2.1\,\text{m}\,\text{s}^{-1}$. The vertical structures of the eddy ϕ_E and ISW ϕ_I are then calculated by solving the eigenvalue problems (16.23) and (16.24), respectively. The single ISW is initialized in the MITgcm by setting the parameters $\eta^0 = 120\,\text{m}$, $V = 2.35\,\text{m}\,\text{s}^{-1}$ and $\Delta = 1160.4\,\text{m}$. For configuration of eddy, L_E is set as 25 km. This scale is consistent with that of the first baroclinic Rossby deformation radius (20–40 km) at a depth of ∼1000 m in the northeastern SCS (Cai *et al.*, 2008) and 20 times larger than the ISW characteristic width. Statistics indicate that mesoscale eddies in the

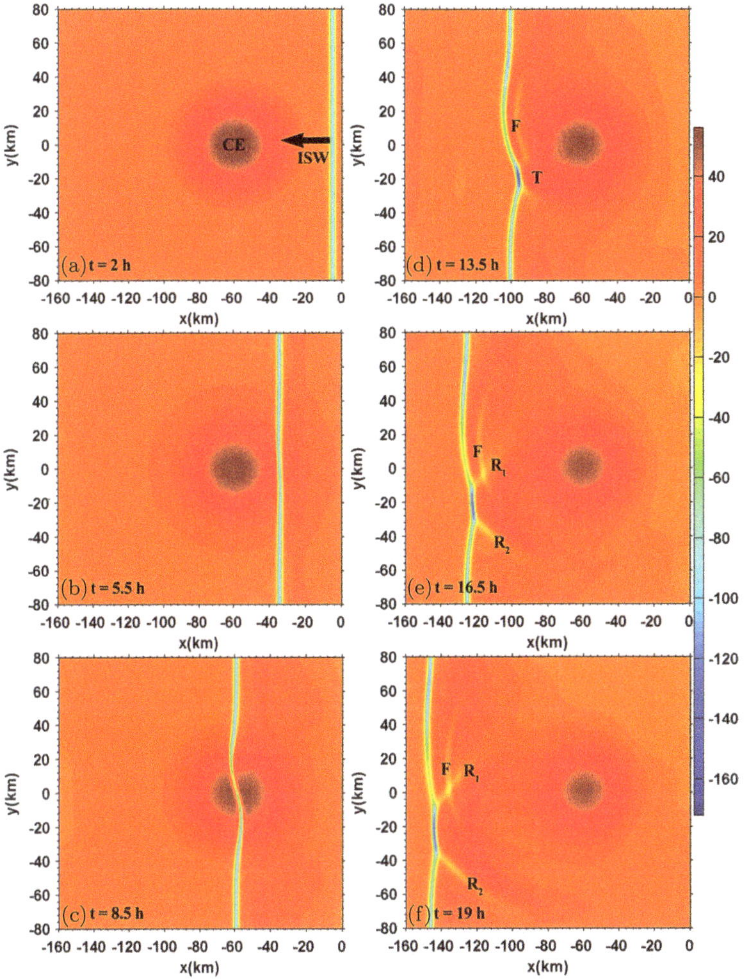

Fig. 16.14. MITgcm simulation results of dynamic processes of a single soliton modulated by a typical CE in the northern SCS. Color codes show displacements of the isopycnal of 1025 kg m^{-3} (relative to a reference depth of 250 m) in m. F, R1, R2 and T represent the newly generated ISWs and wake tail in the x–y plane. *Source*: Taken from Xie *et al.* (2015).

SCS have a mean lifetime of 7–10 weeks and a mean travel speed of 8–10 cm s^{-1} (Chen *et al.*, 2011; Chelton *et al.*, 2011), implying that eddies can be treated as a stationary water body compared to the ISWs.

From the MITgcm simulation results shown in Figure 16.15, one can see that modulation of the depression ISW by the CE become remarkable as running into the dome-shaped pycnocline at $t = 8.5$ h. In the CE current field, the modulated ISW is featured by that (a) the crestline is modified to be S-shaped, (b) the amplitude reduces 30 m from −60 m to −30 m and (c) the characteristic width reduces about 50%. In the downstream far field of CE beyond 1.6 L_E from the CE center after $t = 13.5$ h, the ISW goes into a self-adjusting stage, which is featured by generation of new ISWs F, R_1, R_2 and wake tail T.

Figure 16.15 shows the MITgcm simulation results in the case of AE. One can see that in the AE current field, the modulated

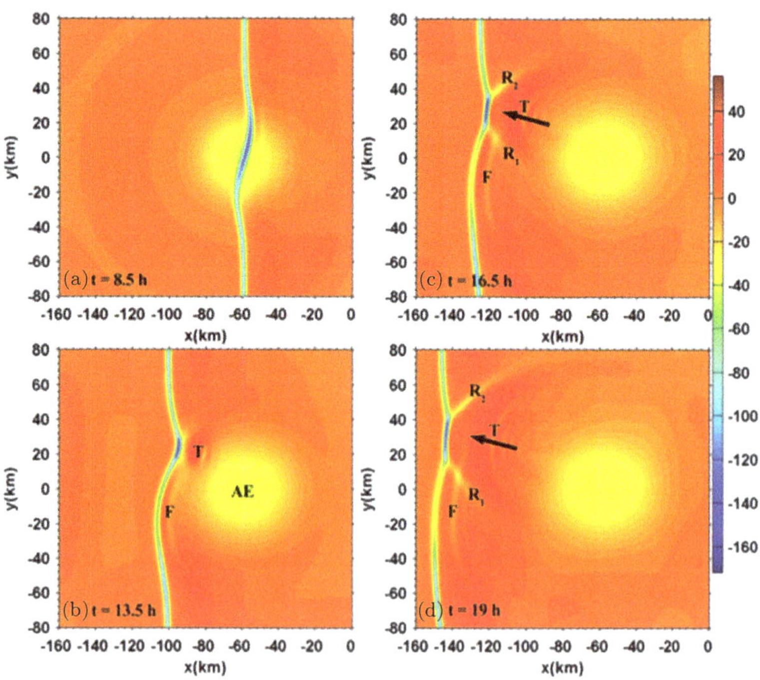

Fig. 16.15. The same as Fig. 16.14, but in the case of AE. Thick black arrows represent the propagation direction of wake trail T.

Source: Taken from Xie *et al.* (2015).

ISW is featured differently by that (a) the crestline is deformed to be opposite S-shaped, (b) the amplitude increases 70 m from −30 m to −100 m and (c) the characteristic width increases about 50%. In the downstream far field of AE beyond $1.6L_E$ from the AE center after $t = 16.5$ h, the ISW also goes into a self-adjusting stage like the case of CE. The signals of all newly generated small waves are stronger than that in the case of CE. The behind mechanisms are worth pursuing.

16.4 Summaries

This chapter summarizes recent progress in the research of mesoscale eddy modulation to the ISWs in the SCS. Satellite and mooring observations show that modulated by spatially varying shear flows of a single eddy or a couple of eddies, (a) ISW crestline patterns are deformed to convex, concave and S-shaped patterns, (b) propagation directions, (c) amplitudes and (d) characteristic widths of the ISWs are variable along the crestlines. Statistics indicate that the modulation events mainly occur in the northern SCS, particularly near the Luzon Strait, where powerful mesoscale eddies and intensive ISWs are concurrent frequently.

To explore the dynamic mechanisms of above observed phenomena and simulate the evolution processes of ISWs modulated by mesoscale eddies in the SCS, theories and numerical models have been developed. From the PKdV equation, the theoretical solution for the problem how does a shoaling thermocline force the ISW amplitude growing is derived. Based on the solution, the behavior of the ISW amplitudes modulated by the thermoclines of normal AE and CE models are predicted. The ISW amplitudes experience a dropping – rising process during passing over a normal AE, while a reverse process during passing over a normal CE. Meanwhile, the solution can be used to construct the ISW crestline patterns during passing over mesoscale eddies.

The nonlinear refraction model (NRM) is used for predicting the evolution of the ISWs as propagating in a spatially varying background current field of mesoscale eddies. The idea is to reduce the horizontal 2D wave motion problem to the 1D problem, through slicing the initial ISW crestline into a group of fine pieces that serve as the radiation ray sources and then use the NRM to predict the behavior of wave radiation along the rays. The new 2D ISW patterns and parameters at any time can be obtained by the data taken from the radiation rays.

The 3D MITgcm model is adopted to examine how the ISW characteristics are modulated by ocean eddies in the case of the ISW crestline length much longer than the horizontal scale of eddies, and the characteristic speed of ISWs much higher than the travel speed of eddies. Simulation results reveal detailed dynamic processes of a single soliton modulated by the typical CE and AE and explain observations in the northern SCS well. In addition, MITgcm simulation results uniquely show the signals of generation of new ISWs and wake tails of small amplitudes in the downstream far field of eddies. The behind mechanisms are worth pursuing.

References

Cai, S., Xie, J. and He, J. (2012). An overview of internal solitary waves in the South China Sea, *Surv. Geophys.*, 33, pp. 927–943, doi:10.1007/s10712-012-9176-0.

Cai, S., Long, X., Wu, R. and Wang, S. (2008). Geographical and monthly variability of the first baroclinic Rossby radius of deformation in the South China Sea, *J. Mar. Syst.*, 74, pp. 711–720, doi:10.1016/j.jmarsys.2007.12.008.

Carton, X. J. (2001). Hydrodynamical modeling of oceanic vortices, *Surv. Geophys.*, 22, pp. 179–263, doi:10.1023/A:1013779219578.

Chelton, D. B., Schlax, M. G. and Samelson, R. M. (2011). Global observations of nonlinear mesoscale eddies, *Progr. Oceanogr.*, 91, pp. 167–216, doi:10.1016/j.pocean.2011.01.002.

Chen, G., Hou, Y. and Chu, X. (2011). Mesoscale eddies in the South China Sea: Mean properties, spatiotemporal variability, and impact on thermohaline structure, *J. Geophys. Res. Oceans*, 116, C06018, doi:10.1029/2010JC006716.

Dunphy, M. and Lamb, K. G. (2014). Focusing and vertical mode scattering of the first mode internal tide by mesoscale eddy interaction, *J. Geophys. Res. Oceans*, 119, pp. 523–536, doi:10.1002/2013JC009293.

Grimshaw, R., Pelinovsky, E., Talipova, T. and Kurkin, A. (2004). Simulation of the transformation of internal solitary waves on oceanic shelves, *J. Phys. Oceanogr.*, 34, pp. 600–606, doi:10.1175/JPO2652.1.

Huang, X., Chen, Z., Zhao, W., Zhang, Z., Zhou, C., Yang, Q. and Tian, J. (2016). An extreme internal solitary wave event observed in the northern South China Sea, *Sci. Rep.*, 6, 30041, doi:10.1038/srep30041.

Huang, X., Wang, Z., Zhang, Z., Yang, Y., Zhou, C., Yang, Q., Zhao, W. and Tian, J. (2018). Role of mesoscale eddies in modulating the semidiurnal internal tide: Observation results in the northern South China Sea, *J. Phys. Oceanogr.*, 48, pp. 1749–1770, doi:10.1175/jpo-d-17-0209.1.

Huang, X., Zhang, Z., Zhang, X., Qian, H., Zhao, W. and Tian, J. (2017). Impacts of a mesoscale eddy pair on internal solitary waves in the northern South China Sea revealed by mooring array observations, *J. Phys. Oceanogr.*, 47, pp. 1539–1554, doi:10.1175/JPO-D-16-0111.1.

Li, Q., Wang, B., Chen, X., Chen, X. and Park, J.-H. (2016). Variability of nonlinear internal waves in the South China Sea affected by the Kuroshio and mesoscale eddies, *J. Geophys. Res. Oceans*, 121, pp. 2098–2118, doi:10.1002/2015JC011134.

Liao, G., Yang, C., Xu, X., Shi, X., Yuan, Y. and Huang, W. (2012). Effects of mesoscale eddies on the internal solitary wave propagation, *Acta Oceanol. Sin.*, 31, pp. 26–40, doi:10.1007/s13131-012-0233-9.

Liu, A. K., Zhao, Y., Tang, T. Y. and Ramp, S. R. (2004). A case study of internal wave propagation during ASIAEX-2001, *IEEE J. Oceanic Eng.*, 29, pp. 1144–1156, doi:10.1109/JOE.2004.841392.

Marshall, J., Adcroft, A., Hill, C., Perelman, L. and Heisey, C. (1997). A finite-volume, incompressible Navier Stokes model for studies of the ocean on parallel computers, *J. Geophys. Res.*, 102, pp. 5753–5766, doi:10.1029/96JC02775.

Nash, J. D., Alford, M. H. and Kunze, E. (2005). Estimating internal wave energy fluxes in the ocean, *J. Atmos. Ocean. Tech.*, 22, pp. 1551–1570, doi:10.1175/JTECH1784.1.

Newell, A. C. (1985). *Soliton in Mathematics and Physics*, Soc. for Ind. and Appl. Math., Philadelphia, PA, pp. 1–246.

Small, J. (2001). A nonlinear model of the shoaling and refraction of interfacial solitary waves in the ocean. Part II: Oblique refraction across a continental slope and propagation over seamount, *J. Phys. Oceanogr.*, 31, pp. 3184–3199, doi:10.1175/15200485(2001)031,3184:ANMOTS.2.0.CO;2.

Sun, L., Zhang, J., Meng, J. and Cui, W. (2022). Influence of mesoscale eddy on internal solitary wave propagation in the northern South China Sea based on remote sensing, *Haiyang Xuebao*, 44, pp. 137–144, doi:10.12284/hyxb2022106.

Wang, J., Flierl, G., LaCasce, J., McClean, J. and Mahadevan, A. (2013). Reconstructing the ocean's interior from surface data, *J. Phys. Oceanogr.*, 43, pp. 1611–1626, doi:10.1175/JPO-D-12-0204.1.

Xie, J., He, Y., Chen, Z., Xu, J. and Cai, S. (2015). Simulations of internal solitary wave interactions with mesoscale eddies in the northeastern South China Sea, *J. Phys. Oceanogr.*, 45, pp. 2959–2978, doi:10.1175/JPO-D-15-0029.1.

Xie, J., He, Y., Lü, H., Chen, Z., Xu, J. and Cai, S. (2016). Distortion and broadening of internal solitary wavefront in the northeastern South China Sea deep basin, *Geophys. Res. Lett.*, 43, pp. 7617–7624, doi:10.1002/2016GL070093.

Xu, J., He, Y., Chen, Z., Zhan, H., Wu, Y., Xie, J., Shang, X., Ning, D., Fang, W. and Cai, S. (2020). Observations of different effects of an anti-cyclonic eddy on internal solitary waves in the South China Sea, *Progr. Oceanogr.*, 188, 102422, doi:10.1016/j.pocean.2020.102422.

Yang, Y., Huang, D., Zhou, C., Zhang, Z., Zhao, W. and Tian, J. (2022). Three-dimensional structures of internal solitary waves in the northern South China sea revealed by mooring array observations, *Progr. Oceanogr.*, 209, 102907, doi:10.1016/j.pocean.2022.102907.

Yuan, Y., Zheng, Q., Dai, D., Hu, X., Qiao, F. and Meng, J. (2006). Mechanism of internal waves in the Luzon Strait, *J. Geophys. Res.*, 111, C11S17, doi:10.1029/2005JC003198.

Zhao, Z. (2014). Internal tide radiation from the Luzon Strait, *J. Geophys. Res. Oceans*, 119, pp. 5434–5448, doi:10.1002/2014jc010014.

Zhao, Z. and Alford, M. H. (2006). Source and propagation of internal solitary waves in the northeastern South China Sea, *J. Geophys. Res.*, 111, C11012, doi:10.1029/2006jc003644.

Zhao, Z., Klemas, V., Zheng, Q. and Yan, X.-H. (2004). Remote sensing evidence for baroclinic tide origin of internal solitary waves in the northeastern South China Sea, *Geophys. Res. Lett.*, 31, L06302, doi:10.1029/2003GL019077.

Zhao, Z., Alford, M. H., MacKinnon, J. A. and Pinkel, R. (2010). Long range propagation of the semidiurnal internal tide from the Hawaiian Ridge, *J. Phys. Oceanogr.*, 40, pp. 713–736, doi:10.1175/2009JPO4207.1.

Zhao, Z., Alford, M. H., Lien, R.-C., Gregg, M. C. and Carter, G. S. (2012). Internal tides and mixing in a submarine canyon with time varying stratification, *J. Phys. Oceanogr.*, 42, pp. 2121–2142, doi:10.1175/JPO-D-12-045.1.

Zheng, Q. (2017). *Satellite SAR Detection of Sub-Mesoscale Ocean Dynamic Processes*, World Scientific, London, pp. 121–178.

Zheng, Q., Yuan, Y., Klemas, V. and Yan, X.-H. (2001). Theoretical expression for an ocean internal soliton SAR image and determination of the soliton characteristic half width, *J. Geophys. Res.*, 106, pp. 31415–31423, doi:10.1029/2000JC000726.

Zheng, Q., Susanto, R. D., Ho, C.-R., Song, Y. T. and Xu, Q. (2007). Statistical and dynamical analyses of generation mechanisms of solitary internal waves in the northern South China Sea, *J. Geophys. Res.*, 112, C03021, doi:10.1029/2006JC003551.

Zheng, Q., Song, Y. T., Lin, H., Hu, X., Meng, J. and Wang, D. (2008). On generation source sites of internal waves in the Luzon Strait, *Acta Oceanol. Sin.*, 27, pp. 38–50.

Zheng, Q., Xie, L., Xiong, X., Hu, X. and Chen, L. (2020). Progress in research of submesoscale processes in the South China Sea, *Acta Oceanol. Sin.*, 39, pp. 1–13, doi:10.1007/s13131-019-1521-4.

Chapter 17

Atmospheric Responses to Ocean Mesoscale Eddies

17.1 Introduction

As an atmospheric underlying boundary, the ocean can modulate the atmospheric motion by heat exchange and momentum coupling. Therefore, the ocean plays an important role in regulating the global climate due to its enormous heat capacity. In warm ocean regions, the marine atmosphere boundary layer (MABL) is directly heated locally by the underlying ocean. Changes in atmospheric temperature in turn alter the turbulent mixing of the air and atmospheric pressure in the MABL, producing higher wind speeds over warmer water and lower wind speeds over cooler water (Chelton, 2013; Frenger *et al.*, 2013).

Mesoscale eddies are the most powerful component of the ocean circulation. Meanwhile, their sea surface temperature (SST) fields directly affect the temperature structure and instability of the overlying atmosphere through air-sea heat exchange (Cabrera *et al.*, 2022). The eddies affect the sea surface winds, i.e., an AE (CE) strengthens (weakens) the sea surface winds. The SST difference alters the lower clouds and precipitation (Chelton, 2013). Using a coupled atmosphere-ocean regional model, Byrne *et al.* (2015) showed that the atmospheric imprint of ocean eddies is driven by the modification of vertical mixing in the MABL, rather than secondary flows driven by horizontal pressure gradients.

The SCS is featured by basin scale distributed powerful mesoscale eddies all year around (see Parts 2–4 of this book for details) and passages of powerful atmospheric processes, such as the seasonally reversed East Asia monsoon and the intensive tropical cyclones (TCs) in summer and fall. Thus, it is of significance to clarify the regional atmospheric responses to mesoscale eddies and possible impacts on the regional weather, ecological systems, oceanic and atmospheric environments, resulting from the interaction processes. For recent years, the research in this field has gained important progresses. For examples, Liu *et al.* (2018) showed that the status of the MABL is influenced remarkably by ocean eddies, using the composite analysis results derived from 438 mesoscale eddies in the SCS from 2000 to 2012. Their results reveal that mesoscale eddies may induce higher than 5% of the total variance of turbulent heat fluxes, surface wind speed and evaporation rate. Sun *et al.* (2023) investigated effects of grouped AEs on the intensity of TC passing over the northern SCS (NSCS), using observed data during passage of TC Hato (2017) and numerical modeling methods. They found that AEs in the outer area (about 300–400 km away from the TC track) suppress TC development, while AEs in the inner area (within about 150 km from the TC track) greatly favor TC development.

This chapter summarizes the recent research results on atmospheric responses to mesoscale eddies in the SCS, including effects of eddy heating, water transport across air-sea boundary, ocean heat energy transformed to atmospheric mechanical energy and TC responses to ocean eddies. Basin-scale air-sea dynamic environments in the SCS are included as the background knowledge for further understanding of occurrence conditions and mechanisms for basin scale air-sea coupling processes. The final goal of this research field aims at further improving the weather system prediction and forecasts in the SCS.

17.2 Basin-scale Air-sea Dynamic Environments in the SCS

17.2.1 *Geography-bottom boundary conditions*

The SCS is a large, semi-enclosed, marginal sea of the western North Pacific (WNP), which extends from the Equator to the Tropic

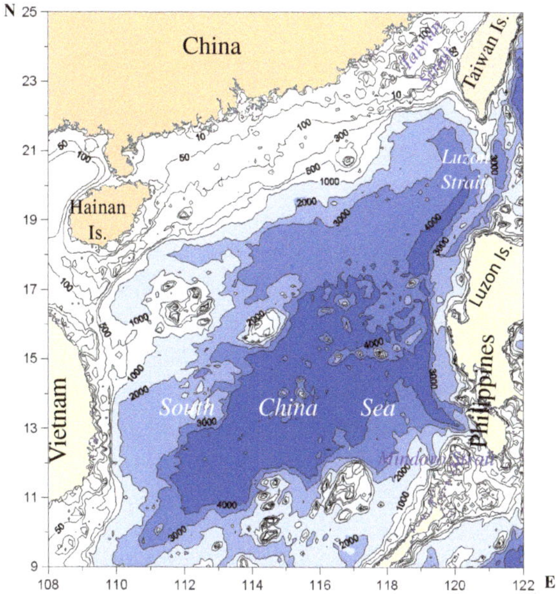

Fig. 17.1. Map of the SCS. Bathymetry is in m.
Source: Taken from Xie *et al.* (2018).

of Cancer and from 99°E to 121°E, occupying an area of about 3.5×10^6 km^2 as shown in Figure 17.1. The SCS is bordered by South China and Indochina Peninsula to its north and west and surrounded by a chain of islands to the east and south, ranging from the Luzon Island in the north to the Borneo Island in the south. The Taiwan Strait, Luzon Strait, Mindoro Strait and Karimata Strait are major channels, through which the SCS connects the East China Sea, the Pacific Ocean, the Sulu Sea, and the Java Sea. The bottom topography of the SCS comprises three parts: the deep basin, the continental slope, and the continental shelf, respectively covering about 15%, 38% and 47% of the total area with an average depth of 1140 m. The rhomboid deep basin extends from NE to SW with a major axis of about 2000 km and a minor axis of about 1000 km, an average depth of 4700 m and a maximum depth of 5559 m (Wang and Li, 2009). These physical features imply that the SCS possesses favorable conditions, including the best geographic location, huge space and water body for generation, evolution and dissipation of multiscale ocean processes and atmosphere-ocean coupling.

17.2.2 *Atmospheric forcing processes*

17.2.2.1 *East Asian monsoon*

The SCS dynamics is dominated by the seasonally reversing East/South Asian monsoons, which induce the corresponding seasonal variations of physical characteristics in the upper ocean of the SCS (Wyrtki, 1961). The stronger winter monsoon lasts nearly six months (November to April), while the weaker summer monsoon lasts nearly 4 months (mid-May to mid-September) (Chu and Wang, 2003). The mean surface wind stress over the SCS is nearly $0.2\,\mathrm{N\,m^{-2}}$, reaching $0.3\,\mathrm{N\,m^{-2}}$ in mid-December and dropping to nearly $0.1\,\mathrm{N\,m^{-2}}$ in June. Figure 17.2 shows an example of monthly average sea surface wind fields in the SCS during the winter and summer monsoons (Chen *et al.*, 2012). One can see that the wind vectors completely reverse from northeasterly winds in winter to southwesterly winds in summer. The SCS plays a role of wind tunnel, in which the wind speeds are remarkably intensified, so that multiscale ocean motions manifest the seasonal variability due to the monsoon modification.

17.2.2.2 *Tropical cyclones*

The SCS seats in the tropics and is connected to the Pacific through the Luzon Strait (Hu *et al.*, 2000). Thus, approximately eight TCs (typhoons) per year pass over the SCS (Pan *et al.*, 2017). Most are

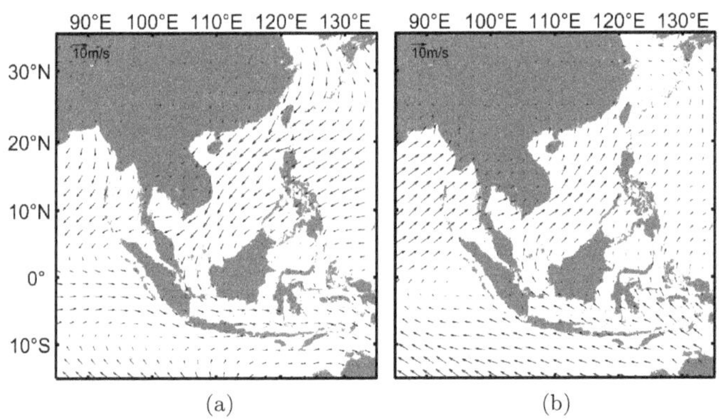

Fig. 17.2. Monthly average sea surface wind fields in the SCS in (a) January and (b) July 1999.
Source: CCMP (Cross-Calibrated, Multi-Platform) wind data.

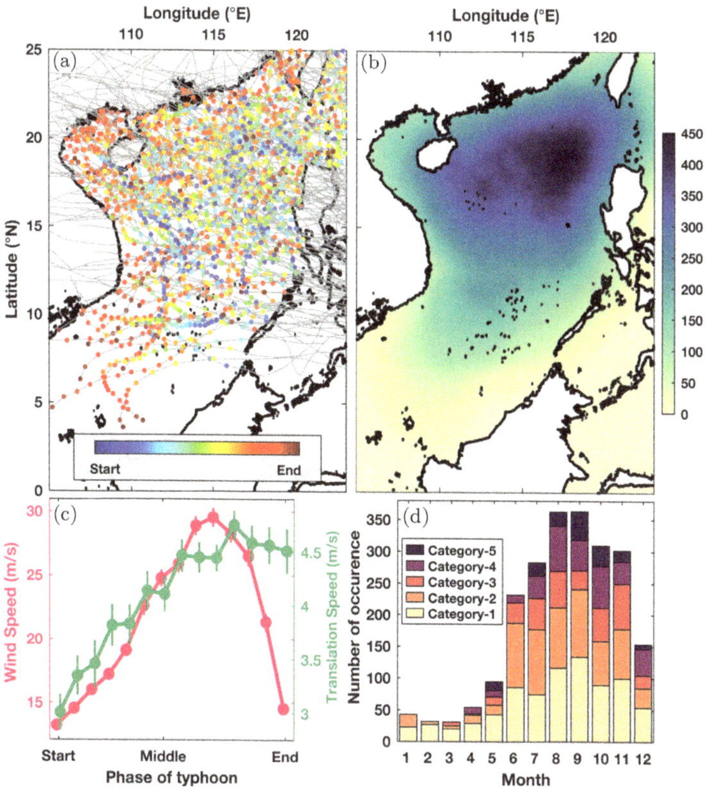

Fig. 17.3. Statistics of the SCS typhoon features derived from the observed data from September 2002 to August 2018. (a) Trajectories of typhoons with color-coded phases of individual typhoons in the SCS. (b) Number of occurrences when typhoon centers are less than 300 km from each location. (c) Averaged wind speed (red curve) and translation speed (green curve) at different phases of typhoons. (d) Monthly number of typhoons (bars) with different categories.
Source: Taken from Wang (2020).

generated in the WNP and travel into the SCS via the Philippines or the Luzon Strait (Lin *et al.*, 2009). The intensities of typhoons vary greatly, ranging from Categories 1 to 5, leading to a significant impact on the regional dynamic environment (Zhao *et al.*, 2008; Wang, 2020). Figure 17.3 shows statistics of the SCS typhoon features derived from the observed data from September 2002 to August 2018 (Wang, 2020). One can see that the typhoon trajectories cover the northern and central SCS and concentrate in the northern deep basin. In the time domain, typhoons may occur all year around, but concentrate in the summer-fall season from June to November. The maximum average wind speed may reach $30\,\mathrm{m\,s^{-1}}$ and the maximum average translation speed reaches $4.6\,\mathrm{m\,s^{-1}}$.

17.2.3 *ENSO events*

The previous studies revealed that the eddy kinetic energy (EKE) in the SCS has interannual variation under the influence of the anomalous wind stress curl induced by the El Niño-Southern Oscillation (ENSO) event as shown in Figure 17.4 (Wu *et al.*, 1999; Wang *et al.*, 2000, 2012). Cheng and Qi (2010) showed that the positive (negative) EKE anomaly in the SCS is significantly dominant during the La Niña (El Niño) events. The patterns of anomalous wind stress curl of the atmosphere tend to form the corresponding swirling mesoscale eddies in the ocean. Therefore, it is expectable that the mesoscale eddy activity would be more active during the La Niña event.

17.2.4 *Upper oceanic dynamic features*

17.2.4.1 *General seasonal circulation in the SCS*

The general circulation is the most fundamental dynamic feature of a marginal sea. The SCS circulation is characterized by that the

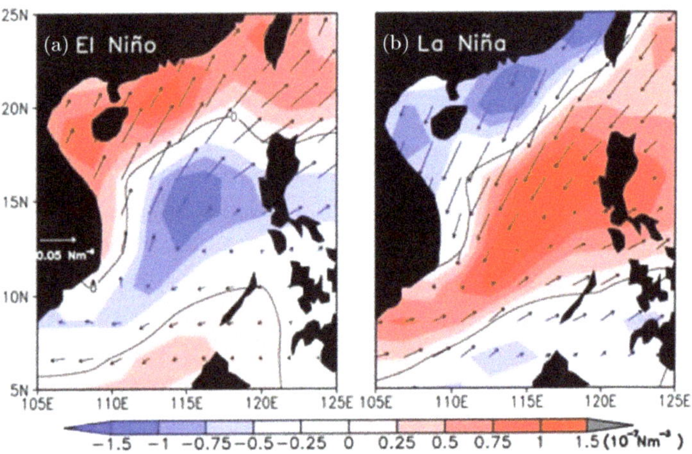

Fig. 17.4. Composite maps of wind stress (black arrows) and wind stress curl anomaly (color codes) in December of (a) El Niño and (b) La Niña events. *Source*: Taken from Cheng and Qi (2010).

surface currents undergo a seasonal reversal of direction driven by monsoon system. The southwest summer monsoon pushes the shelf water northward, so results in compensatory southward movements over the eastern deep basin, while the northeast winter monsoon reverses the flow direction and drives a strong boundary current along the Vietnam coast. Comprehensively analyzed available results, Hu *et al.* (2000) proposed the general upper layer mean circulation patterns of the SCS as shown in Figure 17.5. One can see that the upper layer (0–200 m) mean circulation is cyclonic in winter and anticyclonic in summer. Meanwhile, two seasonal CEs (CE_{w1} and CE_{w2}) exist in winter to the west of the Luzon Island and the southeast of the Zhongnan (Indochina) Peninsula, respectively. In summer, an AE (AE_s) appears to the southeast of the Zhongnan Peninsula, while a CE (CE_{s1}) to the east of Vietnam coast. In addition, it has been suggested that there exists an AE (AE_w in Figure 17.5(a)) in the northwest of the Kalimantan Island in winter, and a CE (CE_{s2} in Figure 17.5(b)) in the northwest of the Luzon Island in summer.

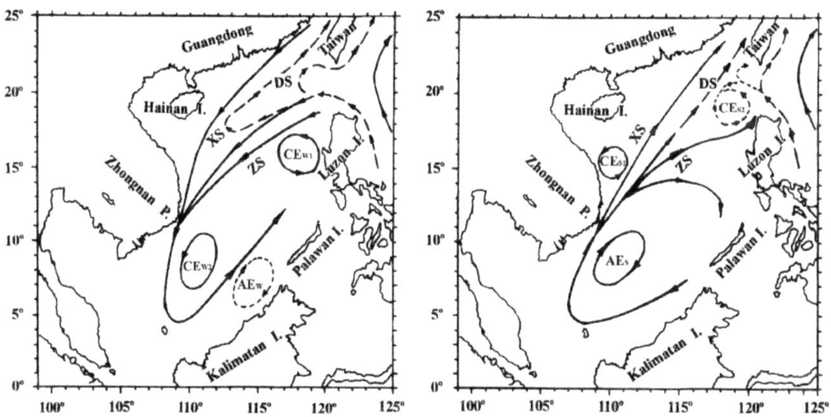

Fig. 17.5. General upper layer mean circulation patterns in the SCS in (a) winter and (b) summer. Long dash arrows schematically show the SCS Warm Current (SCSWC) and the Kuroshio intrusion. CE_{w1}, CE_{w2} and AE_w mark seasonal eddies in winter and CE_{s1}, CE_{s2} and AE_s in summer. Short dash arrows show the unstable Kuroshio intrusion and newly suggested eddies. Abbreviations of islands or atolls: DS-Dongsha, XS-Xisha and ZS-Zhongsha.

Source: Taken and re-labeled from Hu *et al.* (2000).

17.2.4.2 *EKE active regions*

The EKE [$= \frac{1}{2}(u^2 + v^2)$ where u and v are velocity components] has been used as a measure of the strength of mesoscale and sub-mesoscale processes in the ocean including mesoscale eddies (Li *et al.*, 1998; Wang *et al.*, 2000, 2003; Yuan *et al.*, 2007; Nan *et al.*, 2011), filaments (Xie *et al.*, 2003, 2007), fronts (Wang *et al.*, 2001), jets and meanders (Cheng and Qi, 2010). Wang *et al.* (2012) derived seasonal distributions of EKE ($cm^2\,s^{-2}$) in the SCS from satellite altimeter geostrophic current data from 1993 to 2010 as shown in Figure 17.6. One can see that the Luzon Strait (LS) and the western boundary of the SCS deep basin off coast of Vietnam (OCV) are most active

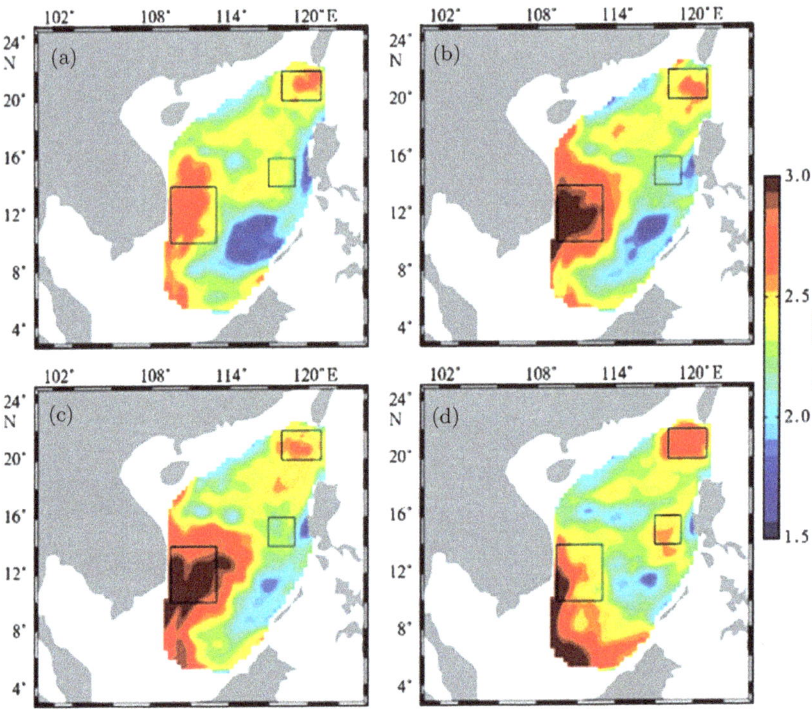

Fig. 17.6. Seasonal distributions of EKE ($cm^2\,s^{-2}$, in $\log_{10}$ scale) in the SCS derived from satellite altimeter geostrophic current data from 1993 to 2010. (a) Spring (MAM), (b) summer (JJA), (c) fall (SON) and (d) winter (DJF). Three boxes represent the Luzon Strait (LS), western boundary of the SCS deep basin off coast of Vietnam (OCV) and west of Luzon Island (WLI).

Source: Taken from Wang *et al.* (2012).

EKE areas all year round, and a small active area appears to west of Luzon Island (WLI) in winter. These EKE active areas are also active areas of air-sea interaction. Active area OCV appears stronger in summer and fall, but weaker in winter and spring, while active area LS appears the strongest in winter and WLI appears only in winter, indicating that the seasonally reverse monsoon winds play a dominant role in regulating basin-scale EKE distribution in the SCS.

17.2.4.3 *Basin-scale distribution of eddies*

Parts 3 and 4 of this book give the detailed statistical features of mesoscale eddies in the SCS. As an example, Figure 17.7 shows a map of dynamic topography (MDT) derived from satellite altimeter AVISO product on 1 August 2007. One can see that mesoscale eddy distribution in the SCS is a basin-scale phenomenon, so that the

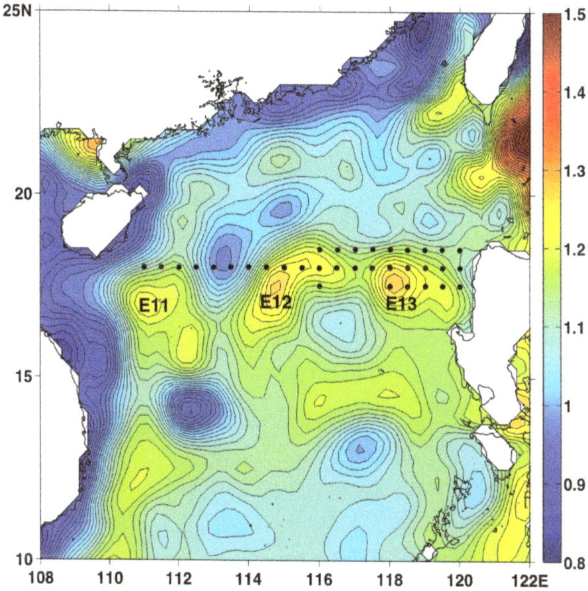

Fig. 17.7. Map of dynamic topography (MDT) of the SCS derived from satellite altimeter AVISO product on 1 August 2007. Eddies E11, E12 and E13 belong to the standing wave event, which was confirmed by cruise observations along sections 17.5°, 17.0° and 17.5°N (dots) conducted from 30 July to 4 August 2007. Color codes are the sea surface dynamic height in cm.

Source: Taken from Xie *et al.* (2018).

atmospheric underlying surface is always in a state by disturbance of eddies. Thus, it is expectable that the atmospheric system would have responses to disturbance from the underlying active ocean.

17.3 Atmospheric Responses to Mesoscale Eddies in the SCS

As an atmospheric underlying surface, the ocean can modulate the atmospheric motion by heat exchange and momentum coupling. Mesoscale eddies are the most powerful component of the ocean circulation; thus, they would play a role in regulating the atmospheric motion. Previous investigators have dedicated the research on two-way interaction between mesoscale eddy and the atmosphere (Renault *et al.*, 2016, 2019; Liu *et al.*, 2018; Cabrera *et al.*, 2022; Seo *et al.*, 2023). Two main mechanisms of atmospheric responses to mesoscale eddies in the SCS have been proposed. One is the vertical mixing mechanism, i.e., the eddy-induced SST anomaly (SSTA) changes the stability of the MABL (Hayes *et al.*, 1989; Wallace *et al.*, 1989). The positive (negative) SSTA of AEs (CEs) makes the MABL unstable (stable), strengthen (weaken) the near-surface winds and increase (decrease) the upward movements in the atmosphere and the possibility of precipitation (Liu *et al.*, 2018; Cabrera *et al.*, 2022). The other is pressure adjustment mechanism, which is caused by modifications of sea level pressure resulting from the anomalous heating of mesoscale eddies and contribute the wind anomaly (Lindzen and Nigam, 1987; Liu *et al.*, 2018). The following subsections will specify the recent advances in the research of the atmospheric responses to mesoscale eddies in the SCS.

17.3.1 *Effects of eddy heating*

The heat exchange is the most direct way for air-sea interaction. Meanwhile, the heat flux across the air-sea boundary depends on the air-sea temperature difference. In the tropics with active heat transport, ocean eddies may significantly affect the spatial distribution of SSTA, which is defined as the difference of SST at each point minus the 2D spatial average value of SST. For instances, Chen *et al.* (2011)

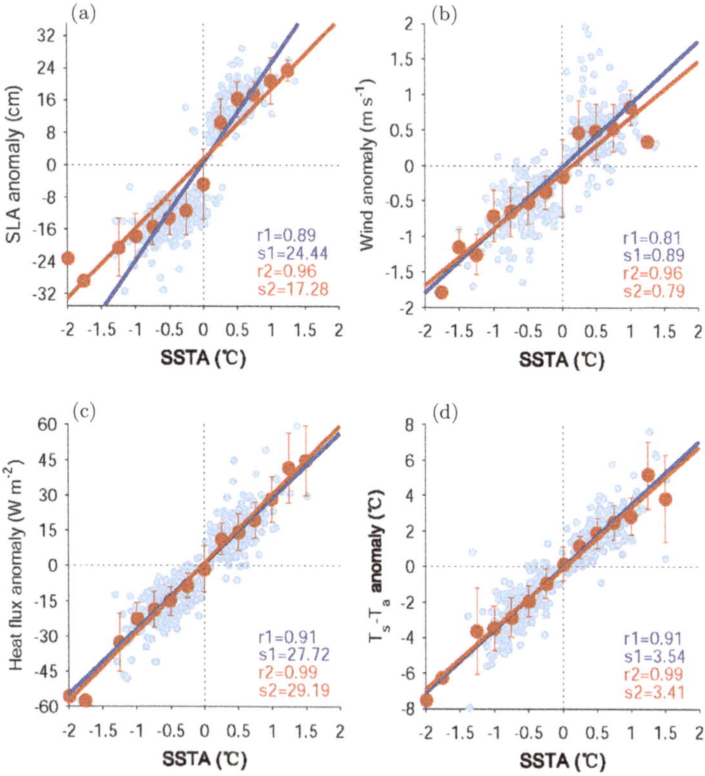

Fig. 17.8. Statistics of linear regression relationships of major air-sea boundary dynamic parameters vs SSTA of mesoscale eddies in the SCS. (a) SLA anomaly, (b) wind anomaly, (c) turbulent heat flux anomaly and (d) difference between SST and air temperature. The correlation coefficients r1 and r2 and slopes s1 and s2 are derived from the linear least-squares fit to the un-binned (blue) and binned (red) data with significance at the 99% confidence level.

Source: Taken from Liu *et al.* (2018).

showed that the AEs contribute 0.4°C to the average SSTA after isolating mesoscale SST signals from the background in the SCS.

Using the composite analysis results derived from 438 mesoscale ocean eddies in the SCS from 2000 to 2012, Liu *et al.* (2018) revealed that MABL is remarkably affected by such eddies. The results showed that mesoscale eddies may induce higher than 5% of the total variance of turbulent heat fluxes, surface wind speed and evaporation rate. Figure 17.8 shows statistics of linear regression relationships of major air-sea boundary dynamic parameters, including SLA anomaly, wind anomaly, turbulent heat flux anomaly and difference between SST and air temperature, vs SSTA of mesoscale eddies. One can see that in all cases the correlation coefficients are higher than 0.8 (Liu *et al.*, 2018).

17.3.2 *Water transport across air-sea boundary*

Anomalous sea surface heating induced by mesoscale eddies may change the characteristics of precipitation, atmospheric water vapor, and cloud liquid water over mesoscale eddies as shown in Figure 17.9(a–f) (Liu *et al.*, 2018). From Figure 17.9(a) and 17.9(d), one can see that a CE center with negative rain rate anomaly is sandwiched by rain bands with positive rain rate anomaly and vice versa for an AE.

Figure 17.9(b) and 17.9(e) show eddy-induced columnar cloud liquid water anomaly, which has similar spatial distribution features to that of the precipitation anomaly. Figure 17.9(c) and 17.9(f) show that distribution features of columnar water vapor also reflect the influence of mesoscale eddies on the entire atmosphere. One can see that the negative columnar water vapor anomaly occupies the CE core, and vice versa in the case of AE.

17.3.3 *Ocean heat energy transformed to atmospheric mechanical energy*

Figure 17.10 shows spatially averaged sea surface wind field response to underlying eddy heating (cooling). From Figure 17.10(a) and 17.10(d), one can see that a CE (an AE) core corresponds to the negative (positive) wind speed anomaly with a minimum (maximum) value of -0.19 (0.11) $\mathrm{m\,s^{-1}}$. Meanwhile, Figure 17.10(b, c) and 17.10(e, f) show that the sea surface wind anomaly over mesoscale eddies further affects the air flow horizontal divergence and vorticity. As Figure 17.10(b) shows that sea surface air flows converge on the left side of the CE and diverge on the right side, forming a dipole pattern. For the AE, a single convergence core is located at its center surrounded by divergence cores (Figure 17.10(e)). For the composite vorticity anomaly as shown in Figure 17.10(c) and 17.10(f), the CE has a negative vorticity anomaly core located at its center surrounded

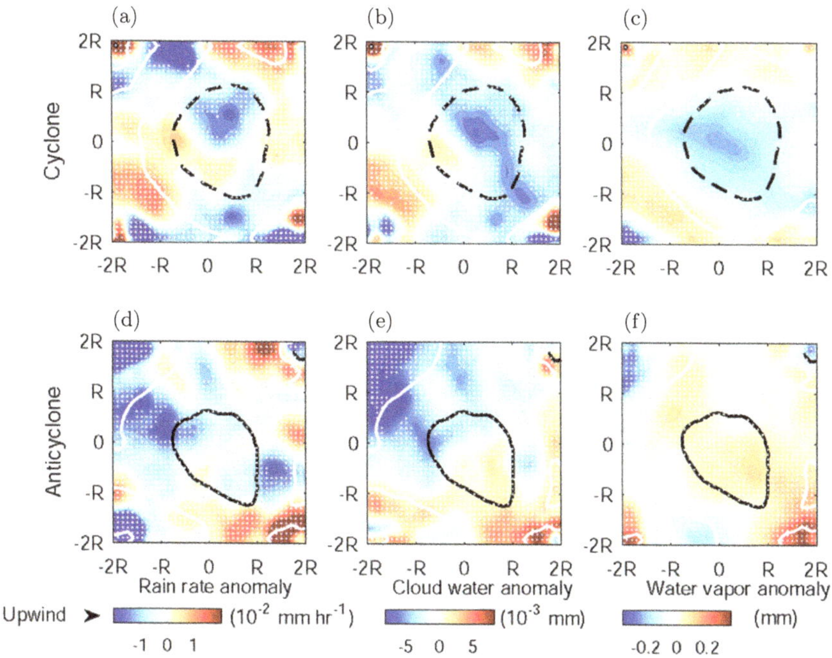

Fig. 17.9. Composite maps of the mean (a, d) rain rate anomaly (color codes), (b, e) columnar cloud liquid water anomaly (color codes), and (c, f) columnar water vapor anomaly (color codes) for (a–c) CE and (d–f) AE.
Source: Taken from Liu *et al.* (2018).

by positive vorticity anomaly centers and vice versa in the case of AE (Liu *et al.*, 2018).

17.3.4 *Tropical cyclone responses to ocean eddies*

Previous investigators have well documented that ocean eddies have significant influence on air–sea interaction during tropical cyclone (TC) passage by either hindering or sustaining air–sea fluxes into the TC core (Lin *et al.*, 2005; Shay *et al.*, 2000; Jaimes and Shay, 2009, 2015; Zheng *et al.*, 2010). Meanwhile, preexisting eddies may cause change in the TC intensity (Wu *et al.*, 2007; Zheng *et al.*, 2008).

Sun *et al.* (2023) investigated effects of grouped AEs on the intensity of TC in the NSCS using observed data and numerical modeling methods. The data observed during passage of TC Hato (2017) were

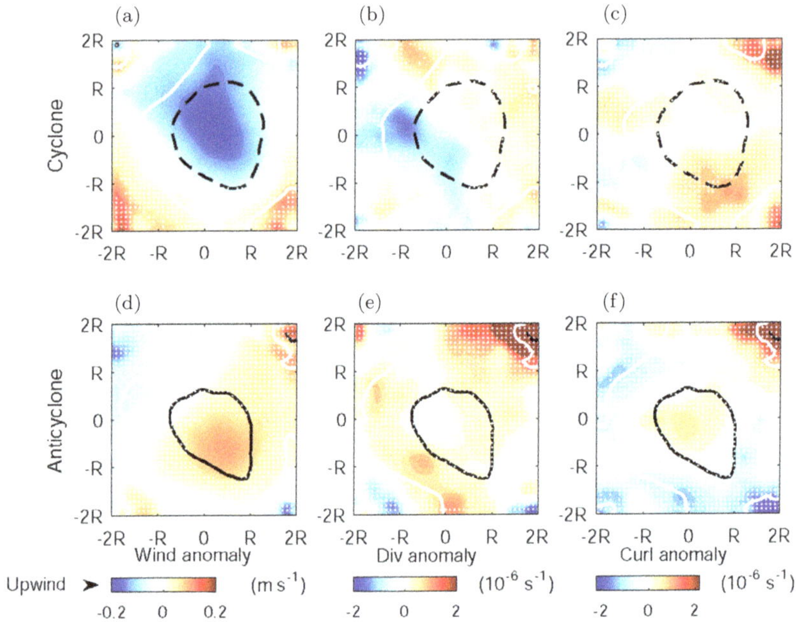

Fig. 17.10. Composite maps of the mean sea surface (a, d) wind speed anomaly, (b, e) divergence anomaly and (c, f) vorticity anomaly for (a–c) CE and (d–f) AE. *Source*: Taken from Liu *et al.* (2018).

used to validate the modeled results. Their study area and numerical modeling domains are shown in Figure 17.11(a) and zoomed Domain 3 shown in Figure 17.11(b). One can see that the core area of Domain 3 is divided into two bands parallel to the TC track, Box-In and Box-Out, which are defined as the inner area and the outer area, respectively. Box-In with a width of three times TC radius of maximum wind speed is centered at the TC track. Box-Out with the same width as that of Box-In is aligned in the south of Box-In. Three AEs were used for validation. AE1 is centered at 20.7°N, 117.5°E on the TC track in Box-In, AE2 centered at 17.2°N, 114.1°E and AE3 centered at 17.5°N, 117.6°E are distributed in Box-Out.

The observed and modeled results show that AEs in the outer area suppress TC development, while AEs in the inner area greatly favor TC development. Further analysis and simulation experiments reveal that the opposite effects primarily result from the different responses of secondary circulation of TC to the AE-induced local

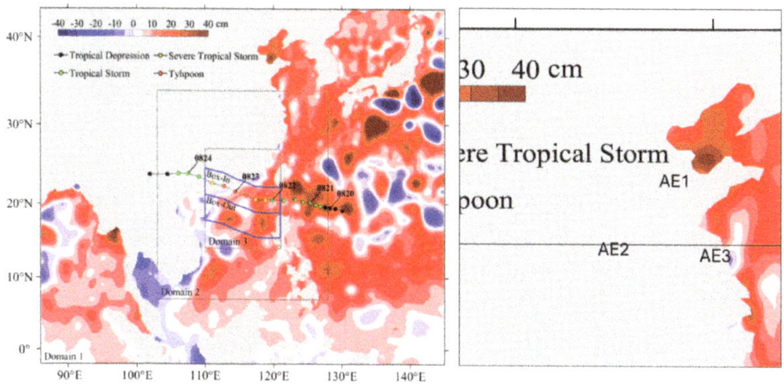

(a) (b)

Fig. 17.11. (a) The model domains overlaid on the SSH anomaly field on 19 August 2017. Blue boxes are used for the sensitivity experiments. Black line with color dots denotes the best track and categories of TC with dates (August/DD, 2017) taken from China Meteorological Administration observations. (b) Zoomed Domain 3. AE1, AE2 and AE3 denote three AEs in the NSCS used for validation. *Source*: Taken and re-drawn from Sun *et al.* (2023).

SSTA in the outer and inner areas. The AE in the outer area triggers a low sea level pressure anomaly and convergent wind toward the AE, so that weakens TC inflow at the lower layers and further weakens TC secondary circulation. Consequently, the energy conversion from ocean heat energy to TC kinetic energy is decreased by the weakened TC secondary circulation. The AE in the outer area also extends the TC eyewall, weakens the TC warm core, and increases precipitation in the outer area. As a result, the TC intensity is weakened. On the other hand, the AE in the inner area causes opposite changes in TC secondary circulation, eyewall, warm core and precipitation, so that favors TC development (Sun *et al.*, 2023).

17.4 Summaries

The ocean mesoscale eddies and the atmospheric system make the tremendous and highly variable air-sea interactions in the SCS. During the processes, the exchange of the heat flux plays an important role. The strengthening or weakening of eddies and near-surface winds depends on the heat flux direction. Section 17.2 in this chapter summaries basin-scale air-sea dynamic environments in the SCS, which serve as the background knowledge for further understanding of occurrence conditions and mechanisms for basin scale air-sea coupling processes. They include geography-bottom boundary conditions, atmospheric forcing processes (East Asian monsoon and tropical cyclone), ENSO events and upper oceanic dynamic features (general seasonal circulation in the SCS, EKE active regions and basin-scale distribution of eddies). Section 17.3 summaries the most recent research results on atmospheric responses to mesoscale eddies in the SCS, including effects of eddy heating, water transport across air-sea boundary, ocean heat energy transformed to atmospheric mechanical energy and tropical cyclone responses to ocean eddies.

The above summarized research results indicate that the basin scale distributed energetical eddies in the SCS make the MABL conditions complicated and variable, which then raise difficulties for accurate prediction and forecasts of the regional weather system. Thus, more comprehensive and advanced understating of the atmospheric responses to ocean eddies needs to be reached by further enhanced observation and research, as well as development of more complete numerical environment models.

References

Byrne, D., Papritz, L., Frenger, I., Münnich, M. and Gruber, N. (2015). Atmospheric response to mesoscale sea surface temperature anomalies: Assessment of mechanisms and coupling strength in a high-resolution coupled model over the South Atlantic, *J. Atmos. Sci.*, 72(5), pp. 1872–1890, doi:10.1175/JAS-D-14-0195.1.

Cabrera, M., Santini, M., Lima, L., Carvalho, J., Rosa, E., Rodrigues, C. and Pezzi, L. (2022). The southwestern Atlantic Ocean mesoscale eddies: A review of their role in the air-sea interaction processes, *J. Mar. Sys.*, 235, 103785, doi:10.1016/j.jmarsys.2022.103785.

Chelton, D. (2013). Mesoscale eddy effects, *Nat. Geosci.*, 6(8), pp. 594–595, doi:10.1038/ngeo1894.

Chen, G., Hou, Y. and Chu, X. (2011). Mesoscale eddies in the South China Sea: Mean properties, spatiotemporal variability, and impact on thermohaline structure, *J. Geophys. Res. Oceans*, 116, C06018, doi:10.1029/2010JC006716.

Chen, H., Tkalich, P., Malanotte-Rizzoli, P. and Wei, J. (2012). The forced and free response of the South China Sea to the large-scale monsoon system, *Ocean Dyn.*, 62(3), pp. 377–393, doi:10.1007/s10236-011-0511-7.

Cheng, X. and Qi, Y. (2010). Variations of eddy kinetic energy in the South China Sea, *J. Oceanogr.*, 66, pp. 85–94, doi:10.1007/s10872-010-0007-y.

Chu, P. and Wang, G. (2003). Seasonal variability of thermohaline front in the central South China Sea, *J. Oceanogr.*, 59, pp. 65–78, doi:10.1023/A:1022868407012.

Frenger, I., Gruber, N., Knutti, R. and Münnich, M. (2013). Imprint of Southern Ocean eddies on winds, clouds and rainfall, *Nat. Geosci.*, 6(8), pp. 608–612, doi:10.1038/ngeo1863.

Hayes, S. P., McPhaden, M. J. and Wallace, J. M. (1989). The influence of sea-surface temperature on surface wind in the eastern equatorial Pacific: Weekly to monthly variability, *J. Climate*, 2(12), pp. 1500–1506, doi:10.1175/1520-0442(1989)002<1500:TIOSST>2.0.CO;2.

Hu, J., Kawamura, H., Hong, H. and Qi, Y. (2000). A review on the currents in the South China Sea: Seasonal circulation, South China Sea Warm Current and Kuroshio Intrusion, *J. Oceanogr.*, 56, pp. 607–624, doi:10.1023/A:1011117531252.

Jaimes, B. and Shay, L. K. (2009). Mixed layer cooling in mesoscale oceanic eddies during Hurricanes Katrina and Rita, *Mon. Weather Rev.*, 137(12), pp. 4188–4207, doi:10.1175/2009MWR2849.1.

Jaimes, B. and Shay, L. K. (2015). Enhanced wind-driven downwelling flow in warm oceanic eddy features during the intensification of Tropical Cyclone Isaac (2012): Observations and theory, *J. Phys. Oceanogr.*, 45(6), pp. 1667–1689, doi:10.1175/JPO-D-14-0176.1.

Li, L., Nowlin Jr, W. D. and Su, J. (1998). Anticyclonic rings from the Kuroshio in the South China Sea, *Deep Sea Res. Part I: Oceanogr. Res. Pap.*, 45(9), pp. 1469–1482, doi:10.1016/S0967-0637(98)00026-0.

Lin, I.-I., Wu, C.-C., Emanuel, K. A., Lee, I.-H., Wu, C.-R. and Pun, I.-F. (2005). The interaction of super typhoon Maemi (2003) with a warm ocean eddy, *Mon. Weather Rev.*, 133(9), pp. 2635–2649, doi:10.1175/MWR3005.1.

Lin, I. I., Pun, I. F. and Wu, C. C. (2009). Upper-ocean thermal structure and the western North Pacific category-5 typhoons. Part II: Dependence on translation speed, *Mon. Weather Rev.*, 137, pp. 3744–3757, doi:10.1175/2009MWR2713.1.

Lindzen, R. S. and Nigam, S. (1987). On the role of sea surface temperature gradients in forcing low-level winds and convergence in the tropics, *J. Atmos. Sci.*, 44(17), pp. 2418–2436, doi:10.1175/1520-0469(1987)044 <2418:OTROSS>2.0.CO;2.

Liu, H., Li, W., Chen, S., Fang, R. and Li, Z. (2018). Atmospheric response to mesoscale ocean eddies over the South China Sea, *Adv. Atmos. Sci.*, 35, pp. 1189–1204, doi:10.1007/s00376-018-7175-x.

Nan, F., Xue, H., Xiu, P., Chai, F., Shi, M. and Guo, P. (2011). Oceanic eddy formation and propagation southwest of Taiwan, *J. Geophys. Res. Oceans*, 116, C12045, doi:10.1029/2011JC007386.

Pan, G., Chai, F., Tang, D. and Wang, D. (2017). Marine phytoplankton biomass responses to typhoon events in the South China Sea based on physical-biogeochemical model, *Ecol. Modelling*, 356, pp. 38–47, doi:10.1016/j.ecolmodel.2017.04.013.

Renault, L., Molemaker, M. J., McWilliams, J. C., Shchepetkin, A. F., Lemarié, F., Chelton, D., Illig, S. and Hall, A. (2016). Modulation of wind work by oceanic current interaction with the atmosphere, *J. Phys. Oceanogr.*, 46(6), pp. 1685–1704, doi:10.1175/JPO-D-15-0232.1.

Renault, L., Marchesiello, P., Masson, S. and Mcwilliams, J. C. (2019). Remarkable control of western boundary currents by eddy killing, a mechanical air-sea coupling process, *Geophys. Res. Lett.*, 46(5), pp. 2743–2751, doi:10.1029/2018GL081211.

Seo, H., O'Neill, L. W., Bourassa, M. A., Czaja, A., Drushka, K., Edson, J. B., Fox-Kemper, B., Frenger, I., Gille, S. T., Kirtman, B. P., Minobe, S., Pendergrass, A. G., Renault, L., Roberts, M. J., Schneider, N., Small, J., Stoffelen, A. and Wang, Q. (2023). Ocean mesoscale and frontal-scale ocean–atmosphere interactions and influence on large-scale climate: A review, *J. Climate*, 36(7), pp. 1981–2013, doi:10.1175/JCLI-D-21-0982.1.

Shay, L. K., Goni, G. J. and Black, P. G. (2000). Effects of a warm oceanic feature on Hurricane Opal, *Mon. Weather Rev.*, 128(5), pp. 1366–1383, doi:10.1175/1520-0493(2000)128<1366:EOAWOF>2.0.CO;2.

Sun, J., Ju, X., Zheng, Q., Wang, G., Li, L. and Xiong, X. (2023). Numerical study of the response of typhoon Hato (2017) to grouped mesoscale eddies in the northern South China Sea, *J. Geophys. Res. Atmos.*, 128, e2022JD037266, doi:10.1029/2022JD037266.

Wallace, J. M., Mitchell, T. P. and Deser, C. (1989). The influence of sea-surface temperature on surface wind in the eastern equatorial Pacific: Seasonal and interannual variability, *J. Climate*, 2(12), pp. 1492–1499, doi:10.1175/1520-0442(1989)002<1492:TIOSST>2.0.CO;2.

Wang, D., Liu, Y., Qi, Y. and Shi, P. (2001). Seasonal variability of thermal fronts in the northern South China Sea from satellite data, *Geophys. Res. Lett.*, 28(20), pp. 3963–3966, doi:10.1029/2001GL013306.

Wang, G., Su, J. and Chu, P. C. (2003). Mesoscale eddies in the South China Sea observed with altimeter data, *Geophys. Res. Lett.*, 30(21), 2121, doi:10.1029/2003GL018532.

Wang, H., Wang, D., Liu, G., Wu, H. and Li, M. (2012). Seasonal variation of eddy kinetic energy in the South China Sea, *Acta Oceanol. Sin.*, 31, pp. 1–15, doi:10.1007/s13131-012-0170-7.

Wang, L., Koblinsky, C. J. and Howden, S. (2000). Mesoscale variability in the South China Sea from the TOPEX/Poseidon altimetry data, *Deep Sea Res. Part I: Oceanogr. Res. Pap.*, 47(4), pp. 681–708, doi:10.1016/S0967-0637(99)00068-0.

Wang, P. and Li, Q. (2009). *The South China Sea, Developments in Paleoenvironmental Research*, 13, Springer Science+Business Media B.V., Berlin, doi:10.1007/978-1-4020-9745-4.

Wang, Y. (2020). Composite of typhoon induced sea surface temperature and chlorophyll-a responses in the South China Sea, *J. Geophys. Res. Oceans*, 125, e2020JC016243, doi:10.1029/2020JC016243.

Wu, C.-C., Lee, C.-Y. and Lin, I.-I. (2007). The effect of the ocean eddy on tropical cyclone intensity, *J. Atmos. Sci.*, 64(10), pp. 3562–3578, doi:10.1175/JAS4051.1.

Wu, C. R., Shaw, P. T. and Chao, S. Y. (1999). Assimilating altimetric data into a South China Sea model, *J. Geophys. Res. Oceans*, 104(C12), pp. 29987–30005, doi:10.1029/1999JC900260.

Wyrtki, K. (1961). *Physical oceanography of the Southeast Asian waters*, NAGA, La Jolla, Calif., Rept. 2, pp. 1–195.

Xie, L., Zheng, Q., Zhang, S., Hu, J., Li, M., Li, J. and Xu, Y. (2018). Rossby wave normal modes in the South China Sea deep basin evidenced by satellite altimetry, *Intl. J. Rem. Sens.*, 39(2), pp. 399–417, doi:10.1080/01431161.2017.1384591.

Xie, S. P., Chang, C. H., Xie, Q. and Wang, D. (2007). Intraseasonal variability in the summer South China Sea: Wind jet, cold filament, and recirculations, *J. Geophys. Res. Oceans*, 112, C10008, doi:10.1029/2007JC004238.

Xie, S. P., Xie, Q., Wang, D. and Liu, W. T. (2003). Summer upwelling in the South China Sea and its role in regional climate variations, *J. Geophys. Res. Oceans*, 108(C8), 3261, doi:10.1029/2003JC001867.

Yuan, D., Han, W. and Hu, D. (2007). Anti-cyclonic eddies northwest of Luzon in summer–fall observed by satellite altimeters, *Geophys. Res. Lett.*, 34, L13610, doi:10.1029/2007GL029401.

Zhao, H., Tang, D. and Wang, Y. (2008). Comparison of phytoplankton blooms triggered by two typhoons with different intensities and translation speeds in the South China Sea, *Mar. Ecol. Progr. Ser.*, 365, pp. 57–65, doi:10.3354/meps07488.

Zheng, Z.-W., Ho, C.-R. and Kuo, N.-J. (2008). Importance of pre-existing oceanic conditions to upper ocean response induced by Super Typhoon Hai-Tang, *Geophys. Res. Lett.*, 35, L20603, doi:10.1029/2008GL035524.

Zheng, Z. W., Ho, C.-R., Zheng, Q., Lo, Y.-T., Kuo, N.-J. and Gopalakrishnan, G. (2010). Effects of preexisting cyclonic eddies on upper ocean responses to Category 5 typhoons in the western North Pacific, *J. Geophys. Res. Oceans*, 115(C9), C09013, doi:10.1029/2009JC005562.

Chapter 18

Mesoscale Eddies in Nearby Marginal Seas

18.1 Introduction

The core contents of this book are concentrated on mesoscale eddies in the South China Sea (SCS). Meanwhile, from Figure 18.1, one can see that the SCS relates to other marginal seas, i.e., the Andaman Sea on the Indian Ocean side, the Sulu Sea, the Celebes (Sulawesi) Sea and the Banda Sea on the Pacific side. Thus, it is of significance to explore the common and different features of mesoscale eddies in those marginal seas compared to that in the SCS, which would help for understanding of genesis and variation mechanisms of eddies.

The basic parameters of the SCS and nearby marginal seas are listed in Table 18.1. From Figure 18.1 and Table 18.1, one can see the common features of the marginal seas, which are favorable for genesis and evolution of mesoscale eddies: (1) all of them are characterized by semi-enclosed water bodies with deep basins deeper than 2000 m located in lower latitudes; (2) their total areas are larger than $10 \times 10^4\,\text{km}^2$, providing the spaces large enough for development of mesoscale eddies and (3) all the Pacific marginal seas sit in Indonesian Throughflow (ITF) pathways (He *et al.*, 2015).

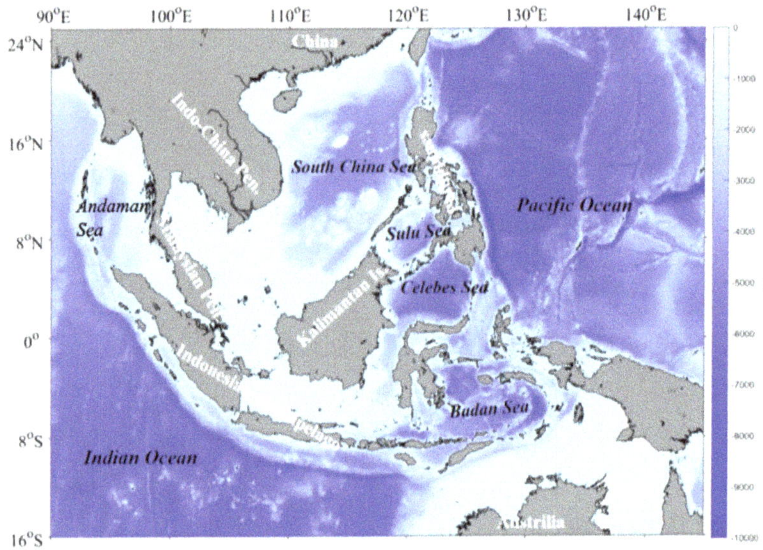

Fig. 18.1. A chart of the SCS and nearby marginal seas. Color codes represent the water depths in meter.

Table 18.1. Basic parameters of the SCS and nearby marginal seas.

Name	Deep basin location	Depth (m)	Total area (km^2)	ITF pathway
South China Sea	10°–19°N, 110°–122°E	2000–5559	350×10^4	West
Andaman Sea	6°–16°N, 92°–98°E	2000–4500	60×10^4	
Sulu Sea	6°–10°N, 119°–123°E	4000–5600	10×10^4	West/central
Celebes Sea	1°–5°N, 118°–125°E	6000–8574	43.5×10^4	Central
Banda Sea	4°–8°S, 123°–132°E	3000–7440	47×10^4	East/central

18.2 Mesoscale Eddies in the Andaman Sea

18.2.1 *Spatial distribution*

The Andaman Sea (AS) is a marginal sea of the northeastern Indian Ocean as shown in Figure 18.1. It connects the Indian Ocean through the Ten Degree Channel and the Gretel Strait on the west side and the SCS through the Strait of Malacca on the southeast side. It is an important navigation channel from the SCS to the Indian Ocean.

Previous investigators have figured out that the Andaman Sea is an ocean area with active mesoscale eddies (Cui *et al.*, 2016; Huang *et al.*, 2020). Using mesoscale eddy trajectory product from the Archiving Validation and Interpolation of Satellite Oceanographic Data (AVISO), Lin *et al.* (2023) analyzed the climatology characteristics and seasonal variability of mesoscale eddies in the AS. Their results show that a total of 328 mesoscale eddies were generated in the AS during 27 years from 1993 to 2019, of which anticyclonic eddies (AEs) and cyclonic eddies (CEs) account for 171 and 157, respectively. The eddies are mainly distributed in the central and western deep basin as shown in Figure 18.2.

From Figure 18.2(a–d), one can see that the spatial distribution patterns of mesoscale eddies in the AS show two remarkable features: (1) eddies of the same polarity are relatively concentrated in certain regions, i.e., AC1-AC6 for AEs and C1-C6 for CEs and (2) eddy polarities in the concentration regions alternate seasonally from summer to winter. Figure 18.2(a) shows that in spring, there are 39 AEs concentrated in regions AC1 in the northwestern basin and AC2 near the Malay Peninsula and 42 CEs in region C1 in the southwestern basin. Figure 18.2(b) shows that in summer, there are total 79 eddies distributed in a CE–AE–CE alternative pattern from north to south, i.e., C2 (34 CEs)- AC3 (22 AEs)- C3 (10 CEs). Figure 18.2(c) shows the case in fall, which is like that in spring. The difference is that the northwestern concentration region alternates from AC1 in spring to CE dominant C4. Figure 18.2(d) shows the case in winter (62 AEs and 26 CEs). The eddy spatial distribution appears the same polarity alternative patterns as that in summer, but the polarities alternate to AE-CE-AE, implying the possibility of monsoon influence.

18.2.2 *Seasonal distribution*

Figure 18.3(a–d) show seasonal distribution histograms of eddy features in the AS. From Figure 18.3(a), one can see that the maximum total eddy number of 88 appears in winter, and the minimum 79 in summer. A remarkable phenomenon is that the AE number is twice as CE number in winter. Figure 18.3(b) shows that the sea level anomaly (SLA) amplitude of AEs appears as a seasonal increase trend from spring to winter, while CEs as a decrease trend. Figure 18.3(c) shows that the eddy radius has the same seasonal variation trends as that of SLA amplitude. Figure 18.3(d) shows the seasonal increase trends of rotating speeds for AEs and CEs.

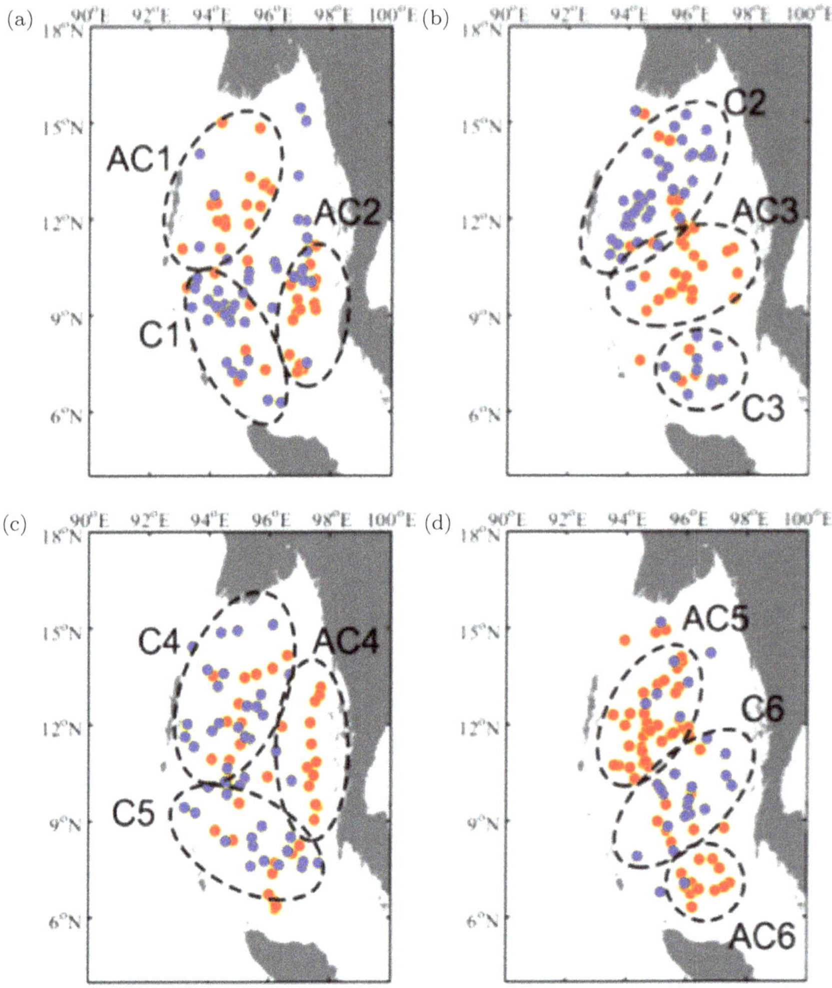

Fig. 18.2. Seasonal variations of the locations of mesoscale eddies in the Andaman Sea from 1993 to 2019. (a) Spring, (b) summer, (c) fall and (d) winter. Codes AC1, C1 and so on mark the cluster regions of AEs (red dots) and CEs (blue dots).

Source: Taken and redrawn from Lin *et al.* (2023).

18.2.3 *Statistics of dynamic parameters*

Figure 18.4(a–d) show probability distribution histograms of eddy dynamic characteristics in the AS. From Figure 18.4(a), one can see

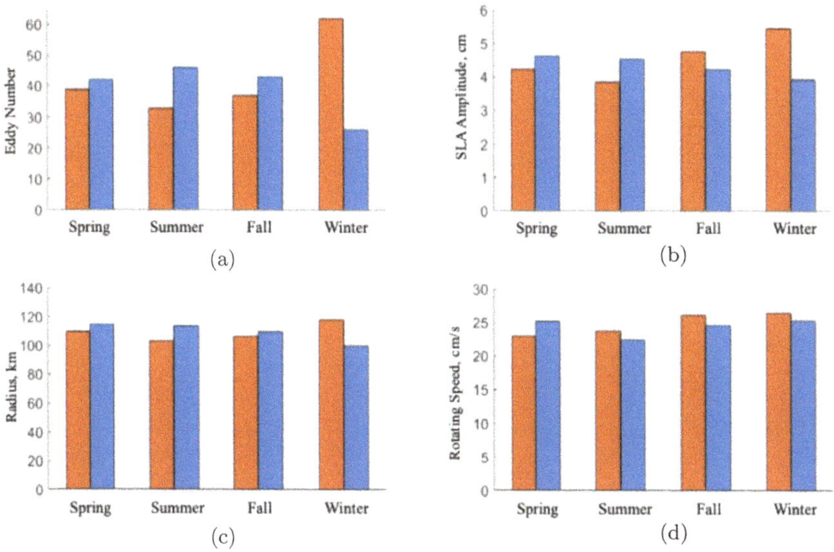

Fig. 18.3. Seasonal distribution histograms of mesoscale AEs (red bars) and CEs (blue bar) in the AS: (a) eddy number, (b) SLA amplitude (cm), (c) radius (km) and (d) maximum rotating speed (cm s^{-1}).

Source: Taken and redrawn from Lin *et al.* (2023).

that the radius distributions of AEs and CEs appear as the same Rayleigh distribution. 75.74% of eddy radii are concentrated within a range of 65–145 km with an average radius of 111.8 km. Figure 18.4(b) shows that the SLA amplitude distributions of AEs and CEs appear as the Rayleigh distribution with an average amplitude of 4.7 cm. The peak values are located at 2–3 cm. Eddies with amplitudes smaller than 7.5 cm account for 85.53% of total number, and strong eddies larger than 8.5 cm for 8.30%. Figure 18.4(c) shows that the rotating speed distributions of AEs and CEs appear as the Gaussian distribution with an average of 24.8 cm s^{-1}. The peak values are located at 23–25 cm s^{-1}. Figure 18.4(d) shows that the propagating speed distributions of AEs and CEs appear as the Rayleigh distribution with an average of 15.0 cm s^{-1}. The peak values are located at 6 cm s^{-1} for AEs and 8 cm s^{-1} for CEs. 58.52% of eddy propagating speeds are concentrated within a range of 4–14 cm s^{-1}. AEs and CEs account for 59.93% and 57.07% of total eddy number.

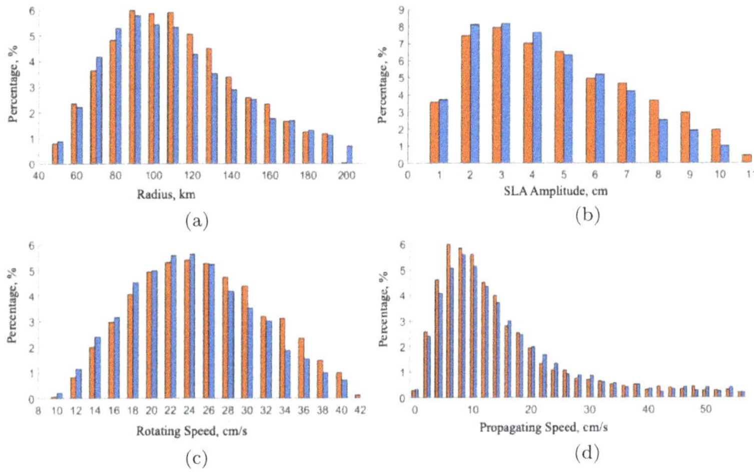

Fig. 18.4. Statistical probability distribution histograms of eddy dynamic characteristics in the AS: (a) radius, (b) SLA amplitude, (c) maximum rotating speed and (d) propagating speed. Red (blue) bars represent AEs (CEs).
Source: Taken and redrawn from Lin *et al.* (2023).

18.2.4 *Statistics of eddy lifetime and propagation distance*

The eddy lifetime is defined as the time interval from the genesis to the disappearance. Figure 18.5(a) shows statistics of the eddy number vs the eddy lifetime in the AS. One can see that the shortest (longest) lifetime is 28 (131) days. The mean lifetime of 45.5 days for AEs is shorter than 47.4 days for CEs. 62.95% of eddy lifetimes are within a range of 25–45 days, which are much shorter than 61.6 days in the SCS.

The eddy propagating distance is defined as the distance from the genesis location to the disappearance location. Figure 18.5(b) shows statistical probability of eddy propagating distance in the AS. One can see that 83.13% of eddy propagating distances are shorter than 350 km. The mean propagating distance of 241.1 km for AEs is longer than 221.5 km for CEs.

18.3 Mesoscale Eddies in the Sulu Sea

The Sulu Sea is a semi enclosed marginal sea of the western North Pacific (WNP) located at $6° - 10°$N, $119° - 123°$E. It is bordering the

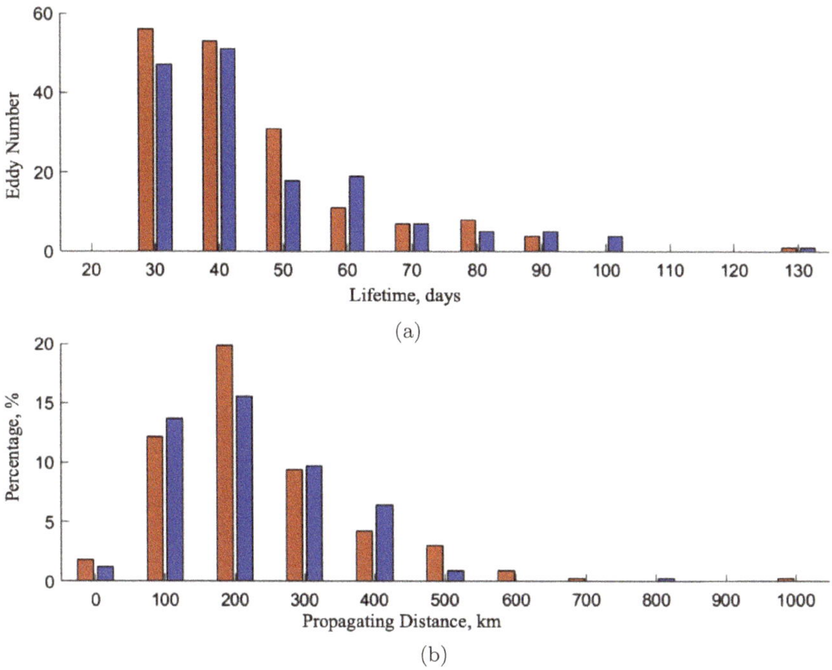

Fig. 18.5. Statistics of (a) eddy lifetime and (b) propagating distance in the AS. Red (blue) bars represent AEs (CEs).
Source: Taken and redrawn from Lin *et al.* (2023).

SCS, the Celebes Sea and the WNP as shown in Figure 18.6(a). Its total area is about $10 \times 10^4\,\mathrm{km}^2$ with a deep basin of 4000–5600 m. The Sulu Sea provides two significant secondary routes for both the Indonesian Throughflow and the extension of the Pacific northern tropic gyre, so that water exchanges of the Sulu Sea with external water bodies have been intensively studied (Cai *et al.*, 2008; Gordon *et al.*, 2011b; Qu and Song, 2009; Sprintall *et al.*, 2012; Tessler *et al.*, 2010; He *et al.*, 2017).

Figure 18.6(b) shows combined spaceborne and shipboard-mounted acoustic Doppler current profiler (ADCP) cruise observations from 10 March to 19 March 2009 during the Philippines Straits Dynamics Experiment (PhilEx) regional cruise (Gordon *et al.*, 2011a). One can see that the Sulu Sea is filled with mesoscale AEs and CEs with a horizontal scale (radius) $O(100)$ km, SLA amplitudes of AEs as high as 10 cm and rotating speed of $0.5\,\mathrm{m\,s^{-1}}$.

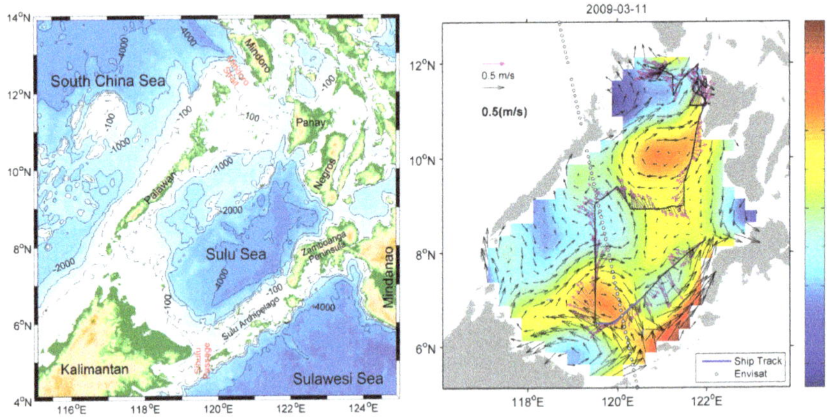

(a) (b)

Fig. 18.6. (a) Bathymetry of the Sulu Sea from ETOPO1. (b) Current vectors (purple arrows) from the shipboard-mounted 150 kHz ADCP cruise within the depths of 25–55 m in March 2009. The cruise track is shown as black and blue solid lines. Geostrophic currents (black arrows) are derived from satellite SLA data (color shading) on 11 March 2009. White circles show the satellite Envisat ground track passing over the Sulu Sea, which is just across the cruise track (blue thick line) on the same day.
Source: Taken from He *et al.* (2017).

18.3.1 *Spatial distribution of eddy genesis and decay events as well as propagation*

From 22 years of SLA data from January 1993 to December 2014 (Chelton *et al.*, 2011), He *et al.* (2017) identified total 6668 CEs and 6357 AEs in the Sulu Sea, i.e., average 1.6 eddies per day. Figure 18.7(b) and 18.7(c) show the geographical distribution of the number of the eddy genesis and decay events at each $1^\circ \times 1^\circ$ latitude-longitude box. The locations of eddy genesis and decay correspond to the initial and last points of an eddy track, respectively. A total of 225 AE tracks and 243 CE tracks are identified. An average of 1.8 eddies generated per month. Two pronounced maxima of eddy genesis are in west of Negros Island and the Zamboanga Peninsula. In general, eddies are more likely generated in the central and northeastern Sulu Sea, while eddies usually decay in the central and southwestern Sulu Sea.

Figure 18.7(d) shows the mean eddy velocity propagation field calculated by a forward difference scheme at each day of eddy centroid

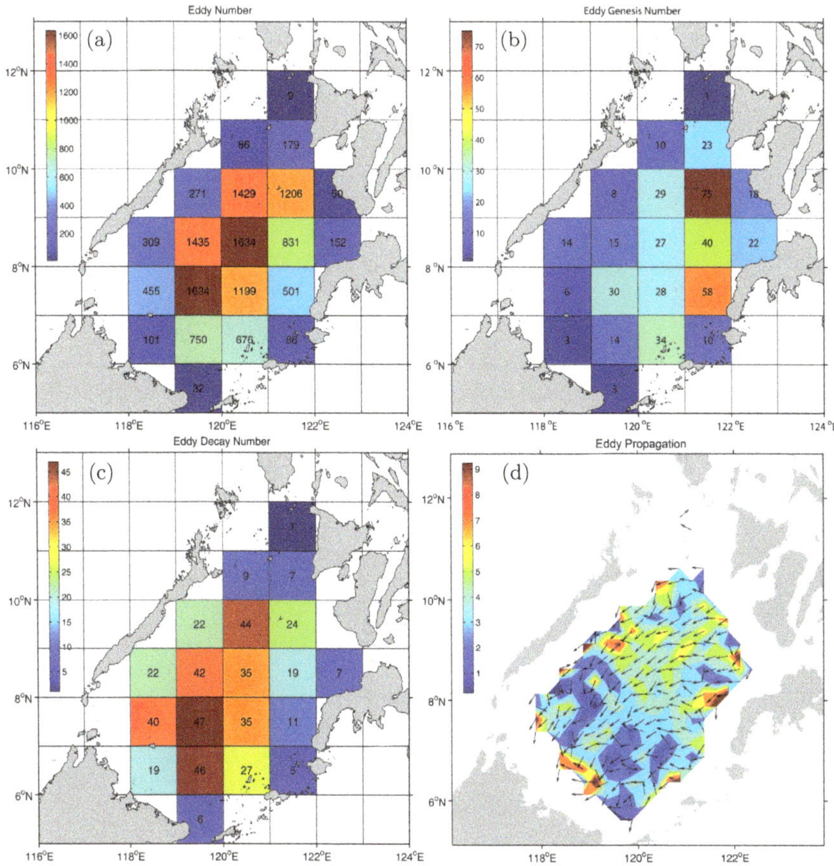

Fig. 18.7. Spatial distribution of (a) number of all identified eddies, (b) number of eddy genesis events (c) number of eddy decay events and (d) eddy propagation velocity vectors (black arrows) and magnitude (color codes in $\text{cm}\,\text{s}^{-1}$).
Source: Taken from He *et al.* (2017).

displacement. To reduce the random noise, the velocities were further filtered by a central moving average of 7 days and then averaged in every $0.25^\circ \times 0.25^\circ$ grid. The Sulu Sea eddy propagation speed is approximately $4.5 \pm 2.7\,\text{cm}\,\text{s}^{-1}$. In the northwestern Sulu Sea, eddies commonly propagate southwestward, which is parallel to the long axis of the Sulu Sea basin and the isobaths. In the eastern Sulu Sea basin, eddies generally propagate westward with small southward migrations (He *et al.*, 2017).

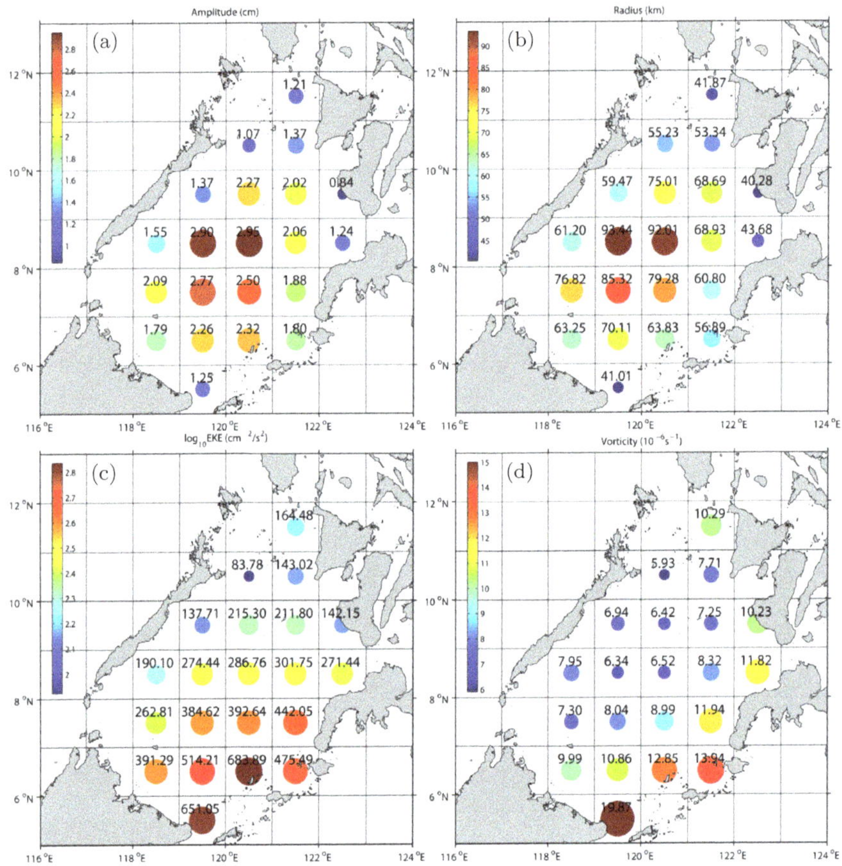

Fig. 18.8. Spatial distribution of mean (a) amplitude (cm), (b) radius (km), (c) EKE ($cm^2\,s^{-2}$) and (d) vorticity ($10^{-6}\,s^{-1}$) of mesoscale eddies in the Sulu Sea derived from SLA data from January 1993 to December 2014.

Source: Taken from He *et al.* (2017).

18.3.2 *Spatial distribution of eddy properties*

Figure 18.8(a–d) show 1° × 1° latitude-longitudinal distributions of mean amplitude, radius, eddy kinetic energy (EKE) and vorticity of mesoscale eddies in the Sulu Sea derived from SLA data from January 1993 to December 2014. One can see that the maximum eddy amplitude of 2.95 cm appears at 8.5°N, 120.5°E in the central Sulu Sea, while the minimum value of 0.84 cm is at 9.5°N, 122.5°E to the west of Negros Island (Figure 18.8(a)). The maximum eddy radius of 93.44 km is at 8.5°N, 119.5°E near the central Sulu Sea, while the minimum 40.28 km at 9.5°N, 122.5°E to the west of Negros Island (Figure 18.8(b)). The maximum mean EKE of 683.89 $cm^2 s^{-2}$ is at 6.5°N, 120.5°E in the southeastern Sulu Sea, while the minimum 83.78 $cm^2 s^{-2}$ at 10.5°N, 120.5°E in the northern Sulu Sea (Figure 18.8(c)). The maximum mean vorticity of $19.87 \times 10^{-6} s^{-1}$ is at 5.5°N, 119.5°E in the south end of the Sulu Sea, while the minimum $5.93 \times 10^{-6} s^{-1}$ at 10.5°N, 120.5°E in the northern Sulu Sea (Figure 18.8(d)).

18.3.3 *Seasonal variability of eddy properties*

Figure 18.9(a–f) show the seasonal variability of mean eddy properties the Sulu Sea derived from SLA data from January 1993 to December 2014. Figure 18.9(a) shows that the seasonal distribution of eddy numbers can be divided into two seasons: boreal summer half year (BSHY) from April to September and boreal winter half year (BWHY) from October to March of next year. In BSHY, the average monthly mean number of AEs is greater than 1, while in BWHY, it is smaller than 0.5. The seasonal variability of CE numbers is antiphase with that of AEs, i.e., in BSHY, the average monthly mean number of CEs is near 0.5, and greater than 1 in BWHY.

Figure 18.9(b) shows that the seasonal distribution of eddy amplitudes can be also divided into two seasons: BSHY and BWHY. In BSHY, the average monthly mean amplitude of AEs is greater than 2.5 cm, while in BWHY, it is smaller than 1.7 cm. The seasonal variability of CE amplitudes is antiphase with that of AEs, i.e., the average monthly mean amplitudes of CEs is near 2.0 cm in BSHY and near 2.5 cm in BWHY.

Figure 18.9(c) shows that the seasonal variability of monthly mean AE radius can be also divided into two seasons: BSHY and BWHY, it reaches a maximum of 91.8 km in August and a minimum of 54.0 km in February. The monthly mean CE radius reaches a maximum of 88.9 km in February, while a minimum of 58.7 km in July.

Figure 18.9(d) shows that the seasonal variability of the mean EKE of AEs and CEs appears as a bimodal pattern. Two peaks of the EKE of AEs appear at 342 and 435 $cm^2 s^{-2}$ in May and October and two troughs of 186 and 310 $cm^2 s^{-2}$ appear in January and July, respectively. An annual maximum EKE of 483 $cm^2 s^{-2}$

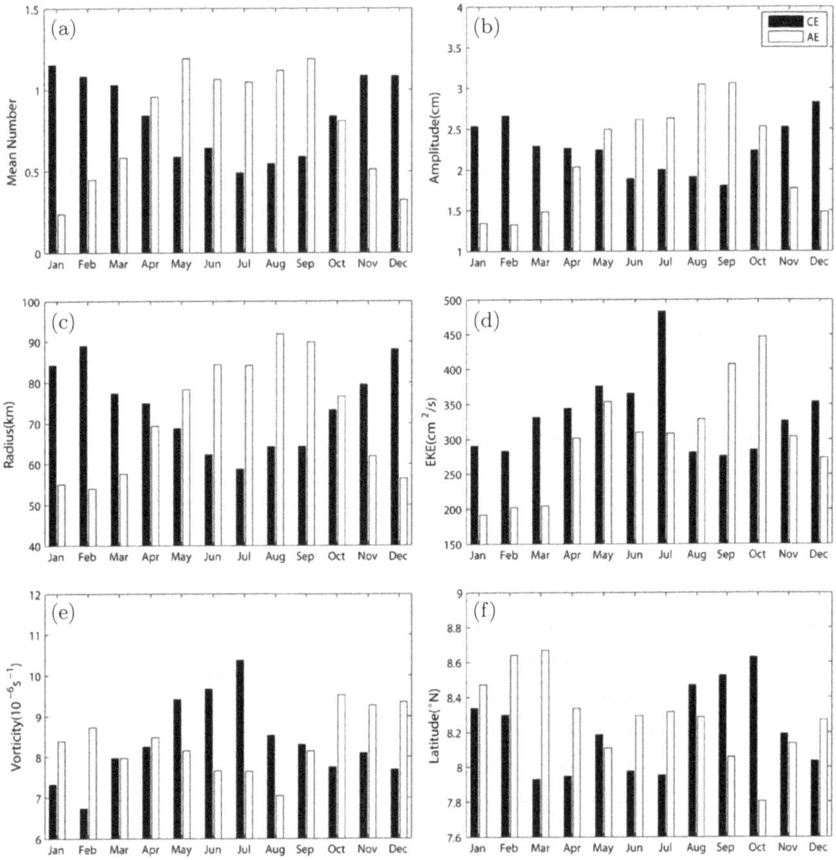

Fig. 18.9. Seasonal variability of properties of AEs (open bars) and CEs (solid bars): (a) number of eddies, (b) amplitude (cm), (c) radius (km), (d) mean EKE ($cm^2\,s^{-2}$), (e) mean vorticity ($10^{-6}\,s^{-1}$) and (f) mean latitude of eddy centers (°N) in the Sulu Sea derived from SLA data from January 1993 to December 2014.

Source: Taken from He *et al.* (2017).

for CEs appears in July, drops dramatically to an annual minimal value of $276\,cm^2\,s^{-2}$ in September, gradually reaches a local maximum of $353\,cm^2\,s^{-2}$ in December and drops to a local minimum of $283\,cm^2\,s^{-2}$ in February of next year.

Figure 18.9(e) shows the seasonal variability of the monthly mean vorticity of AEs and CEs. The AE vorticity is weak in boreal summer and reaches an annual minimum of $-7.0 \times 10^{-6}\,s^{-1}$ in August, and an annual maximum of $-9.5 \times 10^{-6}\,\mathrm{s}^{-1}$ in October. The CE vorticity reaches a minimum of $6.7 \times 10^{-6}\,\mathrm{s}^{-1}$ in February and a maximum of $10.4 \times 10^{-6}\,\mathrm{s}^{-1}$ in July.

Figure 18.9(f) shows the seasonal variability of the monthly mean latitude of AE and CE centers. One can see that the average monthly mean latitudes of AE and CE centers are 8.3°N and 8.2°N with the maximum fluctuation range of ±0.2° latitude.

18.4 Mesoscale Eddies in the Indonesian Seas

Hao *et al.* (2021) investigated the spatiotemporal distributions and generation mechanisms of mesoscale eddies in the Indonesian Seas detected from AVISO satellite altimeter SLA data from January 1992 to December 2019 by the SSH-based automatic identification algorithm developed by Faghmous *et al.* (2015). The Indonesian Seas, from north to south including the Sulu Sea, the Sulawesi (Celebes) Sea, the Maluku Sea and the Banda Sea as shown in Figure 18.10,

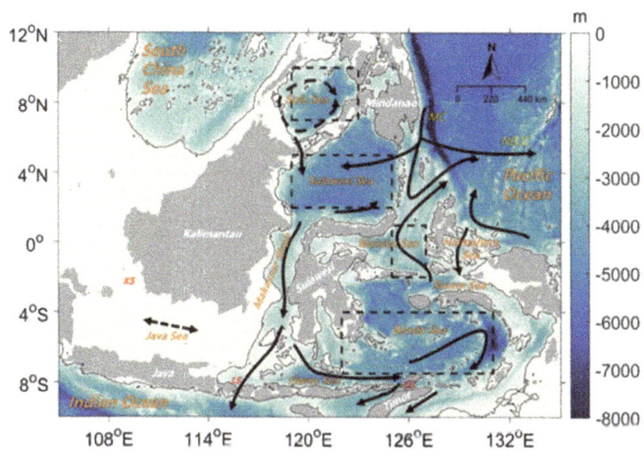

Fig. 18.10. Schematic of the upper-ocean circulation in the Indonesian Seas. Color shading is the bathymetry from ETOPO1. The thin solid lines are isobaths of 200 m. KS, LS and OS in red represent the Karimata Strait, Lombok Strait and Ombai Strait, respectively. MC and NECC in yellow represent the Mindanao Current and the North Equatorial Countercurrent, respectively. Four dashed boxes from north to south mark the Sulu Sea, the Sulawesi Sea, the Maluku Sea and the Banda Sea.

Source: Taken from Hao *et al.* (2021).

provide the sole tropical pathway from the Pacific to the Indian Ocean known as the Indonesian Throughflow (ITF) (Feng *et al.*, 2018; Sprintall *et al.*, 2019), meanwhile, they are the marginal seas with intensive activities of mesoscale eddies. The statistical features of mesoscale eddies in the Sulu Sea have been specified in Section 18.3. Thus, this section will summarize the results of the other seas.

18.4.1 *Spatial distributions of eddies*

In the Indonesian Seas, total 46,676 AEs and 47,004 CEs with amplitudes larger than 2 cm were identified from January 1993 to December 2018. Most identified eddies are concentrated on four larger and deeper basins, i.e., the Sulu Sea, the Sulawesi Sea, the Maluku Sea and the Banda Sea as shown in Figure 18.11(a) and 18.11(b) (Hao *et al.*, 2021).

18.4.2 *Genesis and decay*

Figure 18.12(a–d) show the spatial distributions of the numbers of eddy genesis and decay events for AEs and CEs in the Indonesian Seas from 1993 to 2018. The locations for eddy generation and decay

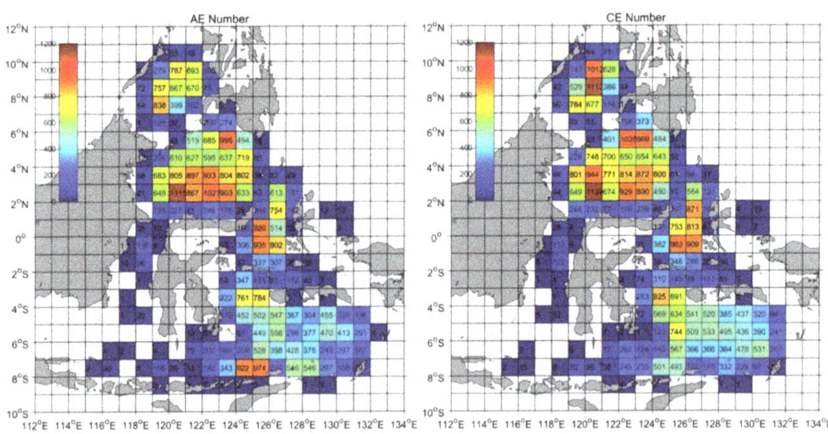

(a) (b)

Fig. 18.11. Spatial distribution of the numbers of identified (a) AEs and (b) CEs in the Indonesian Seas from 1993 to 2018.
Source: Taken from Hao *et al.* (2021).

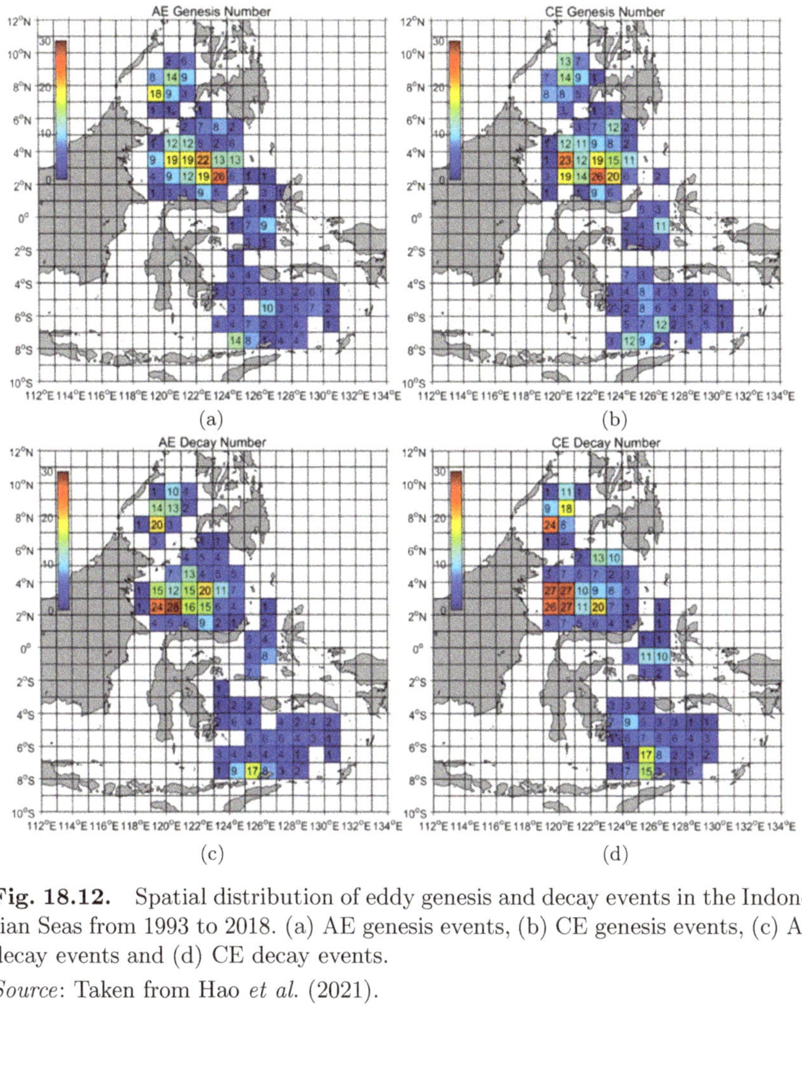

Fig. 18.12. Spatial distribution of eddy genesis and decay events in the Indonesian Seas from 1993 to 2018. (a) AE genesis events, (b) CE genesis events, (c) AE decay events and (d) CE decay events.

Source: Taken from Hao *et al.* (2021).

events are defined as the first and the last points in each eddy track, respectively. Because the minimum resolvable temporal scale of gridded AVISO SLA product is about 10 days (Pujol *et al.*, 2016), those eddies with a lifetime shorter than 10 days are excluded. Consequently, total 469 AE tracks and 500 CE tracks are identified. AE tracks in the Sulu Sea, the Sulawesi Sea, the Maluku Sea and the Banda Sea account for 15%, 53%, 7% and 25% of the total AE tracks, while CE tracks for 15%, 52%, 7% and 27%, respectively (Hao *et al.*, 2021).

18.4.3 *Propagation velocity vectors*

The propagation velocities of eddies are calculated by a forward difference scheme for eddy centroid displacement every day. The velocities of each eddy track were further filtered using a central moving average of 7 days to reduce the random noise and then averaged in each $0.25° \times 0.25°$ grid, following He *et al.* (2017). Figure 18.13(a) and 18.13(b) show the mean propagation velocity fields and the standard deviation (STD) of azimuth of propagation direction relatively to the west for AEs and CEs. One can see that in the Sulu Sea, the mean translation speeds of AEs and CEs are $2.3 \pm 1.0\,\mathrm{cm\,s^{-1}}$ and $0.9 \pm 0.5\,\mathrm{cm\,s^{-1}}$, respectively. AEs move southeastward in the northeastern half basin because of the advection of mean flow from the Mindoro Strait during boreal spring and summer when AEs are

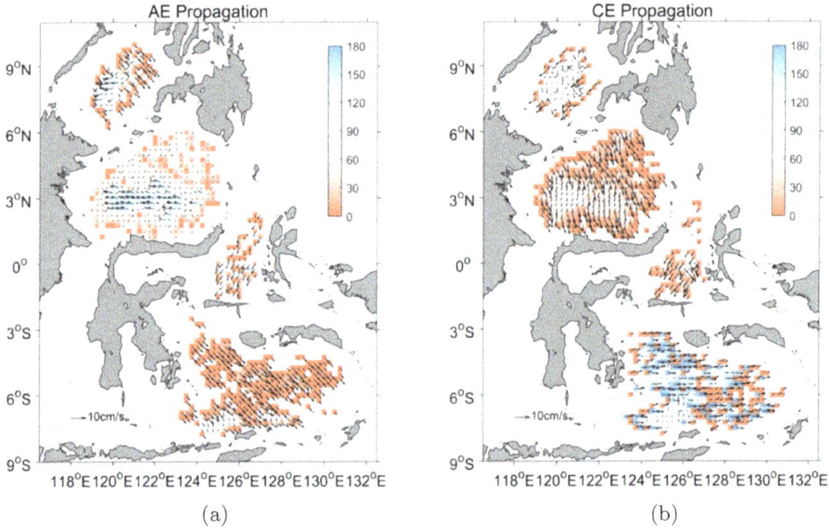

Fig. 18.13. Spatial distribution of propagation velocity vectors (black arrows, in $\mathrm{cm\,s^{-1}}$) for (a) AEs and (b) CEs in the Indonesian Seas from 1993 to 2018. Color codes represent the STD of propagation azimuth direction relative to west at each $0.25° \times 0.25°$ grid (in degree).
Source: Taken from Hao *et al.* (2021).

more frequent. In the Sulawesi Sea, the mean AE translation speed is $2.5 \pm 2.3\,\mathrm{cm\,s^{-1}}$, with westward propagation in the central basin. The CE migrations across the entire basin with a mean speed of $6.3 \pm 2.7\,\mathrm{cm\,s^{-1}}$ are significantly affected by the intense intrusion of MC. In the Maluku Sea, all eddies propagate southwestward, i.e., parallel to the major axis of basin at a mean velocity of $1.9 \pm 0.7\,\mathrm{cm\,s^{-1}}$ for AEs and $1.7 \pm 0.6\,\mathrm{cm\,s^{-1}}$ for CEs. In the Banda Sea, AEs propagate southeastward at a mean velocity of $5.7 \pm 2.5\,\mathrm{cm\,s^{-1}}$ and CEs eastward at $3.9 \pm 2.5\,\mathrm{cm\,s^{-1}}$ (Hao *et al.*, 2021).

18.4.4 *Eddy lifetime*

Figure 18.14(a–d) show the upper-tail cumulative histograms of the eddy lifetime in the Indonesian Seas. Numbers of AEs (CEs) of lifetimes within a range of 10–30 days account for 73% (84%), 85% (86%), 100% (94%) and 86% (80%) in the Sulu Sea, the Sulawesi

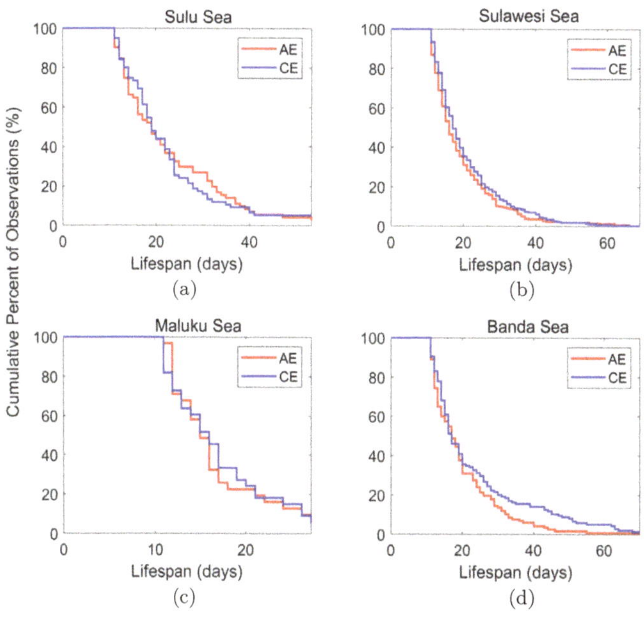

Fig. 18.14. Upper-tail cumulative histograms of the lifetime of AEs (red curve) and CEs (blue curve) in the (a) Sulu Sea, (b) Sulawesi Sea, (c) Maluku Sea and (d) Banda Sea.

Source: Taken from Hao *et al.* (2021).

Sea, the Maluku Sea and the Banda Sea, respectively. The lifetime distributions of AEs and CEs have a generally similar trend with a rapid decrease within 40 days and a slow decrease beyond 40 days except for the Maluku Sea where the eddy lifetime is mostly shorter than 30 days (Hao *et al.*, 2021).

18.4.5 *Spatial distributions of eddy properties*

The spatial distributions of amplitudes and radii of AEs and CEs averaged in $1^\circ \times 1^\circ$ grid in the Indonesian Seas are shown in Figure 18.15(a–d) (Hao *et al.*, 2021). Eddies have typical amplitudes of 2–6 cm and radii of 50–160 km. The similarity between distribution patterns of amplitudes and radii indicates the positive correlation of between the two. Meanwhile, both amplitudes and radii in the interior basin are greater than that near the coast in the Sulu, Sulawesi and Banda Seas, implying the strong energy dissipation near the coast (He *et al.*, 2017). On the other hand, the amplitudes and radii of AEs and CEs in the Maluku Sea and the Seram Sea are smaller due to the small basin scale. The large values of eddy radii are observed in the Banda Sea, where the majority exceed 100 km with the maximum value of 157 km observed in the central basin. Additionally, the amplitudes and radii of CEs are greater than that of AEs in the Sulawesi Sea, where MC intrudes into the sea with a cyclonic loop flow structure (Lukas *et al.*, 1991).

Figure 18.16(a–d) show the spatial distributions of mean EKE and mean relative vorticity of AE and CE averaged in $1^\circ \times 1^\circ$ grid in the Indonesian Seas. There are no significant differences among the spatial patterns of mean EKE of AEs and CEs. The most energetic eddies are observed in the Sulawesi Sea and the Seram Sea with high mean-flow kinematic energy (MKE) (He *et al.*, 2015). The highest eddy amplitude is observed in the central Banda Sea, however, the value of mean EKE is smaller than that in the Sulawesi Sea due to larger eddy area. The spatial distribution of mean relative vorticity of AEs and CEs are generally similar. Eddies with high mean relative vorticity were found in the Sulawesi Sea and the Seram Sea, which is like the distribution of mean EKE. However, compared with other eddy properties, high value of mean vorticity is located near the coast instead of in the interior basin (Hao *et al.*, 2021).

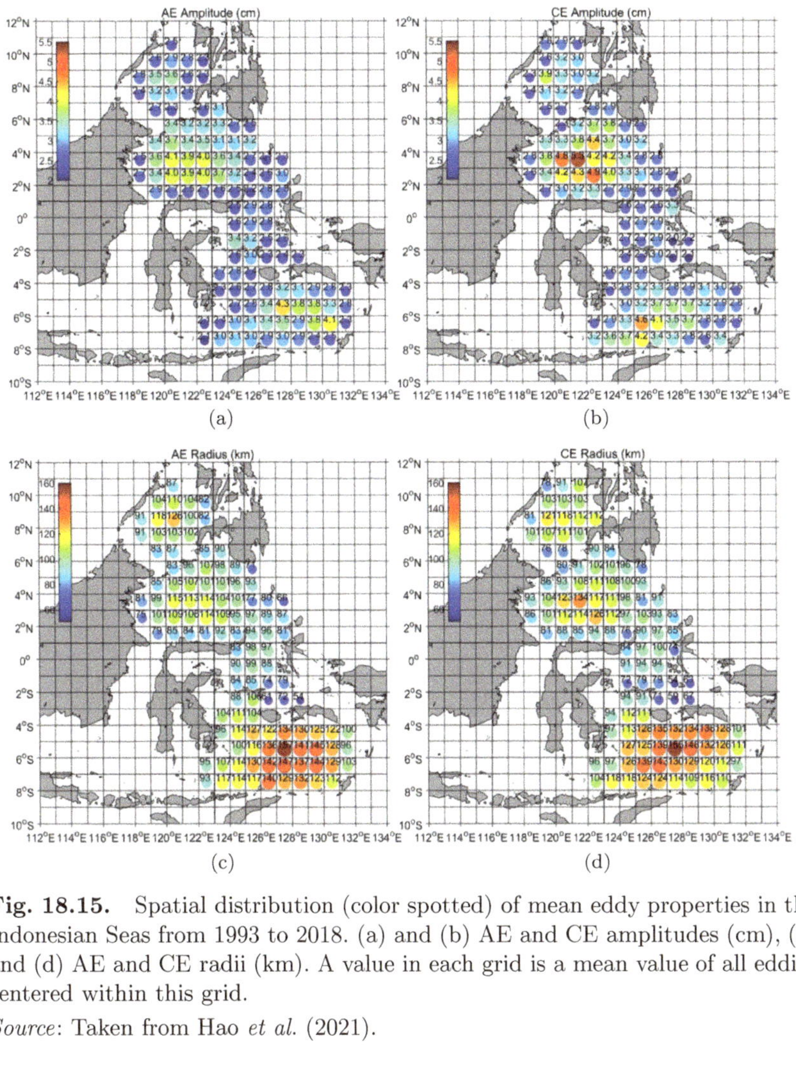

Fig. 18.15. Spatial distribution (color spotted) of mean eddy properties in the Indonesian Seas from 1993 to 2018. (a) and (b) AE and CE amplitudes (cm), (c) and (d) AE and CE radii (km). A value in each grid is a mean value of all eddies centered within this grid.

Source: Taken from Hao *et al.* (2021).

18.4.6 *Seasonal variability of eddy properties*

Figure 18.17(a–f) show the seasonal variability of eddy mean properties in the Sulawesi Sea from 1993 to 2018. One can see that the seasonal variability of mean number, amplitude and radius of AEs and CEs follow the bimodal curves. For AEs, the mean number, amplitude and radius reach annual minimum values of 0.6, 3.2 cm and 97.6 km in September, and annual maximum values of 1.0, 4.1 cm and

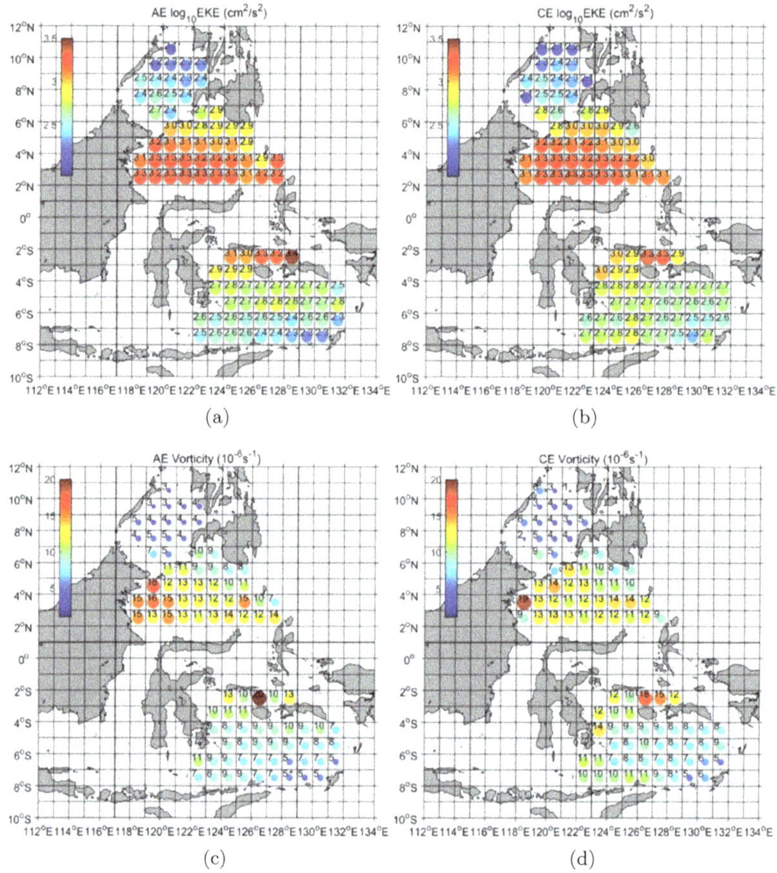

Fig. 18.16. Spatial distribution (color spotted) of mean eddy properties in the Indonesian Seas from 1993 to 2018. (a) and (b) mean AE and CE EKE (cm^2 s^{-1} in logarithm scale), (c) and (d) mean AE and CE vorticities (10^{-6} s^{-1}). A value in each grid is a mean value of all eddies centered within this grid.
Source: Taken from Hao *et al.* (2021).

111.8 km in boreal winter. Then they attain local minimum values of 0.7, 3.5 cm and 99.9 km in May and reach local maximum values of 0.8, 3.6 cm and 105.2 km in July. Meanwhile the mean number, amplitude and radius of CEs vary in antiphase with that of AEs. It is noted that CEs have amplitudes larger than that of AEs through the whole year. The latitudes of eddy centers of AEs and CEs present an opposite tendency, i.e., more AEs concentrated in the southern

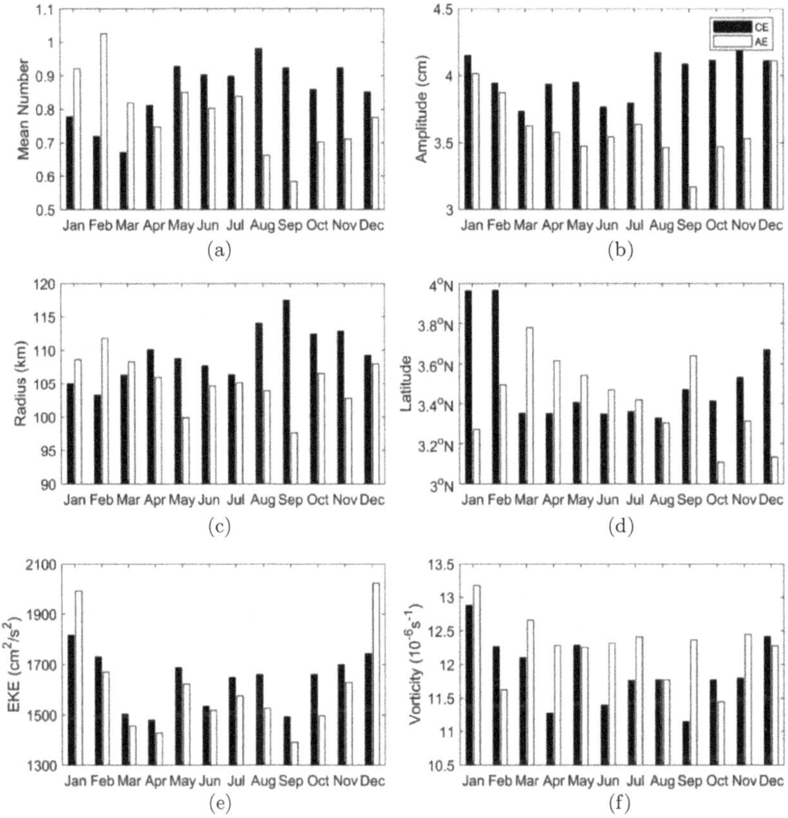

Fig. 18.17. Seasonal variability of properties of AEs (open bars) and CEs (solid bars) in the Sulawesi Sea. (a) Number of eddies, (b) amplitude (cm), (c) radius (km), (d) central latitude (°N), (e) mean EKE ($\mathrm{cm}^2\,\mathrm{s}^{-2}$) and (f) mean vorticity ($10^{-6}\,\mathrm{s}^{-1}$).

Source: Taken from Hao *et al.* (2021).

Sulawesi Sea from March to September, while more CEs in the northern Sulawesi Sea from October to February.

The seasonal variability of mean EKE and mean relative vorticity of CEs have the bimodal structure. Two peak values of $1816\,\mathrm{cm}^2\,\mathrm{s}^{-2}$ and $12.9 \times 10^{-6}\,\mathrm{s}^{-1}$ appear in January and $1688\,\mathrm{cm}^2\,\mathrm{s}^{-2}$ and $12.3 \times 10^{-6}\,\mathrm{s}^{-1}$ in May, while two trough values of $1480\,\mathrm{cm}^2\,\mathrm{s}^{-2}$ and $11.3 \times 10^{-6}\,\mathrm{s}^{-1}$ appear in April and $1493\,\mathrm{cm}^2\,\mathrm{s}^{-2}$ and $11.2 \times 10^{-6}\,\mathrm{s}^{-1}$ in September. The mean EKE of AEs have a similar pattern to that of CEs, with two peaks in December and May and

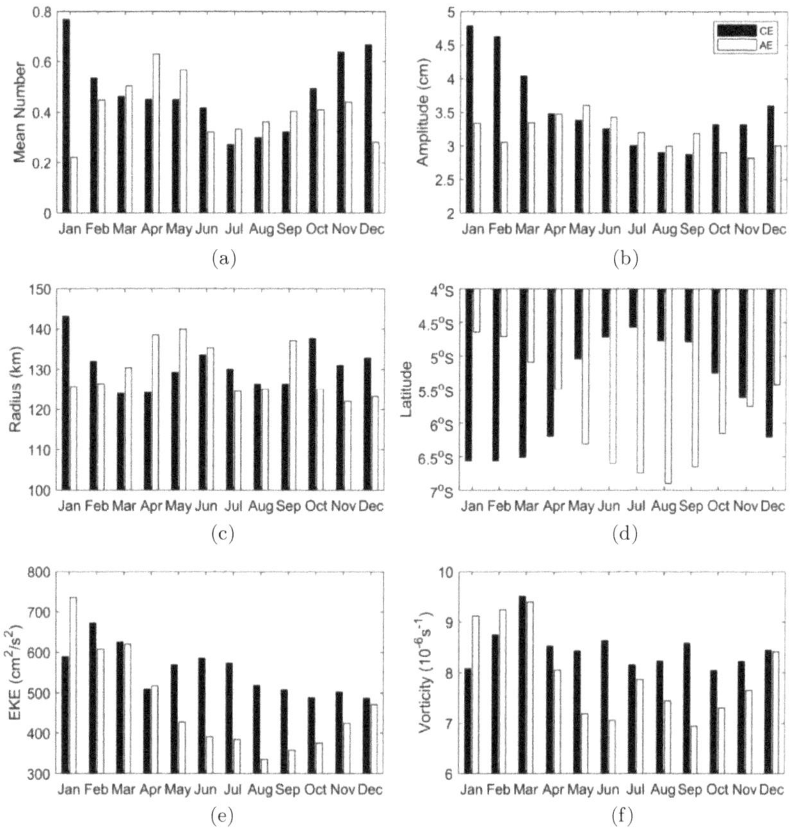

Fig. 18.18. Seasonal variability of properties of AEs (open bars) and CEs (solid bars) in the Maluku Sea. (a) Number of eddies, (b) amplitude (cm), (c) radius (km), (d) central latitude (°N), (e) mean EKE ($cm^2\,s^{-2}$) and (f) mean vorticity ($10^{-6}\,s^{-1}$).

Source: Taken from Hao *et al.* (2021).

two troughs in April and September. No apparent seasonal variations are observed for the mean relative vorticity of AEs with a maximum value of $13.2\times10^{-6}\,s^{-1}$ and a minimum value of $11.5\times10^{-6}\,s^{-1}$ (Hao *et al.*, 2021).

Figure 18.18(a–f) show the seasonal variability of eddy mean properties in the Maluku Sea. The monthly mean number of AEs and CEs are less than 0.2. The mean amplitude of CEs has a bimodal curve with annual maximum of 3.1 cm in February, annual minimum of 2.6 cm in July, local maximum of 2.8 cm in October and local minimum of 2.7 cm in December. Meanwhile, there is no clear seasonal variability of AE amplitude with the maximum of 3 cm in November and the minimum of 2.7 cm in July. The seasonal variability of CE radii is like that of CE amplitudes. Two peak values of 98.7 km and 103.7 km appear in March and October, and two trough values of 88.51 km and 91.4 km in February and May. Like AE amplitudes, the AE radii have no apparent seasonal variability with the maximum of 99.8 km in October and the minimum of 87.9 km in January. AEs prefer to appear in south of the equator from July to November and north of equator from December to June. On the contrary, CEs tend to concentrate in south of the equator in the first half of the year and in north of the equator in the second half of the year (Hao *et al.*, 2021).

Figure 18.19(a–f) show the seasonal variability of eddy mean properties in the Banda Sea, which is located at 4°–8°S, 123°–132°E. One can see that the mean number, amplitude and radius of AEs exhibit bimodal structure, i.e., the annual maximum values of 0.6, 3.6 cm and 139.9 km appearing in austral fall, then dropping to annual minimum values of 0.3, 3.0 cm and 124.6 km, in austral winter. After approaching the minima, they increase to local maximum values of 0.4, 3.2 cm and 137.2 km in austral spring and then decrease to local minimum values of 0.2, 3.1 cm and 122.1 km in austral summer. Meanwhile, the mean number and amplitude of CEs hit peak values of 0.7 and 4.8 cm in austral summer, then decrease gradually to the trough values of 0.3 and 2.9 cm in austral winter. The mean CE radius reaches the maximum value of 143.1 km in January and the minimum 124.0 km in March.

The latitudes of AEs and CEs show antiphase patterns, i.e., AEs tend to sit in the southern Banda Sea during austral winter and in the northern Banda Sea during austral summer, while CEs prefer to concentrate in the southern Banda Sea in austral summer and in the northern Banda Sea in austral winter. In the other words, the polarities of mesoscale eddies observed at the same location reverse seasonally. The mechanisms of polarity distribution reversal (PDR) of mesoscale eddies are attributed to that the basin-scale vorticity, wind stress curl and the meridional shear of zonal current reverse seasonally (Chen *et al.*, 2020).

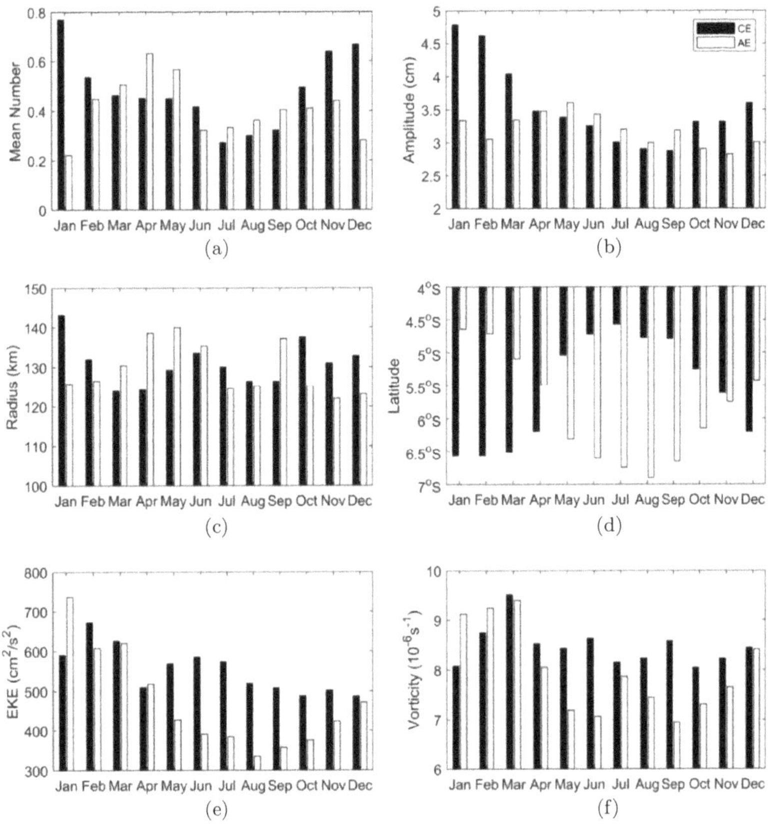

Fig. 18.19. Seasonal variability of monthly mean properties of AEs (open bars) and CEs (solid bars) in the Banda Sea. (a) Number of eddies, (b) amplitude (cm), (c) radius (km), (d) central latitude (°S), (e) EKE ($cm^2\,s^{-2}$) and (f) vorticity ($10^{-6}\,s^{-1}$).

Source: Taken from Hao *et al.* (2021).

Figure 18.19(e) shows that energetic AEs with a peak EKE value of $736\,cm^2\,s^{-2}$ appear in January of austral summer, and the mean minimum EKE of $335.2\,cm^2\,s^2$ in August of austral winter. The seasonal variability of mean EKE of CEs shows a bimodal curve with two peak values of $672.5\,cm^2\ s^{-2}$ and $585.6\,cm^2\,s^{-2}$ in February and June and two trough values of $509.2\,cm^2\,s^{-2}$ and $488.5\,cm^2\ s^{-2}$ in April and October.

Figure 18.19(f) shows that the seasonal variability of mean relative vorticity of CEs is featured by a weak seasonality, i.e., the maximum of $9.5 \times 10^{-6}\,s^{-1}$ in March and the minimum of $8.0 \times 10^{-6}\,s^{-1}$ in October. Meanwhile, the seasonal variability of mean relative vorticity of AEs shows a remarkable bimodal structure, i.e., two peak values of $9.4 \times 10^{-6}\,s^{-1}$ and $7.9 \times 10^{-6}\,s^{-1}$ appear in March and July, and two trough values of $7.1 \times 10^{-6}\,s^{-1}$ and $6.9 \times 10^{-6}\,s^{-1}$ in June and September.

18.5 Summaries

This chapter summarizes statistical features of mesoscale eddies in nearby marginal seas of the SCS, including the Andaman Sea on the Indian Ocean side, the Sulu Sea, the Celebes (Sulawesi) Sea, the Maluku Sea and the Banda Sea on the Pacific Ocean side. These marginal seas have common features as follows. (1) Like the SCS, all nearby marginal seas have active mesoscale eddy fields all year around. (2) Eddy properties, including monthly mean eddy number, amplitude, radius, central latitude, EKE and vorticity, show remarkable seasonal variability, implying modulation by monsoon winds. (3) Enclosed or semi-enclosed boundaries and bottom topography of marginal seas play important roles in formation of spatial distribution patterns of eddies. (4) Background flows serve as generators for eddies and modulators to seasonal variability and polarity reversal of spatial distribution patterns of eddies.

Table 18.2 lists statistical means of eddy parameters in the SCS and nearby seas. One can see that owing to different scales, locations and boundary conditions of the seas, the statistical mean eddy parameters differ from each other. For the WNP marginal seas including the SCS, the Sulu Sea, the Celebes (Sulawesi) Sea and the Banda Sea, the statistical mean SLA amplitudes are within a range of (3.2 ± 0.7) cm, which is 42% of the global mean. For the northeastern Indian Ocean marginal seas, the statistical mean SLA amplitude of the Bay of Bengal reaches 9.2 cm, 19% higher than the global mean, while that of the Andaman Sea is 4.7 cm, 61% of the global mean and 19% higher than that of the SCS located within the same latitudinal range.

Table 18.2. Statistical means of eddy parameters in the SCS and nearby seas.

Name of the Sea (ref(s).)	Basin Scales NS/WE (km)	Eddy Lifetime (d)	Eddy SLA Amplitude (cm)	Eddy Radius (km)	Ratio of (AEs:CEs)
Global Ocean (Tian *et al.*, 2020)	10000/10000	84.9	7.7	82.4	1:1.03 (173245:179127)
The Bay of Bengal (Cui *et al.*, 2016)	2750/1600	55	9.2	124	1:1.45 (389:565)
Andaman Sea (Lin *et al.*, 2023)	450/1000	45.5 (A*) 47.4 (C*)	4.7	112	1:0.92 (171:157)
The SCS (Jin *et al.*, 2024; Chapter 12)	2000/1000	17(A) 18 (C)	3.7 (A) −3.7 (C)	77 (A) 75 (C)	1:1.02
Sulu Sea (Hu *et al.*, 2017; Hao *et al.*, 2021)	790/600	32	2.1 (A) −2.2 (C)	72 (A) 74 (C)	1:1.08 (225:243)
Celebes Sea (Hao *et al.*, 2021)	400/1000	∼30	3.7 (A) −4.0 (C)	105 (A) 110 (C)	1:1.07
Banda Sea (Chen *et al.*, 2020; Hao *et al.*, 2021)	450/1000	40	3.2 (A) −3.3 (C)	129 (A) 131 (C)	1:1.07 (71:76)

Note: *A and C for AE and CE.

References

Cai, S., He, Y., Long, X. and Wang, S. (2008). Model study of the upper circulation in the Sulu Sea and its relation to the South China Sea circulation, *Acta Oceanol. Sin.*, 27, pp. 119–128.

Chelton, D. B., Schlax, M. G. and Samelson, R. M. (2011). Global observations of nonlinear mesoscale eddies, *Progr. Oceanogr.*, 91(2), pp. 167–216, doi:10.1016/j.pocean.2011.01.002.

Chen, B., Xie, L., Zheng, Q., Zhou, L., Wang, L., Feng, B. and Yu, Z. (2020). Seasonal variability of mesoscale eddies in the Banda Sea inferred from altimeter data, *Acta Oceanol. Sin.*, 39(12), pp. 11–20, doi:10.1007/s13131-020-1665-2.

Cui, W., Yang, J. and Ma, Y. (2016). A statistical analysis of mesoscale eddies in the Bay of Bengal from 22-year altimetry data, *Acta Oceanol. Sin.*, 35(11), pp. 16–27, doi:10.1007/s13131-016-0945-3.

Faghmous, J. H., Frenger, I., Yao, Y., Warmka, R., Lindell, A. and Kumar, V. (2015). A daily global mesoscale ocean eddy dataset from satellite altimetry, *Sci. Data*, 2, pp. 1–16, doi:10.1038/sdata.2015.28.

Feng, M., Zhang, N., Liu, Q. and Wijffels, S. (2018). The Indonesian throughflow, its variability and centennial change, *Geosci. Lett.*, 5(1), pp. 1–10, doi:10.1186/s40562-018-0102-2.

Gordon, A. L., Sprintall, J. and Ffield, A. (2011a). Regional oceanography of the Philippine Archipelago, *Oceanography*, 24(1), pp. 14–27, doi:10.5670/oceanog.2011.01.

Gordon, A. L., Tessler, Z. D. and Villanoy, C. (2011b). Dual overflows into the deep Sulu Sea, *Geophys. Res. Lett.*, 38, L18606, doi:10.1029/2011GL048878.

Hao, Z., Xu, Z., Feng, M., Li, Q. and Yin, B. (2021). Spatiotemporal variability of mesoscale eddies in the Indonesian Seas, *Rem. Sens.*, 13(5), 1017, doi:10.3390/rs13051017.

He, Y., Feng, M., Xie, J., Liu, J., Chen, Z., Xu, J. and Cai, S. (2017). Spatiotemporal variations of mesoscale eddies in the Sulu Sea, *J. Geophys. Res. Oceans*, 122(10), pp. 7867–7879, doi:10.1002/2017JC013153.

He, Z., Feng, M., Wang, D. and Slawinski, D. (2015). Contribution of the Karimata Strait transport to the Indonesian Throughflow as seen from a data assimilation model, *Cont. Shelf Res.*, 92, pp. 16–22, doi:10.1016/j.csr.2014.10.007.

Hu, D., Chen, X. and Mao, K. (2017). Statistical characteristics and composed three dimensional structures of mesoscale eddies in the South Indian Ocean, *Haiyang Xuebao*, 39(9), pp. 1–14.

Huang, R., Xie, L., Zheng, Q., Li, M., Bai, P. and Tan, K. (2020). Statistical analysis of mesoscale eddy propagation velocity in the

South China Sea deep basin, *Acta Oceanol. Sin.*, 39(11), pp. 91–102, doi:10.1007/s13131-020-1678-x.

Jin, Y., Jin, M., Wang, D. and Dong, C. (2024). Statistical analysis of multi-year South China Sea eddies and exploration of eddy classification, *Rem. Sens.*, 16, 1818, doi:10.3390/rs16101818.

Lin, Z., Xie, L., Huang, R. and Bai, P. (2023). Seasonal variation of mesoscale eddies in the Andaman Sea, *Haiyang Xuebao*, 45(4), pp. 1–16.

Lukas, R., Firing, E., Hacker, P., Richardson, P. L., Collins, C. A., Fine, R. and Gammon, R. (1991). Observations of the Mindanao Current during the western equatorial Pacific Ocean circulation study, *J. Geophys. Res. Oceans*, 96(C4), pp. 7089–7104, doi:10.1029/91JC00062.

Pujol, M. I., Faugère, Y., Taburet, G., Dupuy, S., Pelloquin, C., Ablain, M. and Picot, N. (2016). DUACS DT2014: The new multi-mission altimeter data set reprocessed over 20 years, *Ocean Sci.*, 12(5), pp. 1067–1090, doi:10.5194/os-12-1067-2016.

Qu, T. and Song, Y. T. (2009). Mindoro Strait and Sibutu Passage transports estimated from satellite data, *Geophys. Res. Lett.*, 36, L09601, doi:10.1029/2009GL037314.

Sprintall, J., Gordon, A. L., Flament, P. and Villanoy, C. L. (2012). Observations of exchange between the South China Sea and the Sulu Sea, *J. Geophys. Res. Oceans*, 117, C05036, doi:10.1029/2011JC007610.

Sprintall, J., Gordon, A. L., Wijffels, S. E., Feng, M., Hu, S., Koch-Larrouy, A. and Setiawan, A. (2019). Detecting change in the Indonesian seas, *Front. Mar. Sci.*, 6, 257, doi:10.3389/fmars.2019.00257.

Tessler, Z. D., Gordon, A. L., Pratt, L. J. and Sprintall, J. (2010). Transport and dynamics of the Panay sill overflow in the Philippine seas, *J. Phys. Oceanogr.*, 40(12), pp. 2679–2695, doi:10.1175/2010jpo4395.1.

Tian, F., Wu, D., Yuan, L. and Chen, G. (2020). Impacts of the efficiencies of identification and tracking algorithms on the statistical properties of global mesoscale eddies using merged altimeter data, *Intl. J. Rem. Sens.*, 41(8), pp. 2835–2860, doi:10.1080/01431161.2019.1694724.

Index

B

C

D

E

H

I

J

K

L

M

N

S

T